# Oral Microbial Ecology

## Current Research and New Perspectives


Edited by

Nicholas S. Jakubovics

Oral Biology
School of Dental Sciences
Newcastle University
Newcastle upon Tyne
UK

and

Robert J. Palmer Jr.

Oral Infection and Immunity Branch
National Institute of Dental and Craniofacial Research
National Institutes of Health
Bethesda, MD
USA



Caister Academic Press

Caister Academic Press
Norfolk, UK

www.caister.com

British Library Cataloguing-in-Publication Data
A catalogue record for this book is available from the British Library

ISBN: 978-1-908230-17-1



Cover design adapted from Figures 4.4, 4.6, and 4.7

Printed and bound in Great Britain

# Contents

# Contributors

**Frank Abbas**
Department of Periodontology
Center for Dentistry and Oral Hygiene
University Medical Center Groningen
University of Groningen
Groningen
Netherlands

f.abbas@umcg.nl

**David Beighton**
Department of Microbiology
Dental Institute
King's College London
London
UK

david.beighton@kcl.ac.uk

**Indranil Biswas**
Department of Microbiology
University of Kansas Medical Center
Kansas City, KS
USA

iBiswas@kumc.edu

**George H. Bowden**
Department of Oral Biology
University of Manitoba
Winnipeg
Canada

bowden@cc.umanitoba.ca

**William H. Bowen**
Center for Oral Biology;
Department of Microbiology and Immunology
University of Rochester Medical Center
Rochester, NY
USA

william_bowen@urmc.rochester.edu

**Henk J. Busscher**
Department of Biomedical Engineering
University Medical Center Groningen
University of Groningen
Groningen
Netherlands

h.j.busscher@umcg.nl

**Michael F. Cole**
Department of Microbiology & Immunology
Georgetown University School of Medicine
Washington, DC
USA

colem@georgetown.edu

**Thuy Do**
Department of Oral Biology
Leeds Dental Institute
Faculty of Medicine Health
University of Leeds
Leeds
UK

n.t.do@leeds.ac.uk

**Megan L. Falsetta**
Center for Oral Biology
University of Rochester Medical Center
Rochester, NY
USA

megan_wood@urmc.rochester.edu

**Christof Godts**
Department of Periodontology
Catholic University Leuven
Leuven
Belgium

christof.godts@med.kuleuven.be

**George Hajishengallis**
Department of Microbiology
University of Pennsylvania
Philadelphia, PA
USA

geoh@upenn.edu

**Hermie J.M. Harmsen**
Department of Medical Microbiology
University Medical Center Groningen
University of Groningen
Groningen
Netherlands

h.j.m.harmsen@umcg.nl

**Chris G. van Hoogmoed**
Department of Biomedical Engineering
University Medical Center Groningen
University of Groningen
Groningen
Netherlands

c.g.van.hoogmoed@umcg.nl

**Nicholas S. Jakubovics**
Oral Biology
School of Dental Sciences
Newcastle University
Newcastle upon Tyne
UK

nick.jakubovics@ncl.ac.uk

**Marlise I. Klein**
Center for Oral Biology
University of Rochester Medical Center
Rochester, NY
USA

marlise_klein@urmc.rochester.edu

**Hyun Koo**
Center for Oral Biology;
Department of Microbiology and Immunology
University of Rochester Medical Center
Rochester, NY
USA

hyun_koo@urmc.rochester.edu

**Jens Kreth**
Department of Microbiology;
Department of Oral Biology
University of Oklahoma Health Sciences Center
Oklahoma City, OK
USA

jens-kreth@ouhsc.edu

**Purnima S. Kumar**
College of Dentistry
The Ohio State University
Columbus, OH
USA

kumar.83@osu.edu

**Gitte Loozen**
Department of Periodontology
Catholic University Leuven
Leuven
Belgium

gitte.loozen@med.kuleuven.be

**Matthew R. Mason**
College of Dentistry
The Ohio State University
Columbus, OH
USA

mason.369@osu.edu

**Henny C. van der Mei**
Department of Biomedical Engineering
University Medical Center Groningen
University of Groningen
Groningen
Netherlands

h.c.van.der.mei@umcg.nl

**Justin Merritt**
Department of Microbiology;
Department of Oral Biology
University of Oklahoma Health Sciences Center
Oklahoma City, OK
USA

justin-merritt@ouhsc.edu

**Saswat Sourav Mohapatra**
Department of Microbiology
University of Kansas Medical Center
Kansas City, KS
USA

saswatsmohapatra@gmail.com

**Annette Moter**
Institute for Microbiology and Hygiene
Charité – Universitätsmedizin Berlin
Berlin
Germany

annette.moter@charite.de

**Sadaf Rasheed Mughal**
Department of Microbiology
Dental Institute
King's College London
London
UK

sadaf.mughal@kcl.ac.uk

**William Nance**
Department of Epidemiology
School of Public Health
University of Michigan
Ann Arbor, MI
USA

wcnance@umich.edu

**Marieke P.T. Otten**
Department of Biomedical Engineering
University Medical Center Groningen
University of Groningen
Groningen
Netherlands

mptotten@gmail.com

**Marc Quirynen**
Department of Periodontology
Catholic University Leuven
Leuven
Belgium

marc.quirynen@med.kuleuven.be

**Alexander H. Rickard**
Department of Epidemiology
School of Public Health
University of Michigan
Ann Arbor, MI
USA

alexhr@umich.edu

**Wim Teughels**
Department of Periodontology
Catholic University Leuven
Leuven
Belgium

Wim.Teughls@med.kuleuven.be

**John G. Thomas**
Department of Pathology;
Department of Periodontology
West Virginia University
Morgantown, WV
USA

jthomas@hsc.wvu.edu

**Adam J. Underwood**
Department of Biological Sciences
Binghamton University
Binghamton, NY
USA

aju8483@gmail.com

**William G. Wade**
Microbiology Unit
King's College London Dental Institute
London
UK

william.wade@kcl.ac.uk

**Katherine A. Wirth**
Department of Microbiology & Immunology
Georgetown University School of Medicine
Washington, DC
USA

katherine.wirth@associates.dhs.gov

**Jin Xiao**
Center for Oral Biology
University of Rochester Medical Center
Rochester, NY
USA;
State Key Laboratory of Oral Diseases
Sichuan University
Chengdu
China

jin_xiao@urmc.rochester.edu

**Janel Yu**
College of Dentistry
The Ohio State University
Columbus, OH
USA

janelyu@gmail.com

**Vincent Zijnge**
Department of Medical Microbiology;
Department of Periodontology
Center for Dentistry and Oral Hygiene
University Medical Center Groningen
University of Groningen
Groningen
Netherlands

vzijnge@hotmail.com

# Preface

From a microbe's perspective, the human mouth is an island. The habitats available to oral micro-organisms are sufficiently distinct from the surrounding tissues (nasopharynx or skin) that only a small number of generalists are able to successfully colonize both environments. For the more specialist microbes, adapted to exist solely on the hard or soft tissues of the oral cavity, opportunities to move to a new niche are few and far between. Like Darwin's finches on the islands of Galapagos, microbial species constantly change and evolve, and each person harbours distinct strains in their mouth. Yet, at the level of species or genus, the diversity between micro-organisms in different peoples' mouths is remarkably small. It is currently estimated that around 700 species-level phylotypes are indigenous to the human oral cavity. Whilst more sensitive detection and identification methods such as pyrosequencing are capable of identifying larger numbers of phylotypes, these must be used carefully if they are to distinguish between the stable microflora and 'contaminants', or transient colonisers, that survive in the mouth for just a short time. Nevertheless, even 700 species provides sufficient complexity to keep microbiologists occupied for many years to come!

The development of oral biofilms is underpinned by specific interactions between bacteria and the host, and between bacteria of different species. The streptococci are particularly well adapted to colonize both hard and soft tissues. Streptococci grow freely in monocultures in standard bacteriological media. In saliva, however, streptococci interact through multiple catabolic activities and signalling networks. Some bacteria are entirely dependent on such interactions, and as such are difficult or impossible to culture in isolation. Pure cultures remain the mainstay of microbiology, and without them little information can be obtained on the physiology of individual organisms. Sensitive methods for community analysis and new approaches to bacterial co-culture are being successfully applied in identification, isolation, and cultivation of fastidious micro-organisms. It may not be long before the term 'unculturable' is rendered obsolete.

The identification of bacterial cell–cell interactions in biofilms requires non-destructive methods that provide information on the spatial distribution of individual species or genera. Culture-independent methods such as fluorescence in situ hybridization (FISH) methods are being enhanced to increase sensitivity, and the detection of multiple targets within a sample is becoming routine thanks to advances in flours and detectors. FISH-based microscopy, especially on samples that have been taken from the mouths of patients, is providing exciting insights into the microbial partnerships that are fundamental to the stability of the biofilm. One limitation of FISH is that it requires dehydration, and the structure of the intercellular matrix is lost. Therefore, other approaches to documenting matrix components are needed. The addition of fluorophore-conjugated dextrans as a substrate for fluorescent extracellular glucan production has provided an elegant approach for revealing the *Streptococcus mutans* exopolysaccharide matrix. However, elucidating the structure of the extracellular matrix of natural dental plaque, which undoubtedly contains proteins and nucleic acids in addition to carbohydrates, is still some way off.

Defined structures in dental plaque such as 'maize cob' and 'bristle brush' formations can be seen in situ, and can be reconstituted in the laboratory by mixing together micro-organisms that bind specifically with one another. This process of coaggregation brings cells into close proximity, where different catabolic capabilities can be combined efficiently to extract the maximum energy from substrates. Microbes constantly sense their local environment, and communication between cells in close proximity leads to gene regulation and the expression of distinct phenotypes. Mutualistic interactions, including the exchange of genetic material, are enhanced in close-knit communities. However, competition is rife, and incompatible species will fight using weapons such as bacteriocins or hydrogen peroxide. Such interactions are critical for maintaining a healthy balance of different species in oral biofilms. Disruption to the microbial homeostasis can lead to the development of microbial communities detrimental to the health of the host.

By far the most common oral diseases of bacterial origin are dental caries and gingivitis/periodontitis. Each of these can be described by the ecological plaque hypothesis, which theorizes that perturbations to the oral environment, for example excess sugar intake, smoking, or disruption to the host's immunity, leads to a shift in the microbial balance away from less pathogenic organisms and the subsequent overgrowth by the more pathogenic species. Once species with pathogenic characteristics start to outgrow other organisms, they can further modify the oral environment by promoting the growth of different fastidious pathogens thereby exacerbating the disease. Oral communities can also act as reservoirs for pathogens that cause diseases at remote sites. One of the most dangerous and costly of these diseases is ventilator-associated pneumonia. Patients on mechanical ventilation in intensive care units are highly susceptible to infections. The endotracheal tube provides an unnatural highway for migration of pathogens from the mouth to the lungs, where they cause life-threatening pneumonia.

Oral hygiene is largely directed towards the removal of dental plaque and the addition of fluoride to inhibit bacteria and strengthen teeth. Interestingly, small amounts of biofilm may actually be beneficial for the delivery of fluoride or antimicrobials. Residual biofilm can act as a reservoir for antimicrobial agents, and release them slowly over time. Such observations support the view that the complete removal of oral biofilms is not necessarily the best strategy for oral health. An alternative approach is to manipulate the existing microflora to promote health and to exclude pathogens. This can be achieved by the addition of beneficial bacteria (probiotics), or by the intake of functional foods (prebiotics), which stimulate the growth of pre-existing beneficial bacteria in the biofilm.

In summary, the oral cavity supports a rich and diverse microbial population. Oral health is dependent on the maintenance of stable microbial communities. Disease occurs when this balance is disturbed and more pathogenic species outgrow the commensals. There is no single caries pathogen or periodontitis pathogen that neatly fulfils Koch's postulates. Instead, health and disease in the mouth are active processes in which the ecology of communities, not of single organisms, is paramount. Here, we provide an update on recent developments in the burgeoning field of oral microbial ecology.

Nick S. Jakubovics and Robert J. Palmer Jr

# Acknowledgements

This book could not have come together without the dedicated efforts of all the authors. We are very grateful for their excellent contributions. We have been kept on track by Hugh Griffin at Horizon Press. We thank Hugh and the publication team at Horizon for their hard work in the publishing process. We have had many fruitful discussions over the years with colleagues in our laboratories and outside. These have been a constant source of ideas and constructive criticism which have broadened our horizons and helped to shape our thinking about oral bacteria and the communities they form. We acknowledge these contributions, which undoubtedly have had a significant bearing on the structure of the book. Finally, NSJ wishes to thank his wife, Kate, for her constant support and encouragement throughout the editorial process.

# Current books of interest

| | |
|---|---|
| Prions: Current Progress in Advanced Research | 2013 |
| RNA Editing: Current Research and Future Trends | 2013 |
| Microbial Efflux Pumps: Current Research | 2013 |
| Cytomegaloviruses: From Molecular Pathogenesis to Intervention | 2013 |
| Bionanotechnology: Biological Self-assembly and its Applications | 2013 |
| Real-Time PCR in Food Science: Current Technology and Applications | 2013 |
| Bacterial Gene Regulation and Transcriptional Networks | 2013 |
| Bioremediation of Mercury: Current Research and Industrial Applications | 2013 |
| *Neurospora*: Genomics and Molecular Biology | 2013 |
| Rhabdoviruses | 2012 |
| Horizontal Gene Transfer in Microorganisms | 2012 |
| Microbial Ecological Theory: Current Perspectives | 2012 |
| Two-Component Systems in Bacteria | 2012 |
| Malaria Parasites: Comparative Genomics, Evolution and Molecular Biology | 2013 |
| Foodborne and Waterborne Bacterial Pathogens | 2012 |
| *Yersinia*: Systems Biology and Control | 2012 |
| Stress Response in Microbiology | 2012 |
| Bacterial Regulatory Networks | 2012 |
| Systems Microbiology: Current Topics and Applications | 2012 |
| Quantitative Real-time PCR in Applied Microbiology | 2012 |
| Bacterial Spores: Current Research and Applications | 2012 |
| Small DNA Tumour Viruses | 2012 |
| Extremophiles: Microbiology and Biotechnology | 2012 |
| *Bacillus*: Cellular and Molecular Biology (Second edition) | 2012 |
| Microbial Biofilms: Current Research and Applications | 2012 |
| Bacterial Glycomics: Current Research, Technology and Applications | 2012 |
| Non-coding RNAs and Epigenetic Regulation of Gene Expression | 2012 |
| *Brucella*: Molecular Microbiology and Genomics | 2012 |
| Molecular Virology and Control of Flaviviruses | 2012 |
| Bacterial Pathogenesis: Molecular and Cellular Mechanisms | 2012 |
| *Bunyaviridae*: Molecular and Cellular Biology | 2011 |
| Emerging Trends in Antibacterial Discovery: Answering the Call to Arms | 2011 |
| Epigenetics: A Reference Manual | 2011 |
| Metagenomics: Current Innovations and Future Trends | 2011 |

# Microbial Populations in Oral Biofilms

1

Michael F. Cole, Katherine A. Wirth and George H. Bowden

**Abstract**

In this chapter we consider the biology of the viridans streptococci in the human oropharynx with a particular focus on the pioneer bacterium *Streptococcus mitis*. We show that, although this species is a constant component of the human oral cavity, each person harbours a unique and diverse population of strains that appear not to be shared within a family and, apparently, are rarely transmitted from mother to neonate. The population of strains of *S. mitis* within the mouth of each individual exhibits turnover perhaps in response to pressure exerted by the mucosal immune system since it has been shown that some secretory immunoglobulin A (SIgA) antibodies are clone-specific. We assert that the strains that are successful in establishing in the mouth are physiologically adapted to occupy their niche within their habitat. While it is clear that *in vitro* experiments and animal models have provided useful information, they are no substitute for studying commensal oral bacteria in their environment, the human oral cavity.

## Introduction

The human oral cavity is a significant but not unique habitat for streptococci. These bacteria have a wide host range and occupy several habitats associated with humans and animals. *Streptococcus* strains are consistently present in the human oral cavity within a day of birth and continue to colonize the mouth throughout life. Ideally, species of oral streptococci, in common with other species in the mouth, survive and maintain their populations on oral surfaces by establishing a non-pathogenic balance between their biological activities and those of the host (Bowden *et al.*, 1998).

Strains of many species of oral streptococci, such as those of *S. salivarius*, are generally commensal in humans. However, strains of other species, including those of *S. mutans* and *S. oralis* (Whiley *et al.*, 1992; Burnie *et al.*, 1996; Ahmed *et al.*, 2003), may be more commonly associated with significant infections. In some cases commensal species like *S. salivarius* may be opportunistic pathogens (Delorme *et al.*, 2007) and can infect normally sterile body sites, while others like *S. mitis* can express defined virulence characteristics and contribute to disease (Mitchell, 2011). Indeed, when one considers dental caries, which has an ecological aetiology (Marsh *et al.*, 2006; Takahashi *et al.*, 2011), a wide range of characteristics including adherence (Banas *et al.*, 2003), bacteriocin production (Hossain *et al.*, 2011) and aciduance (Alam *et al.*, 2000; Matsui *et al.*, 2010) can be included among a wide range of factors contributing to tissue destruction by *S. mutans*.

The biology of the interaction between oral streptococci and their human hosts has formed a rewarding area of study and, as one might expect, extensive research has been carried out on species associated with disease. Again, taking dental caries as an example, research has been directed for decades at the biology of the mutans streptococci and has explored means of controlling their transmission, their colonization of the host, and their role in the disease.

Study of commensal strains may seem less appealing. However, oral commensal streptococci are among the first bacteria to colonize and establish a balance with their host, they contribute to host protection by occupying habitats and producing substances toxic to pathogens, and possibly by stimulating host immunity to antigens shared among them and other clearly pathogenic bacteria. Thus, study of commensal oral streptococci, although not as widespread as study of those associated with disease, should form a significant part of research into the biology of oral bacteria.

Studies of the biology or the natural history of oral streptococcal populations are made difficult by the need to examine the status of the organisms in their natural environment. For example, unless the hosts are part of the experimental system, one cannot analyse transmission or colonization, the host response, or the activities of organisms *in vivo*. This contrasts to modelling the biology of one or more organisms as a suspended culture or biofilm in the laboratory, where one requires only the strains and a more or less complex model system. This clear difference in requirements is reflected in the large amount of data available from *in vitro* biofilm studies compared to that from studies involving human subjects. For example, some workers have cautioned that the role of antimicrobials in bacterial competition in nature may differ from that *in vitro*, and that predictions based on laboratory data should be tested in more natural environments (Hibbing *et al.*, 2010). Moreover, analysis of bacterial genomes has shown that organisms possess the capacity to produce large numbers of small molecules that have no known biological role and are not often produced in the laboratory. These molecules may be involved in interspecies interactions in their natural environments (Gross, 2007; Shank *et al.*, 2009).

Despite such difficulties, some *in vivo* studies of the activities of organisms exist. Very often these models involve use of an enamel chip or a glass surface placed in the mouth for different periods of time. Such devices may be coated with a layer of bacteria prior to insertion into the oral cavity.

With our interest in the biology of oral streptococci and the initiation of the host response, we feel that studies of these bacteria in their natural habitat should form a significant component of research. This would include demonstration in nature of the large variety of signalling and other systems currently known from *in vitro* studies, as well as confirmation *in vivo* of predictions of activities based on laboratory studies. Among the best examples of this approach has been the study of the lantibiotics produced by *Streptococcus salivarius*, in which workers have blended laboratory studies with studies in humans to great success (Tagg, 1983; Ross *et al.*, 1993; Dierksen *et al.*, 2000; Horz *et al.*, 2007; Power *et al.*, 2008).

## Phases of the life cycle of oral streptococci

In order to survive and expand their populations, oral streptococci follow a series of phases in their life cycle. They have to contact their hosts and also possess mechanisms that allow them to survive in the oral environment and colonize new hosts (Fig. 1.1). These organisms are constant components of the oral microbiota, ubiquitous in the human oral cavity, confirming that they are superbly adapted to the human oral environment. The following list shows phases of their life cycle whereby oral streptococci survive in their natural environments. Some aspects of these phases are discussed in more detail elsewhere in this volume and will only be mentioned briefly here for completeness.

Selected phases in the life cycle of oral streptococci are shown below.

1. survival in and transmission among the host population;
2. initial retention on the oral surfaces of the host;
3. persistence as a component of the commensal oral microbiota of the host through incorporation into a biofilm community:
   a. successful competition or interaction with other organisms in the community;
   b. adaptation to variation in the biofilm environment. Expressing diversity to generate phenotypes better suited to a given environment;
   c. survival in a dormant state;
4. avoid or be unaffected by innate and adaptive immunity during and subsequent to initial colonization.

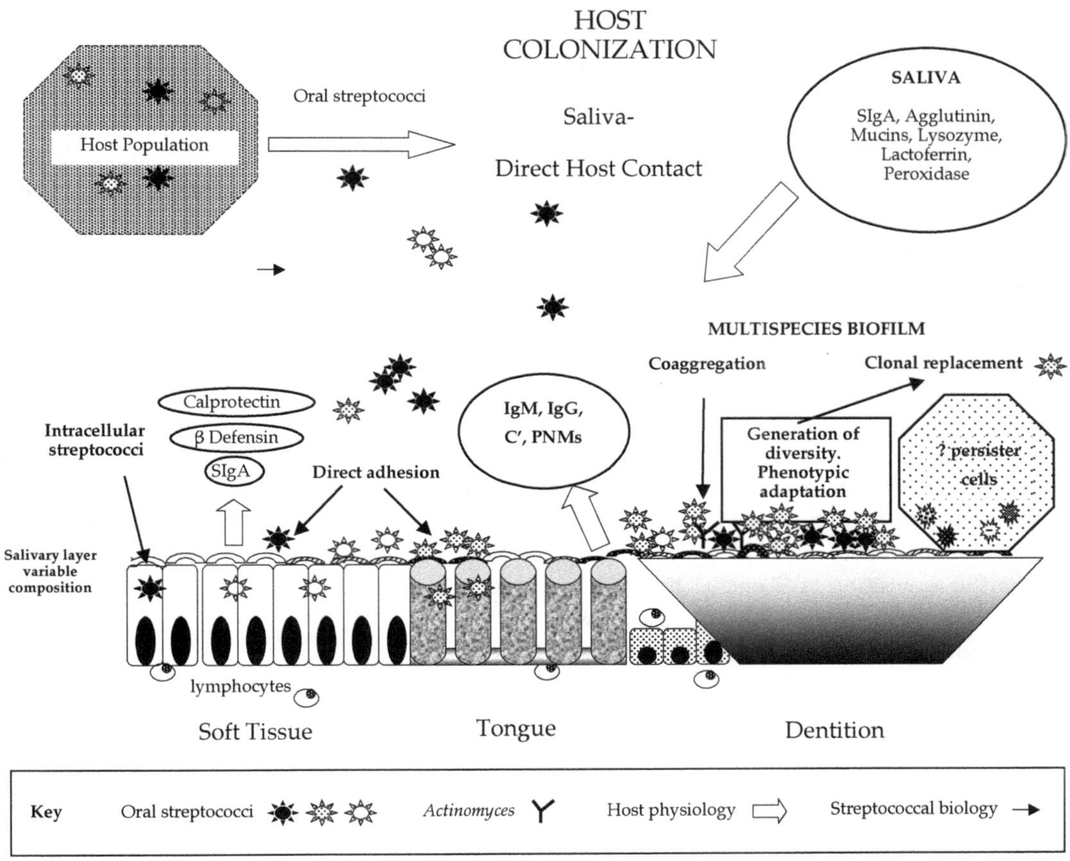


**Figure 1.1** Colonization of the human host by oral streptococci. Oral streptococci colonize via saliva and other contacts. A balance is achieved with host defence mechanisms.

Topic 4 has been a major interest of ours for several years, following the biology of *S. mitis* in infants as a model. Consequently, this area will be dealt with in more detail and include factors such as *S. mitis* strain diversity that may impact on the avoidance of the immune response.

## Transmission among the host population

Generally the different species of oral streptococci seem to be distributed among their human hosts worldwide, independent of the geographic location. This may be the result of their habitat being under strict physiological control, resulting in a closely similar and stable environment among their hosts. It is not known whether strains of the relatively recently described oral species (Kawamura *et al.*, 1998; Willcox *et al.*, 2001; Tong *et al.*, 2003) are so widely distributed among humans. Although *Streptococcus pneumoniae* is carried by animals (van der Linden *et al.*, 2009), other than for members of the mutans group, relatively little is known about animal carriage of the human oral streptococci. However, species of human oral streptococci are isolated from animal infections (Higgins *et al.*, 1984; Lair *et al.*, 1996; Dolente *et al.*, 2000; Nam *et al.*, 2009).

Members of the mutans group are known to colonize the oral cavities of animals and species within this group have been isolated from different animals (Whiley *et al.*, 1998), most recently from pigs (Takada *et al.*, 2007; Igarashi, 2008). *S. ratti* (Whiley *et al.*, 1998) and *S. downei* (Kim *et al.*, 2005) may also be isolated occasionally from humans. Primates in captivity apparently carry a similar microbiota to humans, including oral streptococci (Mashimo *et al.*, 1979; Loftin *et al.*, 1980; Eke *et al.*, 1993) but almost nothing is known of the oral microbiota of animals in nature. Consequently, with the exception of the Mutans Group, humans seem to be the only, or certainly the major, reservoir of the oral streptococci. Furthermore, it seems unlikely that domestic animals would be a significant source of the oral streptococci colonizing humans.

Most of the studies of transmission and early colonization of humans by oral streptococci have focused on transmission among family and other closely associated groups such as children in day care. Generally it is thought that saliva is the main means of transfer among hosts (Könönen *et al.*, 1992, 2000). Studies have centred on *Streptococcus mutans*, given the close association of this organism with dental caries. There are extensive data on the acquisition of *S. mutans* by children from adults, usually their mothers and this will not be dealt with in detail here (Douglass *et al.*, 2008; Mitchell *et al.*, 2009). Suffice it to say that an important feature of colonization by this organism and others, such as *S. sanguinis* (Caufield *et al.*, 2000), that require the presence of hard surfaces is that they establish in the oral cavity after tooth eruption. This is in contrast to *S. mitis*, *S. oralis* and *S. salivarius*, which colonize and persist on oral soft tissue surfaces in newborns (Pearce *et al.*, 1995). Such early colonization suggests that the interaction between the relatively naïve infant host and *S. mitis*, *S. oralis* and *S. salivarius* will be different from that of *S. mutans* and *S. sanguinis* that colonize following eruption of teeth by which time there is an established microbiota in the mouth and elsewhere on the body.

There are relatively few data on the acquisition of those oral streptococci that are generally seen as commensals (Pearce *et al.*, 1995), with the exception of *Streptococcus mitis*. The origin of the strains of *S. mitis* that colonize newborns is not known with any certainty, although these organisms also colonize the nasopharynx and other sites in the respiratory tract (Hohwy *et al.*, 2001). It would seem most likely that in common with other oral organisms *S. mitis* strains are transmitted by saliva, however, studies typing strains find very few clones in common between mothers and their infants. Typing reveals extensive diversity of strains within and between infants and their mothers (Kirchherr *et al.*, 2005). Occasionally however, individual clones of *S. mitis* may persist, sometimes on different surfaces in both the infants' and mothers' mouths. This extensive diversity of *S. mitis* together with some persistence among unique strains within an infant's mouth has implications for the effectiveness of host immune responses and will be dealt with in detail later (see 'Avoid or be unaffected by innate immunity and adaptive immunity during and subsequent to initial colonization' below). Although there is some information on transmission of oral streptococci to infants there is very little or no information on transmission among adults. The small number of data that do exist on transmission of oral organisms among adults appear to be limited to potential periodontal pathogens (van Steenbergen *et al.*, 1993).

## Initial retention on host surfaces

In order to colonize the mouth, bacteria first have to avoid the mechanisms of innate immunity and the mucosal immune response (see 'Avoid or be unaffected by innate immunity and adaptive immunity during and subsequent to initial colonization' below). In addition, they need to resist the removal forces of saliva, mastication and swallowing. In common with most other microorganisms in nature, the oral streptococci have adhesins that interact with exposed surfaces in their environment, mediating their initial retention and allowing them to survive and grow. A recent extensive review covers most aspects of streptococcal adhesion and growth, including that of oral streptococci (Nobbs *et al.*, 2009). Based on selective adherence, oral streptococci exhibit tissue tropisms, some like *S. mutans* and *S. sanguinis* favour the non-shedding surfaces of the teeth, while others such as *S. salivarius* colonize the tongue. *S. mitis* and *S. oralis* are ubiquitous, being isolated from most areas of the oral cavity.

There are two basic types of surface in the mouth, the teeth (hard non-shedding) and soft tissues, the tongue, and the oral mucosa (shedding surfaces). In addition, prosthetic and orthodontic devices provide excellent surfaces for colonization. Also, the primary colonizing bacteria present a variety of surface molecules for interaction with bacteria colonizing at later stages.

It should be noted that all of these surfaces are covered with a fine film of saliva (salivary pellicle/conditioning layer) providing a layer of exposed molecules for interaction with streptococcal adhesins during colonization (Wu *et al.*, 1998; Hamada *et al.*, 2004; Yamaguchi *et al.*, 2004). The salivary film flows towards the pharynx and varies in chemical composition, flow rate, and thickness within the mouth (Dawes, 2008). Consequently, when saliva is thought of in terms of bacterial adherence it cannot be assumed that the film is of the same composition throughout the mouth, suggesting that the salivary film may present different receptors, or different concentrations of receptors for streptococcal adhesion at different oral sites (Carlén *et al.*, 1998). Indeed, streptococci on enamel chips situated at different areas of the mouth varied in their binding of salivary protein (Rudney *et al.*, 1996).

Apart from contributing to the adhesion of oral streptococci, salivary components may also play a role in host protection through promoting bacterial aggregates in suspension, facilitating their removal (Madsen *et al.*, 2010) and also by carrying a range of antibacterial substances (Rudney, 2000; Malamud *et al.*, 2011). It is straightforward to understand how organisms incorporated into a biofilm on a non-shedding surface persist in the mouth. However, the mechanisms whereby organisms colonizing buccal mucosal cells that are regularly shed into saliva survive are less obvious. Specific strains of *S. mitis* in infants' mouths have been isolated from different areas at different times over a given time period, these areas include the tongue and when present, tooth surfaces (Kirchherr *et al.*, 2005). The tongue presents a physically complex habitat with crypts that could aid bacterial retention and biofilms on enamel are also protective. Thus, these two habitats could contribute to the survival of *S. mitis* against cell shedding and salivary flow. Interestingly, oral streptococci have been shown to be dominant among the intracellular bacteria found in buccal epithelial cells (Rudney *et al.*, 2005). Although these cells will be shed, intracellular bacteria could be protected to some extent from removal by saliva.

## Persisting as a component of the commensal oral flora of the host through incorporation into a biofilm community

### Successful competition or interaction with other organisms in the community

Oral streptococci are among the first bacteria to colonize the newborn infant (Pearce *et al.*, 1995) and also among the first to colonize newly erupted teeth (Brailsford *et al.*, 2005) and cleaned tooth surfaces (Li *et al.*, 2004; Diaz *et al.*, 2006). An important characteristic of oral bacteria in relation to biofilm formation is that they exhibit interspecies coaggregation via specific receptors (Yang *et al.*, 2009). This mechanism ensures species diversity within the oral biofilm community. In common with other oral organisms the oral streptococci coaggregate with a variety of species of other genera including *Actinomyces* (Palmer *et al.*, 2003; Ledder *et al.*, 2008). Coaggregation initially measured *in vitro* and subsequently confirmed by some studies in humans, forms the basis for proposals of the development of multispecies oral biofilms on hard non-shedding surfaces. A recent detailed review describes such a model supported by both *in vitro* and in *vivo* observations, of the development of multispecies biofilms in the mouth, where the oral streptococci, *S. mitis*, *S. sanguinis* and *S. oralis*, form the basis for the construct (Kolenbrander *et al.*, 2010). Taken together the reviews by Nobbs *et al.* (2009), Kolenbrander *et al.* (2010) and Kuramitsu *et al.* (2007) provide extensive details both at the molecular level and at the level of cell populations *in vitro* and *in vivo* on the interactions of oral streptococci during the process of biofilm development.

Although there are considerable data on the process of inclusion of oral streptococci into biofilms on hard non-shedding surfaces in the mouth, there are little on biofilms on soft tissue surfaces. There are data on the microbiota of the tongue and oral soft tissues (Aas *et al.*, 2005; Sachdeo *et al.*, 2008; Preza *et al.*, 2009; Papaloannou *et al.*, 2009) that include oral streptococci, however little is known of the development or nature of any biofilms that form on soft tissue surfaces. The tongue should provide a protected habitat, but again the structural relationships of bacteria on the tongue seem not to have been studied.

In the three reviews listed above relatively little mention is made of the extensive genetic diversity that can occur among populations of the oral streptococci and the possibility that such diversity is reflected in phenotypic diversity. Most but not all (Sissons *et al.*, 2007) of the laboratory models of biofilms include standard laboratory strains that may be common among different workers. Use of such strains is an accepted approach but it is worth pointing out that different phenotypes occur among species populations in nature. Given the impact of the local environment on the biology of a biofilm, selection of phenotypes 'best suited' to the environment is an important aspect of the biology of oral streptococci, particularly when dental caries is considered (Marsh, 2006; Takahashi *et al.*, 2011).

## Adaptation to variation in the biofilm environment – expressing diversity to generate phenotypes better suited to a given environment

The response to stress by oral streptococci has generally been studied in *S. mutans,* and involves changes in environmental pH (Crowley *et al.*, 2004; Fozo *et al.*, 2004; Sheng *et al.*, 2007; Welin-Neilands *et al.*, 2007; Matsui *et al.*, 2010), oxygen (Ahn *et al.*, 2009) and other stresses (Svensäter *et al.*, 2000; Deng *et al.*, 2007). In the case of fluoride (Bowden, 1990) individual strains may adapt to grow in higher levels of fluoride than when first isolated (van Loveren *et al.*, 2008) and, upon withdrawal of the fluoride, may revert to their original sensitivity. However, the mechanism of this adaptation is not well understood. Much less is known of the responses of other oral streptococci to stress (Wilkins *et al.*, 2003), although the biological mechanisms used to adapt to environmental changes are probably similar to those of *S. mutans*.

Considerable genetic diversity has been shown among strains of several species of oral streptococci (Bowden *et al.*, 1998; Hohwy *et al.*, 2001; Pan *et al.*, 2001; Kirchherr *et al.*, 2005; Delorme *et al.*, 2007; Nakano *et al.*, 2007; Bek-Thomsen *et al.*, 2008; Do *et al.*, 2009, 2010; Pieralisi *et al.*, 2010; Moser *et al.*, 2010). The impact of such diversity on the survival of oral streptococci in the mouth is not known although it may provide a means to avoid host responses (see 'Avoid or be unaffected by innate immunity and adaptive immunity during and subsequent to initial colonization' below). However, if the diversity is reflected by similar phenotypic diversity then it could, by selection of strains physiologically 'best suited' to the environment, aid in the survival of a species during perturbations of the oral environment.

The generation of diversity within a microbial population is achieved by a variety of mechanisms (Aertsen *et al.*, 2005; Avery *et al.*, 2006; Coenye *et al.*, 2010), several of which are utilized by oral streptococci (Lemos *et al.*, 2005). There is a large body of information on the genetics of the oral streptococci including those of structural and cell surface components, signalling systems and cell interactions that is beyond the scope of this chapter. However, gene transfer among the Mitis group (King *et al.*, 2005; Delorme *et al.*, 2007; Kilian *et al.*, 2008; Do *et al.*, 2009; Johnston *et al.*, 2010) that would tend to increase diversity appears to be relatively common.

Despite the data available concerning the genetics *per se* and genetic diversity that occurs among strains of the oral streptococci, recent information on any phenotypic diversity among strains in nature is scant. Some data are available on the aciduránce of oral isolates, because of the importance of this characteristic in dental caries. Acid tolerant strains (pH 5.0) of several species of oral streptococci including *S. sanguinis* and *S. salivarius* have been isolated from lesions and caries free surfaces in children (Svensäter *et al.*, 2003). Similarly, aciduric streptococci of different species colonize root surfaces with and without caries (Brailsford *et al.*, 2001) and in a detailed study of *S. oralis* genotypes (Alam *et al.*, 2000) an aciduric sub-population was detected. Importantly, strains from this population were transmitted between cohabiting adults. Relatively little is known of other examples of phenotypic diversity among the oral streptococci. The isolation of oral streptococci growing on low-pH media with fluoride from children in an area with water fluoridation has been described previously (Bowden *et al.*, 1982), while individual ribotypes of *S. mutans* showed varying degrees of sensitivity to chlorhexidine (Grönroos *et al.*, 1995). Most recently, an extensive study of strains of *S. mitis* from infants confirmed not only phenotypic diversity among the strains but also antigenic diversity (Kirchherr *et al.*, 2007) (see 'Avoid or be unaffected by innate immunity and adaptive immunity during and subsequent to initial colonization', below).

It is apparent that despite the relative lack of data, oral populations of streptococci are phenotypically diverse. In their natural habitat these populations present a range of strains with distinct characteristics that could be advantageous in a given environment and aid survival of the species. Such advantageous characteristics may be a feature of distinct subpopulations (Alam *et al.*, 2000) while others may result from phenotypic adaptation. In any event, the possibility of this extensive diversity should be borne in mind when approaches are proposed to control populations of oral streptococci in their natural environment.

A significant aspect of phenotypic diversity is the persistence of phenotypes in the environment. One could expect that generation of a specific phenotype that was best suited to the local environment would result in increased cell numbers in the habitat, resulting in the survival of that phenotype. The fate of phenotypes that are not 'best suited' is not known, they may be too few to be detected by cultural methods and, given clonal replacement among some oral streptococci, considered to be 'eliminated'. Both *S. mitis* and *S. oralis* exhibit extensive clonal replacement (Alam *et al.*, 2000; Kirchherr *et al.*, 2005) but this is not the case with *S. mutans*. Although the *S. mutans* population is diverse it apparently does not undergo extensive clonal replacement in its natural habitat (Alaluusua *et al.*, 1994; Grönroos *et al.*, 1995). It is possible that the persistence of a single genotype of *S. mutans* relates to its habitat on non-shedding surfaces. If this were the case then one would expect *S. sanguinis* and *S. gordonii* to show genotypic stability in their hosts. However, nothing is known of the stability of specific genotypes or phenotypes of these species in the mouth.

The phenomenon of persistence, dormancy and cell death has been addressed in other areas of microbiology (Lewis, 2000; Sachidanandham *et al.*, 2009; Coenye *et al.*, 2010; Lewis 2010; Lennon *et al.*, 2011). The possibility that phenotypes of *S. mitis S. oralis* and *S. infantis* perhaps survive in the mouth in a dormant state is supported to some extent by the observation of Bek-Thompsen *et al.* (2008) who, using a non-culture strategy, showed that small numbers of clones of these species survived in the mouth. They suggested that the cells in these populations were too few in number to be detected by culture. However, some may also represent dormant or persister clones. There are some data on the impact of starvation on oral streptococci (Chávez de Paz *et al.*, 2008; Perry *et al.*, 2009; Busuioc *et al.*, 2010) but generally little is known about the nature of dormancy or persistence among the oral streptococci, although this phenomenon would impact significantly on the biology of these bacteria.

## Avoid or be unaffected by innate immunity and adaptive immunity during and subsequent to initial colonization

As has been mentioned earlier the populations of bacteria that comprise the resident microbiotas on the epithelial surfaces of the body are, for the most part, in a state of dynamic equilibrium with the innate and adaptive immune systems that operate there. Therefore, an understanding of the population biology of commensal bacteria at mucosal surfaces must begin with a consideration of the interplay between the adaptive mucosal immune system and commensal bacteria.

The barrier epithelia are, for the most part, delicate monolayers involved in the exchange of gases, the uptake of nutrients, excretion of waste products and other essential physiological functions. Alone, the mucosal surfaces of the alimentary tract of adult humans comprise an area in excess of 400 $m^2$ (Cerutti *et al.*, 2008). Such is the potential susceptibility of these surfaces to colonization and invasion by microbial pathogens that vertebrates have evolved a sophisticated adaptive immune system, termed the mucosal immune system, to protect these surfaces. The mucosal immune system consists of sentinel secondary lymphoid tissue that is present at the portals of entry to the body and is distributed throughout the aero-digestive tract, the genitourinary tract, eyes and mammary glands. Over 80% of the total number of B cells in humans are located in the alimentary canal and are committed to the synthesis of secretory immunoglobulin A (SIgA), the principal immunoglobulin isotype secreted onto mucosal surfaces. SIgA antibodies in the mucous layer and in exocrine secretions are considered to protect mucosal surfaces through immune exclusion, a non-inflammatory mechanism that neutralizes toxins and facilitates the removal of microorganisms by the flow of secretions, which blocks their adhesion to epithelial receptors and/or agglutinates/aggregates microbial cells (see review by Cerutti *et al.*, 2008). The mucosal and systemic immune systems, while integrated, function largely independently. IgA antibodies in blood are produced by plasma cells in the bone marrow, whereas exocrine IgA is produced by plasma cells in the submucosa. IgA exists as two subclasses, 1 and 2. They are virtually identical in their primary structure except that IgA2 lacks a 13 amino acid sequence at the hinge region that renders it resistant to bacterial IgA1 proteases. IgA in blood is almost entirely monomeric and of subclass 1, whereas exocrine IgA is polymeric, largely dimeric, and may comprise as much as 50% of IgA2 (Mestecky *et al.*, 1986).

Exocrine IgA is produced by both T cell-dependent and T cell-independent pathways. In the former, dendritic cells acquire antigen, which can be as large as bacterial cells, via IgA inductive sites such as tonsils, adenoids, Peyer's patches, etc., and present processed antigenic peptides to $CD4^+$ $T_{H2}$ cells which drive $IgM^+IgD^+$ B cells (B2 B cells) to class switch to IgA through CD40L and TGF-β. Epithelial cells, stromal cells and Mast cells may contribute to IgA class switching by providing TGF-β, while dendritic cells provide retinoic acid, IL-6 and nitric oxide. This T cell-dependent pathway is posited to give rise to high affinity, mono-reactive SIgA antibodies that preferentially target pathogens and toxins (Cerutti *et al.*, 2008). In the mouse the T cell-independent IgA pathway is mediated largely by B1 B cells. B1 B cells are primitive, 'innate' lymphocytes that bear, in the mouse, the surface marker CD5. Recently, the human counterpart of the murine CD5 B1 B cell has been identified in cord and peripheral blood as a $CD20^+CD27^+CD43^+CD70^-$ lymphocyte (Griffin *et al.*, 2011). In the mouse, CD5 B1 B cells originate in the peritoneal cavity where they are self-renewing and produce polyreactive 'natural' IgM antibodies. Also, they seed the submucosa. Under the influence of dendritic cells that have captured bacteria from the mucosal surface by extending their dendrites into the lumen, B1 B cells undergo somatic hypermutation and class switching to produce IgA antibodies. It has been suggested that IgA antibodies produced by this pathway are directed largely at the commensal microbiota. Despite the fact that the B1 B cells undergo somatic hypermutation it appears that the mutated sites in the $V_H$ region show no indication of antigenic selection and that the antibodies are polyreactive. In this context, polyreactive SIgA antibodies have been detected in human saliva and colostrum (Quan *et al.*, 1997).

Despite the formidable commitment of the human immune system to SIgA synthesis the role of mucosal IgA in the regulation of both pathogenic and commensal bacteria remains unclear. Selective IgA deficiency is the most common primary immunodeficiency occurring in as few as 1:500 to 1:700 persons yet the majority of these individuals are asymptomatic and the diagnosis is incidental. Despite suggestions from some mouse studies that mucosal IgA can 'shape' the commensal microbiota (Fagarasan *et al.*, 2002; Phalipon *et al.*, 2002; Suzuki *et al.*, 2004; Peterson *et al.*, 2007) other studies have found little difference in the composition of the microbiota between immunodeficient and normal mice and humans (Norhagen *et al.*, 1990; Reinholdt *et al.*, 1993; Marcotte *et al.*, 1996; Sait *et al.*, 2003).

The remainder of this section will focus on our current understanding about the interrelationship between the secretory immune system in the oropharynx and the α-haemolytic *Streptococcus S. mitis*, a pioneer commensal bacterium in the human oral cavity, and its close relative, the mucosal pathogen, *Streptococcus pneumoniae*. It should be understood from the outset that studying the role of the mucosal immune system in the regulation of commensal bacteria and mucosal pathogens requires a prospective study design to determine cause and effect. Conducting a longitudinal study in animal models can be difficult but pales in comparison to the difficulty of performing such a study in humans. Added to the complexity of study design is the fact that measuring antibody in mucosal secretions presents its own set of unique challenges. All too often external secretions are considered to be analogous to blood and the same experimental approaches are employed. However, there are significant differences between the measurement of antibodies in external secretions, such as saliva, and the measurement of antibodies in serum. In contrast to the systemic immune system where blood containing high concentrations of antibodies circulates in a closed vascular system, the secretory immune system is an open system where high volumes of secretions containing low concentrations of antibodies continuously flush mucosal epithelia. For example, about 1.0–1.5 l of saliva containing SIgA antibodies are produce every day that are swallowed and eventually excreted in faeces. Immunoglobulin concentration is inversely related to flow rate in external secretions. For this reason it is imperative to control for differences in flow rate within and between subjects. Analytes in external secretions such as saliva exhibit a circadian rhythm. Thus, whenever possible saliva samples should be collected at the same time of the day and at least an hour after eating, as gustatory stimuli increase flow rate resulting in lower antibody concentrations.

In addition, saliva samples may be 'contaminated' with secretions from other mucosal surfaces or from blood. For example, lacrimal secretions drain into the floor of the nose via the inferior nasal meatus. Therefore, crying results in the contamination of saliva with both lacrimal and nasal secretions. Furthermore, if a neonate/infant has been breast fed shortly before saliva collection it is likely that the saliva will be contaminated with milk, which contains high concentrations of SIgA. Crevicular fluid from the gingival crevice, and gingival exudate, resulting from periodontal inflammation or inflammation of the oral mucosa will contribute blood-derived antibodies, to whole saliva. SIgA in saliva also forms calcium-dependent heterotypic complexes with mucin, lactoferrin and other secretory analytes (Biesbrock *et al.*, 1991; Iontcheva *et al.*, 1997; Soares *et al.*, 2003). Consequently, unless a chelating agent such as EDTA is added to the saliva sample before freezing these heterotypic complexes fall out of solution and form an insoluble precipitate upon thawing. Thus, an indeterminate fraction of SIgA is lost.

Because of the possible contribution of inflammatory exudate and blood to whole saliva it cannot be assumed that all IgA antibody in secretions is exocrine. Antibodies specific for the alpha heavy chain of IgA cannot discriminate between secretory and serum IgA. Thus, measurement of exocrine IgA requires the use of an antibody specific for secretory component. Finally, quantitation of SIgA antibodies in primary units such as micrograms/minute/millilitre of saliva requires use of a secretory IgA standard. A serum IgA standard is not appropriate for this purpose.

Based on the foregoing it is not difficult to appreciate why there are few studies of bacteria–host interactions in the human oral cavity and at other human mucosal surfaces for that matter. A major handicap in studies seeking to determine the specificity of antibodies that react with either commensal or pathogenic bacteria is the lack of species-specific antigens that can unambiguously assign the antibodies as being induced by a particular species. The findings of studies examining the specificity of commensal-reactive salivary SIgA antibodies have been reviewed previously (Smith *et al.*, 1992; Marcotte *et al.*, 1998). It should be noted that almost all published studies employed a single culture collection 'type' strain to determine bacteria-reactive SIgA antibodies. Essentially the studies fall into two categories: those that conclude that the antibody is induced by the bacteria with which they react, and those that conclude that, while a small fraction of antibody is species-specific, the majority of antibodies are directed against cross-reacting antigens. The most likely sources of the cross-reacting antigens include other bacteria in the mouth, commensal bacteria at other mucosal surfaces or even food antigens. As an example of the first category Smith *et al.* in a cross-sectional study (Smith *et al.*, 1992), examined the reactivity of salivary IgA antibodies to culture supernatant antigens of *S. mitis* and *S. salivarius* in groups of infants between 3 and 27 weeks of age. Reactivity with these antigens was observed only after the isolation of the respective streptococcal species from the mouths of the infants from whom the SIgA antibodies were detected. Their data argued that the *S. mitis* and *S. salivarius*-reactive SIgA antibodies were induced by these bacteria and were specific for them.

In a carefully controlled prospective study of human infants from birth to 2 years of age (Cole *et al.*, 1999) we examined the specificity of SIgA antibodies in whole saliva reactive with *S. mitis, S. oralis*, several other species of viridans streptococci and *Enterococcus faecalis* (formerly *Streptococcus faecalis*) that are commensals in the oropharynx and the large bowel, respectively. Cross-absorption showed that antigens common to the streptococci and enterococcus bound a significant amount of the salivary SIgA antibody, although some species-specific antibodies were detected. An example of one such common/shared antigen is phosphorylcholine that is widely distributed among bacteria from the mouth and respiratory tract (Gillespie *et al.*, 1993, 1996). In this context, it is interesting that IgM monoclonal antibodies derived from mice immunized with subgingival plaque from advanced periodontal pockets or with strains of *Aggregatibacter actinomycetemcomitans, Fusobacterium nucleatum, Actinomyces israelii, S. mitis* and *S. oralis* were frequently reactive with phosphorylcholine (Gmür *et al.*, 1999).

*Streptococcus pneumoniae* is extremely closely related to *S. mitis* and *S. oralis* at the 16S ribosomal RNA level (Kawamura *et al.*, 1995; Kilian *et al.*, 2008) and exists in a state of intermittent carriage in the human nasopharynx. The Finnish Otitis Media (FinOM) cohort study is the most comprehensive study to examine naturally occurring salivary IgA antibodies reactive with *S. pneumoniae* antigens (Simell *et al.*, 2001, 2002, 2006). This prospective study followed 329 children from 2 to 24 months during 10 well baby visits (2, 3, 4, 5, 6, 9, 12, 15, 18, 24 months of age). Nasopharyngeal swabs (NPS) for detection of pneumococci were collected at each visit and whole saliva was collected at the 6, 12, 18, and 24 month visits. If the infants attended for a 'sick' visit then a nasopharyngeal aspirate (NPA) and middle ear fluid (MEF), where indicated, were obtained. At 6 months of age 33% of infants had one or more positive NPS, NPA or MEF culture finding. By 24 months 87% of the infants were culture positive for pneumococci. However, the carriage rate of pneumococci from infants who remained healthy (well visits only) throughout the study was 17% at 6 months and 43% at 24 months (Syrjanen *et al.*, 2001). Salivary IgA antibodies reactive with pneumococcal surface adhesin A (PsaA), pneumococcal surface protein A (PspA) and pneumolysin (Ply) were measured at 6-monthly intervals. At 6 months anti-PsaA IgA antibodies were detected in 57%, anti-Ply IgA antibodies in 85%, and anti-PspA IgA antibodies in 25% of the infants. At 24 months the prevalence of antibodies was 74%, 93% and 88% for the three antigens, respectively. As might be expected a higher proportion of IgA-positive samples and a higher antibody concentration was observed in culture-positive infants than in culture-negative infants. However, pneumococci-reactive antibodies were detected in more infants than were carriage-positive for the bacterium. This may reflect the fact that some carriage-negative infants were, in fact, carriage positive. Alternatively, the anti-PsaA and anti-PspA salivary IgA antibodies may have been induced in part by other viridans streptococci (Sampson *et al.*, 1994; Paton *et al.*, 1997).

In our prospective study that examined the salivary SIgA antibody response to *S. mitis* (Cole *et al.*, 1999) we observed that although the proportion of SIgA antibody in whole saliva reactive with the type strain of *S. mitis* biovar 1 increased from birth to 6 months, it declined over the next 18 months of life even though the concentration of SIgA in saliva continued to increase. The finding that, despite being stable at the species level, *S. mitis* exhibits significant clonal diversity, clonal turnover and replacement (Hohwy *et al.*, 1995; Fitzsimmons *et al.*, 1996; Hohwy *et al.*, 2001; Kirchherr *et al.*, 2005) called into question whether the use of a type strain or any single strain for that matter was adequate to probe the subtleties of bacteria–mucosal antibody interactions. Therefore, we decided to study the salivary SIgA antibody response of infants to their own colonizing strains.

The experimental approach (Wirth *et al.*, 2008) was to probe immunoblots of cell surface extracts of clones with saliva samples collected from an infant: (1) before detection of the clone, (2) during colonization by the clone, and (3) after the clone was no longer detected. We hypothesized that the appearance of a clone in the mouth could induce clone-specific SIgA antibodies that may play a role in its subsequent suppression or elimination. Also, we hypothesized that the limited concentration of *S. mitis*-specific SIgA antibody in saliva was the result of the very short exposure of the mucosal immune system to the individual clones of *S. mitis* that appear and rapidly disappear from the mouth. Under these conditions of clonal replacement, it might be expected that the secretory immune system would respond to those antigens that are common to every clone of *S. mitis* because these provide a 'stationary target'. Such antigens may be species specific, genus specific or even extend to related genera.

Our findings (Wirth *et al.*, 2008) showed simple patterns of immunoblots of cell surface extracts developed with saliva samples collected before 2 months of age, suggesting that neonates produce only a narrow repertoire of SIgA antibodies in response to the limited diversity of *S. mitis* colonizing the mouth. It is possible that antigens specific for strains that colonize shortly after birth do not provide a sufficient stimulus to the mucosal immune system. Early-colonizing strains may instead induce antibodies directed against common antigens that contribute to their elimination, precluding induction of SIgA antibodies to unique antigens. With increasing age, saliva samples showed, in most cases, an increase in the numbers of bands recognized by SIgA antibodies. Presumably this reflects the broadening specificities of SIgA antibodies induced by the expanding population of *S. mitis* biovar 1 and other strain-specific antibody.

It is possible that the antibody reactive with these bands was not induced by that clone but by other undetected clones or other related bacteria colonizing the infant. However, these antibodies did not react with bands common among the clones that persisted in an infant. Therefore, they discriminated among the clones colonizing the infant and any influence they had on colonization would affect a specific clone.

Some bands on immunoblots that were unique to a clone were only detected by SIgA antibody in single saliva samples collected at colonization or post colonization. This was the case for three-quarters of the clones examined. It could be argued that colonization by the strain induced SIgA antibodies specific for bands that were unique among the other persistent clones, isolated from the infant, and that these antibodies resulted either in the elimination of the clone from the site or caused its numbers to fall below the ability to detect it. That the antibody to 'unique' bands was very often transient in nature is consistent with the loss of the organism from the site eliminating the antigenic stimulus.

In an attempt to identify the antigens recognized by salivary SIgA antibodies we conducted a preliminary study of two strains of *S. mitis* biovar 1 isolated from the same infant using two-dimensional immunoblots of surface-associated antigens probed with saliva obtained from that infant 6 months of age. The two strains were #6855 (isolated 2 weeks post partum) and #9574 (isolated at 1 month and 2 months post partum). Both strains were isolated from the tongue and had been ribotyped. The PVU II ribotypes shared only 50% similarity. *S. mitis* biovar 1 surface-associated antigens were extracted as described previously (Pearce *et al.*, 1995). Two-dimensional (2D) gel electrophoresis was performed using standard methods (Görg *et al.*, 2004). Following electrophoresis one slab gel of a pair was stained with Coomassie Brilliant Blue R-250 and the other gel was submitted to electrophoretic transfer to a polyvinylidene fluoride (PVDF) membrane. PVDF membranes were stained with Ponceau-S to visualize spots and then completely destained. Blocked membranes were then incubated overnight with saliva samples diluted to contain a concentration of 1.5 pg of SIgA per ml. After washing away unbound saliva, biotin-conjugated rabbit IgG antibody to human α-chains was used to detect bound SIgA. Bound SIgA was detected using horseradish peroxidase-conjugated streptavidin and 3,3-diaminobenzidine tetrahydrochloride/$H_2O_2$ as substrate. Each gel and blot was scanned with a laser densitometer and the images were analysed such that all major spots were outlined, quantified and matched on the gels. In-gel digestion of protein 'spots' was performed and peptides subjected to LC-MS/MS sequencing.

The cell surface extracts of isolate #6855 and isolate #9574 were resolved into 135 and 122 spots, respectively, by 2D gel electrophoresis (Figs. 1.2 and 1.3). No proteins with a pI and an $M_r$ corresponding to lactate dehydrogenase or pyruvate oxidase were observed when the gels were stained with Coomassie Blue R-250, indicating that there had been no significant cell lysis during the preparation of the cell surface extracts. Using subtraction analysis 13 spots were unique to isolate #6855 while 22 spots were unique to isolate #9574 (Table 1.1). Spots that were approximately 3.0-fold higher in each isolate compared to the other are shown in Table 1.2. Blots of each isolate treated with whole saliva from infant #6 and developed with anti-human α-chain antibody revealed two spots on the 2D immunoblot of isolate #6855 (Fig. 1.4) and three on the blot of isolate #9574 (Fig. 1.5) that could be matched with those on the corresponding Coomassie Blue stained gel. Mass spectroscopy revealed that two of the spots in each of the gels were α-enolase and triosephosphate isomerase. The sequence of the third protein from the immunoblot of isolate #9594 could not be identified. The relatively few spots bound by salivary SIgA antibodies was surprising, given the complexity of the 2D Coomassie-stained gels and our detection of 12 reactive bands in one dimensional immunoblotting of the surface extract of isolate #9574 with infant #6 saliva used in this study (Wirth *et al.*, 2008). It may be that the detection method used was not sufficiently sensitive to reveal binding of low affinity SIgA antibodies or the separation in two dimensions resolved spots so that too few epitopes were present.

However, despite the relatively few reactive spots we can say that those detected were either in high abundance or were bound by high affinity SIgA antibody or both and constitute significant antigens in terms of the salivary SIgA antibody response. In particular, the demonstration of antibody to α-enolase is interesting, given that α-enolase and triosephosphate isomerase are thought to be significant in the colonization of surfaces by bacteria. The presence of a glycolytic pathway enzyme on the bacterial surface was first reported for glyceraldehyde 3-phosphate dehydrogenase (GAPDH) on the surface of *Streptococcus pyogenes* (Pancholi *et al.*, 1992). GAPDH was named the streptococcal surface dehydrogenase (SDH) (Pancholi *et al.*, 1992) and has since

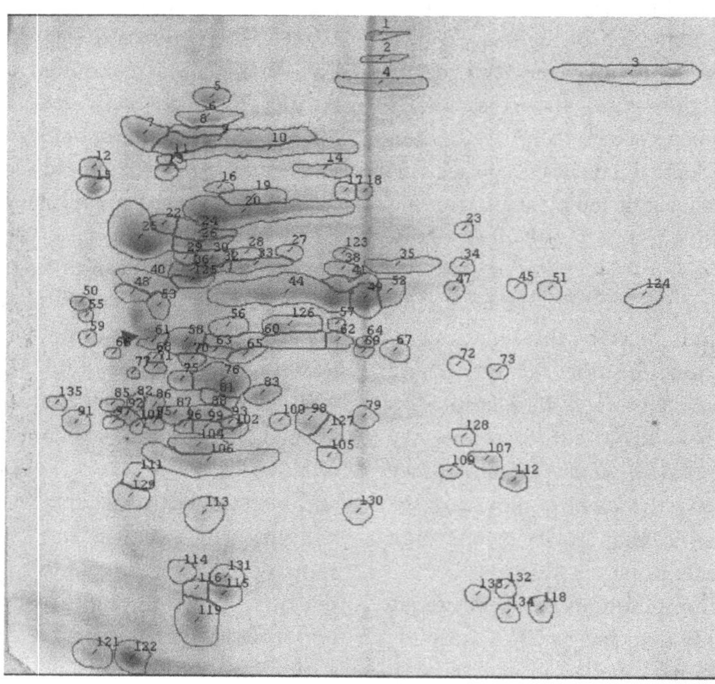

**Figure 1.2** 2D gel of the cell surface extract of isolate #6855 stained with Coomassie Brilliant Blue R250. The polypeptides were resolved into 135 spots.

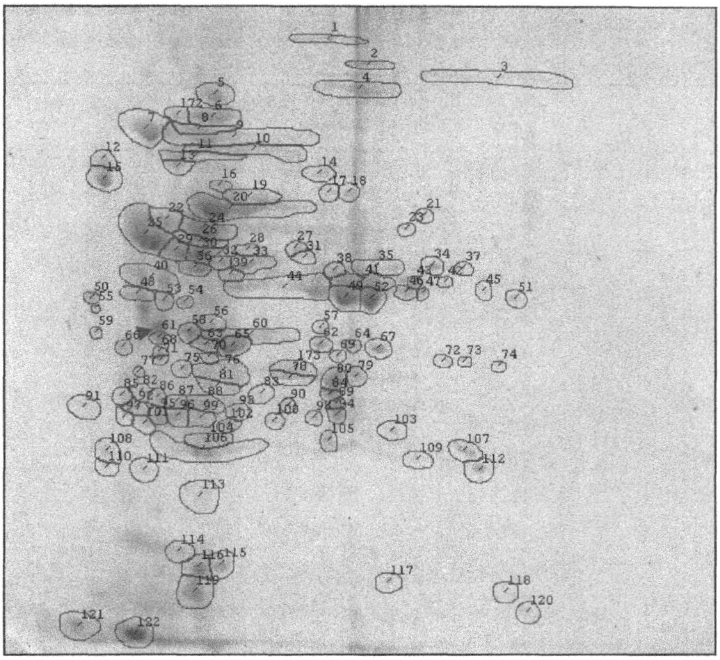

**Figure 1.3** 2D gel of the cell surface extract of isolate #9574 stained with Coomassie Brilliant Blue R250. The polypeptides were resolved into 122 spots.

been found to be a cell surface component of many bacteria and the opportunistic fungus *Candida albicans* (Pancholi *et al.*, 2003). To date, five glycolytic enzymes, GAPDH, α-enolase, phosphoglycerate kinase, phosphoglycerate mutase and triosephosphate isomerase, all sequential enzymes in the second half of the Embden–Myerhof–Parnas pathway, have been detected on the surface of the group A streptococcus, other streptococcal species, other bacteria, fungi, protozoa and parasites (Fischetti *et al.*, 2000). SDH binds fibronectin, lysozyme, myosin, actin and α-enolase (SEN and Eno) appear to be the major plasminogen-binding protein of pathogenic streptococci (Pancholi *et al.*, 1998) and also bind mucin (Ge *et al.*, 2004). Alpha-enolase and triosephosphate isomerase are significant molecules on the surfaces of *S. pneumoniae* and *S. oralis*, both of which are closely related to *S. mitis* (Bergmann *et al.*,2001; Wilkins *et al.*, 2003). Interestingly, triosephosphate isomerase was one of 23 immunogenic proteins recognized by IgG antibodies in the convalescent serum of a patient with disseminated pneumococcal infection (Zysk *et al.*, 2000).

Since commensal viridans streptococci, in particular *S. mitis* and *S. oralis*, share many surface-associated antigens with the pneumococcus and colonize the oropharynx almost immediately post partum, it is conceivable that they could prime the mucosal immune system ahead of colonization of pneumococci much in the same way that *Neisseria lactamica* is thought to contribute to acquired immunity to *Neisseria meningitidis* (Troncoso *et al.*, 2002).

## The significance of clonal diversity, *S. mitis* and survival against the host immune response

Up to this point we have considered the immunological pressures brought to bear on bacteria that attempt to establish on the mucosal surfaces of the body. However, commensal and pathogenic bacteria are not passive participants in this relationship as they respond and adapt continuously. Human mucosal pathogenic bacteria exist by continuous horizontal transfer within a host population and at best may exist in their human host, intermittently, during asymptomatic carriage. Mucosal pathogens have a variety of methods by which to evade SIgA, the principal effector of humoral

**Table 1.1** Reference spot numbering, p1 and MW for polypeptide spots analysed for samples #6855 (gel 2298 #5) and #9574 (gel 2298 #6). Also shown are fold increase or decrease (difference) between the polypeptides for *6855* versus *9574*

| Spot no. | pI | MW | 6855, 2298 #5, Spot% | 9574, 2298 #6, Spot% | 6855 vs. 9574, difference |
|---|---|---|---|---|---|
| 123 | 5.8 | 51,036 | 0.118 | – | +++ |
| 124 | 6.7 | 41,614 | 0.097 | – | +++ |
| 125 | 5.3 | 47,136 | 1.525 | – | +++ |
| 126 | 5.6 | 41,111 | 0.732 | – | +++ |
| 127 | 5.8 | 30,384 | 0.160 | – | +++ |
| 128 | 6.2 | 28,080 | 0.051 | – | +++ |
| 129 | 5.1 | 23,682 | 0.201 | – | +++ |
| 130 | 5.9 | 24,045 | 0.078 | – | +++ |
| 131 | 5.4 | 20,002 | 0.341 | – | +++ |
| 132 | 6.3 | 18,179 | 0.033 | – | +++ |
| 133 | 6.2 | 17,756 | 0.041 | – | +++ |
| 134 | 6.3 | 16,593 | 0.046 | – | +++ |
| 135 | 4.8 | 32,814 | 0.041 | – | +++ |
| 21 | 6.1 | 59,101 | – | 0.067 | – |
| 31 | 5.7 | 50,757 | – | 0.064 | – |
| 37 | 6.2 | 47,894 | – | 0.044 | – |
| 39 | 5.5 | 47,693 | – | 0.172 | – |
| 42 | 6.2 | 45,009 | – | 0.040 | – |
| 43 | 6.1 | 44,898 | – | 0.266 | – |
| 46 | 6.0 | 45,221 | – | 0.400 | – |
| 54 | 5.3 | 44,120 | – | 0.247 | – |
| 74 | 6.3 | 34,589 | – | 0.029 | – |
| 78 | 5.7 | 34,780 | – | 0.366 | – |
| 80 | 5.8 | 33,909 | – | 1.764 | – |
| 84 | 5.8 | 32,283 | – | 0.536 | – |
| 89 | 5.8 | 31,238 | – | 0.315 | – |
| 90 | 5.6 | 31,064 | – | 0.053 | – |
| 94 | 5.8 | 29,960 | – | 0.624 | – |
| 103 | 6.0 | 30,168 | – | 0.136 | – |
| 108 | 5.0 | 26,698 | – | 0.147 | – |
| 110 | 5.0 | 25,331 | – | 0.031 | – |
| 117 | 6.0 | 17,540 | – | 0.093 | – |
| 120 | 6.4 | 15,501 | – | 0.099 | – |
| 172 | 5.3 | 91,310 | – | 0.225 | – |
| 173 | 5.7 | 36,000 | – | 0.370 | – |

The fold differences are calculated from spot percentages (individual spot density divided by total density of all measured spots). Polypeptide spots present only in 6855 are indicated with +++. Polypeptide spots present only in 9574 are indicated with –. A total of 137 spots were analysed.

**Table 1.1** (Continued)

| Spot no. | pI | MW | 6855, 2298 #5, Spot% | 9574, 2298 #6, Spot% | 6855 vs. 9574 difference |
|---|---|---|---|---|---|
| 44 | 5.6 | 45,464 | 7.121 | 2.242 | 3.2 |
| 76 | 5.4 | 34,838 | 5.852 | 0.820 | 7.1 |
| 81 | 5.4 | 33,154 | 0.675 | 0.130 | 5.2 |
| 82 | 5.1 | 32,530 | 0.264 | 0.026 | 10.2 |
| 83 | 5.6 | 32,341 | 0.797 | 0.052 | 15.3 |
| 87 | 5.3 | 31,412 | 0.814 | 0.228 | 3.6 |
| 92 | 5.1 | 30,766 | 0.503 | 0.153 | 3.3 |
| 98 | 5.7 | 31,292 | 0.385 | 0.104 | 3.7 |
| 11 | 5.3 | 77,349 | 0.086 | 0.507 | –5.9 |
| 29 | 5.3 | 51,964 | 0.406 | 1.841 | –4.5 |
| 34 | 6.1 | 48,449 | 0.051 | 0.253 | –5.0 |
| 52 | 5.9 | 44,563 | 1.059 | 3.786 | –3.6 |
| 65 | 5.5 | 37,684 | 0.387 | 2.797 | –7.2 |
| 66 | 5.1 | 37,422 | 0.076 | 0.236 | –3.1 |
| 105 | 5.8 | 27,647 | 0.060 | 0.324 | –5.4 |
| 116 | 5.3 | 18,605 | 0.230 | 1.282 | –5.6 |

Increased polypeptide spots with a fold difference of ≥3 are shown in the upper half of the table by positive numbers, while decreased spots with a fold difference of ≥–3 are shown in the lower part of the table by negative numbers. A total of 137 spots were analysed.

**Figure 1.4** Western blot of surface extract of strain #6855. Two immunoreactive proteins were detected with anti-human α-chain antibody (arrows).

**Figure 1.5** Western blot of surface extract of strain #9574. Blot was developed with anti-human α-chain antibody and three spots were detected (arrows).

immunity at surface epithelia. Such mechanisms include proteases that either specifically or non-specifically degrade SIgA (Kilian *et al.*, 1996), IgA binding proteins that can capture SIgA via the Fc region (Woof, 2002), shedding surface antigens that bind antibody away from the bacterial cell surface (Lee, 1995), or the display of a large number of immunologically distinct virulence factors such as the over 90 capsule types of *S. pneumoniae*, or M proteins of *S. pyogenes*. In addition, pathogenic bacteria may undergo phase and antigenic variation (van der Woude *et al.*, 2004) that enable bacteria to turn on or turn off the expression of one or more surface proteins (phase variation) or to continually change the epitopes on surface proteins, such as pili (antigenic variation).

Since commensal bacteria persist on mucosal surfaces for the lifetime of the individual it stands to reason that they too must have evolved mechanisms by which to evade the constant pressure of secretory immunity. In fact several species of viridans streptococci including the pioneers, *S. mitis* and *S. oralis*, produce IgA1 protease (Kilian *et al.*, 1996) that may inactivate SIgA1 antibodies in saliva. In this context, it is interesting that over 90% of SIgA in the saliva of neonates belongs to subclass 1 of SIgA (Fitzsimmons *et al.*, 1994). Furthermore, viridans streptococci elaborate extracellular polysaccharide (Wilcox *et al.*, 1987), they also bind salivary macromolecules (De Jong *et al.*, 1987) which may mask them from host immunity and most likely shed surface antigens, but their principal mode of evasion appears to be their extraordinary clonal diversity and clonal turnover (Howhy *et al.*, 1995, 2001; Fitzsimmons *et al.*, 1996; Kirchherr *et al.*, 2005; Bek-Thomsen *et al.*, 2008) which may prevent targeting of SIgA antibodies to colonizing clones.

Before discussing the diversity of *S. mitis* an important point must be addressed, which is not limited to the oral streptococci but for which the oral streptococci provide a classic example: accurate taxonomy. A major problem in studying the natural biology of the commensal viridans streptococci, particularly when different studies are compared is the complexity of their taxonomy, which arises in part from their close genetic relationship. The majority of traditional studies of the taxonomy and population biology of viridans streptococci have, often for practical reasons, used small sample sizes that did not reveal the degree of phenotypic diversity among strains. This led to conclusions and identification schemes that subsequently have had to be revised. Consequently, reliably assigning an isolate to a species in order to follow its biology in nature has been extremely difficult, and accurate identification is a cornerstone of population studies of oral streptococci *in vivo*.

*S. mitis* and *S. oralis* are the two members of the 'mitis' group that have been the focus of most investigations because they are pioneers in the human oropharynx and can colonize both soft and hard tissues. The 'mitis' group of viridans *Streptococcus* currently includes the important pathogen *S. pneumoniae* and 12 other validly described species, *S. australis, S. cristatus* (formerly *S. crista*), *S. gordonii, S. infantis, S. mitis, S. oligofermentans, S. oralis, S. parasanguinis* (formerly *S. parasanguis*), *S. peroris, S. pseudopneumoniae, S. sanguinis* (formerly *S. sanguis*) and *S. sinensis*. As outlined above a major problem in studying 'mitis' group streptococci is that they are notoriously difficult to speciate using traditional physiological tests, despite the existence of well respected phenotypic identification schemes (Kilian *et al.*, 1989; Beighton *et al.*, 1991). One problem of drawing conclusions based on small data sets is exemplified by the fact that it has been recently argued that it is not possible to differentiate *S. mitis, S. oralis* and *S. infantis* by anything other than molecular biologic approaches (Bek-Thomsen *et al.*, 2008). The clear inference is that the strains examined in studies of the diversity of *S. mitis* based on the identification of this bacterium by physiological tests probably included strains of *S. oralis* and *S. infantis*. Consequently, it now seems likely that molecular methods will need to be applied to isolates from studies *in vivo* to assign strains to a species. However, phenotypic tests will still need to be applied in order to show phenotypic diversity among these strains.

We appreciated the difficulty in separating *S. mitis*, and *S. oralis* (Kirchherr *et al.*, 2007; Wirth *et al.*, 2008) following analysis of 4,440 recent isolates and traditional physiological tests. The tests included physiological, serological, and DNA–DNA hybridization examination of isolates obtained from four infants during the first year of life. The results of our studies brought into question the suitability of the current 'type' strains of *S. mitis* NCTC 12261 and *S. oralis* ATCC 35037 as representatives of these species. While there is little doubt about the power of non-culture, molecular techniques it is an inescapable fact that one must be able to isolate and cultivate strains if one wishes to determine their interaction with the mucosal immune system. Again, this is a daunting task that, because of the diversity of these bacteria, requires the random collection of large numbers of isolates using non-selective media to avoid selection bias, proper identification of the isolates, and storage to allow further study.

The population dynamics of *S. mitis* bv. 1 has been studied within parents and their infants and within neonates (Howhy *et al.*, 1995, 2001; Fitzsimmons *et al.*, 1996; Kirchherr *et al.*, 2005; Bek-Thomsen *et al.*, 2008). These studies reported extensive diversity within an individual as well as between subjects. We studied the genetic diversity and clonal turnover of *S. mitis* bv. 1 to a large number of isolates collected from infants from birth to 1 year of age (Kirchherr *et al.*, 2005). We examined clonal diversity and turnover of *S. mitis* bv. 1 colonizing the cheeks, tongue, and primary central incisors. Because of the phenotypic heterogeneity discussed earlier we selected a subset of *S. mitis* bv. 1 isolates that produced neuraminidase, β-*N*-acetylglucosaminidase, β-*N*-acetyl-galactosaminidase, and IgA1protease. Strains with this phenotype represented 1/3 of the total number of isolates of *S. mitis* bv. 1 recovered from these infants and this phenotype was similarly numerically significant in their mothers' saliva.

Overall, our findings concerning the clonal diversity and turnover of *S. mitis* are remarkably complementary to those of Kilian's group despite the fact that our studies were conducted on different continents and with slight methodological differences. For example, in our study (Kirchherr *et al.*, 2005) of 859 isolates obtained from the cheeks, tongue, and teeth of these infants, two-thirds represented unique clones. The vast majority of clones were isolated once only, and rarely from more than one surface (Hohwy *et al.*, 1995, 2001; Fitzsimmon *et al.*, 1996). However, a small number of clones were isolated on two or more subsequent visits and were not always found on the same surface from which they were isolated initially, although there was no preference for clones to move from one particular surface to another. Moreover, clones did not appear to exhibit specific tropisms for transfer to any particular sampled surface, showing that they were capable of colonizing both soft and hard tissues. There was little evidence to suggest that clones, either persistent in a habitat or transferring to a habitat, could increase the numbers of their population significantly. These findings support the suggestion of Hohwy *et al.* (2001) that the species niche in the habitat appears to be maintained by a succession of clones rather than by stable strains.

One of the impressive features of *S. mitis* bv. 1 colonization is the rate of clonal turnover/replacement. In our studies (Kirchherr *et al.*, 2005) and those of Kilian's group (Hohwy *et al.*, 1995, 2001) it was rare to recover the same clone at successive visits even when such visits were but a few weeks apart. *S. oralis* which, genetically, is extremely closely related to *S. mitis* (Kawamura *et al.*, 1995) also appears to be highly diverse (Alam *et al.*, 2000; Do *et al.*, 2009). It is interesting that while stability of *S. oralis* was not observed in oral rinse samples over a 12-week period, there was evidence of stability in parallel samples of approximal dental plaque (Alam *et al.*, 2000). However, Hohwy *et al.* (2001) failed to detect persisting genotypes of *S. mitis* on the tooth surfaces of two adults over a 5-year period and, similarly, there was no evidence of persistent dental clones in our study (Kirchherr *et al.*, 2005). Furthermore, there was no evidence that clonal diversity on tooth surfaces was less than that observed on mucosal surfaces, although, admittedly, the period of observation was short. An obvious question remains regarding the source of these diverse genotypes of *S. mitis* bv. 1 that colonize the infant's mouth and appear to be in a constant state of flux.

Studies of *S. mitis* (Hohwy *et al.*, 1995; Kirchherr *et al.*, 2005) show a lack of fidelity between mother and infant clones that does not support the neonatal acquisition of this bacterium from either mother or father. If the large number and frequent turnover of genotypes of *S. mitis* bv. 1 within an infant cannot be explained by frequent acquisition of new exogenous clones from the mother and their subsequent loss, to what can these phenomena be attributed? Other possibilities may include any or all of the following: (i) newly emerging clones arising from other habitats in the respiratory tract colonized by this species (Howhy *et al.*, 2001); (ii) the numbers of particular clones fluxing such that they fall below the level of detection, but are still present in the mouth; (iii) a high rate of genetic mutation; and (iv) horizontal gene transfer. However, Hohwy *et al.* (2001) have discounted recombination *in situ* as playing a major role in clonal diversity and turnover. It remains unclear what factors drive clonal diversity and turnover: dietary, availability of host components for metabolism, interactions between bacteria, or host immunity. We have hypothesized that mucosal immunity contributes to the environmental pressure driving the genetic diversity, and clonal turnover may be a mechanism employed by *S. mitis* to evade immune elimination.

## Concluding remarks

To date we have considerable data on the biology of the oral streptococci but the emphasis has been placed on the genetics of specific strains, particularly in relation to their formation of, and interactions within, biofilms. The data are generated most often from *in vitro* models. This emphasis and approach has been responsible for a dramatic increase in our understanding of the oral streptococci; however we have a relative paucity of knowledge about the biology of these organisms in nature. It is obvious that there are limitations to the type of study that can be undertaken *in vivo* in humans and in experimental animals. However, there is a long history in oral microbiology of experiments *in vivo*. Moreover, modern analytical techniques are being applied to oral bacterial communities collected *in vivo* by use of traditional technology (Aas *et al.*, 2008; Zaura *et al.*, 2009; Kanasi *et al.*, 2010) and these, compared to cultural analysis, have shown the inherent complexity of the oral microbiota in health and disease; for example, the association of organisms other than *S. mutans* with dental caries. However, clonal analysis does not provide viable isolates of the organisms detected, meaning that they cannot be included into *in vitro* models. Consequently, the majority of *in vitro* models seem to be restricted to a limited number of laboratory strains and seldom use recent clinical isolates.

It is becoming apparent that, as our knowledge

of diversity among bacteria increases, it becomes necessary to consider modern approaches to taxonomy in oral microbial ecology, and most of the material in this chapter is oral microbial ecology. It should also be repeated that accurate taxonomy, and a comprehensive understanding of taxonomy, is fundamental to microbial ecology and the biology of the oral streptococci.

Modern taxonomic methods such as multilocus sequence typing (MLST) or multilocus sequence analysis (MLSA) (Gevers *et al.*, 2005; Hanage *et al.*, 2006) can be used to identify recent isolates, although these methods may not identify a specific phenotype. Currently, the application of MLST and MLSA is a topic of keen discussion among taxonomists, who recognize the need to identify strains of ecological significance (Gevers *et al.*, 2005; Hanage *et al.*, 2006; Schleifer *et al.*, 2009).

Some use has also been made of modern technology in examining the formation of dental biofilms, including the role of *Streptococcus* and *Veillonella*. It has been known for many years that strains of these two genera interact (Mikx *et al.*, 1972; McBride *et al.*, 1981). Relatively recently, *in vivo* examinations have been made of this interaction in humans using an enamel chip model (Palmer *et al.*, 2006; Chalmers *et al.*, 2008). The study by Palmer *et al.* (2006) demonstrated changes in the predominant phenotype of *Veillonella* strains during early development of human dental plaque. Although the strains were tested against streptococcal reference strains, the streptococci or other organisms associated with the veillonellae in the dental plaque were not examined. In a following study by Chalmers *et al.* (2008) the microbiota colonizing enamel chips was grown in enrichment medium and reconstructed *in vitro*. The streptococci isolated included strains identified as *S. oralis* and *S. gordonii* and perhaps significantly, the *Veillonella* strain, although shown to be present, could not be cultivated. Subsequently, in a three-species *in vitro* biofilm model, these workers used the laboratory strain *Veillonella* sp. PK1910 a, in place of the uncultivated *Veillonella* strain seen *in vivo*, even though PK1910 was clearly not equivalent to it. The authors suggest that these strains had a nutritional requirement that might be satisfied by growing together with a *Streptococcus* in saliva. This example is provided in this closing section to emphasize the complexity of understanding and analysing the role of oral streptococci *in vivo*.

As is common with chapters on specific bacteria in oral ecology one can forget that these organisms grow in complex communities. The oral streptococci grow in association with hundreds of other species of oral bacteria, they are highly diverse, and they adapt their phenotypes to survive in their local environments. It remains to be seen whether the intense study of these organisms will reveal mechanisms to control dental disease. Certainly opinions differ on ways to approach this problem (Taubman *et al.*, 2006; Takahashi *et al.*, 2011).

## References

Aas, J.A., Paster, B.J., Stokes, L.N., Olsen, I., and Dewhurst, F.E. (2005). Defining the normal bacterial flora of the oral cavity. J. Clin. Microbiol. *43*, 5721–5732.

Aas, J.A., Griffen, A.L., Dardis, S.R., Lee, A.M., Olsen, I., Dewhirst, F.E., Leys, E.J., and Paster, B.J. (2008). Bacteria of dental caries in primary and permanent teeth in children and young adults. J. Clin. Microbiol. *46*, 1407–1417.

Aertsen, A., and Michiels, C.W. (2005). Diversify or die: Generation of diversity in response to stress. Crit. Rev. Microbiol. *31*, 69–78.

Ahmed, R., Hassall, T., and Gray, J. (2003). Viridans *Streptococcus* bacteremia in children on chemotherapy for cancer: an underestimated problem. Pediatr. Hematol. Oncol. *20*, 439–444.

Ahn, S.-J., Ahn, S.-J., Browngardt, C.M., and Burnie, R.A. (2009). Changes in biochemical and phenotypic properties of *Streptococcus mutans* during growth with aeration. Appl. Environ. Microbiol. *75*, 2517–2527.

Alaluusua, S., Alaluusua, S.J., Karjalainen, J., Saarela, M., Holttinen, T., Kallio, M., Höllttä, P., Torkko, H., Relander, P., and Asikainen, S. (1994). The demonstration by ribotyping of the stability of oral *Streptococcus mutans* infection over 5 to 7 years in children. Arch. Oral Biol. *39*, 467–471.

Alam, S., Brailsford, S.R., Adams, S., Allison, C., Sheehy, E., Zoitopoulos, L., and Kidd, E.A. (2000). Genotypic heterogeneity of *Streptococcus oralis* and distinct aciduric subpopulations in human dental plaque. Appl. Environ. Microbiol. *66*, 3330–3336.

Avery, S.V. (2006). Microbial cell individuality and the underlying sources of heterogeneity. Nature Rev. Microbiol. *4*, 577–587.

Banas, J.A., and Vickerman, M.M. (2003). Glucan-binding proteins of the oral streptococci. Crit. Rev. Oral Biol. Med. *14*, 89–99.

Beighton, D., Hardie, J.M., and Whiley, R.A. (1991). A scheme for the identification of viridans streptococci. J. Med. Microbiol. 35, 367–372.

Bek-Thomsen, M., Tettelin, H., Hance, I., Nelson, K.E., and Kilian, M. (2008). Population diversity and dynamics of *Streptococcus mitis*, *Streptococcus oralis* and *Streptococcus infantis* in the upper respiratory tracts of adults determined by a nonculture strategy. Infect. Immun. *76*, 1889–1896.

Bergmann, S., Rohde, M., Chatwal, G.S., and Hammerschmidt, S. (2001). α-enolase of *Streptococcus pneumoniae* is a plasmin(ogen)-binding protein displayed on the bacterial cell surface. Mol. Microbiol. *40*, 1273–1287.

Biesbrock, A.R., Reddy, M.S., and Levine, M.J. (1991). Interaction of a salivary mucin-secretory immunoglobulin A complex with mucosal pathogens. Infect. Immun. *59*, 3492–3497.

Bowden, G.H. (1990). Effects of fluoride on the microbial ecology of dental plaque. J. Dent. Res. *69*, (special issue) 653–659.

Bowden, G.H., and Hamilton, I.R. (1998), Survival of oral bacteria. Crit. Rev. Oral Biol. Med. *9*, 54–85.

Bowden, G.H.W., Odlum, O., Nolette, N., and Hamilton, I.R. (1982). Microbial populations growing in the presence of fluoride at low pH isolated from dental plaque of children living in an area with fluoridated water. Infect. Immun. *36*, 247–254.

Brailsford, S.R., Shah, B., Simons, D., Gilbert, S., Clark, D., Ines, I., Adams, S.E., Allison, C., and Beighton, D. (2001). The predominant aciduric microflora of root caries lesions. J. Dent. Res. *80*, 1822–1833.

Brailsford, S.R., Sheehy, E.C., Gilbert, S.C., Clark, D.T., Kidd, E.A.M., Zoitopoulos, L., Adams, S.E., Visser, J.M., and Beighton, D. (2005). The microflora of the erupting first permanent molar. Caries Res. *39*, 78–84.

Burnie J.P., Brooks, W., Donohoe, M., Hodgets, S., al-Ghamdi, A., and Matthews, R.C. (1996). Defining antibody targets in *Streptococcus oralis* Infection. Infect. Immun. *64*, 1600–1608.

Busuioc, M., Buttaro, B.A., and Piggot, P.J. (2010). The *pdh* operon is expressed in a subpopulation of stationary-phase bacteria and is important for survival of sugar-starved *Streptococcus mutans*. J. Bacteriol. *192*, 4395–4402.

Carlén, A., Börjesson, A.C., Nikdel, K., and Olsson, J. (1998). Composition of pellicles formed *in vivo* on tooth surfaces in different parts of the dentition, and in vitro on hydroxyapatite. Caries Res. *32*, 447–455.

Caulfield, P.W., Dasanayake, A.P., Li, Y., Pan, Y., Hsu, J., and Hardin, M. (2000). Natural history of *Streptococcus sanguinis* in the oral cavity of infants: Evidence for a discrete window of infectivity. Infect. Immun. *68*, 4018–4023.

Cerutti, A., and Rescigno, M. (2008). The biology of intestinal immunoglobulin A responses. Immunity *28*, 740–750.

Chalmers, N.I., Palmer, R.J., Jr., Cisar, J.O., and Kolenbrander, P.E. (2008). Characterisation of a *Streptococcus* sp. *Veillonella* sp. community micromanipulated from dental plaque. J. Bacteriol. *190*, 8145–8154.

Chávez de Paz, L.E., Hamilton, I.R., and Svensäter, G. (2008). Oral bacteria in biofilms exhibit slow reactivation from nutrient deprivation. Microbiology *154*, 1927–1938.

Childers, N. (2009). Maternal transmission of mutans streptococci in severe-early childhood caries. Pediatric. Dent. *31*, 193–201.

Coenye, T. (2010). Response of sessile cells to stress from changes in gene/s expression to phenotypic adaptation. FEMS Immunol. Med. Microbiol. *59*, 239–252.

Cole, M.F., Bryan, S., Evans, M.K., Pearce, C.L., Sheridan, M.J., Sura, P.A., Wientzen, R.L., and Bowden, G.H. (1999). Humoral immunity to commensal oral bacteria in human infants: salivary secretory immunoglobulin A antibodies reactive with *Streptococcus mitis* biovar 1, *Streptococcus oralis, Streptococcus mutans,* and *Enterococcus faecalis* during the first two years of life. Infect. Immun. *67*, 1878–1886.

Crowley, P.J., Svensäter, G., Snoep, J.L., Bleiweis, A.S., and Brady, L.J. (2004). An *ffh* mutant of *Streptococcus mutans* is viable and able to physiologically adapt to low pH in continuous culture. FEMS Microbiol. Lett. *234*, 315–324.

Dawes, C. (2008). Salivary flow patterns and the health of hard and soft oral tissues. J. Am. Dent. Assoc. *139*, 18S-24S.

De Jong, M.H., and Van Der Hoeven. J.S. (1987). The growth of oral bacteria on saliva. J. Dent. Res. *66*, 498–505.

Delorme, C., Poyart, C., Dusko Ehrlich, S., and Renault, P. (2007). Extent of horizontal gene transfer in evolution of streptococci of the salivarius group. J. Bacteriol. *189*, 1330–1341.

Deng. D.M., ten Cate, J.M., and Crielaard, W. (2007). The adaptive response of *Streptococcus mutans* towards oral care products: involvement of the ClpP serine protease. Eur. J. Oral Sci. *115*, 363–370.

Diaz, P.I., Chalmers. N.I., Rickard, A.H., Kong, C., Milburn, C.L., Palmer, R.J., and Kolenbrander, P.E. (2006). Molecular characterization of subject-specific oral mcroflora during Initial colonization of enamel. Appl. Environ. Microbiol. 72, 2837–2848.

Dierksen K.P., and Tagg, J.R. (2000). The influence of indigenous bacteriocin- producing *Streptococcus salivarius* on the acquisition of *Streptococcus pyogenes* by primary school children in Dunedin, New Zealand. In Streptococci and Streptococcal Diseases Entering the New Millenium, Martin, D.R., and Tagg, J.R., eds. (Securacopy, Aukland, New Zealand), pp. 81–85.

Do, T., Jolley, K.A., Maiden, M.C.J., Gilbert, S.C., Clark, D., Wade, W.G., and Beighton, D. (2009). Population structure of *Streptococcus oralis*. Microbiology *155*, 2593–2602.

Do. T., Gilbert, S.C., Clark, D., Farida, A., Fatturi Parolo C.C., Maltz, M., Russell, R.R., Wade, W.G., and Beighton, D. (2010). Generation of diversity in *Streptococcus mutans* genes demonstrated by MLST. PLoS ONE *5*, e9073. doi:10.1371/journal.pone.0009073.

Dolente, B.A., Seco, O.M., and Lewis, M.L. (2000). Streptococcal toxic shock in a horse. J. Am. Vet. Med. Assoc. *217*, 64–67.

Douglass, J.M., Li, Y., and Tinanoff, N. (2008). Association of mutans streptococci between caregivers and their children. Pediatr. Dent. *30*, 375–387.

Eke, P.I., Braswell, L., Arnold, R., and Fritz, M. (1993). Sub-gingival microflora in *Macaca mulata* species of rhesus monkey. J. Periodont. Res. *28*, 72–80.

Fagarasan, S., Muramatsu, M., Suzuki, K., Nagaoka, H., Hiai, H., and Honjo, T. (2002). Critical roles of activation-induced cytidine deaminase in the homeostasis of gut flora. Science *298*, 1424–1427.

Fischetti, V.A. (2000). Surface proteins on Gram-positive bacteria. In Gram-Positive Pathogens, Fischetti, V.A., Novick, R.P., Ferretti, J.J., and Portnoy, D.A., eds. (ASM Press, Washington, DC), pp. 11–24.

Fitzsimmons, S.P., Evans, M.K., Pearce, C.L., Sheridan, M.J., Wientzen, R.L., and Cole, M.F. (1994). Immunoglobulin A subclasses in infants' saliva and in saliva and milk from their mothers. J. Pediatr. *124*, 556–573.

Fitzsimmons, S., Evans, M., Pearce, C., Sheridan, M.J., Wientzen, R., Bowden, G., and Cole, M.F. (1996). Clonal diversity of *Streptococcus mitis* biovar 1 isolates from the oral cavity of human neonates. Clin. Diagn. Lab. Immunol. *3*, 517–522.

Fozo, E.M., Kaifasz, J.K., and Quivey, R.G., Jr. (2004). Low pH induced membrane fatty acid alterations in oral bacteria. FEMS Microbiol. Lett. *238*, 291–295.

Ge, J., Catt, D.M., and Gregory, R.L. (2004). *Streptococcus mutans* surface α-enolase binds salivary mucin MG2 and human plasminogen. Infect. Immun. *72*, 6748–6752.

Gevers, D., Cohan, F.M., Lawrence, J.G., Spratt, B.G., Coeyne, T., Feil, E.J., Stackebrandt, E., Van de Peer, Y., Vandamme, P., Thompson, F.L., *et al.* (2005). Re-evaluating prokaryotic species. Nature Rev. Microbiol. *3*, 733–739.

Gillespie, S.H., McWhinney, P.H.M., Patel, S., Raynes, J.G., McAdam, K.P.W.J., Whiley, R.A., and Hardie, J.M. (1993). Species of alpha-hemolytic streptococci possessing a C-polysaccharide phosphorylcholine-containing antigen. Infect. Immun. *61*, 3076–3077.

Gillespie, S.H., Ainsclough, S., Dickens, A., and Lewin, J. (1996). Phosphorylcholine-containing antigens in bacteria from the mouth and respiratory tract. J. Med. Microbiol. *44*, 35–40.

Gmür, R., Thurnheer, T., and Guggenhein, B. (1999). Dominant cross-reactive antibodies generated during the response to a variety of oral bacterial species detect phosphorylcholine. J. Dent. Res. *78*, 77–85.

Görg, A., Weiss, W., and Dunn, M.J. (2004). Current two-dimensional electrophoresis technology for proteomics. Proteomics *4*, 3665–3865.

Griffin D.O., Holodick, N.E., and Rothstein, T.L. (2011). Human B1 cells in umbilical cord blood and adult peripheral blood express the novel phenotype $CD20^+$ $CD27^+$ $CD43^+CD70^-$. J. Exp. Med. *208*, 67–80.

Grönroos, L., Mättö, J., Saarela, M., Luoma, A.-R., Luoma, H., Jousimies-Somer, H., Pyhälä, L., Asikainen, S., and Alaluusua, S. (1995). Chlorhexidine susceptibilities of Mutans streptococcal serotypes and ribotypes. Antimicrob. Agents Chemother. *39*, 894–898.

Gross, H. (2007). Strategies to unravel the function of orphan biosynthesis pathways: recent examples and future prospects. Appl. Microbiol. Biotechnol. *75*, 267–277.

Hamada, T., Kawashima, M., Watanabe, H., Tagami, J., and Senpuku, H. (2004). Molecular Interactions of surface protein peptides of *Streptococcus gordonii* with human salivary components. Infect. Immun. *72*, 4819–4826.

Hanage, W.P., Fraser, C., and Spratt, B.G. (2006). Sequences, sequence clusters and bacterial species. Phil. Trans. R Soc B *361*, 1917–1927.

Hibbing, M.E., Fuqua, C., Parsek, M.R., and Brook Peterson, S. (2010). Bacterial competition surviving and thriving in the microbial jungle. Nat. Rev. Microbiol. *8*,15–25.

Higgins, R., Biberstein, E.L., and Jang, S. (1984). Nutritionally variant streptococci from corneal ulcers in horses. J. Clin Microbiol. *20*, 1130–1134.

Hohwy, J., and Kilian, M. (1995). Clonal diversity of the *Streptococcus mitis* biovar 1 population in the human oral cavity and pharynx. Oral Microbiol. Immunol. *10*, 19–25.

Hohwy, J., Reinholdt, J., and Kilian, M. (2001). Population dynamics of *Streptococcus mitis* in its natural habitat. Infect. Immun. *69*, 6055–6063.

Horz, H.P., Meinelt, A., Houben, B., and Conrads, G. (2007). Distribution and persistence of probiotic *Streptococcus salivarius* K12 in the human oral cavity as determined by real-time quantitative polymerase chain reaction. Oral Microbiol. Immunol. *22*, 126–130.

Hossain, M.S., and Biswas, I. (2011). Mutacins from *Streptococcus mutans* UA159 are active against multiple streptococcal species. Appl. Environ. Microbiol. *77*, 2428-2434.

Igarashi, M. (2008). Identification of *Streptococcus ferus* and the distribution of three kinds of mutans streptococci from pig tooth surface. Int. J. Oral Med. Sci. *7*, 7–11.

Iontcheva, I., Oppenheim, F.G., and Troxler, R.F. (1997). Human salivary mucin MG1 selectively forms heterotypic complexes with amylase, proline-rich proteins, statherin, and histatins. J. Dent. Res. *76*, 734–743.

Johnston, C., Hinds, J., Smith, A., van der Linden, M., Eldere, J.V., and Mitchell, T.J. (2010). Detection of large numbers of pneumococcal virulence genes in streptococci of the Mitis group. J. Clin. Microbiol. *48*, 2762–2769.

Kanasi, E.M., Dewhirst, F.E., Chalmers, N.I., Kent, R., Jr., Moore, A., Hughes, C.V., Pradhan, N., Loo, C.Y., and Tanner, A.C.R. (2010). Clonal analysis of the microbiota of severe early childhood caries. Caries Res. *44*, 485–497.

Kawamura, Y., Hou, X.G., Sultana, F., Miura, H., and Ezaki, T. (1995). Determination of 16S rRNA sequences of *Streptococcus mitis* and *Streptococcus gordonii* and phylogenetic relationships among members of the genus *Streptococcus*. Int. J. Syst. Bacteriol. *45*, 406–408.

Kawamura, Y., Hou, X.-G., Todome, Y., Sultana, F., Hirose, K., Shu, S.-E., Ezaki, T., and Ohkuni, H. (1998). *Streptococcus peroris* sp.nov. and *Streptococcus infantis* sp.nov., new members of the *Streptococcus mitis* group, isolated from human clinical specimens. Int. J. Syst. Bacteriol. *48,* 921–927.

Kilian, M., Mikkelsen, L., and Henrichsen, J. (1989). Taxonomic study of viridans streptococci: description of *Streptococcus gordonii* sp. nov. and emended description of *Streptococcus sanguis* (White and Niven 1946), *Streptococcus oralis* (Bridge and Sneath 1982), and *Streptococcus mitis* (Andrews and Horder 1906). Int. J. Syst. Bacteriol. *39,* 471–484.

Kilian, M., Reinholdt, J., Lomholt, H., Poulsen, K., and Frandsen, E.V. (1996). Biological significance of IgA1 proteases in bacterial colonization and pathogenesis: critical evaluation of experimental evidence. APMIS *104,* 321–338.

Kilian, M., Poulsen, K., Blomqvist, T., Havarstein, L.S., Bek-Thomsen, M., Tettelin, H., and Sorensen, U.B.S. (2008). Evolution of *Streptococcus pneumoniae* and its close commensal relatives. PLoS One 3, e2683.

Kim, K.-J., Lim, S.-H., Kim, K.-W., Yoo, S.W., Hwang, H.-K., Min, B.-M., Choe, S.-J., and Kook, J.-K. (2005). First isolation of *Streptococcus downei* from human dental plaques. FEMS Microbiol. Lett. *249,* 323–326.

King, S.A., Whatmore, A.M., and Dowson, C.G. (2005). NanA a neuriminidase from *Streptococcus pneumoniae,* shows high levels of sequence diversity at least in part through recombination with *Streptococcus oralis.* J. Bacteriol. *187,* 5376–5386.

Kirchherr, J.L., Bowden, G.H., Richmond, D.A., Sheridan, M.J., Wirth, K.A., and Cole, M.F. (2005). Clonal diversity and turnover of *S. mitis* bv. 1 on shedding and non-shedding oral surfaces of human infants during the first year of life. Clin. Diag. Lab. Immunol. *12,* 1184–1190.

Kirchherr, J.L., Bowden, G.H., Cole, M.F., Kawamura, Y., Richmond, D.A., Sheridan, M.J., and Wirth, K.A. (2007). Physiological and serological variation in *Streptococcus mitis* biovar 1 from the human oral cavity during the first year of life. Arch. Oral Biol. *52,* 90–99.

Kolenbrander, P.E., Palmer, R.J., Jr., Periasamy, S., and Jakubovics, N.S. (2011). Oral multispecies biofilm development and the key role of cell–cell distance. Nature Rev. Microbiol. *8,* 471–480.

Könönen, E. (2000). Development of oral bacterial flora in young children. Ann. Med. *32,* 107–112.

Könönen, E., Jousimies-Somer, H., and Asikainen, S. (1992). Relationship bctween oral gram-negative anaerobic bacteria in saliva of the mother and the colonization of her edentulous infant. Oral Microbiol. Immunol. *7,* 273–276.

Kuramitsu, H.K., He, X., Lux, R., Anderson, M.H., and Shi, W. (2007). Interspecies interactions within oral microbial communities. Microbiol. Mol. Biol. Rev. *71,* 653–670.

Lair, S., Chapais, B., Higgins, R., Mirkovic, R., and Marineau, D. (1996). Myeloencephalitis associated with a viridans group *Streptococcus* in a colony of Japanese macaques (*Macaca fuscata*). Vet. Pathol. *33,* 99–103.

Ledder, R.G., Timperley, A.S., Friswell, M.K., Macfarlane, S., and McBain, A.J. (2008). Coaggregation between and among human intestinal and oral bacteria. FEMS Microbiol. Ecol. *66,* 630–636.

Lee, S.F. (1995). Active release of bound antibody by *Streptococcus mutans.* Infect. Immun. *63,* 1940–1946.

Lemos, J.A.C., Abranches, J., and Burne, R.A. (2005). Responses of cariogenic streptococci to environmental stresses. Curr. Issues Mol. Biol. *7,* 95–108.

Lennon, J.T., and Jones, S.E. (2011). Microbial seed banks: the ecological and evolutionary implications of dormancy. Nature Rev. Microbiol. *9,* 119–130.

Lewis, K. (2000). Progammed death in bacteria. Microbiol. Mol. Biol. Rev. *64,* 503–514.

Lewis, K. (2010). Persister cells. Annu. Rev. Microbiol. *64,* 357–372.

Li, J., Helmerhorst, E.J., Leone, C.W., Troxler, R.F., Yaskell, T., Haffajee, A.D., Socransky, S.S., and Oppenheim, F.G. (2004). Identification of early microbial colonizers in human dental biofilm. J. Appl. Microbiol. *97,* 1311–1318.

van der Linden, M., Al-Lahham, A., Nicklas, W., and Reinert, R.R. (2009). Molecular characterization of pneumococcal isolates from pets and laboratory animals. PLoS One *4(12),* e8286. doi:10.1371/journal.pone.0008286.

Loftin, K.C., Brown, L.R., and Levy, B.M. (1980). Comparison of the predominant cultivable microflora in the dental plaque of *Macaca mulatta* (Rhesus) and *Macaca fascicularis* (Cynomologous). J. Dent. Res. *59,* 1606–1612.

van Loveren, C., Hoogenkamp, M.A., Denr, D.M., and ten Cate, J.M. (2008). Effects of different kinds of fluorides on enolase and ATPase activity of a fluoride-sensitive and fluoride-resistant strain of *Streptococcus mutans.* Caries Res. *42,* 429–434.

McBride, B.C., and van der Hoeven, J.S. (1981). Role of interbacterial adherence in colonization of the oral cavities of gnotobiotic rats infected with *Streptococcus mutans* and *Veillonella alcalescens.* Infect. Immun. *33,* 467–472.

Madsen, J., Mollenhauer, J., and Holmskov, U. (2010). Gp-340/DMBT1 in mucosal innate immunity. Innate Immun. *16,* 160–167.

Malamud, D., Abrams, W.R., Barber, C.A., Weissman, D., Rehtanz, M., and Golub, E. (2011). Antiviral activities in human saliva. Adv. Dent. Res. *23,* 34–37.

Marcotte, H., and Lavoie, M.C. (1996). No apparent influence of immunoglobulins on indigenous oral and intestinal microbiota of mice. Infect. Immun. *64,* 4692–4699.

Marcotte, H., and Lavoie, M.C. (1998). Oral microbial ecology and the role of salivary immunoglobulin A. Micro. Mol. Biol. Rev. *62,* 71–109.

Marsh, P.D. (2006). Dental plaque as a biofilm and a microbial community – implications for health and disease. BMC Oral Health *6*(Suppl. 1), S14. doi:10.1186/1472-6831-6-S1-S14.

Mashimo, P.A., Ellison, S.A., and Slots, J. (1979). Microbial composition of monkey dental plaque (*Macaca arctoides* and *Macaca fascicularis*). Scand. J. Dent. Res, *87,* 24–31.

Matsui, R., and Cvitkovitch, D. (2010). Acid tolerance mechanisms utilized by *Streptococcus mutans.* Future Microbiol. *5,* 403–417.

Mestecky, J., and Russell, M.W. (1986). IgA subclasses. Monogr. Allergy *19*, 277–301.

Mikx, F.H., van der Hoeven, J.S., Konig, K.G., Plasschaert, A.J., and Guggenheim, B. (1972). Establishment of defined microbial ecosystems in germ-free rats 1. The effect of interactions of *Streptococcus mutans* or *Streptococcus sanguis* with *Veillonella alkalescens* on plaque formation and caries activity. Caries Res. *6*, 211–223.

Mitchell, J. (2011). *Streptococcus mitis*: walking the line between commensalism and pathogenesis. Mol. Oral Microbiol. *26*, 89–98.

Mitchell, S.C., Ruby, J.D., Moser, S., Momeni, S., Smith, A., Osgood, R.O., Litaker, M., and Childers, N. (2009). Maternal transmission of mutans streptococci in severe-early childhood caries. Pediatric Dent *31*, 193–201.

Moser, S.A., Mitchell, S.C., Ruby, J.D., Momeni, S., Osgood, R.C., Whiddon, J., and Childers, N.K. (2010). Repetitive extragenic palindromic PCR for study of *Streptococcus mutans* diversity and transmission in human populations. J. Clin. Microbiol. *48*, 599–602.

Nakano, K., Lapirattanaki, J., Nomura, R., Nemoto, H., Alaluusua, S., Grönroos, L., Vaara, M., Hamada, S., Ooshima, T., and Nakagawa, I. (2007). *Streptococcus mutans* clonal variation revealed by multilocus sequence typing. J. Clin. Microbiol. *45*, 2616–2625.

Nam, H.-M., Lim, S.-K., Kang, H.-M., Kim, J.-M., Moon, J.-S., Jang, K.-C., Joo, Y.-S., Kung, M.-I., and Jung, S.-J. (2009). Antimicrobial resistance of streptococci isolated from mastitic bovine milk samples in Korea. J. Vet. Diagn. Invest. *21*, 698–701.

Nobbs, A.H., Lamont, R.J., and Jenkinson, H.F. (2009). *Streptococcus* adherence and colonization. Microbiol. Mol. Biol. Rev. 73, 407–450.

Norhagen, G.E., Engstrom, P.-E., Hammerstrom, L., and Nord, C.E. (1990). Oral and intestinal microflora in individuals with different immunoglobulin deficiencies. Eur J Clin. Microbiol. Infect. Dis. *9*, 631–633.

Palmer, R.J., Jr., Gordon, S.M., Cisar, J.O., and Kolenbrander, P.E. (2003). Coaggregation-mediated interactions of streptococci and actinomyces detected in human dental plaque. J. Bacteriol. *185*, 3400–3409.

Palmer, R.J., Jr., Diaz, P.I., and Kolenbrander, P.E. (2006). Rapid succesion within the *Veillonella* population of a developing human oral biofilm *in situ*. J. Bacteriol. *188*, 4117–4124.

Pan, Y.P., and Caufield, P.W. (2001). Phenotypic and genotypic diversity of *Streptococcus sanguis* in infants. Oral Microbiol. Immunol. *16*, 235–242.

Pancholi, V., and Chhatwal, G.S. (2003). Housekeeping enzymes as virulence factors for pathogens. Int. J. Med. Microbiol. *293*, 391–401.

Pancholi, V., and Fischetti, V.A. (1992). A major surface protein on group A streptococci is a glyceraldehyde-3-phosphate-dehydrogenase with multiple binding activity. J. Exp. Med. *176*, 415–426.

Pancholi, V., and Fischetti, V.A. (1998). α-enolase, a novel strong plasmin(ogen) binding protein on the surface of pathogenic streptococci. J. Biol. Chem. *273*, 14503–14515.

Papaloannou, W., Gizani, S., Hafagee, A.D., Quirynen, M., Mamai-Homata, E., and Papagiannoulis, L. (2009). The microbiota on different oral surfaces in healthy children. Oral Microbiol. Immunol. *24*, 183–189.

Paton, J.C., Berry, A.M., and Lock, R.A. (1997). Molecular analysis of putative pneumococcal virulence proteins. Microb. Drug Resist. *3*, 1–10.

Pearce, C., Bowden, G.H., Evans, M., Fitzsimmons, S.P., Johnson, J., Sheridan, M.J., Wientzen, R., and Cole, M. (1995). Identification of pioneer viridans streptococci in the oral cavity of human neonates. J. Med. Microbiol. *42*, 67–72.

Perry, J.A., Cvitkovitch, D.G., and Lévesque, C.M. (2009). Cell death in *Streptococcus mutans* biofilms: a link between CSP and extracellular DNA. FEMS Microbiol. Lett. *299*, 261–266.

Peterson, D.A., McNulty, N.P., Guruge, J.L., and Gordon, J.I. (2007). IgA response to symbiotic bacteria as a mediator of gut homeostasis. Cell Host Microbe *2*, 328–339.

Phalipon, A., Cardona, A., Kraehenbuhl, J.P., Edelman, L., Sansonetti, P.J., and Corthesy, B. (2002). Secretory component: A new role in secretory IgA-mediated immune exclusion *in vivo*. Immunity *17*, 107–115.

Pieralisi, F.J.S., Rodrigues, M.R., Segurs, V.G., Maciel, F.B.., Garcia, J.E., and Poli-Frederico, R.C. (2010). Genotypic diversity of *Streptococcus mutans* in caries-free and caries-active preschool children. Int. J. Dent. *2010*, 824976. doi:10.1155/2010/824976.

Power, D.A., Burton, J.P., Chilcott, C.N., Dawes, P.J., and Tagg, J.R. (2008). Preliminary investigations of the colonisation of upper respiratory tract tissues of infants using a paediatric formulation of the oral probiotic *Streptococcus salivarius* K12. Eur. J. Clin. Microbiol. Infect. Dis. *27*, 1261–1263.

Preza, D., Olsen, I., Willumsen, T., Grinde, B., and Paster, B.J. (2009). Diversity and site-specificity of the oral microflora in the elderly. Eur. J. Clin. Microbiol. Infect. Dis. *28*, 1033–1040.

Quan, C.P., Berneman, A., Pires, R., Avrameas, S., and Bouvet, J.-P. (1997). Natural polyreactive secretory immunoglobulin A autoantibodies as a possible barrier to infection in humans. Infect. Immun. *65*, 3997–4004.

Reinholdt J., Firman, V., and Kilian, M. (1993). Similar proportions of immunoglobulin A1 (IgA1) protease-producing streptococci in initial dental plaque of selectively IgA-deficient and normal individuals Infect. Immun. *61*, 3993–4000.

Ross, K.F., Ronson, C.W., and Tagg, J.R. (1993). Isolation and characterization of the lantibiotic salivaricin A and its structural gene salA from *Streptococcus salivarius* 20P3. Appl. Environ. Microbiol. *59*, 2014–2021.

Rudney, J.D. (2000). Saliva and dental plaque. Adv. Dent. Res. *14*, 29–39.

Rudney, J.D., Ji, Z., and Larson, C.J. (1996). Saliva protein binding to streptococcal layers placed at different sites in 48 persons. J. Dent. Res. *75*, 1789–1797.

Rudney, J.D., Chen, R., and Zhang, G. (2005). Streptococci dominate the diverse flora within buccal cells. J. Dent. Res. *84*,1165–1171.

Sachdeo, A., Haffajee, A.D., and Socransky, S.S. (2008). Biofilms in the edentulous oral cavity. J. Prosthodont. *17*, 348–356.

Sachidanandham, R., and Yew-Hoong Gin, K. (2009). A dormancy state in non-spore forming bacteria. Appl. Microbiol. Biotechnol. *81*, 927–941.

Sait, L., Galic, M., Strugnell, R.A., and Janssen, P.H. (2003). Secretory antibodies do not affect the composition of the bacterial microbiota in the terminal ileum of 10-week-old mice. Appl. Envir. Microbiol., *69*, 2100–2109.

Sampson, J.S., O'Conner, S.P., Stinson, A.R., Tharpe, J.A., and Russel, H. (1994). Cloning and nucleotide sequence of *psaA*, the *Streptococcus pneumoniae* gene encoding a 37-kilodalton protein homologous to previously reported *Streptococcus* sp. adhesins. Infect. Immun. *62*, 319–324.

Schleifer, K.H. (2009). Classification of Bacteria and Archaea: Past, present and future. Syst. Appl. Microbiol. *32*, 533–542.

Shank, E.A., and Kolter, R. (2009). New developments in microbial interspecies signalling. Curr. Opin. Microbiol. *12*, 205–214.

Sheng, J., and Marquis, R.E. (2007). Malolactic fermentation by *Streptococcus mutans*. FEMS Microbiol. Lett. *272*, 196–201.

Simell, B., Korkella, M., Pursianen, H., Kilpi, T.M., and Käyhty, H. (2001). Pneumococcal carriage and otitis media induce salivary antibodies to pneumococcal surface adhesin A, pneumolysin, and pneumococcal surface protein A in children. J. Infect. Dis. *183*, 887–896.

Simell, B., Kilpi, T.M., and Käyhty, H. (2002). Pneumococcal carriage and otitis media induce salivary antibodies to pneumococcal capsular polysaccharides in children. J. Infect. Dis. *186*, 1106–1114.

Simell, B., Kilpi, T., and Käyhty, H. (2006). Subclass distribution of natural salivary IgA antibodies against pneumococcal capsular polysaccharide of type 14 and pneumococcal adhesin A (PsaA) in children. Clin. Exp. Immunol. *143*, 543–549.

Sissons, C.H., Anderson, S.A., Wong, L., Colemen, M.J., and White, D.C. (2007). Microbiota of plaque microcosm biofilms: Effect of three times daily sucrose pules in different simulated oral environments. Caries Res. *41*, 413–422.

Smith, D.J., and Taubman, M.A. (1992). Ontogeny of immunity to oral microbiota in humans. Crit. Rev. Oral Bio. Med. 3, 109–133.

Soares, R.V., Siqueira, C.C., Bruno, L.S., Oppenheim. F.G., Offner, G.D., and Troxler, R.F. (2003). MG2 and lactoferrin form a heterotypic complex in salivary secretions. J. Dent. Res. *82*, 471–475.

van Steenbergen, T.J.M., Petit, M.D.A., Scholte, L.H.M., van der Velden, U., and de Graff, J. (1993). Transmission of *Porphyromonas gingivalis* between spouses. J. Clin. Periodontol. *20*, 340–345.

Suzuki, K., Meek, B., Doi, Y., Muramatsu, M., Chiba, T., Honjo, T., and Fagarasan, S. (2004). Abberant expansion of segmented filamentous bacteria in IgA-deficient gut. PNAS *101*, 1981–1986.

Svensäter, G., Sjögreen, B., and Hamilton, I.R. (2000). Multiple stress responses in *Streptococcus mutans* and the induction of general and stress-specific proteins. Microbiology *146*, 107–117.

Svensäter, G., Borgström, M., Bowden, G.H.W., and Edwardsson, S. (2003). The acid-tolerant microbiota associated with plaque from initial caries and healthy tooth surfaces. Caries Res. *37*, 395–403.

Syrjanen, R., Kilpi, T.M., Kaijalainen, T.H., Herva, E.E., and Takala, A.K. (2001). Nasopharyngeal carriage of *Streptococcus pneumoniae* in Finnish children younger than 2 years old. J. Infect. Dis. *184*, 451–459.

Tagg, J.R. (1983). Application of inhibitor typing in a study of the transmission and retention in the human mouth of the bacterium *Streptococcus salivarius*. Arch. Oral Biol. *28*, 911–915.

Takada, K., and Hirasawa, M. (2007). *Streptococcus orisuis* sp. nov., isolated from the pig oral cavity. Int. J. Syst. Evol. Microbiol. *57*, 1272–1275.

Takahashi, N., and Nyvad, B. (2011). The role of bacteria in the caries process: ecological perspectives. J. Dent. Res. *90*, 294–303.

Taubman, M.A., and Nash, D.A. (2006). The scientific and public-health imperative for a vaccine against dental caries. Nature Rev. Immunol. *6*, 555–563.

Tong, H., Gao, X., and Dong, X. (2003). *Streptococcus oligofermentans* sp. nov., a novel oral isolate from caries-free humans. Int. J. Syst. Evol. Microbiol. *53*, 1101–1104.

Troncoso, G., Sanchez, S., Criado, M.T., and Ferreiros, C.M. (2002). Analysis of *Neisseria lactamica* antigens putatively implicated in acquisition of natural immunity to *Neisseria meningitidis*. FEMS Immunol. Med. Microbiol. *34*, 9–15.

Welin-Neilands, J., and Svensäter, G. (2007). Acid tolerance of biofilm cells of *Streptococcus mutans*. Appl. Environ. Microbiol. *73*, 5633–5638.

Whiley, R.A., and Beighton, D. (1998). Current classification of the oral streptococci. Oral Microbiol. Immunol. *13*, 195–216.

Whiley, R.A., Beighton, D., Winstanley, T.G., and Fraser, H.Y. (1992). *Streptococcus intermedius, Streptococcus constellatus,* and *Streptococcus anginosus* (the *Streptococcus milleri* group): association with different body sites and clinical infections. J. Clin. Microbiol. *30*, 243–244.

Wilkins, J.C., Beighton, D., and Homer, K.A. (2003). Effect of acid pH on expression of surface proteins of *Streptococcus oralis*. Appl. Environ. Microbiol. *69*, 5290–5296.

Willcox, D.P., and Drucker, D.B. (1987). A simple method for determining extracellular polysaccharide-producing ability of oral streptococci. Microbios *51*, 175–181.

Willcox, M.P.D., Zhu, H., and Know, K.W. (2001). *Streptococcus australis* sp. nov. a novel oral streptococcus. Int. J. Syst. Evol. Microbiol. *51*, 1277–1281.

Wirth, K.A., Bowden, G.H., Kirchherr, J.L., Richmond, D.A., Sheridan, M.J., and Cole, M.F. (2008). Humoral immunity to commensal oral bacteria in human infants: evidence that *Streptococcus mitis* biovar 1 colonization induces strain-specific salivary immunoglobulin A antibodies. ISME J. *2*, 728–738.

Wirth, K.A., Bowden, G.H., Richmond, D.A., Sheridan, M.J., and Cole, M.F. (2008). Antibody binding to *Streptococcus mitis* and *Streptococcus oralis* cell fractions. Arch. Oral Biol. *53*, 141–149.

Woof, J.M. (2002). The human IgA-Fc α receptor interaction and its blockade by streptococcal IgA-binding proteins. Biochem. Soc. Trans. *30*, 491–494.

van der Woude, M.W., and Bäumler, A.J. (2004). Phase and antigenic variation in bacteria. Clin. Micro. Rev. *17*, 581–611.

Wu, H., Mintz, K.P., Ladha, M., and Fives Taylor, P.M. (1998). Isolation and Characterization of Fap1, fimbriae associated adhesin of *Streptococcus parasanguis* FW213. Mol. Microbiol. *28*, 487–500.

Yamaguchi, T. (2004). Purification of saliva agglutinin of *Streptococcus intermedius* and its association with bacterial aggregation and adherence Arch. Microbiol. *181*, 106–111.

Yang, J., Ritchey, M., Yoshida, Y., Allen Bush, C., and Cisar, J.O. (2009). Comparative structural and molecular characterization of ribitol-5-phosphate containing *Streptococcus oralis* coaggregation receptor polysaccharides. J. Bacteriol. *191*, 1891–1900.

Zaura, E., Keijser, B.J.F., Huse, S.M., and Crielaard, W. (2009). Defining the healthy 'core microbiome' of oral communities. BMC Microbiol. *9*, 259. doi:10.1186/1471-2180-9-259.

Zysk, G., Bongaerts, R.J.M., Ten Thoren, E., Bethe, G., Hakenbeck, R., and Heinz, H.-P. (2000). Detection of 23 immunogenic pneumococcal proteins using convalescent-phase serum. Infect. Immun. *68*, 740–743.

# Detection and Culture of Novel Oral Bacteria

# 2

William G. Wade

## Abstract

The oral microbiota is highly diverse and includes fungi, protozoa, viruses and bacteria. Both domains of prokaryotes, *Archaea* and *Bacteria* are present. Representatives of the *Archaea* are restricted to a few taxa in the genus *Methanobrevibacter*, while there are over 600 species of *Bacteria*, from at least 12 phyla. The full diversity of bacterial populations in the mouth has been recognized following the application of culture-independent methods of analysis, based on 16S rRNA gene sequence comparisons. Because oral bacteria are typically slow-growing and fastidious, and around half cannot be grown in the laboratory at all, the taxonomic process of classifying and naming bacterial species is ongoing and over 100 cultivable taxa have still to be named. In recent years, attempts have been made to culture the not-yet-cultured portion of the microbiota. There are a number of reasons why certain taxa are uncultivable and these include a need for a specific nutrient, extreme oxygen sensitivity and dependence on other organisms. The inter-dependence among members of the oral microbial community may relate to cooperative degradation of natural substrates for growth or the need to participate in signalling networks that control growth rate and resuscitation from dormancy. Novel culture media and methods are being developed that reproduce the in-vivo environment and thus encourage previously uncultured organisms to grow in the laboratory.

## Introduction

The human mouth harbours a diverse collection of microorganisms, including bacteria, fungi, viruses and protozoa. For the purposes of this chapter, the generic term 'bacteria' will be used to include members of the prokaryotic domains *Bacteria* and *Archaea*. Much diversity among oral bacteria remains uncharacterised. Novel oral bacterial species have been described as a result of improvements in taxonomic approaches. Some bacterial species have been shown to be heterogeneous and therefore split into multiple new taxa. Alternatively, groups of strains not recognizable as existing species have been proposed as novel taxa. In addition, culture-independent methods have confirmed the observation that a substantial proportion of oral bacteria cannot be cultured using conventional culture media and incubation conditions. Numerous as-yet-uncultured bacterial phylotypes are now recognized and listed (along with cultivable species) in the Human Oral Microbiome Database, an on-line resource which describes the bacterial taxa making up the oral microbiota (Chen *et al.*, 2010; Dewhirst *et al.*, 2010). The aim of this chapter is to summarize recent taxonomic changes and to document how molecular methods have transformed our knowledge of the oral microbiota. To fully understand the metabolism and virulence potential of a species, in-vitro culture is essential. The strategies employed to culture previously uncultivated oral bacteria will be reviewed and future prospects will be assessed.

## The impact of molecular phylogeny on oral bacterial taxonomy

Classification and identification of bacteria has been revolutionized by the adoption of molecular methods, particularly by comparison of small subunit (16S) ribosomal RNA genes (Lane *et al.*, 1985) as the basis for phylogenetic analysis. Oral bacteria are particularly difficult to identify by conventional means because the majority are obligate anaerobes, are typically slow-growing, and are unreactive in biochemical tests. The adoption of 16S rRNA methods has allowed far more precise classification, and the relative ease of automated DNA sequencing methods has enabled quick and reliable strain identification from a partial sequence.

Application of molecular methodologies has led to substantial changes in the taxonomy of oral bacteria. Table 2.1 summarizes the new genera alone proposed in the last 20 years; a far greater

**Table 2.1** Recently described bacterial genera with oral representatives (since 1990)

| Phylum | Genus | Reference |
|---|---|---|
| Actinobacteria | *Actinobaculum* | Lawson *et al.* (1997) |
| | *Atopobium* | Collins and Wallbanks (1992) |
| | *Cryptobacterium* | Nakazawa *et al.* (1999) |
| | *Kocuria* | Stackebrandt *et al.* (1995) |
| | *Olsenella* | Dewhirst *et al.* (2001) |
| | *Parascardovia* | Jian and Dong (2002) |
| | *Scardovia* | Jian and Dong (2002) |
| | *Slackia* | Wade *et al.* (1999) |
| | *Tropheryma* | La Scola *et al.* (2001) |
| Bacteroidetes | *Bergeyella* | Vandamme *et al.* (1994) |
| | *Prevotella* | Shah and Collins (1990) |
| | *Tannerella* | Sakamoto *et al.* (2002) |
| Firmicutes | *Abiotrophia* | Kawamura *et al.* (1995) |
| | *Anaerococcus* | Ezaki *et al.* (2001) |
| | *Anaeroglobus* | Carlier *et al.* (2002) |
| | *Bulleidia* | Downes *et al.* (2000) |
| | *Catonella* | Moore and Moore (1994) |
| | *Dialister* | Moore and Moore (1994) |
| | *Filifactor* | Collins *et al.* (1994) |
| | *Finegoldia* | Murdoch and Shah (1999) |
| | *Granulicatella* | Collins and Lawson (2000) |
| | *Johnsonella* | Moore and Moore (1994) |
| | *Mogibacterium* | Nakazawa *et al.* (2000) |
| | *Parvimonas* | Tindall and Euzeby (2006) |
| | *Peptoniphilus* | Ezaki *et al.* (2001) |
| | *Pseudoramibacter* | Willems and Collins (1996) |
| | *Schwartzia* | van Gylswyk *et al.* (1997) |
| | *Shuttleworthia* | Downes *et al.* (2002) |
| | *Solobacterium* | Kageyama and Benno (2000) |
| Proteobacteria | *Lautropia* | Gerner-Smidt *et al.* (1994) |
| | *Suttonella* | Dewhirst *et al.* (1990) |
| Synergistetes | *Jonquetella* | Jumas-Bilak *et al.* (2007) |
| | *Pyramidobacter* | Downes *et al.* (2009) |

number of novel species have been described. The extent of the ongoing nomenclatural revision has attracted criticism from other workers in the field who have found it confusing that familiar genus and species names are changed. The benefits, however, outweigh the disadvantages because of the clarity afforded by molecular-based taxonomy. In particular, some previously heterogeneous taxa have been split into well-defined species with distinct disease associations that are clearly biologically relevant. The nomenclatural revision process is far from complete. At the time of writing, HOMD includes 113 un-named cultivated species-level taxa; a number of oral bacterial families and genera, particularly in the *Firmicutes*, are polyphyletic and will require re-naming (Dewhirst *et al.*, 2010).

## Detection and culture of oral bacteria

It is well known that among the first bacteria visualized by Leeuwenhoek were oral bacteria scraped from the plaque on teeth (Porter, 1976). Similarly, when bacteria were first cultured on solid media by Pasteur, Koch and others, oral bacteria were among the first studied, thanks largely to the pioneering work of WD Miller (Miller, 1890; Ring, 2002). The aim of culturing is to try to recreate the conditions in which bacteria live *in vivo*. Culture media typically consist of enzymatic digests of meat or vegetable proteins, augmented with a source of vitamins and other growth factors such as yeast extract, and a buffer system normally designed to give a final neutral pH. Media formulated for the growth of pathogens typically also include whole blood or serum to provide additional mammalian host factors. When originally introduced in the nineteenth century, this type of medium was extraordinarily successful and the majority of the principal pathogens of man were successfully cultivated. Indeed, the cause of syphilis, *Treponema pallidum*, is one of the few examples of an as yet uncultured bacterial species that is responsible for a 'classical' infectious diseases (Peeling and Mabey, 2004).

It was quickly recognized that, in addition to supplying nutrients, the appropriate conditions had to be provided in order to grow oral bacteria in the laboratory, including the need for anaerobic incubation. Studies which failed to use anaerobic conditions produced results which led to erroneous associations between species and oral disease states and, in particular, the overemphasis of the role of aerobic and facultatively anaerobic species such as streptococci and staphylococci (Medalia, 1913). This shows the importance of taking into account all possible sources of bias, whatever the methods used, when attempting to characterize complex microbial communities. Anaerobiosis is an important consideration for oral bacterial metabolism. As oral plaque biofilms develop, they become anaerobic extremely quickly (Kenney and Ash, 1969). It is because of this that the majority of oral bacteria are obligate anaerobes; maintenance of anaerobiosis during sample transport, handling and incubation is important. Jars in which anaerobic conditions were created by the catalytic combination of oxygen with hydrogen are convenient and relatively effective, but suffer from the disadvantage of a lag time before anaerobic conditions are established. The Hungate roll tube technique was developed to overcome this and involved inoculating samples on agar around the inside of a tube filled with anaerobic gas. The later development of anaerobic gloveboxes and workstations where anaerobic conditions are permanently maintained, and samples and plates are passed in and out via an airlock, has been shown to be as effective as the roll tube technique for the culture of strictly anaerobic species (Aranki *et al.*, 1969). In addition, efforts have been made to improve culture media. In general, media designed for the culture of obligate anaerobes, particularly the inclusion of growth promoting substances such as haemin, Vitamin K and arginine, give the highest recoveries of bacteria from oral samples (Slots, 1975; Heginbothom *et al.*, 1990). Whatever media and incubation methods are used, however, a substantial proportion of the oral microbiota remains uncultured.

## Uncultivable oral bacteria

It has long been recognized that not all oral bacteria can be cultured in the laboratory. Microscopic counts of bacteria in oral samples are typically twice those of total viable counts. Advances in gene amplification by PCR and cloning have made it possible to analyse complex bacterial communities without culture. Originally developed for environmental samples (Giovannoni *et al.*, 1990; Ward *et al.*, 1990), the amplification, cloning and sequencing of 16S rRNA genes is now widely used to determine the composition of oral samples (Dymock *et al.*, 1996; Kroes *et al.*, 1999; Paster *et al.*, 2001; Munson *et al.*, 2002, 2004; Aas *et al.*, 2005). The results of these and other studies have confirmed that there are a large number of as-yet-uncultured bacterial species among the oral microbiome. These include representatives of the phyla *Bacteroidetes, Firmicutes, Spirochaetes* and *Synergistetes,* as well as entire phyla/divisions such as SR1 and TM7 which have no confirmed cultivable representatives.

High-throughput sequencing methods have been used to perform deep sequencing of oral samples with the aim of revealing the so-called 'rare biosphere' within the oral microbiome (Lazerevic *et al.*, 2011). Virtually all naturally occurring bacterial communities are highly positively skewed in the frequency distribution of taxa with a substantial part of the microbiota made up of a large number of rarely occurring taxa. The mouth presents an additional problem, being a naturally open environment and exposed to bacteria in air, water and beverages and food. In practice, common air- and water-borne bacterial species are not typically found among the predominant oral microbiota although they frequently contaminate PCR reagents (Tanner *et al.*, 1998). This is probably because their concentrations on entering the mouth are relatively low in comparison to the indigenous bacterial community. The use of deep sequencing methods, however, is likely to make it difficult to determine if such organisms are merely transients or members of the resident community, albeit present in low numbers. Initial studies using high-throughput sequencing (Keijser *et al.*, 2008) demonstrated estimates of species richness for the oral microbiota orders-of-magnitude higher than those reported previously. It was subsequently shown that pyrosequencing is prone to sequence error and artefacts particularly at homopolymeric regions of DNA (Quince *et al.*, 2009). The analysis of large sequence datasets is problematic, not least because of the sheer volume of data. This makes it difficult to assign sequences to operational taxonomic units (OTU) at defined levels of sequence identity. As mentioned, this approach led to the identification of spurious novel taxa, because sequencing error had created sequences sufficiently different from known taxa to be described by the analysis algorithms as novel.

## Visualization

The high level of diversity among oral bacteria revealed by 16S rRNA-based methods has been questioned by some workers who suggested that DNA from remnants of dead bacteria or environmental contaminants might be present. In order to demonstrate that intact, viable cells from novel phylotypes are present in oral samples, the rRNA sequence data have been used to design oligonucleotide probes specific for uncultivated taxa. These have then been used with fluorescent labels to stain smears or tissue samples and viewed by conventional or confocal microscopy (Amann *et al.*, 2001). FISH detection of bacteria targeting 16S rRNA relies on hybridization with rRNA molecules in the ribosomes. These will only be present in intact cells that are metabolically active and a positive FISH signal thus indicates that the target cells are viable in the samples, and thus actively growing *in vivo* (Wallner *et al.*, 1993).

Oral members of the uncultured Division TM7 were found to exhibit a filamentous morphotype (Brinig *et al.*, 2003), confirming earlier observations of TM7 cells from a waste reactor (Hugenholtz *et al.*, 2001). These organisms were found to be ubiquitous in nature as well as being found in 96% of samples from periodontal health and disease (Brinig *et al.*, 2003). In a similar way, FISH probes for *Tannerella* phylotypes BU045 and BU063 have been used to detect these phylotypes in oral samples (Zuger *et al.*, 2007). *Tannerella* BU063 is of particular interest, because it is a close relative of *Tannerella forsythia*, a putative periodontal pathogen (Socransky *et al.*, 1998) strongly associated with adult periodontitis (Tran *et al.*, 2001), yet BU063 is associated with gingival health (Kumar *et al.*, 2003). The FISH analyses showed that *Tannerella* BU063 and its related phylotype BU045 are elongated, thin bacilli with a segmented structure (Zuger *et al.*, 2007).

The recently described phylum *Synergistetes* (Jumas-Bilak *et al.*, 2009) includes two cultivable oral species, *Jonquetella anthropi* and *Pyramidobacter piscolens*, but the majority of oral phylotypes detected belong to Cluster A which until recently, had no cultivable representatives (Vartoukian *et al.*, 2007). Group-level FISH probes for Cluster A, as well as specific probes for individual phylotypes, revealed that members of this group are large, motile bacilli that prefer subgingival plaque and have increased numbers in periodontitis (Vartoukian *et al.*, 2009). Cluster A *Synergistetes* have been shown to form a palisade layer in the top level of subgingival plaque and to have specific associations with host neutrophils, with other bacteria, as well as forming aggregates with themselves (Zijnge *et al.*, 2010).

## Why are some bacteria uncultivable?

As already mentioned, the fundamental reason why so many bacteria cannot be cultured in the laboratory is because the conditions to which they have been accustomed *in vivo* have not been reproduced. There are a number of possible explanations for this. Many bacteria are likely to require specific nutrients that are found naturally in their normal habitat but which they are unable to synthesize for themselves. For example, the growth of the common oral bacteria *Abiotrophia* and *Granulicatella*, formerly known as the nutritionally variant streptococci, is markedly stimulated by the addition of pyridoxal or L-cysteine to the medium (Ruoff, 1991), while *T. forsythia* requires N-acetylmuramic acid for growth (Wyss, 1989). Although culture media developed for the growth of fastidious bacterial species typically contain a wide range of additives that are known to stimulate bacterial growth, the diversity of bacterial communities is such that it would be extremely difficult to determine the individual requirements of every member of the community. Other environmental factors such as temperature and pH are clearly important although they should be relatively easy to reproduce *in vitro*.

Despite these advances, a significant proportion of the oral microbiota has remained recalcitrant to culture. It has been recognized that the natural habitat of bacteria is typically a multispecies community in a biofilm. Thus bacteria have evolved to interact with other members of their community and, indeed, may have become dependent on such interactions for growth. These may include the sharing of nutrients, the cooperative degradation of natural substrates, or cell–cell signalling to regulate growth rates and the induction of, and emergence from, dormancy.

Although the precise mechanisms for these interdependencies have yet to be elucidated, a key approach to mimicking them *in vitro* is the establishment of mixed culture systems where bacteria can grown, and the composition of the community monitored. A variety of systems have been described, primarily for environmental bacteria. Members of the ubiquitous SAR11 marine bacterioplankton group were successfully cultured (Rappe *et al.*, 2002) by diluting sea water samples so that on average 22 bacterial cells were placed into microtitre tray wells in a low-nutrient medium. The presence of dividing cells was detected by the non-specific DNA-binding stain 4¢,6-diamidino-2-phenylindole (DAPI) and preliminary identification performed using a panel of specific FISH probes. A variant of this method is to grow bacteria on agar in contact with a semipermeable membrane the other side of which is the natural substrate (Kaeberlein *et al.*, 2002). In a further refinement, mixtures of bacteria are encapsulated in aqueous microdroplets dispersed in a continuous oil phase, enabling a large number of pair-wise symbiotic relationships to be tested simultaneously (Park *et al.*, 2011).

## Identification of growth stimulatory factors

Successful maintenance of a mixed culture that includes an 'uncultivable' taxon is only the first step towards obtaining a pure culture for physiological studies and as a source of DNA for genome sequencing. Purification has been achieved by physical separation through laser dissection or flow cytometry (Huber *et al.*, 1995; Podar *et al.*, 2007). Alternatively, 'domestication' of the culture can be attempted whereby after prolonged culture with co-culture partners, the organism adapts to the in-vitro environment and is able to grow either independently or in close proximity to other organisms. An example of the latter approach was the successful domestication of a member of the Cluster A *Synergistetes* (Vartoukian *et al.*, 2010). This was achieved by repeated passaging of a subgingival plaque sample on blood agar with enrichment of the target organism by localizing their position on the plates using colony hybridization with specific probes. After several passages, the composition of the enriched community consisted of the target *Synergistetes* and four other species. By cross-streaking the culture with these four species, additional growth enhancement was obtained and, after eight passages, a few small colonies of *Synergistetes* Cluster A were seen a few millimetres from a *Parvimonas micra* donor streak. Presumably, a growth-stimulating factor was diffusing from the *P. micra* donor and enabling independent growth of the *Synergistetes* Cluster A strain, which was named as SGP1 and identified as belonging to HOMD oral taxon 363. Sufficient cells were harvested to allow DNA to be extracted and, after whole genomic amplification, the genome was fully sequenced (accession no. FP929056).

The identity of the growth-stimulating factor remains unknown. It has long been known that one bacterial species can stimulate the growth of another, a process termed satellitism: the observation of 'satellites' that grow around donor organisms. One of the oldest examples is *Haemophilus influenzae*, which in primary culture plates can be seen to produce satellite colonies around other bacteria, typically staphylococci (Davis, 1921). The basis for growth-stimulation in this case was shown to result from *H. influenzae*'s requirement for two factors, X and Y, which were later shown to be haemin (Fildes, 1921) and NAD (Lwoff and Lwoff, 1937).

The diffusion chamber described above (Kaeberlein *et al.*, 2002) has been used to isolate previously uncultured organisms that were dependent on helper strains for growth. These pairs of strains were then used to identify the substance responsible for growth enhancement. The growth of *Psychrobacter* strain MSC33 was induced by a 5-amino-acid peptide (Nichols *et al.*, 2008) at levels too low for the peptide to be used as a nutrient. This suggests that the peptide was functioning as a signalling molecule. Resuscitation promotion factor (Rpf), a protein which revived *Micrococcus* cells from dormancy at picomolar concentrations, was described as a 'bacterial cytokine' because of the similarities in action to mammalian cytokines, and it is also thought to act as a signalling molecule (Mukamolova *et al.*, 1998). Interestingly, Rpf was later shown to have a strong structural similarity to lysozyme (Cohen-Gonsaud *et al.*, 2005) and to have the ability to cleave peptidoglycan. It was hypothesized that the action of Rpf on the bacterial cell wall might be to create peptidoglycan fragments that act as signalling molecules (Keep *et al.*, 2006). In parallel work on sporulation in *Bacillus subtilis*, muropeptide fragments were shown to act as signalling molecules by binding to a serine/threonine kinase, PrkC, on the cell surface (Shah *et al.*, 2008). A specific muropeptide is required: a disaccharide–tripeptide with a meso-diaminopimelic-acid residue, found primarily in Gram-positive bacteria, in the third position of the stem peptide. It is possible therefore that the action of Rpf is to produce muropeptides from peptidoglycan which then bind to cell-surface receptors and promote resuscitation/growth-induction. Given the ubiquity of peptidoglycan fragments in bacterial biofilms, it is possible that this is a general mechanism of growth regulation.

Another group of molecules with the ability to induce the growth of 'uncultivable' bacteria are siderophores (D'Onofrio *et al.*, 2010). Because of its genetic tractability, *E. coli* was screened for its ability to induce the growth of uncultivable organisms in order to facilitate the identification of the factors responsible. *E. coli* was found to induce the growth of *Marinibacter polysiphoniae*, while mutant strains unable to produce the siderophore enterobactin did not. Adding siderophores to primary cultures of marine water allowed a range of previously uncultured organisms to grow (D'Onofrio *et al.*, 2010). The relevance of this work to culture of previously uncultured human oral bacteria is unclear because iron is presumed to be relatively abundant in the mouth. However the principle of multiple bacterial species in a community sharing molecules essential for metabolism is of relevance. *E. coli* was also found to stimulate the growth of the recently described novel species,

*Poriferícola rhodea* (Yoon *et al.*, 2011). *P. rhodea* was isolated on bait-streaked agar, an isolation medium for predatory bacteria where *E. coli* is used as 'bait' (Reichenbach and Dworkin, 1992). In this case it is not clear whether the novel species is truly predatory or merely growing in co-culture with *E. coli*.

## Future perspectives

Metagenomic techniques have proved extremely useful in determining the composition and genetic potential of organisms colonizing specific habitats. It is clear, however, that understanding the regulation of gene expression and interactions between organisms requires far greater knowledge than that which is available from gene sequences alone. Individual cells and species have evolved to play a specific role in the complex communities in which they are found and, to understand how these communities function, isolation and culture of the organisms is required. The relative ease in culturing the primary exogenous pathogens may have give risen to complacency – indeed the reason that pathogens grow relatively easy in culture may be because they are pioneer organisms: ones adapted to colonizing new and hostile environments. In contrast, members of the commensal microbiota have evolved to live in multispecies biofilms where interactions with other species and the mammalian host are the normal context for growth. Deprived of these interactions, many are unable to grow. Work to date in reproducing the in-vivo habit *in vitro* has been limited, and only a small number of previously uncultured organisms have been successfully cultured. Future studies need to be on larger scale and must take a systematic high-throughput approach. A number of potential growth-stimulating factors have now been identified and could be added to primary culture media. Co-culture devices and systems could be used in conjunction with molecular detection systems to screen for the growth of previously uncultured organisms. Finally, it is important that oral bacterial systematics remains a vital discipline. As novel species are cultivated, they should be characterized and formally described and proposed as new species, for the benefit of the entire research community.

## References

Aas, J.A., Paster, B.J., Stokes, L.N., Olsen, I., and Dewhirst, F.E. (2005). Defining the normal bacterial flora of the oral cavity. J. Clin. Microbiol. *43*, 5721–5732.

Amann, R., Fuchs, B.M., and Behrens, S. (2001). The identification of microorganisms by fluorescence in situ hybridisation. Curr. Opin. Biotechnol. *12*, 231–236.

Arank, A., Syed, S.A., Kenney, E.B., and Freter, R. (1969). Isolation of anaerobic bacteria from human gingiva and mouse cecum by means of a simplified glove box procedure. Appl. Microbiol. *17*, 568–576.

Brinig, M.M., Lepp, P.W., Ouverney, C.C., Armitage, G.C., and Relman, D.A. (2003). Prevalence of bacteria of division TM7 in human subgingival plaque and their association with disease. Appl. Environ. Microbiol. *69*, 1687–1694.

Carlier, J.P., Marchandin, H., Jumas-Bilak, E., Lorin, V., Henry, C., Carriere, C., and Jean-Pierre, H. (2002). *Anaeroglobus geminatus* gen. nov., sp. nov., a novel member of the family *Veillonellaceae*. Int. J. Syst. Evol. Microbiol. *52*, 983–986.

Chen, T., Yu, W.H., Izard, J., Baranova, O.V., Lakshmanan, A., and Dewhirst, F.E. (2010). The Human Oral Microbiome Database: a web accessible resource for investigating oral microbe taxonomic and genomic information. Database (Oxford) *2010*, baq013.

Cohen-Gonsaud, M., Barthe, P., Bagneris, C., Henderson, B., Ward, J., Roumestand, C., and Keep, N.H. (2005). The structure of a resuscitation-promoting factor domain from *Mycobacterium tuberculosis* shows homology to lysozymes. Nat. Struct. Mol. Biol. *12*, 270–273.

Collins, M.D., and Lawson, P.A. (2000). The genus *Abiotrophia* (Kawamura *et al.*) is not monophyletic: proposal of *Granulicatella* gen. nov., *Granulicatella adiacens* comb. nov., *Granulicatella elegans* comb. nov. and *Granulicatella balaenopterae* comb. nov. Int. J. Syst. Evol. Microbiol. *50(Pt 1)*, 365–369.

Collins, M.D., and Wallbanks, S. (1992). Comparative sequence analyses of the 16S rRNA genes of *Lactobacillus minutus*, *Lactobacillus rimae* and *Streptococcus parvulus*: proposal for the creation of a new genus *Atopobium*. FEMS Microbiol. Lett. *74*, 235–240.

Collins, M.D., Lawson, P.A., Willems, A., Cordoba, J.J., Fernandez-Garayzabal, J., Garcia, P., Cai, J., Hippe, H., and Farrow, J.A. (1994). The phylogeny of the genus *Clostridium*: proposal of five new genera and eleven new species combinations. Int. J. Syst. Bacteriol. *44*, 812–826.

D'Onofrio, A., Crawford, J.M., Stewart, E.J., Witt, K., Gavrish, E., Epstein, S., Clardy, J., and Lewis, K. (2010). Siderophores from neighboring organisms promote the growth of uncultured bacteria. Chem. Biol. *17*, 254–264.

Davis, D.J. (1921). The accessory factors in bacterial growth: V. The value of the satellite (or symbiosis) phenomenon for the classification of hemophilic bacteria. J. Infect. Dis. *29*, 187–189.

Dewhirst, F.E., Paster, B.J., La Fontaine, S., and Rood, J.I. (1990). Transfer of *Kingella indologenes* (Snell and Lapage 1976) to the genus *Suttonella* gen. nov. as *Suttonella indologenes* comb. nov.; transfer of *Bacteroides nodosus* (Beveridge 1941) to the genus *Dichelobacter* gen. nov. as *Dichelobacter nodosus* comb. nov.; and assignment of the genera *Cardiobacterium, Dichelobacter*, and *Suttonella* to *Cardiobacteriaceae* fam. nov. in the gamma division of *Proteobacteria* on the basis of 16S rRNA sequence comparisons. Int. J. Syst. Bacteriol. *40*, 426–433.

Dewhirst, F.E., Paster, B.J., Tzellas, N., Coleman, B., Downes, J., Spratt, D.A., and Wade, W.G. (2001). Characterization of novel human oral isolates and cloned 16S rDNA sequences that fall in the family *Coriobacteriaceae*: description of olsenella gen. nov., reclassification of *Lactobacillus uli* as *Olsenella uli* comb. nov. and description of *Olsenella profusa* sp. nov. Int. J. Syst. Evol. Microbiol. *51*, 1797–1804.

Dewhirst, F.E., Chen, T., Izard, J., Paster, B.J., Tanner, A.C., Yu, W.H., Lakshmanan, A., and Wade, W.G. (2010). The human oral microbiome. J. Bacteriol. *192*, 5002–5017.

Downes, J., Olsvik, B., Hiom, S.J., Spratt, D.A., Cheeseman, S.L., Olsen, I., Weightman, A.J., and Wade, W.G. (2000). *Bulleidia extructa* gen. nov., sp. nov., isolated from the oral cavity. Int. J. Syst. Evol. Microbiol. *50(Pt 3)*, 979–983.

Downes, J., Munson, M.A., Radford, D.R., Spratt, D.A., and Wade, W.G. (2002). *Shuttleworthia satelles* gen. nov., sp. nov., isolated from the human oral cavity. Int. J. Syst. Evol. Microbiol. *52*, 1469–1475.

Downes, J., Vartoukian, S.R., Dewhirst, F.E., Izard, J., Chen, T., Yu, W.H., Sutcliffe, I.C., and Wade, W.G. (2009). *Pyramidobacter piscolens* gen. nov., sp. nov., a member of the phylum '*Synergistetes*' isolated from the human oral cavity. Int. J. Syst. Evol. Microbiol. *59*, 972–980.

Dymock, D., Weightman, A.J., Scully, C., and Wade, W.G. (1996). Molecular analysis of microflora associated with dentoalveolar abscesses. J. Clin. Microbiol. *34*, 537–542.

Ezaki, T., Kawamura, Y., Li, N., Li, Z.Y., Zhao, L., and Shu, S. (2001). Proposal of the genera *Anaerococcus* gen. nov., *Peptoniphilus* gen. nov. and *Gallicola* gen. nov. for members of the genus *Peptostreptococcus*. Int. J. Syst. Evol. Microbiol. *51*, 1521–1528.

Fildes, P. (1921). The nature of the effect of blood-pigment upon the growth of *B. influenzae*. Br. J. Exp. Pathol. *2*, 16–25.

Gerner-Smidt, P., Keiser-Nielsen, H., Dorsch, M., Stackebrandt, E., Ursing, J., Blom, J., Christensen, A.C., Christensen, J.J., Frederiksen, W., Hoffmann, S., *et al.* (1994). *Lautropia mirabilis* gen. nov., sp. nov., a Gram-negative motile coccus with unusual morphology isolated from the human mouth. Microbiology *140(Pt 7)*, 1787–1797.

Giovannoni, S., Britschgi, T., Moyer, C., and Field, K. (1990). Genetic diversity in Sargasso Sea bacterioplankton. Nature *345*, 60–63.

van Gylswyk, N.O., Hippe, H., and Rainey, F.A. (1997). *Schwartzia succinivorans* gen. nov., sp. nov., another ruminal bacterium utilizing succinate as the sole energy source. Int. J. Syst. Bacteriol. *47*, 155–159.

Heginbothom, M., Fitzgerald, T.C., and Wade, W.G. (1990). Comparison of solid media for cultivation of anaerobes. J. Clin. Pathol. *43*, 253–256.

Huber, R., Burggraf, S., Mayer, T., Barns, S.M., Rossnagel, P., and Stetter, K.O. (1995). Isolation of a hyperthermophilic archaeum predicted by in situ RNA analysis. Nature *376*, 57–58.

Hugenholtz, P., Tyson, G.W., Webb, R.I., Wagner, A.M., and Blackall, L.L. (2001). Investigation of candidate division TM7, a recently recognized major lineage of the domain *Bacteria* with no known pure-culture representatives. Appl. Environ. Microbiol. *67*, 411–419.

Jian, W., and Dong, X. (2002). Transfer of *Bifidobacterium inopinatum* and *Bifidobacterium denticolens* to *Scardovia inopinata* gen. nov., comb. nov., and *Parascardovia denticolens* gen. nov., comb. nov., respectively. Int. J. Syst. Evol. Microbiol. *52*, 809–812.

Jumas-Bilak, E., Carlier, J.P., Jean-Pierre, H., Citron, D., Bernard, K., Damay, A., Gay, B., Teyssier, C., Campos, J., and Marchandin, H. (2007). *Jonquetella anthropi* gen. nov., sp. nov., the first member of the candidate phylum '*Synergistetes*' isolated from man. Int. J. Syst. Evol. Microbiol. *57*, 2743–2748.

Jumas-Bilak, E., Roudiere, L., and Marchandin, H. (2009). Description of '*Synergistetes*' phyl. nov. and emended description of the phylum '*Deferribacteres*' and of the family *Syntrophomonadaceae*, phylum '*Firmicutes*'. Int. J. Syst. Evol. Microbiol. *59*, 1028–1035.

Kaeberlein, T., Lewis, K., and Epstein, S.S. (2002). Isolating 'uncultivable' microorganisms in pure culture in a simulated natural environment. Science *296*, 1127–1129.

Kageyama, A., and Benno, Y. (2000). Phylogenic and phenotypic characterization of some *Eubacterium*-like isolates from human feces: description of *Solobacterium moorei* gen. nov., sp. nov. Microbiol. Immunol. *44*, 223–227.

Kawamura, Y., Hou, X.G., Sultana, F., Liu, S., Yamamoto, H., and Ezaki, T. (1995). Transfer of *Streptococcus adjacens* and *Streptococcus defectivus* to *Abiotrophia* gen. nov. as *Abiotrophia adiacens* comb. nov. and *Abiotrophia defectiva* comb. nov., respectively. Int. J. Syst. Bacteriol. *45*, 798–803.

Keep, N.H., Ward, J.M., Cohen-Gonsaud, M., and Henderson, B. (2006). Wake up! Peptidoglycan lysis and bacterial non-growth states. Trends Microbiol. *14*, 271–276.

Keijser, B.J., Zaura, E., Huse, S.M., van der Vossen, J.M., Schuren, F.H., Montijn, R.C., ten Cate, J.M., and Crielaard, W. (2008). Pyrosequencing analysis of the oral microflora of healthy adults. J. Dent. Res. *87*, 1016–1020.

Kenney, E.B., and Ash, M.M., Jr. (1969). Oxidation reduction potential of developing plaque, periodontal pockets and gingival sulci. J. Periodontol. *40*, 630–633.

Kroes, I., Lepp, P.W., and Relman, D.A. (1999). Bacterial diversity within the human subgingival crevice. Proc. Natl. Acad. Sci. U.S.A. *96*, 14547–14552.

Kumar, P.S., Griffen, A.L., Barton, J.A., Paster, B.J., Moeschberger, M.L., and Leys, E.J. (2003). New bacterial species associated with chronic periodontitis. J. Dent. Res. *82*, 338–344.

La Scola, B., Fenollar, F., Fournier, P.E., Altwegg, M., Mallet, M.N., and Raoult, D. (2001). Description of *Tropheryma whipplei* gen. nov., sp. nov., the Whipple's disease bacillus. Int. J. Syst. Evol. Microbiol. *51*, 1471–1479.

Lane, D.J., Pace, B., Olsen, G.J., Stahl, D.A., Sogin, M.L., and Pace, N.R. (1985). Rapid determination of 16S ribosomal RNA sequences for phylogenetic analyses. Proc. Natl. Acad. Sci. U.S.A. *82*, 6955–6959.

Lawson, P.A., Falsen, E., Akervall, E., Vandamme, P., and Collins, M.D. (1997). Characterization of some *Actinomyces*-like isolates from human clinical specimens: reclassification of *Actinomyces suis* (Soltys and Spratling) as *Actinobaculum suis* comb. nov. and description of *Actinobaculum schaalii* sp. nov. Int. J. Syst. Bacteriol. *47*, 899–903.

Lazerevic, V., Whiteson, K., Francois, P., and Schrenzel, J. (2011). The salivary microbiome assessed by a high-throughput and culture-independent approach. J. Integrated OMICS *1*, 28–35.

Lwoff, A., and Lwoff, M. (1937). Studies on codehydrogenases. I. Nature of growth factor 'V'. Proc. R. Soc. Lond., Ser. B: Biol. Sci. *122*, 352–359.

Medalia, L.S. (1913). Chronic alveolar osteomyelitis: its causes and treatment with vaccines. Dental Cosmos *55*, 24–36.

Miller, W.D. (1890). Micro-organisms of the Human Mouth (The S.S. White Dental Manufacturing Company, Philadelphia, PA).

Moore, L.V., and Moore, W.E. (1994). *Oribaculum catoniae* gen. nov., sp. nov.; *Catonella morbi* gen. nov., sp. nov.; *Hallella seregens* gen. nov., sp. nov.; *Johnsonella ignava* gen. nov., sp. nov.; and *Dialister pneumosintes* gen. nov., comb. nov., nom. rev., Anaerobic Gram-negative bacilli from the human gingival crevice. Int. J. Syst. Bacteriol. *44*, 187–192.

Mukamolova, G.V., Kaprelyants, A.S., Young, D.I., Young, M., and Kell, D.B. (1998). A bacterial cytokine. Proc. Natl. Acad. Sci. U.S.A. *95*, 8916–8921.

Munson, M.A., Pitt Ford, T., Chong, B., Weightman, A.J., and Wade, W.G. (2002). Molecular and cultural analysis of the microflora associated with endodontic infections. J. Dent. Res. *81*, 761–766.

Munson, M.A., Banerjee, A., Watson, T.F., and Wade, W.G. (2004). Molecular analysis of the microflora associated with dental caries. J. Clin. Microbiol. *42*, 3023–3029.

Murdoch, D.A., and Shah, H.N. (1999). Reclassification of *Peptostreptococcus magnus* (Prevot 1933) Holdeman and Moore 1972 as *Finegoldia magna* comb. nov. and *Peptostreptococcus micros* (Prevot 1933) Smith 1957 as *Micromonas micros* comb. nov. Anaerobe 5, 555–559.

Nakazawa, F., Poco, S.E., Ikeda, T., Sato, M., Kalfas, S., Sundqvist, G., and Hoshino, E. (1999). *Cryptobacterium curtum* gen. nov., sp. nov., a new genus of Gram-positive anaerobic rod isolated from human oral cavities. Int. J. Syst. Bacteriol. *49(Pt 3)*, 1193–1200.

Nakazawa, F., Sato, M., Poco, S.E., Hashimura, T., Ikeda, T., Kalfas, S., Sundqvist, G., and Hoshino, E. (2000). Description of *Mogibacterium pumilum* gen. nov., sp. nov. and *Mogibacterium vescum* gen. nov., sp. nov., and reclassification of *Eubacterium timidum* (Holdeman *et al.* 1980) as *Mogibacterium timidum* gen. nov., comb. nov. Int. J. Syst. Evol. Microbiol. *50(Pt 2)*, 679–688.

Nichols, D., Lewis, K., Orjala, J., Mo, S., Ortenberg, R., O'Connor, P., Zhao, C., Vouros, P., Kaeberlein, T., and Epstein, S.S. (2008). Short peptide induces an 'uncultivable' microorganism to grow in vitro. Appl. Environ. Microbiol. *74*, 4889–4897.

Park, J., Kerner, A., Burns, M.A., and Lin, X.N. (2011). Microdroplet-enabled highly parallel co-cultivation of microbial communities. PLoS One *6*, e17019.

Paster, B.J., Boches, S.K., Galvin, J.L., Ericson, R.E., Lau, C.N., Levanos, V.A., Sahasrabudhe, A., and Dewhirst, F.E. (2001). Bacterial diversity in human subgingival plaque. J. Bacteriol. *183*, 3770–3783.

Peeling, R.W., and Mabey, D.C. (2004). Syphilis. Nat. Rev. Microbiol. *2*, 448–449.

Podar, M., Abulencia, C.B., Walcher, M., Hutchison, D., Zengler, K., Garcia, J.A., Holland, T., Cotton, D., Hauser, L., and Keller, M. (2007). Targeted access to the genomes of low-abundance organisms in complex microbial communities. Appl. Environ. Microbiol. *73*, 3205–3214.

Porter, J.R. (1976). Antony van Leeuwenhoek: tercentenary of his discovery of bacteria. Bacteriol. Rev. *40*, 260–269.

Quince, C., Lanzen, A., Curtis, T.P., Davenport, R.J., Hall, N., Head, I.M., Read, L.F., and Sloan, W.T. (2009). Accurate determination of microbial diversity from 454 pyrosequencing data. Nat. Methods *6*, 639–641.

Rappe, M.S., Connon, S.A., Vergin, K.L., and Giovannoni, S.J. (2002). Cultivation of the ubiquitous SAR11 marine bacterioplankton clade. Nature *418*, 630–633.

Reichenbach, H., and Dworkin, M. (1992). The myxobacteria. In The Prokaryotes, Balows, A., Truper, H.G., Dworkin, M., Harder, W., and Schleifer, K.-H., eds. (Springer-Verlag, Berlin), pp. 3416–3487.

Ring, M.E. (2002). W. D. Miller. The pioneer who laid the foundation for modern dental research. NY State Dent. J. *68*, 34–37.

Sakamoto, M., Suzuki, M., Umeda, M., Ishikawa, I., and Benno, Y. (2002). Reclassification of *Bacteroides forsythus* (Tanner *et al.* 1986) as *Tannerella forsythensis* corrig., gen. nov., comb. nov. Int. J. Syst. Evol. Microbiol. *52*, 841–849.

Shah, H.N., and Collins, D.M. (1990). Prevotella, a new genus to include *Bacteroides melaninogenicus* and related species formerly classified in the genus *Bacteroides*. Int. J. Syst. Bacteriol. *40*, 205–208.

Shah, I.M., Laaberki, M.H., Popham, D.L., and Dworkin, J. (2008). A eukaryotic-like Ser/Thr kinase signals bacteria to exit dormancy in response to peptidoglycan fragments. Cell *135*, 486–496.

Slots, J. (1975). Comparison of five growth media and two anaerobic techniques for isolating bacteria from dental plaque. Scand. J. Dent. Res. *83*, 274–278.

Socransky, S.S., Haffajee, A.D., Cugini, M.A., Smith, C., and Kent, R.L., Jr. (1998). Microbial complexes in subgingival plaque. J. Clin. Periodontol. *25*, 134–144.

Stackebrandt, E., Koch, C., Gvozdiak, O., and Schumann, P. (1995). Taxonomic dissection of the genus *Micrococcus*: *Kocuria* gen. nov., *Nesterenkonia* gen. nov., *Kytococcus* gen. nov., *Dermacoccus* gen. nov., and *Micrococcus* Cohn 1872 gen. emend. Int. J. Syst. Bacteriol. *45*, 682–692.

Tanner, M.A., Goebel, B.M., Dojka, M.A., and Pace, N.R. (1998). Specific ribosomal DNA sequences from diverse environmental settings correlate with experimental contaminants. Appl. Environ. Microbiol. *64*, 3110–3113.

Tindall, B.J., and Euzeby, J.P. (2006). Proposal of *Parvimonas* gen. nov. and *Quatrionicoccus* gen. nov. as replacements for the illegitimate, prokaryotic, generic names *Micromonas* Murdoch and Shah 2000 and *Quadricoccus* Maszenan *et al.* 2002, respectively. Int. J. Syst. Evol. Microbiol. *56*, 2711–2713.

Tran, S.D., Rudney, J.D., Sparks, B.S., and Hodges, J.S. (2001). Persistent presence of *Bacteroides forsythus* as a risk factor for attachment loss in a population with low prevalence and severity of adult periodontitis. J. Periodontol. *72*, 1–10.

Vandamme, P., Bernardet, J.-F., Segers, P., Kersters, K., and Holmes, B. (1994). New perspectives in the classification of the *Flavobacteria*: description of *Chryseobacterium* gen. nov., *Bergeyella* gen. nov., and *Empedobacter* nom. rev. Int. J. Syst. Evol. Microbiol. *44*, 827–831.

Vartoukian, S.R., Palmer, R.M., and Wade, W.G. (2007). The division '*Synergistes*'. Anaerobe *13*, 99–106.

Vartoukian, S.R., Palmer, R.M., and Wade, W.G. (2009). Diversity and morphology of members of the phylum '*Synergistetes*' in periodontal health and disease. Appl. Environ. Microbiol. *75*, 3777–3786.

Vartoukian, S.R., Palmer, R.M., and Wade, W.G. (2010). Cultivation of a *Synergistetes* strain representing a previously uncultivated lineage. Environ. Microbiol. *12*, 916–928.

Wade, W.G., Downes, J., Dymock, D., Hiom, S.J., Weightman, A.J., Dewhirst, F.E., Paster, B.J., Tzellas, N., and Coleman, B. (1999). The family *Coriobacteriaceae*: reclassification of *Eubacterium exiguum* (Poco *et al.* 1996) and *Peptostreptococcus heliotrinreducens* (Lanigan 1976) as *Slackia exigua* gen. nov., comb. nov. and *Slackia heliotrinireducens* gen. nov., comb. nov., and *Eubacterium lentum* (Prevot 1938) as *Eggerthella lenta* gen. nov., comb. nov. Int. J. Syst. Bacteriol. *49(Pt 2)*, 595–600.

Wallner, G., Amann, R., and Beisker, W. (1993). Optimizing fluorescent in situ hybridization with rRNA-targeted oligonucleotide probes for flow cytometric identification of microorganisms. Cytometry *14*, 136–143.

Ward, D., Weller, R., and Bateson, M. (1990). 16S rRNA sequences reveal numerous uncultured microorganisms in a natural community. Nature *345*, 63–65.

Willems, A., and Collins, M.D. (1996). Phylogenetic relationships of the genera *Acetobacterium* and *Eubacterium sensu stricto* and reclassification of *Eubacterium alactolyticum* as *Pseudoramibacter alactolyticus* gen. nov., comb. nov. Int. J. Syst. Bacteriol. *46*, 1083–1087.

Yoon, J., Oku, N., Park, S., Kasai, H., and Yokota, A. (2011). *Porifericola rhodea* gen. nov., sp. nov., a new member of the phylum *Bacteroidetes* isolated by the bait-streaked agar technique. Antonie Van Leeuwenhoek *100*, 145–153.

Zijnge, V., van Leeuwen, M.B., Degener, J.E., Abbas, F., Thurnheer, T., Gmur, R., and Harmsen, H.J. (2010). Oral biofilm architecture on natural teeth. PLoS One *5*, e9321.

Zuger, J., Luthi-Schaller, H., and Gmur, R. (2007). Uncultivated *Tannerella* BU045 and BU063 are slim segmented filamentous rods of high prevalence but low abundance in inflammatory disease-associated dental plaques. Microbiology *153*, 3809–3816.

# Bacterial Catabolism of Salivary Substrates

3

David Beighton, Sadaf Rasheed Mughal and Thuy Do


## Abstract

The oral biofilm proliferates in the mouth primarily by utilizing components of saliva because dietary foods are rapidly cleared. The complex microbial community functions in a concerted manner to obtain nutrients, sugars and amino acids, from salivary components including mucins, by the production of a range of glycosidic enzymes including sialidase, β-galactosidase, *N*-acetylglucosaminidases, α-fucosidase and mannosidases and exo- and endo-proteolytic activities. Degradation of glycans occurs sequentially and *in vitro* studies indicate that liberated sugars are rapidly transported though evidence of cross-feeding between species, utilizing liberated sugars, is evident. *Streptococcus oralis* is a species with the greatest ability to deglycosylate both N- and O-linked glycans and has been used extensively in model systems. New research should take advantage of modern high throughput sequencing techniques to determine the biofilm transcriptome of humans receiving defined diet, including fasting, to ascertain the response of the biofilm to *in vivo* conditions.


## Carbon sources in saliva

The oral biofilm contains a large number of different bacteria species (Dewhirst *et al.*, 2010) which, for the most part interact, survive and proliferate, in the absence of the host's diet. Dietary foods are rapidly cleared from the mouth (Keene *et al.*, 1996) so that the major nutrient source for the oral microbiome are components of the secretions from the various salivary glands, and serum components entering the mouth through the gingival crevice as gingival crevicular fluid (GCF). In essence though, the components of saliva and GCF may be regarded overall as quite similar, differing only in the proportions of their constituent parts. Thus, proteins and glycoproteins are present in both fluids and a majority contain N- or O-linked glycans. In an early study using 2D-gel electrophoresis in conjunction with mass spectroscopy, 16 proteins were identified in saliva and pooled pellicle samples (Yao *et al.*, 2003). However using a more sensitive approach, LC/MS/MS, a total of 266 proteins, proteins were identified in whole saliva which, when coupled with the number identified using a 2D-gel electrophoretic approach, yielded 309 distinct proteins in whole saliva (Hu *et al.*, 2005). The proteome of saliva and salivary pellicles is very diverse and provides the oral microbiota with a wide and complex variety of substrates for growth.

The predominant glycosylated proteins in saliva are the mucins which are present at >15% of the protein content of saliva (Schenkels *et al.*, 1995). The O-linked glycans, those of salivary mucins, have the more complex structures but the principal sugars in both O- and N-linked glycans are neuraminic acid (sialic acid), *N*-acetylglucosamine, N-acetylgalactosamine, galactose, mannose and fucose (Fig. 3.1). Some sugars may also be sulfated. Mucins are highly glycosylated proteins with the attached sugars constituting at least 50% and up to 90% of the dry weight of the molecule. The O-linked glycan sidechains can vary in length from 1 to more than 20 sugar residues usually attached by O-glycosidic linkages of N-acetyl-galactosamine to serine or threonine (Klein *et al.*, 1992). Thus mucins may be regarded as the

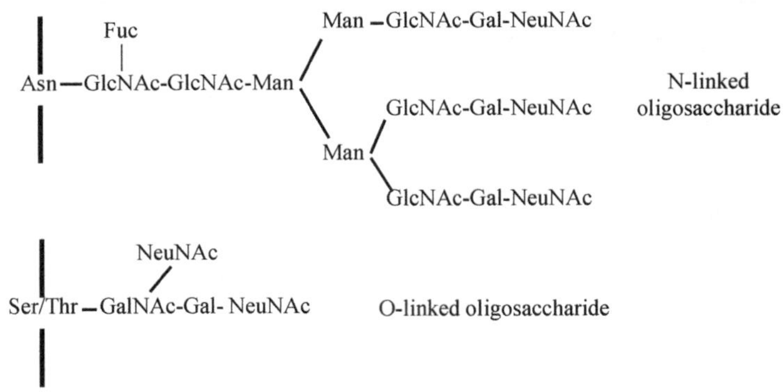


**Figure 3.1** Simplified representative structures of O- and N-linked glycans. Ser/Thr and Asn are serine/threonin and asparagines, respectively. Fuc, GlcNAc, GalNAc, NeuNAc, Man and Gal are fucose, *N*-acetylglucosamine, N-acetylgalactosamine, sialic acid, mannose and galactose, respectively.

principal and most available source of sugars for bacterial maintenance and growth.

It is worth noticing that glucose, the sugar so often included in laboratory media for the growth of bacteria, is not present in these glycans and that the glucose concentration of unstimulated saliva is < 0.3 mM (Soares *et al.*, 2009). In order for the oral biofilm to utilize the salivary glycoproteins they must be degraded into components that are transportable into the bacterial cell. To achieve this members of the oral biofilm produce a variety of exo- and endo-glycosidases to remove either individual sugars or more complex sugars from glycans which may be catabolized by bacteria. Similarly a wide range of proteases is produced by the oral biofilm to release peptides and amino acids which may be used by the oral biofilm for growth.

## Evidence for bacterial growth in the absence of the host diet

It was believed for many years that the host's diet 'fed' the bacteria on the teeth enabling them to grow but the pioneering studies by Beckers and van der Hoeven (1982) clearly demonstrated that individual bacterial species grew on the teeth in the absence of host dietary intake. Thus when germfree Osborne–Mendel rats were mono-associated with *Actinomyces viscosus* or *Streptococcus mutans*, the exponential growth rates of *S. mutans* and *A. viscosus* were 0.63 per hour and 0.24 per hour, respectively: equivalent to a doubling in population size every 1.1 hours for *S. mutans* and 2.9 hours for *A. viscosus*. What was surprising from these studies was that the provision of a sucrose-containing diet did not significantly influence the exponential growth rate of either organism.

In a study in which the growth rates of bacteria in the natural biofilm in the developmental grooves of first deciduous molar teeth of macaque monkeys were measured (Beighton *et al.*, 1986), the median doubling times of the streptococci, lactobacilli, actinomyces and *Neisseria mucosa* were 3.32, 3.45, 2.88 and 4.54 hours respectively. The availability or composition of the diet provided to the monkeys did not influence significantly the bacterial growth rates, in agreement with the previous studies of Beckers and van de Hoeven (1982). Further the addition of 0.5% (w/v) glucose to the drinking water of fasted monkeys also did not influence the size of this bacterial population 18 hours after cleaning. The numbers of bacteria in the grooves on the teeth of fed or fasted monkeys were not significantly different, but a significant increase in the proportion of '*Streptococcus mitior*' (now *Streptococcus oralis* and *Streptococcus mitis*) did occur. In an *in vitro* study in which bacteria in pooled human plaque samples were grown in human saliva, the predominant organism was found to be '*S. mitior*' and salivary mucins and proteins were extensively deglycosylated and the protein cores degraded (de Jong and van der Hoeven, 1987). To determine if this effect was apparent in humans, we examined the oral microbiota of bone marrow transplant patients during a period of restricted dietary intake. We again found that the proportion of the '*Streptococcus oralis* group' (*Streptococcus mitis* and *S. oralis*) increased significantly ($P < 0.003$) from a baseline 12.1% to 48.4% at 7 days post transplantation (Lucas *et al.*, 1997). So we have a situation where the availability of the diet and its composition made no significant difference to the rate of growth of bacteria on the dentition, although *S. oralis* appears to proliferate better than other members of the oral microbiota in the absence of the host's diet using saliva as the main source of nutrient. It is also of interest that the oral streptococci identified as either *S. oralis* or *S. mitis* are often isolated from systemic infections (Hoshino *et al.*, 2005).

## Degradation of O- and N-linked glycans

Glycans may be removed from glycoproteins by bacteria for a number of reasons. Certainly to obtain nutrients for growth may be the most apparent and perhaps the most important reason, and a consequence of this is the liberation of nutrients that cross-feed other bacteria. Thus, sialic acid is utilized for growth by bacteria that do not possess sialidase activity (Byers *et al.*, 1996): *Streptococcus sanguinis* does not produce sialidase but is able to use sialic acid as the major carbon source. However, for bacteria that primarily utilize amino acids for growth the removal of the glycans increases the exposure of the protein backbone to proteolysis so increasing its availability to the bacteria. Additionally some bacteria may use the liberated sugars to decorate their cell surfaces to 'hide' from the host's immune system (Jones *et al.*, 1986; Vimr *et al.*, 2000).

The degradation of the glycans may occur as the result of endo-glycosidic activity by which entire or almost complete glycans are released, as exemplified by *Streptococcus pneumoniae* (Kadioglu *et al.*, 2008) and *Enterococcus faecalis* (Roberts *et al.*, 2000). Although endoglycosidic genes may be identified in the genomes of many oral bacteria, no evidence of their activity against glycoproteins has been reported. Previously we found that N-linked glycans of $\alpha_1$-acid glycoprotein and the high-mannose glycan-containing glycoprotein, ribonuclease B, are degraded sequentially by *S. oralis*, and all liberated sugars, with the exception of fucose, were metabolized (Tarelli *et al.*, 1998; Byers *et al.*, 1999). Many oral bacteria produce exo-glycosidic enzymes with the putative ability to remove individual sugars from glycans; such activities have been used widely to assist in the identification of oral bacteria (Kilian *et al.*, 1989; Moncla *et al.*, 1990, 1991; Beighton *et al.*, 1991; Maiden *et al.*, 1996). However, the major exo-glycosidic activities required for the degradation of N- and O-linked glycans are sialidase (neuraminidase) and β-galactosidase, β-*N*-acetylglucosaminidase, α-fucosidase and α- and β- mannosidase. It is of interest that α- and β- mannosidase activities in *S. oralis* and other viridans streptococci were not detectable using synthetic mannosidase substrates, but only being evident by the use of MALDI-ToF-MS to examine the residual masses of the model glycoprotein ribonuclease B. Likewise, α-fucosidase was not detected in *S. oralis* using synthetic substrates, yet the organism exhibited this activity because it liberated fucose from $\alpha_1$-acid glycoprotein. However, the liberated fucose moieties were not used as a carbon source during growth.

## Bacteria act in concert

It is now accepted that, in the oral biofilm, bacteria act in concert to degrade salivary and serum components. One of the first studies to examine this interaction was reported by ter Steeg *et al.* (1987) who grew samples of subgingival plaque in human serum and found that all measured serum proteins including immunoglobulins, haptoglobin, transferrin and complement C3c were completely degraded after a long period of incubation, whereas partial degradation of immunoglobulins, haptoglobin, transferrin, albumin, alpha-l-antitrypsin and complement C3c and C4 was observed after 48 hours of growth. Partial degradation involved proteolytic degradation and removal of glycans. In each independent culture the composition of the final bacterial population was different, indicating that many different consortia have the same catalytic potential. Using mixtures of mutans streptococci (organisms not regarded as extensively proteolytic) and other oral bacteria, it was possible to demonstrate synergistic degradation of albumin. Notably, mixtures of *S. mutans* or *S. sobrinus* in conjunction with *S. oralis* or *F. nucleatum* (Homer and Beighton, 1992) demonstrated this activity.

Bradshaw *et al.* (1996) reported the use of porcine gastric mucin, which have a similar glycan structures to human mucins (Herp *et al.*, 1979), to grow increasingly complex mixtures of oral bacteria in a chemostat. The organisms and their order of introduction into the chemostat were selected on the basis of their known glycosidic and proteolytic activities. A five member mixed culture of species having little glycosidase and protease activity (*S. mutans, S. gordonii, N. subflava, V. dispar,* and *F. nucleatum*), was established in a chemostat ($D = 0.1/h$) in which hog gastric mucin was the major carbon and energy source. Introduction of additional species with novel enzyme activities [e.g. sialidase (*A. viscosus* and *S. oralis*), α-fucosidase (*L. casei*) and endopeptidases (*P. gingivalis* and *P. nigrescens*])] led to their establishment within the community and resulted in an increase in the total viable counts of the microbial consortium (Fig. 3.2). This increase in viable count was made up of the numbers of the newly added species but most significantly from a rise in the numbers of the existing community members. This result

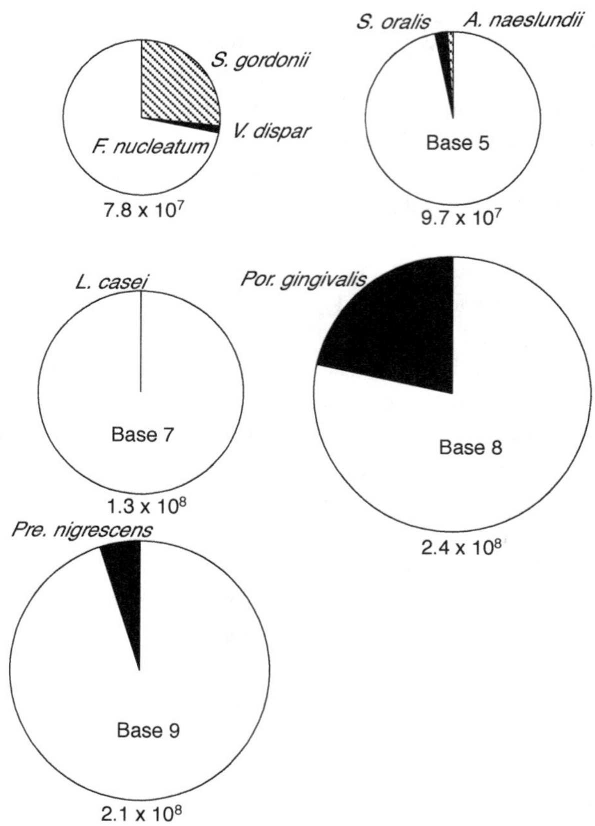


**Figure 3.2** The total viable count within a microbial community increases as the diversity increases. The area of each pie represents the total viable count of the entire community (number below the pie), whereas the area of each slice represents the viable count for a specific organism. Upper left: the initial five-species mixture (*Streptococcus gordonii*, *S. mutans*, *Fusobacterium nucleatum*, *Viellonella dispar*, and *Neisseria sublflava*). Counts of *N. subflava* and *S. mutans* are too small to be visible in the pie. Upper right: after addition of *S. oralis* and *Actinomyces naeslundii*; the white slice represents the summed count of the previous five-member community. Middle left: after addition of *Lactobacillus casei* to the previous seven-member community. Middle right: after addition of *Porphyromonas gingivalis* to the previous eight-member community. Lower left: after addition of *Prevotella nigrescens* to the previous nine-member community.

suggested that glycoprotein catabolism involved the synergistic and concerted action of several species with overlapping patterns of enzyme activity. Previously, van der Hoeven and Camp (1991) reported the synergistic degradation of the glycan side chains on hog gastric mucin by the combination of *S. sanguis* Ny 584 and *S. oralis* strain Ny 586 where the stimulation of the growth of *S. sanguis* Ny 584 was probably due to the generation of non-fucose-ending oligosaccharide chains by fucosidase activity from *S. oralis* Ny 586. Neither organism utilized the liberated fucose which was free in the culture supernatant.

## *In vivo* expression of glycosidic and proteolytic activities

In studies using macaque monkeys we determined the effect of diet on the production of exo-glycosidic and proteolytic enzyme activities by the oral biofilm. The following glycosidase activities were detected, using synthetic substrates, at high levels in supra-gingival plaque: α-galactosidase, β-galactosidase, β-*N*-acetyl-D-galactosaminidase, β-*N*-acetyl-D-glucosaminidase, α-L-fucosidase and neuraminidase, but only low levels of α-*N*-acetyl-D-galactosaminidase and α-*N*-acetyl-D-glucosaminidase (Beighton *et al.*, 1986). The same model was used to assess the effects of fasting on the production of relevant glycosidases by the oral biofilm. In this study fasting monkeys were provided with drinking water either unsupplemented or supplemented with glucose or *N*-acetylglucosamine (Beighton and Smith, 1986). Fasting resulted in significant increases in the plaque levels of α-fucosidase, β-*N*-acetylglucosaminidase, β-*N*-acetylgalactosaminidase and sialidase (Fig. 3.3). These glycosidic enzyme levels, except for sialidase, were also significantly raised when the drinking water was supplemented with 0.025 M glucose. However when the drinking water was supplemented with 0.025 M *N*-acetylglucosamine the plaque level of each glycosidic activity was not significantly different, but generally lower, from those measured in the plaque of fed monkeys.

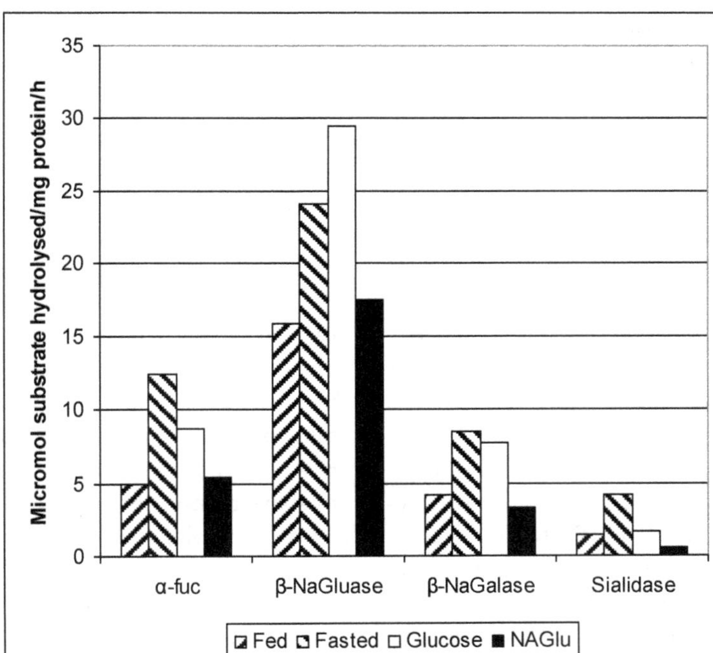


**Figure 3.3** Changes in exoglycosidic levels (α-fuc, α-fucosidase; β-NaGluase, β-*N*-acetylglucosaminidase; β-NaGalase, β-*N*-acetylgalactosaminidase) in dental plaque collected from the smooth tooth surfaces of 12 monkeys either fed, fasted and provided with drinking water, or fasted and provided with drinking water supplemented with 0.025 M glucose (Glu) or 0.025 M *N*-acetylglucosamine (NAGlu). Data from Beighton and Smith (1986).

These observations demonstrate that the biofilm has a preference for amino-sugars rather than simple sugars such as glucose: the biofilm is not simply carbon-limited.

Proteolytic activities of supragingival plaque have also been determined in fed and in fasted monkeys using a range of synthetic peptidase substrates (Smith and Beighton, 1987). After fasting the monkeys for 24 hours, the rates of hydrolysis of *N*-leucyl-, *N*-alanyl-, *N*-isoleucyl- and *N*-tyrosyl-2-naphthylamines, glycyl-L-proline 4-nitroanilide, *N*¢,-benzoyl-αL-arginine-2-naphthylamine and L-leucylglycine were significantly increased. Changes in these peptidase activities may serve to enhance the mobilization of amino acids and peptides, and they may explain why the growth rates of dental plaque bacteria in *vivo* are generally unaffected by the availability of the host's diet. *In vitro* simulations using mixed oral biofilm bacteria grown in chemostats with saliva as the major nutrient source demonstrate the production of similar sets of glycosidic and proteolytic activities (de Jong and van der Hoeven, 1986).

## Recent studies of bacteria interactions with saliva

Wickström and Svensater (2008) used a purified mucin preparation, MUC5B, as a substrate for bacterial growth. They demonstrated, using fluorescent staining for viability and metabolic activity, that the plaque samples, examined using confocal microscopy, were metabolically active in the presence of MUC5B, deglycosylating the mucin, while individual species of *S. oralis* and *S. gordonii* isolated from the sample plaque samples were inactive. Two preparations of MUC5B (solMUC5B and gelMUC5B) were isolated by density-gradient centrifugation and exhibited different carbohydrate content and surface-adsorbing characteristics. Each was tested for its biological activity by incubating them individually with freshly isolated dental plaque and assaying for metabolic using the fluorescent substrate 5-cyano-2,3-ditolyl tetrazolium chloride (Wickström *et al.*, 2009). The two MUC5B preparations elicited different responses from the plaque samples. Higher metabolism was found with the solMUC5B preparation in solution, with 79.4% active cells compared to the controls without mucin (9.6% of cells metabolically active). GelMUC5B showed only 48.2% active cells with the same plaque population. When the same mucins were adhered to a surface, significantly lower metabolic response occurred, with surface-associated plaque cells showing only 12.1% active cells with solMUC5B and 29.2% with gelMUC5B. The authors suggested that the metabolism by the plaque cells adsorbed to surface-associated mucins was down regulated compared to the same cells suspended in mucin solution. These data are the first to show differential responses of oral bacteria to purified mucins; the solMUC5B and gelMUC5B preparations exhibited different biological activity when assayed with freshly plaque bacteria in suspension and in a biofilm.

Periasamy *et al.* (2009) reported that, when saliva was provided as the sole nutrient source, a strain of *F. nucleatum* would not form biofilms alone nor with an *S. oralis* isolate, but it did form biofilms with an *A. naeslundii* strain. However, enhanced growth of the *F. nucleatum* was evident in a three-species biofilm composed of the *F. nucleatum*, *S. oralis* and *A. neaslundii*, indicating that some mixed bacterial interactions with saliva may not necessarily be beneficial to all taxa involved. A similar observation regarding the specificity of the bacterial interactions involving other bacterial species was also reported by the same group (Periasamy *et al.*, 2010). However such observations involving restricted numbers of species and strains need to be interpreted with caution since the diversity of the oral biofilm is much greater than used in these studies and all members of the same species may not exhibit the same phenotype. In this latter report (Periasamy *et al.*, 2010) a central role for *Veillonella* spp. in the establishment of the oral biofilm was suggested owing to its ability to metabolize of lactate, arising from the catabolism of salivary glycans, by mixed multispecies biofilms. These observations extend the role of *Veillonella* spp. in the oral biofilm beyond organisms with the ability to utilize lactate as an energy source. These more recent studies though collectively suggest that the more complex the community, the greater is the supply of nutrients to support the growth of all members of the biofilm; results akin to those reported by Bradshaw *et al.* (1994).

## Catabolism of mucin-derived sugars

The catabolism of the mucin-derived sugars has been studied in viridans streptococci (Byers *et al.*, 1996). Sialic acid, *N*-acetylglucosamine and mannose catabolism by these bacteria demonstrates a convergence in functions required to utilize these diverse sugars. Sialic acid is imported into these bacteria by an inducible transport system. Once internalized, sialic acid is acted upon by *N*-acetylneuraminate pyruvate lyase to produce pyruvate and *N*-acetylmannosamine which may be converted to *N*-acetylglucosamine by the action of *N*-acetylhexosamine isomerase, or phosphorylated to form *N*-acetylmannosamide-6-phosphate by the action of *N*-acetylhexosamine phosphorylase. The *N*-acetylglucosamine and mannosamine-6-phosphate may each be converted to *N*-acetylglucosamine-6-phosphate by the actions of *N*-acetylhexosamine or N-acetylhexosamine-6-phosphate isomerase, respectively. This is the form in which *N*-acetylglucosamine, transported via the glucose PTS, enters the catabolic cycle so the catabolism of both amino sugars proceeds along the same pathway. Mannose, like *N*-acetylglucosamine is also transported via the glucose PTS with the formation of mannose-6-phosphate which is isomerized and enters glycolysis. The metabolism of galactose involves the internalization of the sugar via either the glucose or lactose PTS in *S. mutans* after which it may enter the tagatose or Leloir pathways as galactose-6-phosphate (Zeng *et al.*, 2010).

## Effect of *N*-acetylglucosamine on the growth of mutans streptococci

Mutans streptococci, which are acidogenic and aciduric, are associated with the initiation and progression of dental caries. Of this group *Streptococcus sobrinus* has, *in vitro*, the most potent cariogenic properties yet in the mouth the numbers and proportion of this species are low in comparison to those of *S. mutans* (Beighton, 2005). The exception is in the case of individuals with excessive sugar consumption and extensive caries where *S. sobrinus* prevalence is greatly increased. Through a simple set of differentiating tests, *S. sobrinus* was shown to be unable to ferment *N*-acetylglucosamine (Kraal and Daneo-Moore, 1981), a trait that would seem to be detrimental to the proliferation of this species in the mouth. However, when we examined the basis for the inability of *S. sobrinus* to ferment this mucin-derived amino sugar we found that *N*-acetylglucosamine inhibited the growth of *S. sobrinus* in the presence of glucose (Homer *et al.*, 1993). When *S. mutans* and *S. sobrinus* were grown together in the presence of glucose and *N*-acetylglucosamine, *S. mutans*, which is able to utilize *N*-acetylglucosamine, out-competed *S. sobrinus*. When grown together on glucose alone, *S. sobrinus* out-competed *S. mutans*. It was demonstrated that *S. sobrinus* transports *N*-acetylglucosamine via the glucose phosphoenolpyruvate-dependent phosphotransferase system to form *N*-acetylglucosamine-6-phosphate. However, the intracellular levels of *N*-acetylglucosamine-6-phosphate deactylase and glucosamine-6-phosphate deaminase, required for the catabolism of *N*-acetylglucosamine-6-P, are low, resulting in reduced intracellular levels of phosphoenolpyruvate and reduced cell growth. So uptake of *N*-acetylglucosamine by *S. sobrinus* is detrimental to its proliferation in the oral biofilm via a mechanism analogous to that involved in the metabolic inhibition of streptococci mediated by xylitol.

## Identification of a lacto-*N*-biose pathway in *A. oris*

It is clear that there are many enzymes involved in the degradation of glycans and the determination of near-complete bacterial genome sequences will enable the genes coding for potential degradative proteins to be identified. Of course, it is essential that the bacteria being sequenced are accurately identified and the genes be accurately annotated. As an example of how genome sequencing can assist in the identification of novel genes and some of the problems and difficulties that arise with the annotation of genes I will present some recent data we have obtained on the identification of a lacto-*N*-biose operon in *Actinomyces oris*. We (Henssge *et al.*, 2009) have presented description of *A. naeslundii*, *A. oris* and *A. johnsonii* previously described as genospecies of *A. naeslundii*. Three strains of *A. oris* have been sequenced wholly or partially but one, MG1, is described as *A. naeslundii*. Another, C505, is described as *Actinomyces viscosus*. Only *A. oris* K20 is correctly identified. We have undertaken partial genome sequencing of *A. naeslundii* and *A. oris* strains and have found some specific differences between the genome sequences of these two oral species. We found that genes, ANA_1637, ANA_1638, ANA_1639 and ANA_1640, in the MG-1 genome (http://cmr.jcvi.org/cgi-bin/CMR/CmrHomePage.cgi) were annotated in the CMR database as conserved hypothetical protein, solute binding protein of ABC transporter for sugars, permease of ABC transporter for sugars (permease) and permease of ABC transporter for sugars (permease), respectively. However, when we annotated our data using RAST (Rapid Annotations based on Subsystem Technology; Aziz *et al.*, 2008) the genes were annotated as lacto-*N*-biose phosphorylase, galacto-*N*-biose-/lacto-*N*-biose I ABC transporter (periplasmic substrate-binding protein), galacto-*N*-biose-/lacto-N-biose I ABC transporter (permease component 1) and galacto-*N*-biose-/lacto-*N*-biose I ABC transporter (permease component 2). These annotations were confirmed by BLAST searching and so we have found that *A. oris* strains possess a genes coding for proteins required to metabolize either galacto- or lacto-*N*-biose. In bifidobacteria lacto-N-biose phosphorylase plays a role in nutrition by enabling the utilization of galacto-N-biose liberated from mucin sugars and lacto-*N*-biose derived from human milk oligosaccharides (González, *et al.*, 2008). We have also found in subsequent growth experiments that *A. oris* strains grew significantly better on lacto-*N*-biose than did *A. naeslundii* strains and might be the basis for the predominance of *A. oris* over *A. naeslundii* in the oral biofilm. Gene deletion experiments are under way to confirm the functions of these genes. These observations also demonstrate the problems that might be faced by investigators attempting to use genomic data to understand the potential extracellular enzymic activities of bacteria and their potential in the catabolism of salivary proteins.

## Summary

From these various studies it becomes apparent that mixed microbial populations, either contrived *in vitro*, or present *in vivo*, exhibit the ability to degrade salivary components and to utilize the components for maintenance, growth and proliferation. There can be no doubt that different species, and probably different phenotypes of the same species, compete for the liberated nutrients even if they do not produce the degradative enzymes responsible for the protein or glycoprotein degradation. The oral biofilm may harbour in excess of 1200 phylotypes, based on 16S rRNA sequencing. However this is a significant underestimation of the diversity of the oral microbiota as some genera, including *Streptococcus, Veillonella, Actinomyces, Lactobacillus,* cannot be differentiated satisfactorily using 16S rRNA sequencing and every species that has been investigated is comprised of an undetermined number of genotypes which may exhibit different phenotypes (Alam *et al.*, 2000). This diversity and inbuilt redundancy in function is important to the survival and productivity of the oral biofilm. Yachi and Loreau (1999) propose the insurance hypothesis to explain the effect of biodiversity in a fluctuating environment. The oral cavity is certainly a fluctuating environment with respect to oral hygiene practices and dietary intake. They suggested a model to assess the effects of species richness on the expected temporal mean and variance of ecosystem processes such as productivity, based on individual species' productivity responses to environmental fluctuations. They reported two major insurance effects of species richness on ecosystem productivity. Firstly a buffering effect evident as a reduction in the temporal variance of productivity, and secondly a performance-enhancing effect so that productivity increases over time. As the oral biofilm has evolved with us we must assume that the diversity of the oral biofilm and its ability to proliferate as already maximal yet our understanding of the interactions evidenced in the forgoing remains somewhat limited beyond the simplistic.

## Future research

While some interactions of the oral biofilm and protein and glycoprotein components of saliva have been elucidated, the fundamental interactions between mixed bacterial populations and mucins requires more research including specifically the application of meta-transcriptomic methodologies. Such studies are required to ascertain the bacterial response to exposure to defined mucins or glycoproteins and will extend the studies simply investigating either degradation of substrate, production of enzymes or bacterial cell numbers. New research should take advantage of modern high-throughput sequencing techniques to determine the biofilm transcriptome of humans receiving defined diet, including fasting, to ascertain the response of the biofilm to *in vivo* conditions. It is now possible to move beyond simple *in vitro* experiments utilizing one or a few species and determine the real changes in gene expression when the oral biofilm is subject to specific challenges, diet withdrawal or exposure to specific foods including defined sugars, proteins and pre- or pro-biotics.

These more sophisticated approaches may lead to a better understanding of the genetic basis of the ability of mixed bacterial populations to proliferate on mucins and may enable methods to be established which, while not eliminating the oral biofilm, should inhibit the rate of accumulation. This may be affected by reducing the ability of the bacteria to utilize the mucins or increase the maintenance energy of bacteria such that more of the energy derived from the catabolism of mucins is used to repair bacteria and maintain their viability than is spent on proliferation.

## References


Alam, S., Brailsford, S.R., Adams, S., Allison, C., Sheehy, E., Zoitopoulos, L., Kidd, E.A., and Beighton, D. (2000). Genotypic heterogeneity of *Streptococcus oralis* and distinct aciduric subpopulations in human dental plaque. Appl. Environ. Microbiol. *66*, 3330–3336.

Aziz, R.K., Bartels, D., Best, A.A., DeJongh, M., Disz, T., Edwards, R.A., Formsma, K., Gerdes, S., Glass, E.M., Kubal, M., *et al.* (2008). The RAST Server: rapid annotations using subsystems technology. BMC Genomics. *9*, 75.

Beckers, H.J., and van der Hoeven, J.S. (1982). Growth rates of *Actinomyces viscosus* and *Streptococcus mutans* during early colonization of tooth surfaces in gnotobiotic rats. Infect. Immun. *35*, 583–587.

Beighton, D. (2005). The complex oral microflora of high-risk individuals and groups and its role in the caries process. Community Dent. Oral Epidemiol. *33*, 248–255.

Beighton, D., and Smith, K. (1986). The modulation of exoglycosidic enzymes in the supragingival plaque of macaque monkeys. FEMS Microbiol. Lett. *34*, 319–322.

Beighton, D., Smith, K., and Hayday, H. (1986). The growth of bacteria and the production of exoglycosidic enzymes in the dental plaque of macaque monkeys. Arch. Oral Biol. *31*, 829–835.

Beighton, D., Hardie, J.M., and Whiley, R.A. (2005). A scheme for the identification of viridans streptococci. J. Med. Microbiol. *35*, 367–372.

Bradshaw, D.J., Homer, K.A., Marsh, P.D., and Beighton, D. (1994). Metabolic cooperation in oral microbial communities during growth on mucin. Microbiology *140*, 3407–3412.

Byers, H.L., Homer, K.A., and Beighton, D. (1996). Utilization of sialic acid by viridans streptococci. J. Dent. Res. *75*, 1564–1571.

Byers, H.L., Tarelli, E., Homer, K.A., and Beighton, D. (1999). Sequential deglycosylation and utilization of the N-linked, complex-type glycans of human alpha1-acid glycoprotein mediates growth of *Streptococcus oralis*. Glycobiology *9*, 469–479.

Dewhirst, F.E., Chen, T., Izard, J., Paster, B.J., Tanner, A.C., Yu, W.H., Lakshmanan, A., and Wade, W.G. (2010). The human oral microbiome. J. Bacteriol. *192*, 5002–5017.

González, R., Klaassens, E.S., Marinen, E., de Vos, W.M., and Vaughan, E.E. (2008). Differential transcriptional response of *Bifidobacterium longum* to human milk, formula milk, and galactooligosaccharide. Appl. Environ. Microbiol. *74*, 4686–4694.

Henssge, U., Do, T., Radford, D.R., Gilbert, S.C., Clark, D., and Beighton, D. (2009). Emended description of *Actinomyces naeslundii* and descriptions of *Actinomyces oris* sp. nov. and *Actinomyces johnsonii* sp. nov., previously identified as *Actinomyces naeslundii* genospecies 1, 2 and WVA 963. Int. J. Syst. Evol. Microbiol. *59*, 509–516.

Herp, A., Wu, A.M., and Moschera, J. (1979). Current concepts of the structure and nature of mammalian salivary mucous glycoproteins. Mol. Cell. Biochem. *23*, 27–44.

van der Hoeven, J.S., and Camp, P.J. (1991). Synergistic degradation of mucin by *Streptococcus oralis* and *Streptococcus sanguis* in mixed chemostat cultures. J. Dent. Res. *70*, 1041–1044.

Homer, K.A., and Beighton, D. (1992). Synergistic degradation of bovine serum albumin by mutans streptococci and other dental plaque bacteria. FEMS Microbiol. Lett. *69*, 259–262.

Homer, K.A., Patel, R., and Beighton, D. (1993). Effects of *N*-acetylglucosamine on carbohydrate fermentation by *Streptococcus mutans* NCTC 10449 and *Streptococcus sobrinus* SL-1. Infect. Immun. *61*, 295–302.

Hoshino, T., Fujiwara, T., and Kilian, M. (2005). Use of phylogenetic and phenotypic analyses to identify nonhemolytic streptococci isolated from bacteremic patients. J. Clin. Microbiol. *43*, 6073–6085.

Hu, S., Xie, Y., Ramachandran, P., Ogorzalek Loo, R.R., Li, Y., Loo, J.A., and Wong, D.T. (2005). Large-scale identification of proteins in human salivary proteome by liquid chromatography/mass spectrometry and two-dimensional gel electrophoresis-mass spectrometry. Proteomics. *5*, 1714–1728.

de Jong, M.H., and van der Hoeven, J.S. (1987). The growth of oral bacteria on saliva. J. Dent. Res. *66*, 498–505.

Kadioglu, A., Weiser, J.N., Paton, J.C., and Andrew, P.W. (2008).The role of *Streptococcus pneumoniae* virulence factors in host respiratory colonization and disease. Nat. Rev. Micro. *6*, 288–301.

Kral, T.A., and Daneo-Moore, L. (1981). Biochemical differentiation of certain oral streptococci. J. Dent. Res. *60*, 1713–1718.

Keene, H.J., Coykendall. A.L., and Lahmeyer, H.A. (1996). Oral glucose clearance time in caries-active and caries-resistant naval recruits. J. Dent. Res. *45*, 409.

Kilian, M., Mikkelsen, L., and Henrichsen, J. (1989). Taxonomic studies of viridans streptococci: description of *Streptococcus gordonii* sp. nov. and emended descriptions of *Streptococcus sanguis* (White and Niven 1946), *Streptococcus oralis* (Bridge and Sneath 1982), and *Streptococcus mitis* (Andrewes and Horder 1906). Int. J. Syst. Bacteriol. *39*, 471–484.

Klein, A., Carnoy, C.H., Wieruszeski, J.M., Strecker, G., Strang, A.M., and van Halbeek H. (1992). The broad diversity of neutral and sialylated oligosaccharides derived from human salivary mucins. Biochem. *31*, 6152–6165.

Jones, A.H., Lee, C.C., Moncla, B.J., Robinovitch, M.R., and Birdsell, D.C. (1986). Surface localization of sialic acid on *Actinomyces viscosus*. J. Gen. Microbiol. *132*, 3381–3391.

Lucas, V.S., Beighton, D., Roberts, G.J., and Challacombe, S.J. (1997). Changes in the oral streptococcal flora of children undergoing allogeneic bone marrow transplantation. J. Infect. *35*, 135–141.

Maiden, M.F., Tanner, A., and Macuch, P.J. (1996). Rapid characterization of periodontal bacterial isolates by using fluorogenic substrate tests. J. Clin. Microbiol. *34*, 376–384.

Moncla, B.J., Braham, P., and Hillier, S.L. (1990). Sialidase (neuraminidase) activity among Gram-negative anaerobic and capnophilic bacteria. J. Clin. Microbiol. *28*, 422–425.

Moncla, B.J., Braham, P., Rabe, L.K., and Hillier, S.L. (1991). Rapid presumptive identification of black-pigmented Gram-negative anaerobic bacteria by using 4-methylumbelliferone derivatives. J. Clin. Microbiol. *29*, 1955–1958.

Periasamy, S., and Kolenbrander, P.E. (2010). Central role of the early colonizer *Veillonella* sp. in establishing multispecies biofilm communities with initial, middle, and late colonizers of enamel. J. Bacteriol. *192*, 2965–2972.

Periasamy, S., Chalmers, N.I., Du-Thumm, L., and Kolenbrander, P.E. (2009). *Fusobacterium nucleatum* ATCC 10953 requires *Actinomyces naeslundii* ATCC 43146 for growth on saliva in a three-species community that includes *Streptococcus oralis* 34. Appl. Environ. Microbiol. *75*, 3250–3257.

Roberts, G., Tarelli, E., Homer, K.A., Philpott-Howard, J., and Beighton, D. (2000). Production of an endo-beta-*N*-acetylglucosaminidase activity mediates growth of *Enterococcus faecalis* on a high-mannose-type glycoprotein. J. Bacteriol. *182*, 882–890.

Schenkels, L.C., Veerman, E.C., and Nieuw Amerongen, A.V. (1995). Biochemical composition of human saliva in relation to other mucosal fluids. Crit. Rev. Oral Biol. Med. *6*, 161–175.

Soares, M.S., Batista-Filho, M.M., Pimentel, M.J., Passos, I.A., and Chimenos-Küstner, E. (2009). Determination of salivary glucose in healthy adults. Med. Oral Patol. Oral Cir. Bucal. *14*, e510–e513.

Smith, K., and Beighton, D. (1987). Proteolytic activities in the supragingival plaque of monkeys (*Macaca fascicularis*). Arch. Oral Biol. *32*, 473–476.

ter Steeg, P.F., van der Hoeven, J.S., de Jong, M.H., van Munster, P.J., and Jansen, M.J. (1987). Enrichment of subgingival microflora on human serum leading to accumulation of *Bacteroides* species, Peptostreptococci and Fusobacteria. Antonie Van Leeuwenhoek *53*, 261–272.

Tarelli, E., Byers, H.L., Homer, K.A., and Beighton, D. (1998). Evidence for mannosidase activities in *Streptococcus oralis* when grown on glycoproteins as carbohydrate source. Carbohydr. Res. *312*, 159–164.

Vimr, E., Lichtensteiger, C., and Steenbergen, S. (2000). Sialic acid metabolism's dual function in *Haemophilus influenzae*. Mol. Microbiol. *36*, 1113–1123.

Wickström, C., and Svensäter, G. (2008). Salivary gel-forming mucin MUC5B – a nutrient for dental plaque bacteria. Oral Microbiol. Immunol. *23*, 177–182.

Wickström, C., Hamilton, I.R., and Svensäter, G. (2009). Differential metabolic activity by dental plaque bacteria in association with two preparations of MUC5B mucins in solution and in biofilms. Microbiology *155*, 53–60.

Yachi, S., and Loreau, M. (1999). Biodiversity and ecosystem productivity in a fluctuating environment: the insurance hypothesis. Proc. Natl. Acad. Sci. U.S.A. *96*, 1463–1468.

Yao, Y., Berg, E.A., Costello, C.E., Troxler, R.F., and Oppenheim, F.G. (2003). Identification of protein components in human acquired enamel pellicle and whole saliva using novel proteomics approaches. J. Biol. Chem. *278*, 5300–5308.

Zeng, L., Das, S., and Burne, R.A. (2010). Utilization of lactose and galactose by *Streptococcus mutans*: transport, toxicity, and carbon catabolite repression. J. Bacteriol. *192*, 2434–2344.

# Structural Organization of Oral Biofilms in Supra- and Subgingival Environments

4

Vincent Zijnge, Annette Moter, Frank Abbas and Hermie J.M. Harmsen

## Abstract

The more we learn about the oral microbiota, the more we have to accept its high degree of complexity and diversity. The composition of the oral microbial community depends on a number of patient-specific factors; the immunology and genetic background of the patient, underlying diseases (chronic or acute, e.g. diabetes, infections), hormone status (e.g. pregnancy), medication (e.g. immunosuppressive therapy or antibiotics) and behaviour of the patient (e.g. nutrition, oral hygiene, smoking habits). It even differs within the oral cavity of a subject due to the ecological niches influenced by different variables such as oxygen level, nutrition, physical stress and saliva flow.

All these factors influence the combination and multitude of bacterial species that are finally responsible for maintaining health or causing disease within the oral ecosystem. However, bacteria do not function as isolated entities but live together in complex sessile communities called oral biofilms. The efficiency of oral biofilms depends on coordinated, metabolically integrated, spatially organized networks. Therefore, it is crucial to understand these highly dynamic communities and to observe and analyse their composition and architecture. In this chapter we first discuss the methodology used in oral biofilm structure research and subsequently discuss the current status of the research as well as our concept of supra- and subgingival biofilm organization.

## Methods to study oral biofilm structure

### Composition of oral biofilms

A prerequisite for studying the structure of oral plaque is knowledge of the bacteria present in the oral cavity. A substantial number of studies have investigated the prevalence and association of bacterial species with age, health, or disease. The inventory of bacterial species that inhabit the human mouth has been growing constantly with the advent of culture-independent molecular methods. Cloning and sequencing of genes coding for ribosomal RNA have elucidated a number of yet-to-be cultured taxa (Choi *et al.*, 1994; Kumar *et al.*, 2003; Aas *et al.*, 2005; Vartoukian *et al.*, 2007; Dewhirst *et al.*, 2010). A recent work analysed the oral microbiome based on more than 36,034 16S rRNA-clone sequences, resulting in 1,179 taxa, of which 68% were uncultivated phylotypes (Dewhirst *et al.*, 2010). The majority of taxa belong to six major phyla; *Firmicutes, Bacteroidetes, Proteobacteria, Actinobacteria, Spirochaetes,* and *Fusobacteria*. The remaining phyla, *Euryarchaeota, Chlamydia, Chloroflexi,* SR1, *Synergistetes, Tenericutes,* and TM7, contain only 4% of the taxa.

This comprehensive dataset of oral sequences has recently been made available to the public in an online data-bank, the Human Oral Microbiome Database (HOMD at http://www.homd.org/). Next-generation sequencing of saliva and supragingival plaque yielded several thousand phylotypes, whose relevance has yet to be determined (Keijser *et al.*, 2008). In addition, the human mouth harbours fungi [as compiled in the so called oral 'Mycobiome' (Ghannoum *et al.*, 2010)], viruses (Slots, 2010a,b) and even *Archaea* (Lepp *et al.*, 2004).

The association of a number of bacterial species with disease has been investigated using molecular techniques like real-time PCR, species-specific amplification, or polymerase chain reaction (PCR) followed by oligonucleotide capture (de Lillo *et al.*, 2004; Moter *et al.*, 2006; Riep *et al.*, 2009). Chequerboard DNA–DNA hybridization allowed for profiling of species communities in high numbers of samples and resulted in the characterization of distinct bacterial complexes associated with health or disease (Socransky *et al.*, 1998). Also, fingerprinting techniques such as terminal restriction fragment length polymorphism (tRFLP) (Sakamoto *et al.*, 2003) and denaturing gradient gel electrophoresis (DGGE) (Zijnge *et al.*, 2006) have been successfully used to compare oral bacterial community patterns. The resulting community profiles led to the design of high throughput microarray assays for fast and comprehensive sample analysis (Colombo *et al.*, 2009).

All of the above-mentioned techniques have substantially contributed to our understanding of physiology and pathology in the human mouth. However, it is beyond the scope of this chapter to give a thorough review of these 'destructive' molecular techniques that use extracted DNA without regard for spatial resolution. From this point on, we will focus on techniques for investigation of oral biofilm structure and architecture.

## *In vitro* systems

It is obvious, that the diversity and complexity of oral biofilms is difficult to model *in vitro*. Nevertheless, *in vitro* systems have been of utmost importance in finding key players in biofilm formation and architecture. *In vitro* models range from single species biofilms to sophisticated systems such as the artificial mouth in which a plaque microcosm is cultivated from saliva in a complex multistation culture device (Sissons *et al.*, 1991; Angker *et al.*, 2011).

A number of cultured oral bacterial species have been shown to physically interact, adhere to each other, and clump together when mixed in liquid media. This is known as coaggregation and displays specific interbacterial recognition patterns (Kolenbrander, 1988; Whittaker *et al.*, 1996). To observe these interactions microscopically, flow-cell systems were developed that allowed analysis of initial bacterial attachment and multispecies biofilm growth (Palmer Jr. and Caldwell, 1995). Also, phenotypes of *Porphyromonas gingivalis* biofilms were observed by confocal laser scanning microscopy (CLSM) showing the differences of biofilm development in a static system as compared to a flow chamber system. In the latter one, constant flow of nutrients led to thicker and more heterogeneous biofilms than in the static systems (Davey, 2006). *In vitro* models of supragingival plaque have also been used to investigate effects of antimicrobial agents by cultivation of oral bacteria, or of undefined plaque samples, in saliva-based medium on hydroxyapatite discs (Hope and Wilson, 2004; Takeuchi *et al.*, 2007). Similarly, the structural organization of *Streptococcus mutans* exopolysaccharide matrix (EPS) was observed on saliva-coated hydroxyapatite discs using CLSM (Jeon *et al.*, 2009; Xiao and Koo, 2010).

The *in vitro* systems offer several advantages: standardized conditions, non-destructive monitoring over time, and reproducibility. Naturally, all *in vitro* models require cultivated species. Also, influences of the immune system, of genetic background and of oral ecological niches are difficult to incorporate. Recently, an *in vitro* model of biofilm-host cell interactions was developed, that combined human gingival cells with nine subgingival plaque bacteria including fastidious strains (Guggenheim *et al.*, 2009). The authors successfully generated mature biofilms as measured by culture and molecular techniques, and they showed that apoptotic epithelial cells and pro-inflammatory cytokines increased. Sophisticated models like this are promising tools to study the role of oral biofilms in the inflammation process at the biofilm–host interface. However, at some point, results obtained in experimental settings have to be reproduced *in vivo* to elucidate their relevance to the physiology or pathogenicity of the biofilm.

## Analysis of biofilms *in vivo*

Microscopic analyses still remain at the forefront of investigations into *in vivo* grown biofilms. The first reports on microscopic observation of oral bacteria date back to the seventeenth century, when the Dutch scientist Antonie van Leeuwenhoek reported his findings on 'animalcules' to the Royal Society of London. Using a rudimentary microscope, he observed single-celled organisms from oral 'scrapings' – the first visual documentation of bacteria.

## Sampling

One crucial issue common to all microscopic techniques is undisturbed sampling. The most commonly used sampling method is the use of curettes, or of sterile paper points inserted into the periodontal pocket which absorb the biofilm bacteria over a defined period of time. Representative samples can be obtained that allow analysis of smaller biofilm sections. However, orientation and localization of the biofilm's spatial origin is difficult. Furthermore, sampling using paper points might preferentially collect the looser superficial parts of the biofilm and might miss microorganisms at the base of the biofilm or within the tissue.

To sample cohesive oral biofilms *in vivo*, devices have been developed that allow biofilm formation at the location of interest in the mouth for a defined time. Supragingival plaque has been obtained using 'retrievable enamel chips' (Diaz *et al.*, 2006), and subgingival biofilms have been recovered using carriers that were inserted into periodontal pockets and left in place for 3–7 days (Wecke *et al.*, 2000; Schlafer *et al.*, 2010). These sampling strategies allow analysis of undisturbed *in vivo* grown biofilms formed from patient's unique microbiota under the influence of the patient's immunology and genetics. Size, and substratum of the device as well as duration of biofilm development can be defined and varied, and the same location can be sampled several times. Interestingly, no difference in biofilm growth was seen with respect to different device materials (dentin, gold foil, or polytetrafluoroethylene), but colonization did require a dense pellicle on the surface of the device. These *in vivo* grown biofilms might differ from naturally grown biofilms because they are only few days old and thus lack layers of dead or less-active bacteria. Nevertheless, they seem to be representative in terms of biofilm architecture of the active layers of naturally grown biofilms (Schlafer *et al.*, 2010).

In terms of tissue biopsies, natural oral biofilms can be obtained only during tooth extraction or periodontal surgery. These samples are singular snap shots and do not allow for serial investigation, but they are valuable for biofilm analysis and investigation of biofilm–host interaction. Recently, Zijnge *et al.* succeeded in sectioning entire teeth and analysed the attached biofilm (Zijnge *et al.*, 2010). The most difficult part was the tooth preparation: sectioning, embedding and cutting the enamel and bone without destroying the biofilm structure. Elaborate procedures for decalcification and re-embedding, not suitable for routine analysis, were required.

## How to look at oral biofilms

EM studies provided convincing evidence that the communities found on teeth had distinct structures and regularities, and that colonization followed certain rules (Listgarten, 1976; Listgarten, 1994). The influence of surface roughness and charge on early colonizers was investigated (Listgarten *et al.*, 1973; Lie, 1978; Berthold, 1979). Limitations of EM analysis are that a relatively small microscopic field can be investigated and that the solely morphological data does not allow for species identification. Nevertheless, EM data yielded essential insights into the ultrastructure of oral biofilms, and they revealed formations in mature biofilms called corncobs or test-tube brushes. In combination with immuno-gold labelling, species including *P. gingivalis*, *Treponema denticola* or *Aggregatibacter actinomycetemcomitans* were detected (Quirynen *et al.*, 1993; Kigure *et al.*, 1995).

In recent years, molecular *in situ* techniques have enabled insights into the architecture of *in vivo* grown oral biofilms. In particular FISH has become an invaluable tool for biofilm research. FISH is a method for visualization and simultaneous identification of bacteria within the histological context using oligonucleotide probes that bind specifically to the ribosomal ribonucleic acids (rRNA). FISH probes can be designed according to the rRNA sequences at all taxonomic levels. When labelled with different fluorochromes, several species or genera can be visualized simultaneously. Diaz *et al.* used FISH and confocal laser scanning microscopy (CLSM) to determine the species participating in the initial colonization of the supragingival enamel chip model (Diaz *et al.*, 2006). To provide evidence that the coaggregation observed *in vitro* between bacterial species also occurs *in vivo*, immunofluorescence targeting coaggregation mediating factors, such as receptor polysaccharide (RPS), was used in the chip model (Newman, 1976). This approach was further improved by the use of quantum dot (QD)-based primary immunofluorescence combined with micromanipulation and subsequent identification of captured bacteria (Palmer *et al.*, 2003; Chalmers *et al.*, 2007).

For examination of complex, multilayered subgingival biofilms, physical sectioning prior to FISH was employed for several species associated with periodontitis. The large number of yet -to -be cultured species could be shown, predominant species could be specified, and the participants of bacterial aggregates described by EM, such as corncobs, could be identified (Diaz *et al.*, 2006; Wecke *et al.*, 2000; Schlafer *et al.*, 2010). Using sectioned natural teeth, a comprehensive FISH study on plaque provided substantial insights into the architecture of oral biofilms (Schlafer *et al.*, 2010). The combination of different molecular techniques, such as epidemiological detection using amplification-based methods, with FISH of *in vivo* grown biofilms can elucidate the role of a particular species in periodontitis (Wecke *et al.*, 2000; Chalmers *et al.*, 2008). Recently, the number of FISH probes that can be distinguished within a single image has been increased using combinatorial labelling and spectral imaging (CLASI-FISH) (Drescher *et al.*, 2010; Valm *et al.*, 2011). This technique allowed differentiation of 15 different phylotypes in an artificial mixture of laboratory-grown microbes and in dental plaque. Digital image analysis further allowed spatial relations between species to be examined.

## Methods to come

Recent developments in microscopic techniques, like cryo-EM, might further our understanding of adhesion and co-aggregation (Villena *et al.*, 2010). Fluorescence lectin-binding analysis could elucidate the composition of extracellular polymeric substances (EPS) (Neu *et al.*, 2001). The combination of nuclear magnetic resonance and confocal microscopy has already been used on *S. mutans* biofilms grown on hydroxyapatite discs, and this approach might further allow analysis of dynamic metabolic processes *in situ* (McLean *et al.*, 2008). Likewise, enhanced element labelling-catalysed reporter deposition FISH (EL-FISH) combined with nanometre-scale secondary-ion mass spectrometry (NanoSIMS) was used on oral biofilms. This technique allows species identification and simultaneous analysis of metabolic activity of single cells through stable isotope labels (Behrens *et al.*, 2008). Finally, scanning transmission X-ray microscopy (STXM) provides chemical and biochemical information with spatial resolution. STXM using near-edge X-ray absorption fine structure/spectroscopy (NEXAFS) allows mapping of e.g. metals, or chlorhexidine in river biofilms (Dynes *et al.*, 2006). However, this sophisticated technique requires access to a synchrotron and has not yet been applied to oral biofilms.

## Limitations of the *in situ* techniques

Because FISH depends on probe penetration into the bacterial cell, it requires sample fixation which kills the cells. Therefore it gives excellent spatial resolution, but is limited for temporal studies. The target phylotype is determined by the sequence of the FISH probe. However, differences in virulence that exist between strains of a particular species would be missed by this technique.

For FISH as well as EM analysis, sample preparation requires dehydration steps that can result in changes of the EPS and of characteristic biofilm features like water channels or voids. Another problem is that high individuality and variation between oral biofilms, even within a single sample, make it difficult to determine a universal biofilm architecture. On the other hand, consistent patterns and structural elements have been described by different techniques and investigators. Digital image analysis will provide objective data that can be evaluated statistically.

However, the various *in situ* techniques have verified that oral biofilms do have a distinct and conserved organization. Our current understanding of oral biofilm structure has resulted from *in vitro* and *in vivo* analyses both.

## Biofilm structure

### The stepwise formation of biofilms

The human mouth is an environment that provides differently sized surface areas, a wide diversity of nutrients, and optimal temperature and humidity for bacterial growth, all of which make biofilm formation inevitable in a broad spectrum of different micro-niches. Here we limit ourselves to the supragingival enamel and subgingival root surfaces. Already in the 1970s, Loë and coworkers showed in an experimental gingivitis model that biofilm formation is a process in which different steps can be recognized (Loë *et al.*, 1965). More recently Stoodley *et al.* (2002) presented a general model for the stepwise development of biofilms. This model describes five different stages: Stage 1 is initial attachment of cells to a surface. Stage 2 is the transformation of cells from a planktonic into a biofilm mode of growth. In this stage, genes optimal for planktonic growth (like genes for motility, recognition and adhesion to substrates) are turned off and genes more suited for the growth and formation of biofilm, such as for the production of extracellular polysaccharides (EPS), become actively expressed; the adhered bacteria become firmly anchored into the biofilm. Stage 3 is biofilm growth and shaping of biofilm architecture with the controlled division of cells and programmed cell death, through which structures like watery channels within the biofilm and upwards growing piles of cells (often termed 'mushrooms') are created. Stage 4 is maturation of the biofilm architecture, wherein the biofilm reaches a sort of steady state: a balance between bacterial growth and the sloughing off of biofilm parts by external forces such as chewing, tooth brushing of tongue movements. Stage 5 is regulated dispersion of cells or cell aggregates from the biofilm. Whether this is a separate step or occurs already during the maturation step is a matter of definition. Since the timing and importance of steps 4 and 5 are unclear, we combine these last two steps into a final step of maturation and dispersion.

The proposed model for biofilm formation is mainly based on single-species laboratory experiments with model organisms such as *Pseudomonas aeruginosa*. Several studies indicate that these events also take place during biofilm formation in oral microorganisms, and that they also play a role in the formation of biofilms on tooth surface. A more detailed description in the context of oral biofilm follows.

#### Initial attachment to tooth surface

After initial coating of a pristine tooth surface with a pellicle of polymers and proteins from food and saliva (Scannapieco, 1994), Gram-positive bacteria such as streptococci and *Actinomyces* spp. are among the first colonizers (Dige *et al.*, 2009a). Streptococci are especially numerous, followed by *Actinomyces* spp., although other bacteria, such as *Veillonella, Haemophilus* and *Neisseria,* may also play an initial role (Nyvad and Kilian, 1987). They adhere covalently to the tooth surface via specialized adhesins on the bacterial cell surface that recognize proteins in the pellicle (Kolenbrander and London, 1993; Jenkinson and Lamont, 1997); however, non-specific physico/chemical interactions also play a role (Bos *et al.*, 1999). The adhesion process is not homogeneous and results in a patchy distribution of clumps of cells over the surface (Dige *et al.*, 2007; Al-Ahmad *et al.*, 2009)

#### Differential gene expression

Genes for motility, recognition and adhesion to substrates are suppressed and genes for biofilm formation, such as for the production of extracellular polysaccharides (EPS) are induced (Huang *et al.*, 2008; Lo *et al.*, 2009; Mitchell *et al.*, 2010). In this phase the early colonizers, such as streptococci and *Actinomyces* spp. change from a planktonic to a biofilm mode of growth (Jefferson, 2004; Paddick *et al.*, 2006). Differential gene expression ensures heterogeneity for specialized function within the biofilm.

#### Biofilm growth and functional shaping

Biofilms mature by a combination of processes: cell division and growth (Lai *et al.*, 1975; Listgarten, 1994) and by seeding off from the biofilm on new surfaces, attachment of new cells from the environment, either directly to the existing biofilm (co-adhesion) or first through co-aggregation followed by attachment to the biofilm (e.g. corncob and rosette structures (Listgarten *et al.*, 1973). In this stage the biofilm will also attract and bind other colonizers, such as *Capnocytophaga* and anaerobes such as fusobacteria, *Prevotella* and *Propionibacterium* (Kolenbrander and London, 1993). Numerous pairs of bacterial species have been tested for their ability to co-aggregate by the group of Kolenbrander. The outcome of these co-aggregation experiments led to the description of a well-known model, described by Kolenbrander and London (Kolenbrander and London, 1993). In this model early colonizers bind to receptors in the acquired pellicle on the surface. Each newly adherent cell type becomes in turn the nascent surface and bridge for additional adhering cell types. In this model the authors differentiated early colonizers from late colonizers and proposed that *Fusobacterium nucleatum* served as a bridge between early and late colonizers. The early colonizers are primarily Gram-positive species belonging to the dental plaque of healthy individuals. The late colonizers are mostly Gram-negative periodontal pathogens, like *A. actinomycetemcomitans, P. gingivalis, Prevotella denticola* and *Treponema* spp. (Kolenbrander and London, 1993).

## Maturation of the biofilm

Redistribution of the biomass occurs through growth and death of bacterial cells, through development of open spaces including gas and water channels, and through the development of so-called mushroom structures or, in the presence of strong crevicular fluid flow, a more filamentous structure (Netuschil *et al.*, 1998; Wood *et al.*, 2000; Auschill *et al.*, 2001). Once matured, a steady-state may develop in which bacterial growth and adhesion is balanced by sloughing off of biofilm particles through mechanical processes in the mouth. Intriguingly, bacteria within biofilms can decide to escape from the biofilm in response to inter- or intra-bacterial communication, starvation or physical–chemical signals. *A. actinomycetemcomitans* for example has been shown to produce the extracellular enzyme Dispersin B that hydrolyses beta-1,6-*N*-acetyl-D-glucosamine (PNAG). PNAG is a biofilm matrix component comparable to EPS. Secretion of Dispersin B results in liberation and dispersion of cells clumps from the biofilm that may then colonize new surfaces (Kaplan, 2010).

The model of Stoodley *et al.* (2002) provides a general dynamic model for the formation of biofilms that, according to *in vitro* observations on oral bacteria, can be extrapolated to oral biofilms. Besides the fact that oral biofilms are built by a variety of different species, the oral cavity presents enormous technological challenges for the study of these dynamic temporal processes *in vivo*. Alternatively, we would apply an archaeological approach to dissect mature biofilms formed *in vivo* at different times during oral biofilm formation. The observations are single time points that are used to investigate the different steps that take place during biofilm formation. Changes in the composition of the microbiological populations have been described in some detail by Loë *et al.* in 1965 (Loë *et al.*, 1965) and Listgarten in 1976 (Listgarten, 1976); the latter states that 'the flora from a periodontitis sample was essentially similar to that seen on the subgingival parts of artificial crowns after two months'. We support the concept that subgingival biofilms are a subgingival continuation of matured supragingival biofilms. The terms 'supra' and 'subgingival' describe the location whereas the terms 'old' and 'mature' reflect the developmental state of the biofilm. The maturing biofilm that is in close proximity to the gingival epithelium results in the onset of inflammation and the concomitant clinical changes in the tissues such as swelling and redness. The biofilm becomes covered by the gingiva and thereby protected from mechanical disturbances. Continuation of the inflammatory responses eventually results in the progression of the matured biofilm apically (Saglie, 1977). The layered structure of the supragingival biofilm continues subgingivally and characteristic microbial complexes are seen in both biofilms (Haffajee *et al.*, 2008), as depicted in the Fig. 4.1. With this drawing we hope to illustrate this model by describing the layered structures in the transition from supra- to subgingival biofilm. Molecular techniques that identify single bacterial cells to the species level can be used to confirm the Kolenbrander model and to visually annotate the beautiful images obtained by Listgarten.

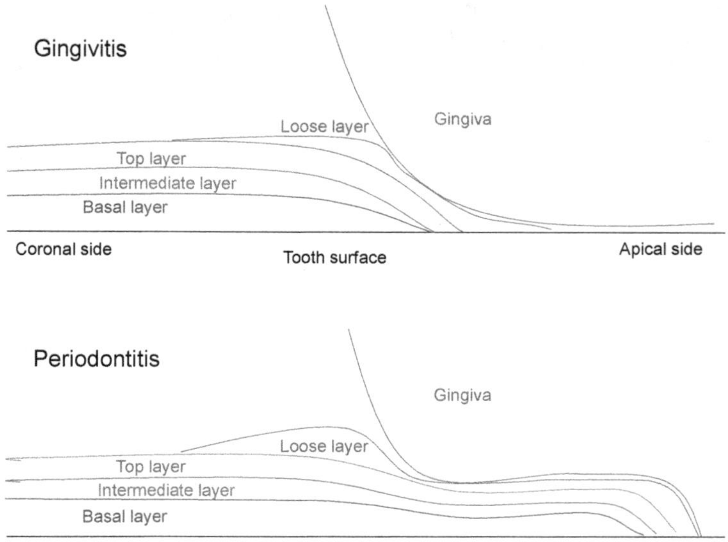


**Figure 4.1** Schematic representation of the progression from a structured supragingival biofilm to subgingival biofilm.

## Biofilm development in time

### Basal layer

The initial adhesion of bacteria to the pellicle-covered surface of the teeth starts with a random seeding of predominantly streptococci and *Actinomyces* sp. This was most elegantly demonstrated by Dige *et al.* (2009b) when they studied the initial adhesion of bacteria on a pristine tooth surface with FISH and CLSM. They observed that the initial colonization resulted in a patchy distribution of *Actinomyces* sp. and streptococci over the surface. This may reflect the randomness of the initial adhesion steps that are mainly the result of physicochemical interactions and cell–cell interactions (Bos *et al.*, 1999; Kolenbrander *et al.*, 2006; Hannig *et al.*, 2007). Other genera, such as *Neisseria, Granulicatella, Corynebacterium* and *Rothia,* play a less dominant role during these first steps in oral biofilm formation, probably due to their lower numbers in saliva. The attached bacteria form a 20-μm-thick biofilm within 48 hours. The growing biofilm is partly the result of newly adhering cells from the surroundings, which is more selective than the initial random events due to specific inter-bacterial adhesion characteristics. In addition to adhesion, the nutrients in saliva stimulate bacterial growth that results in the selective outgrowth of certain species. The importance of growth has been clearly demonstrated by the pyramidal outgrowth of adhered streptococci, which reflects the exponential doubling of bacteria during growth (Fig. 4.2) (Lai *et al.*, 1975). Metabolic interactions between oral bacteria have described in more detail (Bowden and Li, 1997). *Veillonella* sp., for example, scavenges on the lactic acid produced by streptococci that ferment sugars. Experiments on bovine enamel slabs placed in the human mouth (Al-Ahmad *et al.*, 2007) have indeed shown that both species occur in close proximity in biofilms. Despite the importance of nutritional influences on biofilm formation, few

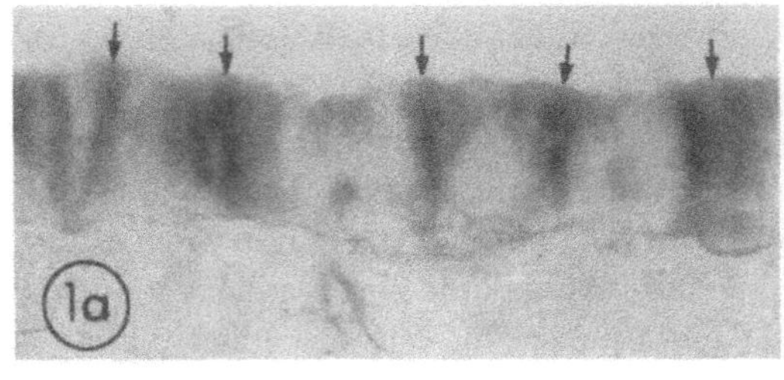


**Figure 4.2** Light microscopic image of a 30 μm thick plaque on an epoxy resin strip. The black pyramidal columns indicate the presence of *Streptococcus sanguis* (now *S. sanguinis*). Reprinted from Lai *et al.* (1975).

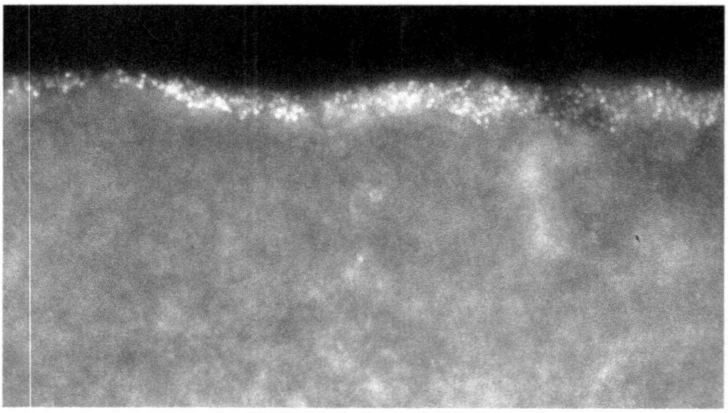

**Figure 4.3** An example of supragingival plaque with a metabolic active layer of streptococci (yellow-green) on top of a less active older biofilm composed of rods and cocci. The red colour is the signal from the red bacterial probe and the yellow-green and orange from a double labelling of a red bacterial probe and a green streptococcal probe (Zijnge *et al.*, 2010). A colour version of this figure is available in the plate section at the back of the book.

studies have addressed this. It can however be speculated that factors such as composition of the available nutrients and growth rate of the different species determine the outcome of biofilm formation. A snap shot of how growth rate and the availability of nutrients may affect biofilm formation can be seen in Fig. 4.3. On top of an inactive layer of bacteria a small band of active streptococci is visible. The activity of the layer is assessed by the relative abundance of the rRNA content. The amount of rRNA is positively correlated to the growth rate, and therefore, as rRNA is the target of FISH, it gives a higher fluorescence signal intensity if cells are active. Whereas the inactive layer shows the presence of both cocci and rods (possibly *Actinomyces*) it is clear that streptococci from the active layer dominate the part of the biofilm exposed to saliva and thereby to salivary nutrients. This observation shows that biofilms may grow successively on top of old and less active biofilms, comparable to corals, whereas the availability and competition for nutrients may play a determining outcome in biofilm composition and architecture.

While biofilm formation continues away from the tooth surface, the cells that adhered first and directly to tooth surface become excluded from nutrients and become less active. Observations show that *Actinomyces* may scavenge on these cellular remnants although the exact metabolic mechanisms need elucidation (Yeung, 1999).

### Intermediate layer

The aforementioned example of a biofilm dominated by streptococci, which have a competitive advantage, may reflect a so-called cariogenic biofilm that develops in an environment rich in sugars. In the absence of such extreme metabolic advantages, biofilm formation may continue along a different route. When the conditions within the biofilm change towards a more anaerobic environment, Gram-negative filaments appear in the biofilm. As observed by culture-based studies on biofilms at different stages of their development (Loë *et al.*, 1965) and, more recently, by molecular techniques (Socransky *et al.*, 1994). Listgarten observed these filaments on epoxy crowns, in agreement with the coaggregation experiments Kolenbrander that suggested *Fusobacterium nucleatum* as the potential dominant organism in this intermediate layer. Recently the presence of *F. nucleatum* has been confirmed (Zijnge *et al.*, 2010). Additionally it appeared that some of the filaments were *Tannerella forsythia* (Fig. 4.4). According to their abundance and localization *T. forsythia* can be considered an important contributor to the structural framework of a maturing biofilm. Because of their fastidious characteristics and dependence on N-acetyl muramic acid, a bacterial wall component, *T. forsythia* has been overlooked for a long time which may explain its absence in the model of Kolenbrander. Its location

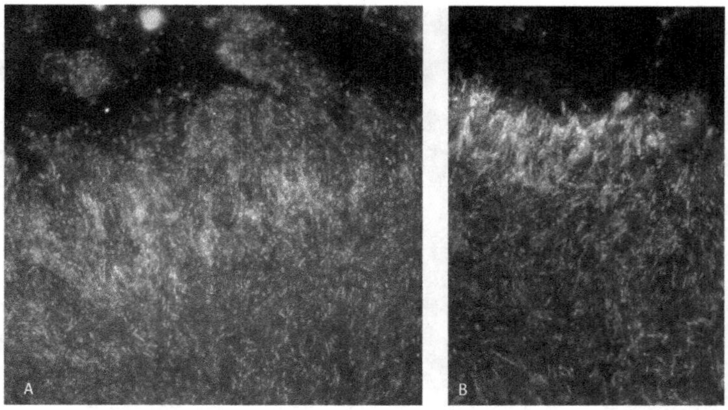


**Figure 4.4** The intermediate layer of the biofilm is dominated by fusobacteria (A) and *Tannerella forsythia* (B). In both panels the yellow-orange bacteria are of interest, which is a double labelling of a green bacterial probe and a red fusobacterial probe (A), or a red *Tannerella* probe (B). Modified from Zijnge *et al.* (2010). A colour version of this figure is available in the plate section at the back of the book.

in the intermediate layer of biofilms concurs with reports of a positive synergy in the development of mixed biofilms between *T. forsythia* and *F. nucleatum* (Sharma *et al.*, 2005). *In vitro* experiments showed that cell–cell interactions are essential for this synergy. *In vivo* observations confirmed this, and together these species seem to form a coherent intermediate layer between the basal and the top layer, a structural layer that forms a basis for colonization of the pathogens identified as late colonizers (Kolenbrander *et al.*, 2006). These late colonizers lack the ability to co-adhere to the initial colonizers. *F. nucleatum* adheres to both Gram-negative and Gram-positive bacteria and is therefore postulated to be a bridging organism. Whether *T. forsythia* exhibits the same capacities remains unknown.

### Top layer

Several days after a biofilm is established, especially near the gingiva, morphological changes take place in the host tissues resembling early steps of inflammation. In addition, nutrients change in quantity and composition. The increased flow of sulcus fluid and leakage of interstitial fluids provide the growing biofilm with an excess of protein. This may provide selective advantages to proteolytic bacteria, mostly Gram-negative and from the *Bacteroidetes* phylum. The predominantly Gram-negative character of the bacteria in the top layer has already been identified through culturing by Listgarten in 1976 (Listgarten, 1976) through culturing, and molecular techniques convincingly proved their dominance. A striking difference in contribution to biofilm architecture exists between *F. nucleatum* and *T. forsythia*, on the one hand, and specific periodontal pathogens such as *P. gingivalis* and *Prevotella* spp. on the other. Whereas *F. nucleatum* and *T. forsythia* can be considered as basic elements of the architectural structure, the putative periodontal pathogens are present in micro-colonies in the top layer of the biofilm (Fig. 4.5) as if they are pathogenic features attached to the architectural structure. *Parvimonas* and *Prevotella* have also been observed attached to the top layer (Zijnge *et al.*, 2010). Recently, *Synergistes* spp. were also identified as a elements of the architectural structure of the biofilm. The palisade lining described by Listgarten were shown to be *Synergistes* cells on the top of the biofilm as described by Zijnge (Listgarten, 1976; Zijnge *et al.*, 2010). Their role is very intriguing: they seem to be in a constant battle with the hosts' immune system. They form a palisade-like lining 'over' the biofilm, being in close contact to eukaryotic cells resembling polymorphonuclear leucocytes (PMNs) (Fig. 4.6).

### Outside layer (fuzzy layer)

After approximately three weeks of undisturbed biofilm formation, a peculiar outside

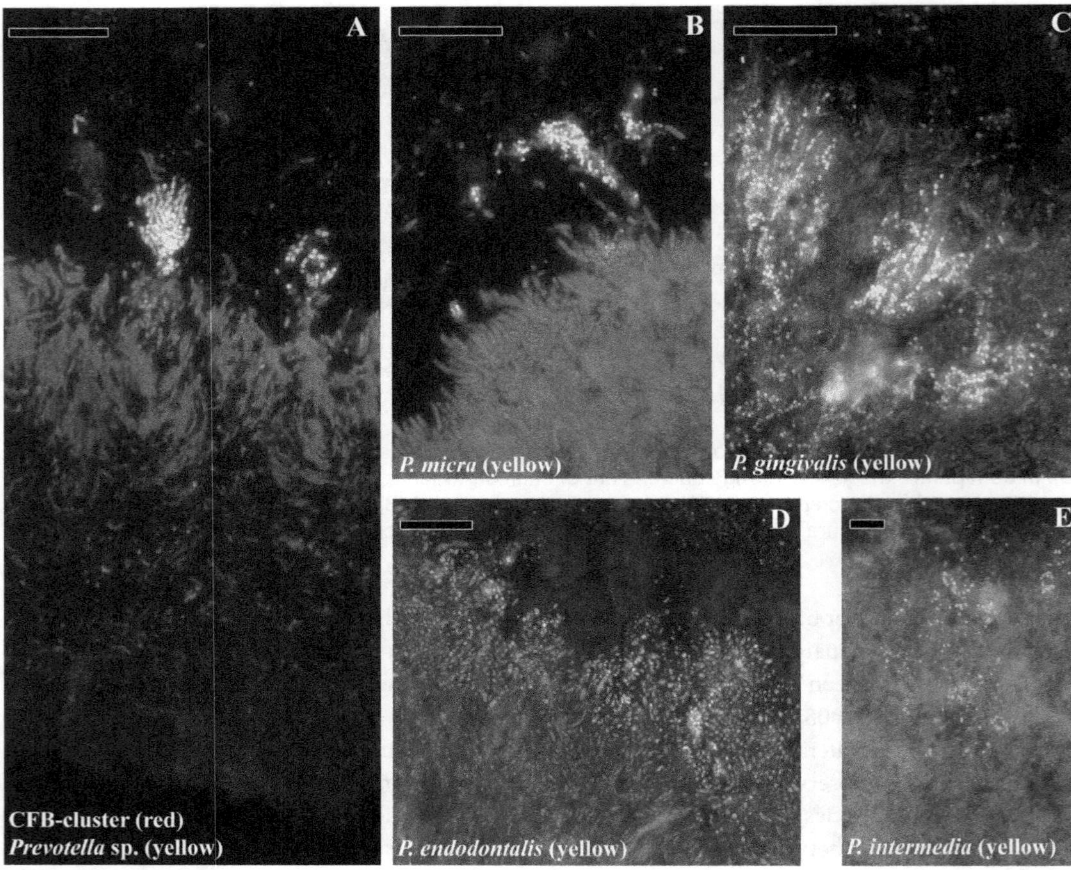


**Figure 4.5** Localization of species associated with periodontitis. (A) Overview of the subgingival biofilm with CFB-cluster species (red) and *Prevotella* sp. (yellow). Since *Prevotella* sp. are part of the CFB-cluster of bacteria, cells appear in yellow. (B) Top of the biofilm with a micro-colony of *P. micra* (yellow). (C) Micro-colonies of *P. gingivalis* (yellow) in the top layer. (D) Micro-colonies of *P. endodontalis* (yellow) in the top layer. (E) Microcolonies of *P. intermedia* in the top layer. Bars are 10 µm. A colour version of this figure is available in the plate section at the back of the book.

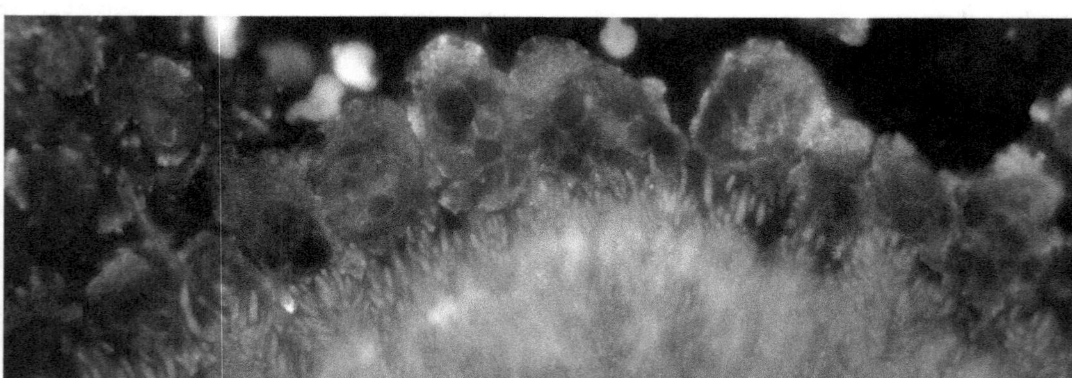

**Figure 4.6** Detail showing a subgingival biofilm with bacteria (red) and eukaryotic cells (green) (probably PMN). Note the suggestive palisade lining of bacteria and the numerical difference between bacteria and immune cells. A colour version of this figure is available in the plate section at the back of the book.

layer was observed by Listgarten. It can even be questioned whether this layer is really part of the attached biofilm. In contrast to the attached biofilm, this outside or fuzzy layer is not clearly organized and seems to float on top of the attached biofilm. The components that provide some kind of structure to this layer are unknown. It has recently been shown that this outside layer is dominated by enormous amounts of spirochetes like *Treponema denticola* (Zijnge *et al.*, 2010). Dispersed through this layer are structures that have been named corncobs (Lancy *et al.*, 1983; Listgarten *et al.*, 1973; Mouton *et al.*, 1980; Zijnge *et al.*, 2010) and test-tube brushes (Listgarten, 1994) (Fig. 4.7). These structures are in general composed of a layer of morphologically similar bacteria organized around a central long rod-shaped bacterium. Corncob structures were identified in the 19th century (Vincentini, 1987) and ever since then it has been a matter of debate which species form these structures. *In vitro* studies put forward *F. nucleatum* as the core rod (Lancy *et al.*, 1983) although a Gram-positive organism (Listgarten *et al.*, 1973; Walter and Frank, 1985) and even hyphae of the yeast *Candida* have been suggested to be the core (Walter and Frank, 1985; Zijnge *et al.*, 2010). Consensus exists that streptococci are the coccoid shapes cells that surround the central rod (Fig. 4.8). An intriguing question is the function of test-tube brushes and corncobs. Should they be considered scouts or settlers? The result of active detachment processes in analogy Dispersin B-linked dispersion of *A. actinomycetemcomitans*? And why are these specific organisms there? The fuzzy layer provides a barrier between host tissues and the attached biofilm. Cellular components of the host immune-system have to migrate through this outside layer where they seem to ignore the bacteria (mainly spirochetes).

## Open questions

Specific questions arise from these studies on biofilm structure. The first question is the impact of growth on biofilm formation and the influence of the availability of nutrients. Because of the complex mixture of nutrients available in saliva and crevicular fluids, and the presence

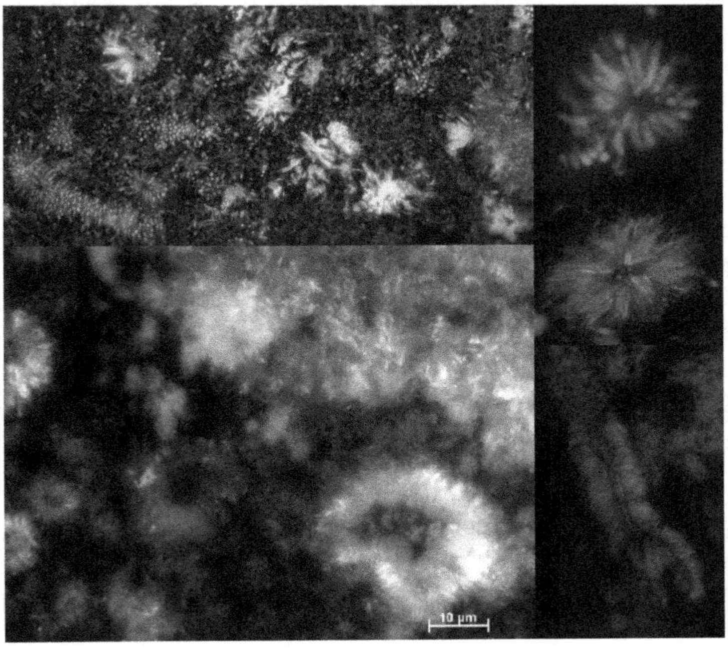


**Figure 4.7** Examples of test tube brushes in the outer layer of subgingival biofilms. A colour version of this figure is available in the plate section at the back of the book.

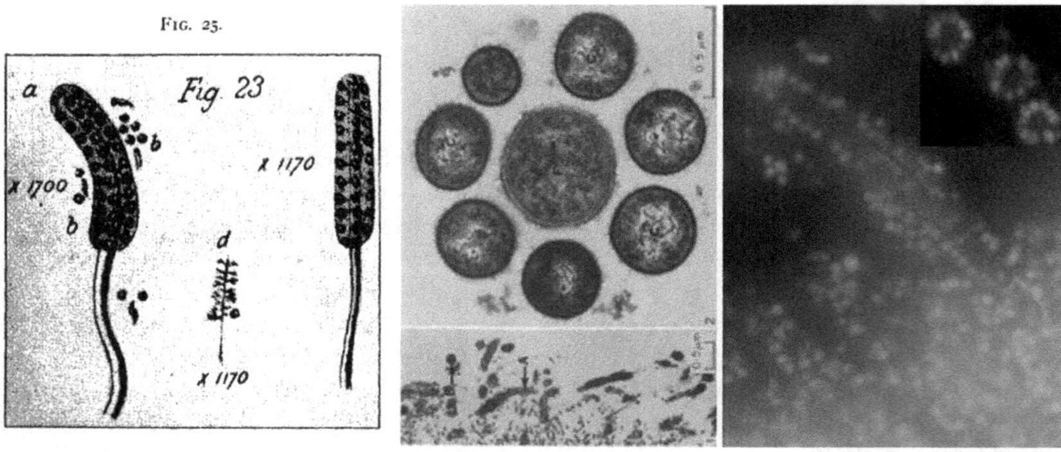


**Figure 4.8** Descriptions of corncob structures through the centuries. The left panel is a graphical representation by Vincentini in 1890 (Vincentini, 1987), the middle panel has been obtained by Listgarten in 1973 (Listgarten *et al.*, 1973) and the right panel shows corncobs as observed by FISH in 2010 (Zijnge *et al.*, 2010). A colour version of this figure is available in the plate section at the back of the book.

of different bacterial species, this topic needs further elucidation for understanding biofilm formation and its control. Second question is the intriguing presence of isolated structures such as corncobs and test-tube brushes. These test-tube brushes are mainly made up from central lactobacilli surrounded with *Tannerella,* fusobacteria, *Filifactor* and *Synergistes* species. These structures seem to float in the loose outer layer, in a dense mass of treponemes; it is easy to imagine that these structures drift around as rafts of pathogens that can disperse through the mouth and spread disease. The third question is the role that *Synergistes* spp. play as structural components of the subgingival plaque by forming a palisade lining. Is the suggestion that they function as a defence correct? And what are the underlying mechanisms?

## Outlook

Our concept of a subgingival biofilm is illustrated in Fig. 4.9: a three-layered biofilm structure with a loose fuzzy layer between the biofilm and the connective tissue of a periodontal pocket. The subgingival biofilm is a continuation of a supragingival biofilm that basically has the same structure but, because the environmental conditions change, the species composition and morphology change. The main pathogenicity lies entrapped in the loose layer, consisting of spirochaetes and structures with possible pathogenic features, like test-tube brushes. Particular pathogens that colonize in or on the top layer (such as *P. gingivalis*) produce enzymes that deteriorate the epithelial tissue. The development of the loose layer and the colonization of the particular pathogens is facilitated by *Fusobacterium* and *Tannerella* species that are present on top of the basal layer, which is itself composed of actinobacteria and bacterial cellular remnants. These species create the architecture and the low oxygen environment for colonization by pathogenic species. This basic construction of a biofilm facilitates colonization of pathogens. This starts supragingivally where the Gram-negative anaerobes colonize the basal layer of inactive and dead material with mainly streptococci, *Veillonella* and filamentous *Actinomycetes*. Here the pathogens compete with streptococci and others oral bacteria for nutrition and space. Subgingival biofilm formation starts supragingivally, and therefore prevention of biofilm formation begins with good oral hygiene.

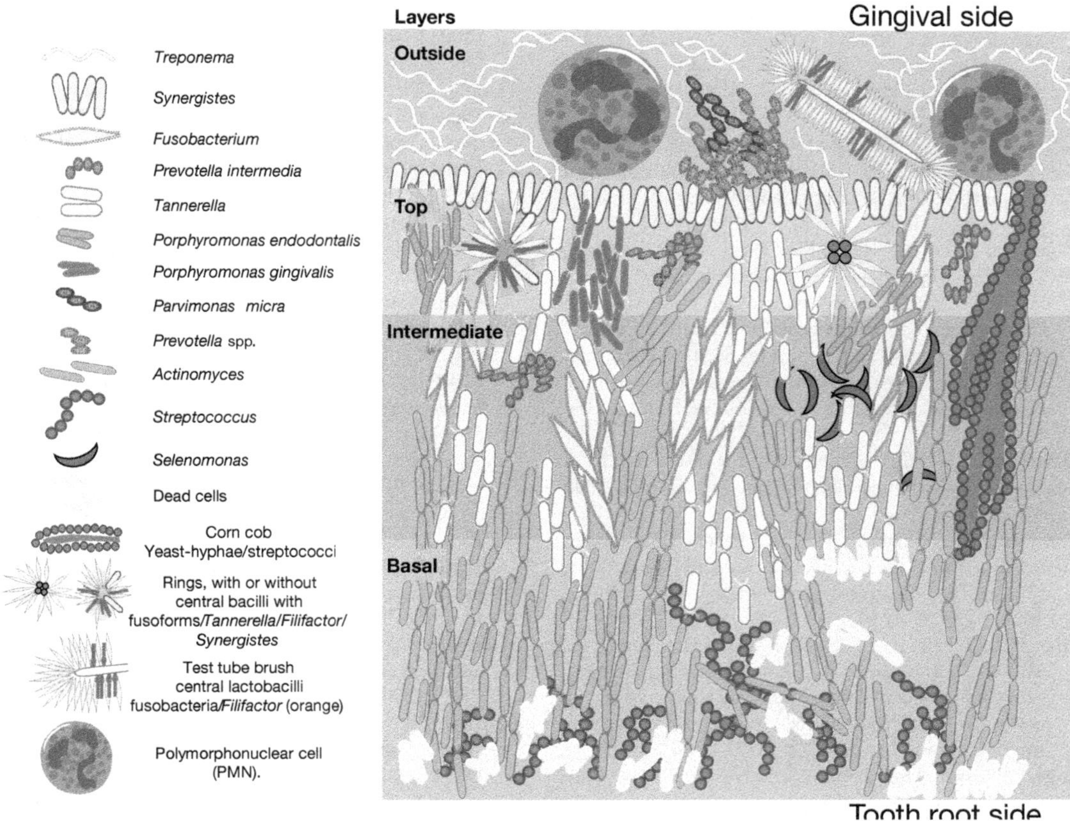


**Figure 4.9** A schematic model of a mature biofilm architecture. This drawing schematically represents the observations that are described in recent literature. It shows the three-layer architecture, with the additional loose outer layer and shows the key players in the architecture of the biofilm. A colour version of this figure is available in the plate section at the back of the book.

## References

Aas, J.A., Paster, B.J., Stokes, L.N., Olsen, I., and Dewhirst, F.E. (2005). Defining the normal bacterial flora of the oral cavity. J. Clin. Microbiol. *43,* 5721–5732.

Al-Ahmad, A., Wunder, A., Auschill, T.M., Follo, M., Braun, G., Hellwig, E., and Arweiler, N.B. (2007). The in vivo dynamics of *Streptococcus* spp., *Actinomyces naeslundii, Fusobacterium nucleatum* and *Veillonella* spp. in dental plaque biofilm as analysed by five-colour multiplex fluorescence in situ hybridization. J. Med. Microbiol. *56,* 681–687.

Al-Ahmad, A., Follo, M., Selzer, A.C., Hellwig, E., Hannig, M., and Hannig, C. (2009). Bacterial colonization of enamel in situ investigated using fluorescence in situ hybridization. J. Med. Microbiol. *58,* 1359–1366.

Angker, L., Swain, M.V., Wong, L., and Sissons, C. (2011). The effects of fluoride and mineralising treatments on plaque microcosm Ca, P and F, pH responses and cariogenicity. N. Z. Dent. J. *107,* 12–18.

Auschill, T.M., Artweiler, N.B., Netuschil, L., Brecx, M., Reich, E., and Sculean, A. (2001). Spatial distribution of vital and dead microorganisms in dental biofilms. Arch. Oral Biol. *46,* 471–476.

Behrens, S., Losekann, T., Pett-Ridge, J., Weber, P.K., Ng, W.O., Stevenson, B.S., Hutcheon, I.D., Relman, D.A., and Spormann, A.M. (2008). Linking microbial phylogeny to metabolic activity at the single-cell level by using enhanced element labeling-catalyzed reporter deposition fluorescence in situ hybridization (EL-FISH) and NanoSIMS. Appl. Environ. Microbiol. *74,* 3143–3150.

Berthold, P. (1979). Formation of salivary coating and dental plaque on two different supporting materials. An electron microscopic study. J. Periodontol. *50,* 397–405.

Bos, R., van der Mei, H.C., and Busscher, H.J. (1999). Physico-chemistry of initial microbial adhesive interactions – its mechanisms and methods for study. FEMS Microbiol. Rev. *23,* 179–230.

Bowden, G.H., and Li, Y.H. (1997). Nutritional influences on biofilm development. Adv. Dent. Res. *11,* 81–99.

Chalmers, N.I., Palmer, R.J., Du-Thumm, L., Sullivan, R., Shi, W., and Kolenbrander, P.E. (2007). Use of quantum dot luminescent probes to achieve single-cell resolution of human oral bacteria in biofilms. Appl. Environ. Microbiol. *73,* 630–636.

Chalmers, N.I., Palmer, R.J., Jr., Cisar, J.O., and Kolenbrander, P.E. (2008). Characterization of a *Streptococcus* sp.-*Veillonella* sp. community micromanipulated from dental plaque. J. Bacteriol. *190,* 8145–8154.

Choi, B.K., Paster, B.J., Dewhirst, F.E., and Gobel, U.B. (1994). Diversity of cultivable and uncultivable oral spirochetes from a patient with severe destructive periodontitis. Infect. Immun. *62,* 1889–1895.

Colombo, A.P.V., Boches, S.K., Cotton, S.L., Goodson, J.M., Kent, R., Haffajee, A.D., Socransky, S.S., Hasturk, H., Van Dyke, T.E., Dewhirst, F., *et al.* (2009). Comparisons of subgingival microbial profiles of refractory periodontitis, severe periodontitis, and periodontal health using the human oral microbe identification microarray. J. Periodontol. *80,* 1421–1432.

Davey, M.E. (2006). Techniques for the growth of *Porphyromonas gingivalis* biofilms. Periodontol. 2000 *42,* 27–35.

Dewhirst, F.E., Chen, T., Izard, J., Paster, B.J., Tanner, A.C.R., Yu, W., Lakshmanan, A., and Wade, W.G. (2010). The human oral microbiome. J. Bacteriol. *192,* 5002–5017.

Diaz, P.I., Chalmers, N.I., Rickard, A.H., Kong, C., Milburn, C.L., Palmer, R.J.,Jr, and Kolenbrander, P.E. (2006). Molecular characterization of subject-specific oral microflora during initial colonization of enamel. Appl. Environ. Microbiol. *72,* 2837–2848.

Dige, I., Nilsson, H., Kilian, M., and Nyvad, B. (2007). In situ identification of streptococci and other bacteria in initial dental biofilm by confocal laser scanning microscopy and fluorescence in situ hybridization. Eur. J. Oral Sci. *115,* 459–467.

Dige, I., Nyengaard, J.R., Kilian, M., and Nyvad, B. (2009a). Application of stereological principles for quantification of bacteria in intact dental biofilms. Oral Microbiol. Immunol. *24,* 69–75.

Dige, I., Raarup, M.K., Nyengaard, J.R., Kilian, M., and Nyvad, B. (2009b). *Actinomyces naeslundii* in initial dental biofilm formation. Microbiology *155,* 2116–2126.

Drescher, J., Schlafer, S., Schaudinn, C., Riep, B., Neumann, K., Friedmann, A., Petrich, A., Göbel, U.B., and Moter, A. (2010). Molecular epidemiology and spatial distribution of *Selenomonas* spp. in subgingival biofilms. Eur. J. Oral Sci. *118,* 466–474.

Dynes, J.J., Lawrence, J.R., Korber, D.R., Swerhone, G.D., Leppard, G.G., and Hitchcock, A.P. (2006). Quantitative mapping of chlorhexidine in natural river biofilms. Sci. Total Environ. *369,* 369–383.

Ghannoum, M.A., Jurevic, R.J., Mukherjee, P.K., Cui, F., Sikaroodi, M., Naqvi, A., and Gillevet, P.M. (2010). Characterization of the oral fungal microbiome (mycobiome) in healthy individuals. PLoS Pathog. *6,* e1000713.

Guggenheim, B., Gmur, R., Galicia, J.C., Stathopoulou, P.G., Benakanakere, M.R., Meier, A., Thurnheer, T., and Kinane, D.F. (2009). In vitro modeling of host–parasite interactions: the 'subgingival' biofilm challenge of primary human epithelial cells. BMC Microbiol. *9,* 280.

Haffajee, A.D., Socransky, S.S., Patel, M.R., and Song, X. (2008). Microbial complexes in supragingival plaque. Oral Microbiol. Immunol. *23,* 196–205.

Hannig, C., Hannig, M., Rehmer, O., Braun, G., Hellwig, E., and Al-Ahmad, A. (2007). Fluorescence microscopic visualization and quantification of initial bacterial colonization on enamel in situ. Arch. Oral Biol. *52,* 1048–1056.

Hope, C.K., and Wilson, M. (2004). Analysis of the effects of chlorhexidine on oral biofilm vitality and structure based on viability profiling and an indicator of membrane integrity. Antimicrob. Agents Chemother. *48,* 1461–1468.

Huang, M., Meng, L., Fan, M., Hu, P., and Bian, Z. (2008). Effect of biofilm formation on virulence factor secretion via the general secretory pathway in *Streptococcus mutans*. Arch. Oral Biol. *53,* 1179–1185.

Jefferson, K.K. (2004). What drives bacteria to produce a biofilm? FEMS Microbiol. Lett. *236,* 163–173.

Jeon, J.G., Klein, M.I., Xiao, J., Gregoire, S., Rosalen, P.L., and Koo, H. (2009). Influences of naturally occurring agents in combination with fluoride on gene expression and structural organization of *Streptococcus mutans* in biofilms. BMC Microbiol. *9,* 228.

Jenkinson, H.F., and Lamont, R.J. (1997). Streptococcal adhesion and colonization. Crit. Rev. Oral Biol. Med. *8,* 175–200.

Kaplan, J.B. (2010). Biofilm dispersal: mechanisms, clinical implications, and potential therapeutic uses. J. Dent. Res. *89,* 205–218.

Keijser, B.J., Zaura, E., Huse, S.M., van der Vossen, J.M., Schuren, F.H., Montijn, R.C., ten Cate, J.M., and Crielaard, W. (2008). Pyrosequencing analysis of the oral microflora of healthy adults. J. Dent. Res. *87,* 1016–1020.

Kigure, T., Saito, A., Seida, K., Yamada, S., Ishihara, K., and Okuda, K. (1995). Distribution of *Porphyromonas gingivalis* and *Treponema denticola* in human subgingival plaque at different periodontal pocket depths examined by immunohistochemical methods. J. Periodontal. Res. *30,* 332–341.

Kolenbrander, P.E. (1988). Intergeneric coaggregation among human oral bacteria and ecology of dental plaque. Annu. Rev. Microbiol. *42,* 627–656.

Kolenbrander, P.E., and London, J. (1993). Adhere today, here tomorrow: oral bacterial adherence. J. Bacteriol. *175,* 3247–3252.

Kolenbrander, P.E., Palmer, R.J., Rickard, A.H., Jakubovics, N.S., Chalmers, N.I., and Diaz, P.I. (2006). Bacterial interactions and successions during plaque development. Periodontol. 2000 *42,* 47–79.

Kumar, P.S., Griffen, A.L., Barton, J.A., Paster, B.J., Moeschberger, M.L., and Leys, E.J. (2003). New bacterial species associated with chronic periodontitis. J. Dent. Res. *82,* 338–344.

Lai, C.H., Listgarten, M.A., and Rosan, B. (1975). Immunoelectron microscopic identification and localization of *Streptococcus sanguis* with peroxidase-labeled antibody: localization of *Streptococcus* sanguis in intact dental plaque. Infect. Immun. *11,* 200–210.

Lancy, P., Jr, Dirienzo, J.M., Appelbaum, B., Rosan, B., and Holt, S.C. (1983). Corncob formation between

*Fusobacterium nucleatum* and *Streptococcus sanguis*. Infect. Immun. *40,* 303–309.

Lepp, P.W., Brinig, M.M., Ouverney, C.C., Palm, K., Armitage, G.C., and Relman, D.A. (2004). Methanogenic *Archaea* and human periodontal disease. Proc. Natl. Acad. Sci. U.S.A. *101,* 6176–6181.

Lie, T. (1978). Ultrastructural study of early dental plaque formation. J. Periodontol. Res. *13,* 391–409.

de Lillo, A., Booth, V., Kyriacou, L., Weightman, A.J., and Wade, W.G. (2004). Culture-independent identification of periodontitis-associated *Porphyromonas* and *Tannerella* populations by targeted molecular analysis. J. Clin. Microbiol. *42,* 5523–5527.

Listgarten, M.A. (1976). Structure of the microbial flora associated with periodontal health and disease in man. A light and electron microscopic study. J. Periodontol. *47,* 1–18.

Listgarten, M.A., Mayo, H., and Amsterdam, M. (1973). Ultrastructure of the attachment device between coccal and filamentous microorganisms in 'corn cob' formations of dental plaque. Arch. Oral Biol. *18,* 651–656, IN5.

Listgarten, M.A. (1994). The structure of dental plaque. Periodontol. 2000 *5,* 52–65.

Lo, A.W., Seers, C.A., Boyce, J.D., Dashper, S.G., Slakeski, N., Lissel, J.P., and Reynolds, E.C. (2009). Comparative transcriptomic analysis of *Porphyromonas gingivalis* biofilm and planktonic cells. BMC Microbiol. *9,* 18.

Loë, H., Theilade, E., and Jensen, S.B. (1965). Experimental gingivitis in man. J. Periodontol. *36,* 177–187.

McLean, J.S., Ona, O.N., and Majors, P.D. (2008). Correlated biofilm imaging, transport and metabolism measurements via combined nuclear magnetic resonance and confocal microscopy. ISME J. *2,* 121–131.

Mitchell, H.L., Dashper, S.G., Catmull, D.V., Paolini, R.A., Cleal, S.M., Slakeski, N., Tan, K.H., and Reynolds, E.C. (2010). *Treponema denticola* biofilm-induced expression of a bacteriophage, toxin–antitoxin systems and transposases. Microbiology *156,* 774–788.

Moter, A., Riep, B., Haban, V., Heuner, K., Siebert, G., Berning, M., Wyss, C., Ehmke, B., Flemmig, T.F., and Gobel, U.B. (2006). Molecular epidemiology of oral treponemes in patients with periodontitis and in periodontitis-resistant subjects. J. Clin. Microbiol. *44,* 3078–3085.

Mouton, C., Reynolds, H.S., and Genco, R.J. (1980). Characterization of tufted streptococci isolated from the 'corn cob' configuration of human dental plaque. Infect. Immun. *27,* 235–245.

Netuschil, L., Reich, E., Unteregger, G., Sculean, A., and Brecx, M. (1998). A pilot study of confocal laser scanning microscopy for the assessment of undisturbed dental plaque vitality and topography. Arch. Oral Biol. *43,* 277–285.

Neu, T., Swerhone, G.D., and Lawrence, J.R. (2001). Assessment of lectin-binding analysis for in situ detection of glycoconjugates in biofilm systems. Microbiology *147,* 299–313.

Newman, H.N. (1976). The apical border of plaque in chronic inflammatory periodontal disease. Br. Dent. J. *141,* 105–113.

Nyvad, B., and Kilian, M. (1987). Microbiology of the early colonization of human enamel and root surfaces in vivo. Scand. J. Dent. Res. *95,* 369–380.

Paddick, J.S., Brailsford, S.R., Rao, S., Soares, R.F., Kidd, E.A., Beighton, D., and Homer, K.A. (2006). Effect of biofilm growth on expression of surface proteins of *Actinomyces naeslundii* genospecies 2. Appl. Environ. Microbiol. *72,* 3774–3779.

Palmer R.J., Jr., and Caldwell, D.E. (1995). A flowcell for the study of plaque removal and regrowth. J. Microbiol. Methods *24,* 171–182.

Palmer, R.J., Jr, Gordon, S.M., Cisar, J.O., and Kolenbrander, P.E. (2003). Coaggregation–mediated interactions of streptococci and actinomyces detected in initial human dental plaque. J. Bacteriol. *185,* 3400–3409.

Quirynen, M., van der Mei, H.C., Bollen, C.M., Schotte, A., Marechal, M., Doornbusch, G.I., Naert, I., Busscher, H.J., and van Steenberghe, D. (1993). An in vivo study of the influence of the surface roughness of implants on the microbiology of supra- and subgingival plaque. J. Dent. Res. *72,* 1304–1309.

Riep, B., Edesi-Neuss, L., Claessen, F., Skarabis, H., Ehmke, B., Flemmig, T.F., Bernimoulin, J.P., Gobel, U.B., and Moter, A. (2009). Are putative periodontal pathogens reliable diagnostic markers? J. Clin. Microbiol. *47,* 1705–1711.

Saglie, R. (1977). A scanning electron microscopic study of the relationship between the most apically located subgingival plaque and the epithelial attachment. J. Periodontol. *48,* 105–115.

Sakamoto, M., Takeuchi, Y., Umeda, M., Ishikawa, I., and Benno, Y. (2003). Application of terminal RFLP analysis to characterize oral bacterial flora in saliva of healthy subjects and patients with periodontitis. J. Med. Microbiol. *52,* 79–89.

Scannapieco, F.A. (1994). Saliva–bacterium interactions in oral microbial ecology. Crit. Rev. Oral Biol. Med. *5,* 203–248.

Schlafer, S., Riep, B., Griffen, A., Petrich, A., Hubner, J., Berning, M., Friedmann, A., Gobel, U., and Moter, A. (2010). *Filifactor alocis* – involvement in periodontal biofilms. BMC Microbiol. *10,* 66.

Sharma, A., Inagaki, S., Sigurdson, W., and Kuramitsu, H.K. (2005). Synergy between *Tannerella forsythia* and *Fusobacterium nucleatum* in biofilm formation. Oral Microbiol. Immunol. *20,* 39–42.

Sissons, C.H., Cutress, T.W., Hoffman, M.P., and Wakefield, J.S. (1991). A multi-station dental plaque microcosm (artificial mouth) for the study of plaque growth, metabolism, pH, and mineralization. J. Dent. Res. *70,* 1409–1416.

Slots, J. (2010a). Herpesviral–bacterial interactions in periodontal diseases. Periodontol. 2000 *52,* 117–140.

Slots, J. (2010b). Human viruses in periodontitis. Periodontol. 2000 *53,* 89–110.

Socransky, S.S., Smith, C., Martin, L., Paster, B.J., Dewhirst, F.E., and Levin, A.E. (1994). 'Checkerboard' DNA–DNA hybridization. BioTechniques *17,* 788–792.

Socransky, S.S., Haffajee, A.D., Cugini, M.A., Smith, C., and Kent, R.L., Jr. (1998). Microbial complexes in subgingival plaque. J. Clin. Periodontol. *25,* 134–144.

Stoodley, P., Sauer, K., Davies, D.G., and Costerton, J.W. (2002). Biofilms as complex differentiated communities. Annu. Rev. Microbiol. *56,* 187–209.

Takeuchi, Y., Guggenheim, B., Filieri, A., and Baehni, P. (2007). Effect of chlorhexidine/thymol and fluoride varnishes on dental biofilm formation in vitro. Eur. J. Oral Sci. *115,* 468–472.

Valm, A.M., Welch, J.L.M., Rieken, C.W., Hasegawa, Y., Sogin, M.L., Oldenbourg, R., Dewhirst, F.E., and Borisy, G.G. (2011). Systems-level analysis of microbial community organization through combinatorial labeling and spectral imaging. Proc. Natl. Acad. Sci. U.S.A. *108,* 4152–4157.

Vartoukian, S.R., Palmer, R.M., and Wade, W.G. (2007). The division '*Synergistes*'. Anaerobe *13,* 99–106.

Villena, G.K., Fujikawa, T., Tsuyumu, S., and Gutierrez-Correa, M. (2010). Structural analysis of biofilms and pellets of *Aspergillus niger* by confocal laser scanning microscopy and cryo scanning electron microscopy. Bioresour. Technol. *101,* 1920–1926.

Vincentini, F. (1987). Bacteria of the Spura and Cryprogamic Flora of the Mouth (Bailliere, Tindall & Cox, London).

Walter, B., and Frank, R.M. (1985). Ultrastructural relationship of denture surfaces, plaque and oral mucosa in denture stomatitis. Journal De Biologie Buccale *13,* 145–166.

Wecke, J., Kersten, T., Madela, K., Moter, A., Gobel, U.B., Friedmann, A., and Bernimoulin, J. (2000a). A novel technique for monitoring the development of bacterial biofilms in human periodontal pockets. FEMS Microbiol. Lett. *191,* 95–101.

Whittaker, C.J., Klier, C.M., and Kolenbrander, P.E. (1996). Mechanisms of adhesion by oral bacteria. Annu. Rev. Microbiol. *50,* 513–552.

Wood, S.R., Kirkham, J., Marsh, P.D., Shore, R.C., Nattress, B., and Robinson, C. (2000). Architecture of intact natural human plaque biofilms studied by confocal laser scanning microscopy. J. Dent. Res. *79,* 21–27.

Xiao, J., and Koo, H. (2010). Structural organization and dynamics of exopolysaccharide matrix and microcolonies formation by *Streptococcus mutans* in biofilms. J. Appl. Microbiol. *108,* 2103–2113.

Yeung, M.K. (1999). Molecular and genetic analyses of *Actinomyces* spp. Crit. Rev. Oral Biol. Med. *10,* 120–138.

Zijnge, V., Welling, G.W., Degener, J.E., van Winkelhoff, A.J., Abbas, F., and Harmsen, H.J. (2006). Denaturing gradient gel electrophoresis as a diagnostic tool in periodontal microbiology. J. Clin. Microbiol. *44,* 3628–3633.

Zijnge, V., van Leeuwen, M., Barbara, M., Degener, J.E., Abbas, F., Thurnheer, T., Gmür, R., and Harmsen, H.J.M. (2010). Oral biofilm architecture on natural teeth. PLoS One *5,* e9321.

# The Role of Extracellular Polysaccharides Matrix in Virulent Oral Biofilms

# 5

Marlise I. Klein, Megan L. Falsetta, Jin Xiao, William H. Bowen and Hyun Koo

## Abstract

Virulent biofilms are responsible for a range of infections, including those occurring in the mouth. The disease dental caries is a prime example of the consequences arising from interactions of host, bacteria and diet (sugars), resulting in the assembly of pathogenic biofilms on susceptible tooth-surfaces. All biofilms harbour a microbial-derived extracellular matrix. The exopolysaccharides (EPS) formed on the tooth pellicle and on the bacterial surfaces provide binding sites for microorganisms; eventually the accumulated EPS enmeshes the microbial cells. The metabolic activity of the bacteria embedded in EPS-rich and diffusion-limited matrix leads to acidification of the milieu, and subsequent onset of the disease through acid dissolution of enamel (expressed clinically as cavities). This chapter focuses on the mechanisms through which the EPS produced by *Streptococcus mutans* modulates the assembly of a spatially complex 3D matrix during biofilm morphogenesis within a mixed-bacterial species community. The architecture of the matrix influences the development of pH heterogeneity in the 3D environment of intact biofilms. The matrix facilitates the formation of structured acidic-microenvironments in close proximity to the apatite surface, which are essential for the evolution of virulent biofilms formed on the tooth. These observations may have relevance beyond the mouth, as matrix is inherent to all biofilms.

## Introduction

Biofilms are highly dynamic, organized, and structured communities of microbial cells that are covered and enmeshed in a three-dimensional (3D) extracellular matrix. The density, composition, and structure of the biofilm matrix may be as variable as the microbial constituents, all of which are influenced by local environmental conditions and all of which can change with time (Branda *et al.*, 2005; Flemming and Wingender, 2010; Bowen and Koo, 2011). Simply put, there is no biofilm development without a matrix (Flemming and Wingender, 2010). Exopolysaccharides (EPS), proteins, lipids, nucleic acids (e.g. DNA), and even lipooligosaccharides have been identified in the matrices of a variety of biofilms formed in nature and at various sites in the human body (Steichen *et al.*, 2008; Karatan and Watnick, 2009; Ma *et al.*, 2009; Mann *et al.*, 2009; Schooling *et al.*, 2009). EPS are key components in the matrix of virulent oral biofilms, clinically known as dental plaque (Bowen and Koo, 2011). The formation of EPS on an initially bacteria-free tooth pellicle surface provides binding sites for oral microorganisms; the polymers are also formed on bacterial surfaces. The accumulated EPS *in situ* forms a matrix that eventually enmeshes the microbial cells, providing structural integrity, cohesion, and protection against antimicrobials and/or other environmental assaults. Furthermore, EPS influences the diffusion of substances into and out of the biofilm (Lewis, 2001; Stewart and Franklin, 2008). The sugar-based metabolic activity of the microorganisms clustered within the matrix leads to the acidification of the milieu, despite the presence of surrounding saliva and alkali generation in the plaque. These biological changes have enormous impact in the pathogenesis of dental caries by creating a favourable microenvironment for the survival of aciduric and EPS-binding pathogens, and highly concentrated acidic regions that enhance demineralization of the adjacent tooth enamel. Despite the recognized importance of the EPS-matrix in biofilm virulence, most studies have focused on the enumeration of pathogenic bacteria and/or the identification of new acidogenic/aciduric organisms, while the biological events occurring in the matrix have received limited attention. Understanding the assembly of and changes that occurs within the EPS-rich matrix over time, and how these modulate biofilm 3D architecture, shifts in microbial population, and the microenvironmental pH could further elucidate biological processes that mediate the transition from non-pathogenic to virulent biofilms. In this chapter, we focus on the role of the extracellular polysaccharide matrix in the development of virulent biofilms associated with dental caries, the single most common and costly oral infectious disease worldwide (Marsh, 2003; Dye *et al.*, 2007).

## Extracellular matrix in dental plaque

Human dental plaque is certainly a highly complex biofilm system that is constantly exposed to environmental changes and may serve as a general model for understanding the process of matrix formation, its role in biofilm biology (Foster *et al.*, 2003), and expression of virulence (Bowen and Koo, 2011). In the mouth, biofilms are initiated on a proteinaceous film present on the tooth surface clinically known as the pellicle, which is derived from mammalian and most importantly bacterial sources [e.g. glucosyltransferases (Gtfs) and fructosyltransferases (Ftfs)]; a small group of organisms (mostly Streptococci and *Actinomyces* spp.) can adhere to this pellicle in low numbers (Nobbs *et al.*, 2009; Valm *et al.*, 2011). The mouth is a dynamic habitat where the highly diverse resident bacteria must adapt to constant variations in the local environmental conditions dictated by extrinsic and intrinsic factors (e.g. nutrients from diet, changes in pH) (Lemos and Burne, 2008; Klein *et al.*, 2010). For example, exposure to dietary sucrose dramatically influences the composition of the pellicle by providing a substrate for extracellular polysaccharides production by Gtf and Ftf enzymes present on surfaces. Streptococcal glucosyltransferases are constituents of the pellicle, and are capable of synthesizing glucans *in situ* from sucrose, which provide bacterial binding sites for subsequent colonization and local accumulation. Gtfs also bind many oral bacteria, converting them into glucan producers. The bacterial cells then form clusters that are enmeshed in an EPS-rich matrix, which are tightly adherent to the saliva-coated apatitic surfaces. Thus, a bacterial product in the pellicle and on microbial surfaces provides the early foundation of the matrix (and microbial binding sites), thereby dramatically influencing the trajectory of virulent biofilm formation (see 'Role of *S. mutans* exoenzymes in the development of EPS-rich matrix' section).

The broad composition of the plaque matrix comprises a mixture of polysaccharides, inorganic ions, DNA, microbial and mammalian (salivary) proteins, and lipids. The content of glucan (homopolymer of glucose) ranges from 10% to 40% of the dry weight of dental plaque and approximately 1–10% of the weight is comprised of fructan (homopolymer of fructose); these proportions vary depending at least in part on the duration since the last intake of food and type of dietary carbohydrate consumed (Critchley *et al.*, 1967; Gibbons, 1968; Wood, 1969; Mayhall and Butler, 1976; Emilson *et al.*, 1984). Furthermore, the biofilm matrix contains variable amounts of Ca, P, Mg, F, and even lipoteichoic acids (Hotz *et al.*, 1972; Cole and Bowen, 1974; Rolla *et al.*, 1980; Cury *et al.*, 2000; Paes Leme *et al.*, 2006; Pessan *et al.*, 2008). Dental plaque is heavily hydrated, harbouring up to 80% water, which may be associated with the presence of fructan (Wilson and Ashley, 1990), a polymer with a high water regain value.

Histochemistry and transmission electron microscopy of plaque biofilms reveals microorganisms embedded in a matrix of polysaccharide (McDougall, 1963; Reese and Guggenheim, 2007), which appears to increase following exposure to sucrose (Birkhed and Rosell, 1975). Fig. 5.1 depicts a microscopic image of human plaque collected from

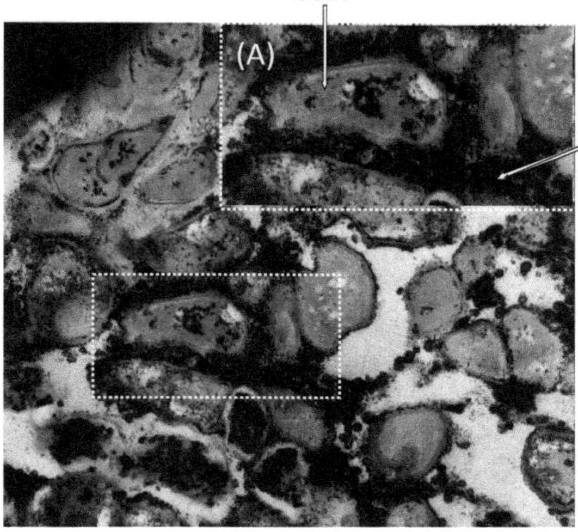


**Figure 5.1** Micrograph of human dental plaque collected from caries-active children. (A) Close-up view of the selected area showing bacterial cells enmeshed in extracellular polysaccharides.

caries-active children showing clusters of bacterial cells surrounded by extracellular polysaccharides. Polysaccharides within dental plaque are not evenly distributed, and the density and structure of plaque is highly variable, changing over time (Guggenheim and Burckhardt, 1974; Birkhed and Rosell, 1975; Reese and Guggenheim, 2007). The development, structure (i.e. branching of EPS), and composition (e.g. proportions of water-soluble and water-insoluble polymers) of the EPS-matrix can directly influence biofilm morphogenesis, ecological development, and the microenvironment (Bowen and Koo, 2011).

In cariogenic oral biofilms, EPS is a key component of the matrix and is associated with the pathogenesis of dental caries (Tanzer *et al.*, 1974, 1985; Munro *et al.*, 1991; Yamashita *et al.*, 1993; Burne *et al.*, 1996; Paes Leme *et al.*, 2006; Bowen and Koo, 2011) as mutant strains defective in glucan and fructan synthesis display a dramatic reduction of virulence expression *in vivo* (Tanzer, 1985; Munro *et al.*, 1991; Yamashita *et al.*, 1993; Burne *et al.*, 1996). The EPS in the matrix of dental plaque is a product of the interaction between microbial glucosyltransferases (Gtfs) and fructosyltransferases (Ftfs) with dietary sucrose (Vacca-Smith *et al.*, 1996a; van Hijum *et al.*, 2006; Bowen and Koo, 2011). Furthermore, the presence of starch hydrolysates greatly influences the production and structure of glucans because they can serve as acceptor sites for polymer formation (Vacca-Smith *et al.*, 1996a). Thus, EPS content and composition are directly dependent on the host diet.

The microbial exoenzymes synthesize polysaccharides comprised of either glucose sugar residues (glucans) or fructose residues (fructans) using primarily sucrose as a substrate. Gtfs synthesize water-soluble (rich in α-1,6 linkages) and insoluble glucans (high proportion of α-1,3 linkages). The profile differs depending on the type of enzyme and its physical state (in solution or adsorbed on surfaces) (van Hijum *et al.*, 2006; Bowen and Koo, 2011). Ftfs catalyse synthesis of fructans composed of a high proportion of β-2,1 (inulin-type) or β-2,6 (levan-type) linkages; inulin is significantly less soluble than levan (Ebisu *et al.*, 1975; van Hijum *et al.*, 2006). Levan is the most commonly detected fructan in dental plaque, although inulin is occasionally found in low concentrations. One of the most intriguing aspects about these enzymes is that Gtfs and Ftfs are detected in elevated amounts in the pellicle, even at the earliest stages of development (Vacca Smith and Bowen, 2000; Hannig *et al.*, 2008), and more importantly in an active form synthesizing glucans and fructans *in situ* (Vacca Smith and Bowen, 2000). *Streptococcus mutans* appears to be the main source of Gtfs in the pellicle, while the exact microbial source of Ftfs remains to be determined; both enzymes are found in whole saliva. The interaction between Ftf and Gtfs, which compete for the same substrate, is certainly worthy of additional exploration.

## Oral microorganisms and EPS

Although distinct species of organisms are found in plaque-biofilms, most of them do not contribute to the assembly of an insoluble EPS-matrix, until they are coated by Gtfs (see later) (Bowen and Koo, 2011). Among oral microorganisms, streptococci, lactobacilli, and *Actinomyces* spp. express Gtfs and Ftfs, which utilize dietary sucrose to produce glucans and fructans (van Hijum *et al.*, 2006; Bowen and Koo, 2011) as summarized in Table 5.1. The Gtfs also utilize starch hydrolysates as acceptors during glucan synthesis. In addition, eukaryotic organisms (e.g. *Candida albicans*) and some periodontopathogens also produce EPS. However, their role in the development of virulent biofilms warrants further investigation. The ongoing sequencing of the oral microbiome (Dewhirst *et al.*, 2010) may reveal additional organisms that encode enzymes for EPS production. In this chapter, we focus on EPS produced by *S. mutans* Gtfs (while at the same time recognizing the possible contribution of fructan to expression of virulence) because of the proven role of Gtfs in the formation of the extracellular matrix and in the development of virulent biofilms associated with dental caries.

## Role of *S. mutans* exoenzymes in the development of EPS-rich matrix

Cariogenic biofilms are comprised of mixed flora *in vivo*; the organisms are embedded and/or covered by EPS, and are usually potent producers of acid and are highly acid tolerant. Most of them do not participate directly in the assembly of the insoluble matrix. In contrast, mutans streptococci, particularly *S. mutans*, are the primary developers of the EPS-rich matrix in cariogenic biofilms (Bowen and Koo, 2011). *Streptococcus mutans* produces at least six different enzymes, which catalyse the synthesis and breakdown of high-molecular-weight polymers of glucose and fructose. Three Gtfs (encoded by *gtfB*, *gtfC* and *gtfD*) and an endodextranase (Dex; product of *dexA*) participate in the synthesis and degradation of glucans (Guggenheim and Burckhardt, 1974; Hayacibara *et al.*, 2004). GtfB synthesizes mostly insoluble glucans (rich in α-1,3-linked glucose). GtfC forms a mixture of insoluble and soluble glucans (mostly α-1,6-linked glucose). GtfD produces soluble glucans, and Dex hydrolyses α-1,6-linked glucose. In addition, *S. mutans* produces a fructosyltransferase (Ftf) and a fructan hydrolase (Fru), the products of *ftf* and *fruA* genes respectively (Shiroza and Kuramitsu, 1988; Burne and Penders, 1992), which catalyse the synthesis and breakdown of fructans composed predominantly of β-2,1 linkages (Ebisu *et al.*, 1975). The concerted actions of these exoenzymes are responsible for the continuous formation and remodelling of the biofilm EPS matrix (Bowen and Koo, 2011).

The exoenzymes and their polysaccharide products have been implicated in various roles in plaque biofilm formation and in the pathogenesis of dental caries. The insoluble glucans produced by surface-adsorbed GtfB and GtfC are essential for the construction of a 3D extracellular matrix, whereas fructans and water-soluble glucans (primarily produced by GtfD) are used as an extracellular carbohydrate reservoir (Burne *et al.*, 1996; Bowen and Koo, 2011). The digestion of soluble fructans and glucans by fructanase and dextranase provide low-molecular-weight carbohydrates, which are readily metabolized by a range of oral microorganisms during periods of nutrient deprivation (Burne *et al.*, 1996; Bowen and Koo, 2011). The availability of fermentable sugars from EPS digestion may account at least in part for the low pH values detected in 'starved' plaque from caries active persons (Stephen and Jenkins, 1971; Walker *et al.*, 1981). In addition, the partially digested glucans may serve as primer for insoluble glucan synthesis by GtfB. Fructan, which has high water regain value, contributes to the gelatinous nature of plaque-biofilm and helps to keep plaque hydrated (Wilson and Ashley, 1990). Moreover, the deletion of *gtfB* and *gtfC* genes results in greater than 90% reduction in the incidence and severity of carious lesions *in vivo* (Tanzer, 1985; Yamashita *et al.*, 1993). In addition, *S. mutans* strains defective in the production of fructans are less cariogenic than their parent strains (Burne *et al.*, 1996; Vacca-Smith *et al.*, 1996b). Clearly, the dynamic interplay of the enzymes responsible for synthesis and cleavage of the exopolysaccharides directly influences the synthesis and chemical structure of the matrix, and consequently the development of cariogenic biofilm. The constant provision of additional acceptor sites leads to decreased EPS solubility, and subsequently an increased resistance to effects of dextranases, while more binding sites are created as the plaque biofilm matures (Bowen and Koo, 2011).

## Surface-adsorbed Gtfs mediate the initial assembly of EPS *in situ*

The Gtfs secreted by *S. mutans* become an integral part of the pellicle in an enzymatically active form (Rolla *et al.*, 1983; Schilling and Bowen, 1988, 1992; Vacca-Smith *et al.*, 1996b; Vacca Smith and Bowen, 2000), rapidly producing glucans *in situ* (Fig. 5.2), which provides additional non-mammalian sites for colonization. Among the Gtfs, GtfC has the greatest affinity for the pellicle (Venkitaraman *et al.*, 1995; Vacca-Smith and Bowen, 1998; Vacca Smith and Bowen, 2000). The glucans synthesized in the pellicle are produced primarily by GtfC (Vacca-Smith and Bowen, 1998; Vacca Smith and Bowen, 2000; Bowen and Koo, 2011). They provide distinct binding sites with high avidity for oral microorganisms, particularly for *S. mutans*, which does not bind to the pellicle as effectively as do other early colonizers (thereby enhancing its competitiveness). The binding of *S. mutans* to glucans formed *in situ* is mediated by cell-associated Gtf enzymes and several non-Gtf glucan binding proteins (Gbps), particularly GbpC (and possibly GbpB) (Banas and Vickerman, 2003). Interestingly, Gtfs (mostly GtfB) also bind to many oral microorganisms (McCabe and Donkersloot, 1977; Hamada *et al.*, 1978; Vacca-Smith and Bowen, 1998; Gregoire *et al.*, 2011), even to those that do not produce Gtfs, producing glucans on microbial surfaces (Fig. 5.2). *Actinomyces viscosus, Lactobacillus casei*, and *S. mutans* all bind GtfB with great avidity, thereby converting them into *de facto* glucan producers (Vacca-Smith and Bowen, 1998), which can enhance binding to each other and to tooth-surfaces, facilitating the colonization and accumulation of these organisms during the biofilm development process. Furthermore, GtfB binds to *C. albicans* yeast-cells in an active form that synthesizes glucans in the presence of sucrose (Gregoire *et al.*, 2011). More importantly, *S. mutans* cells then bind in high numbers to the glucans formed on the *C. albicans* surface. It is noteworthy that *S. mutans* binds poorly to *C. albicans* cells that are not coated with glucan. Thus, the presence of surface-formed glucans dramatically enhances the cross-kingdom interactions between *S. mutans* and *C. albicans* (Gregoire *et al.*, 2011), which are commonly found associated with plaque collected from subjects with severe early childhood caries (de Carvalho *et al.*, 2006).

It is noteworthy that starches are usually consumed together with sucrose. The concomitant presence of both main dietary carbohydrates dramatically influences the synthesis and structure of glucans on surfaces, as well as subsequent formation of the initial EPS-matrix (Vacca-Smith *et al.*, 1996a; Klein *et al.*, 2009, 2010). Starches are gradually digested in the mouth through the action of salivary α-amylase, releasing a myriad of hydrolysates, such as maltose, maltotriose and small oligosaccharides. The hydrolysates serve as acceptors sites for further glucan synthesis, constituting an integral part of the polymer molecule (Fukui and Moriyama, 1983; Koga *et al.*, 1988; Fu and Robyt, 1991; Vacca-Smith *et al.*, 1996a; Klein *et al.*, 2010). The combination of starch hydrolysates and sucrose with surface-adsorbed GtfB results in enhanced formation of predominantly insoluble glucan and an increase in the number of α-1,4- and α-1,3-linkages with 3,4-linked glucose as the major branch point (Kopec *et al.*, 1997). These features confer enhanced resistance to digestion by dextranase and mutanase (Kopec *et al.*, 1997). Furthermore, *Actinomyces* spp. and *S. mutans* bind in greater numbers to glucan formed in the presence of starch hydrolysates as compared to that formed in the absence of starch (Vacca-Smith *et al.*, 1996a).

Clearly, insoluble glucans produced *in situ* on the pellicle-covered tooth and on microbial

**Table 5.1** EPS-producing oral microorganisms

| Organisms | Type of EPS[a] | Gene | Function[b] | References |
|---|---|---|---|---|
| *Actinomyces naeslundii* | Fructan, levan-type (β-2,6-linked) polysaccharides, units joined through multiple (1,2,6)-linked fructosyl branch units | *ftf* | Storage, water retention | Allen and Bowen (1990), Bergeron *et al.* (2000) |
| *Streptococcus mutans* | Water insoluble glucans, mostly α-1,3-linked glucose<br>Surface-formed glucans contain greater amounts of 3-linked glucose (and 3-linked branch points) than glucans formed in solution<br>Surface-formed glucans in presence of sucrose and starch hydrolysates also display 4-linked glucose (and α-3,4 linked branch points) | *gtfB* | Adhesion/cohesion, structural integrity and virulence | Hamada *et al.* (1984), Aoki *et al.* (1986), Shiroza *et al.* (1987), Munro *et al.* (1991, 1995), Shimamura *et al.* (1994), Vacca-Smith *et al.* (1996b), Kopec *et al.* (1997), Koo *et al.* (2010) |
| | Water-soluble and insoluble glucans, mixture of α-1,3 linked glucose and α-1,6 linked glucose<br>Surface-formed glucans display mostly 3-linked glucose, while glucans in solution have higher proportion of 6-linked glucose | *gtfC* | Adhesion/cohesion, structural integrity, and virulence | Hamada *et al.* (1984), Ueda *et al.* (1988), Munro *et al.* (1991), Yamashita *et al.* (1993), Shimamura *et al.* (1994), Munro *et al.* (1995), Kopec *et al.* (1997), Koo *et al.* (2010) |
| | Water-soluble glucans, mostly α-1,6 linked glucose<br>Surface-formed glucans contain greater amounts of 6-linked glucose than glucans formed in solution | *gtfD* | Adhesion/cohesion, primer, and virulence | Hamada *et al.* (1984), Hanada and Kuramitsu (1989), Honda *et al.* (1990), Munro *et al.* (1991, 1995), Yamashita *et al.* (1993), Shimamura *et al.* (1994), Kopec *et al.* (1997) |
| | Fructan, inulin-type, mostly β-2,1 linked | *ftf* | Adhesion/cohesion, storage, and virulence | Shiroza and Kuramitsu (1988), Burne (1991), Munro *et al.* (1991), Burne *et al.* (1996), Rozen *et al.* (2004), Shemesh and Steinberg (2006) |
| *Streptococcus sobrinus* (an infrequent inhabitant) | Water insoluble glucans (mutan) mostly α-1,3-linked glucose | *gtfI* | Adhesion/cohesion | Loesche (1986), Hanada and Takehara (1987), Abo *et al.* (1991) |
| | Water-soluble glucans mostly α-1,6-linked glucose, linear (primer independent) | *gtfS* | Adhesion/cohesion | Yamashita *et al.* (1988), Hayakawa *et al.* (1993) |
| | Water-soluble glucans mostly α-1,6-linked glucose (primer independent) | *gtfT* | Adhesion/cohesion | Hanada *et al.* (1993) |
| | Water-soluble glucans mostly α-1,6 linked glucose (primer dependent) | *gtfU* | Adhesion/cohesion | Hanada *et al.* (2002) |
| | Fructan, levan-type, mostly β-6,1-linked | | Adhesion/cohesion, water retention and storage | Corrigan and Robyt (1979) |
| *Streptococcus sanguinis* | Water-soluble glucans (primer independence) | *gtfP* | Adhesion/cohesion | Hoshino *et al.* (2004), Xu *et al.* (2007) |
| *Streptococcus gordonii* | Water-soluble glucans, mostly α-2,6 linked glucose | *gtfG* | Adhesion/cohesion | Haisman and Jenkinson (1991), Vickerman *et al.* (1996, 1997, 2007) |

| | | | | |
|---|---|---|---|---|
| *Streptococcus oralis* | Water-soluble glucans, mostly α-1,6-linked glucose (no primer dependency) | *gtfR* | Adhesion/cohesion | Fujiwara *et al.* (2000) |
| *Streptococcus salivarius* | Water-insoluble glucans (mutan) mostly α-1,3-linked glucose (primer dependent) | *gtfJ* | Adhesion/cohesion | McCabe and Donkersloot (1977), Tsumuraya and Misaki (1979), Eifuku *et al.* (1989), Pitty and Jacques (1989), Giffard *et al.* (1991) |
| | Water-soluble glucans (dextran) mostly α-1,6-linked glucose (primer dependent) | *gtfK* | Adhesion/cohesion | McCabe and Donkersloot (1977), Pitty and Jacques (1989), Giffard *et al.* (1991), Simpson *et al.* (1995) |
| | Water-insoluble glucans (mutan) mostly α-1,3 linked glucose (primer independent) | *gtfL* | Adhesion/cohesion | Simpson *et al.* (1995) |
| | Water-soluble glucans (dextran) mostly α-1,6 linked glucose (primer independent) | *gtfM* | Adhesion/cohesion | Simpson *et al.* (1995) |
| | Fructan, levan-type, mostly β-2,6 linked | *ftf* | Storage and water retention | Jacques (1984), Simms *et al.* (1990), Song and Jacques (1999a,b) |
| *Streptococcus downei* | Water-insoluble glucan, mostly α-1,3-linked glucose | *gtfI* | Adhesion/cohesion | Gilpin *et al.* (1985), Ferretti *et al.* (1987) |
| | Water-soluble glucan (primer independent); α-1,6-linked linear oligo-isomaltosaccharides without any branching sites | *gtfS* | Adhesion/cohesion | Gilmore *et al.* (1990), Russell *et al.* (1990) |
| *Candida albicans* | β-1,3- and β-1,6-linked glucan, and mannan | *fks1p*<br>Various proteins identified in secreted EPS | Biofilm-specific drug resistance<br>Structural integrity and virulence | Thomas *et al.* (2006), Lal *et al.* (2010), Nett *et al.* (2010) |
| *Aggregatibacter actinomycetemcomitans* | PGA, a β-1,6-linked *N*-acetyl-D-glucosamine (GlcNAc) polymer | *pgaABCD* | Mediates intercellular adhesion and resistance to detergents<br>Adhesion/cohesion, and structural integrity | Kaplan *et al.* (2004), Izano *et al.* (2008) |
| *Capnocytophaga ochracea* | Mannose | Remains to be elucidated | Protection from attack by human innate immune system<br>Virulence | Dyer and Bolton (1985), Yamanaka *et al.* (2009) |
| *Prevotella intermedia* | Mannose | Remains to be elucidated | Protection from attack by human innate immune system<br>Virulence | Yamanaka *et al.* (2009) |
| *Prevotella nigrescens* | Mannose | Remains to be elucidated | Protection from attack by human innate immune system<br>Virulence | Yamane *et al.* (2005), Yamanaka *et al.* (2009) |

**Table 5.1** Continued

| Organisms | Type of EPS[a] | Gene | Function[b] | References |
|---|---|---|---|---|
| *Porphyromonas gingivalis* | Mainly mannose with low levels of rhamnose, glucose, galactose, and 2-acetamido-2-deoxy-D-glucose | *ltp1* | Involved in immune modulation and evasion Virulence | Farquharson *et al.* (2000), Comstock and Kasper (2006), Arndt (2009) |
| *Tannerella forsythia* | Glycoprotein | *wecC, neuC,* and *gtf* | Surface attachment and biofilm formation Adhesion/cohesion | Honma *et al.* (2007) |
| *Lactobacillus fermentum* | Water-soluble glucan; mainly dextran α-1,6 linked | *gtfKg3* | Adhesion/cohesion, and structural integrity | Hammond (1969), Kralj *et al.* (2004), Badel *et al.* (2011) |
| *Lactobaccillus reuteri* | Mainly α-1,6-linked dextran | *gtf180* | Storage, bacterial adherence/cohesion, structural integrity, and water retention | Kralj *et al.* (2005), Badel *et al.* (2011) |
| | Mainly α-1,3-linked mutan | *gtfML1* | | Kralj *et al.* (2005), Badel *et al.* (2011) |
| | Mainly α-1,4-linked reuteran | *gtfA* | | Kralj *et al.* (2005), Badel *et al.* (2011) |
| | Mainly β-2,6-linked levan | *lev* | | van Hijum *et al.* (2006), Badel *et al.* (2011) |
| | Mainly β-2,1-linked inulin | *Inu* | | Ozimek *et al.* (2006), Badel *et al.* (2011) |
| *Lactobacillus salivarius* | Mainly galactose and glucose with small amounts of xylose, arabinose, and mannose | Remains to be elucidated | Storage, bacterial adherence/cohesion, and structural integrity | Frengova *et al.* (2000), Badel *et al.* (2011) |
| *Lactobacillus rhamnosus* | Glucose and galactose | Remains to be elucidated | Storage, bacterial adherence/cohesion, and structural integrity | Vanhaverbeke *et al.* (1998), Badel *et al.* (2011) |
| *Lactobacillus casei* | Galactose and rhamnose | Attributed in part to activity of UDP-galactose 4-epimerase enzyme | Storage, bacterial adherence/cohesion, and structural integrity | Hammond (1969), Kojic *et al.* (1992), Mozzi *et al.* (2003), Maze *et al.* (2010), Badel *et al.* (2011) |
| | Water-soluble glucan; mainly dextran α-1,6 | *dexB* | | |
| *Lactobacillus delbrueckii* | Galactose, glucose, and rhamnose | Remains to be elucidated | Storage, bacterial adherence/cohesion, and structural integrity | Faber *et al.* (2001), Marshall *et al.* (2001), Harding *et al.* (2005), Badel *et al.* (2011) |

[a]Name of the polymer and types of linkage.
[b]There are five major functions: binding sites (adherence and cohesion), virulence, storage (nutrient source), structural integrity/stability, and primer (for EPS synthesis). For comprehensive review, refer to Bowen and Koo (2011) and van Hijum *et al*. (2006).
Note: Databases GenBank/NCBI (http://www.ncbi.nlm.nih.gov/genbank/) and Human Oral Microbiome Database (http://www.homd.org/) were searched for new gene/protein sequences of glucosyltransferases (glucansucrases) and fructosyltransferases (fructansucrases) (on 5 October 2011), but the output did not reveal uncharacterized contributors to EPS matrix.

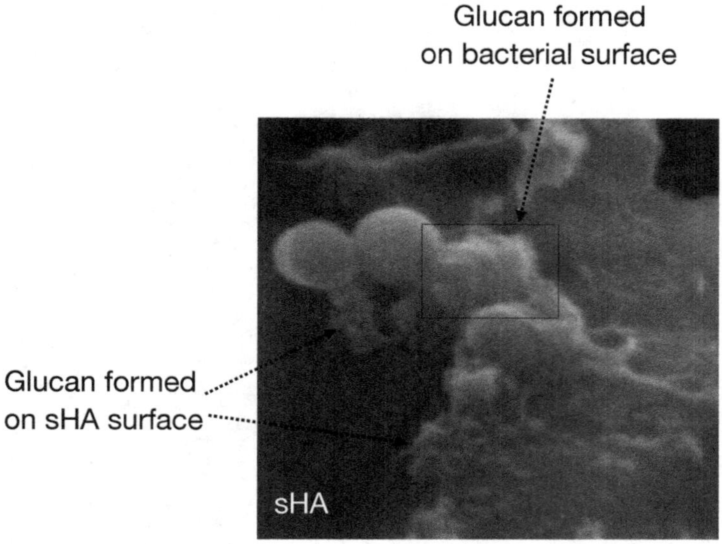


**Figure 5.2** Glucans formed *in situ* by surface-adsorbed glucosyltransferases. Surface-formed glucans provide distinctive binding sites with high avidity to oral bacteria, such as *S. mutans* (shown). sHA, saliva-coated hydroxyapatite.

surfaces by the actions of GtfC and GtfB are critical for the development of the initial EPS matrix (Koo *et al.*, 2010). Concomitantly, these polymers enhance the cohesion of bacterial cells and adherence to the saliva-coated apatitic surfaces, allowing for the development of highly localized and dense cell clusters enmeshed in EPS (termed EPS–microcolony complexes; Fig. 5.3) (Tamesada *et al.*, 2004; Kreth *et al.*, 2008; Koo *et al.*, 2010; Xiao and Koo, 2010). These phenomena may even explain the microscopic and histochemical preparations of early plaque-biofilms, which reveal microorganisms (even normally non-EPS-producers) embedded in a highly insoluble polysaccharide matrix; thereby, influencing the evolution of virulent biofilms even at early stages of development.

## Persistent EPS synthesis modulates the formation of the body of the matrix in the biofilm

If the biofilm is exposed to sucrose frequently (and to starch hydrolysates), EPS will be continuously produced *in situ*, thereby providing mechanical stability and a continuum of diverse binding sites. In addition, Gtfs may also bind to the glucan already formed within biofilms (Tamesada *et al.*, 2004). The continuous formation of EPS helps to assemble the body of the matrix, which further encases the cells to provide structural support for the development of microcolonies of varying size, which form a highly complex and compartmentalized 3D architecture (see details in Fig. 5.4). Furthermore, the structure of EPS is altered over time by the presence of mutanase and dextranases (and possibly other enzymes) in the plaque (Guggenheim and Burckhardt, 1974; Walker *et al.*, 1981; Hayacibara *et al.*, 2004).

During the biofilm maturation and EPS-matrix development, the acidogenic members of the biofilm community also metabolize sucrose, starch hydrolysates, and intracellular polysaccharides to form acids, accumulation of which leads to the acidification of the milieu (Marsh, 2003). *S. mutans* up-regulates the genes encoding GtfB, GtfC, and dextranase in response to acidic pH in the environment and carbohydrate availability (Li and Burne, 2001; Klein *et al.*, 2010). The simultaneous synthesis of glucans by GtfB and GtfC on surfaces contributes to the continuous production and accumulation of EPS-rich matrix. Furthermore, the presence of starch hydrolysates also induces the expression of *gtfB* (Klein *et al.*, 2009, 2010). The up-regulation of dextranase

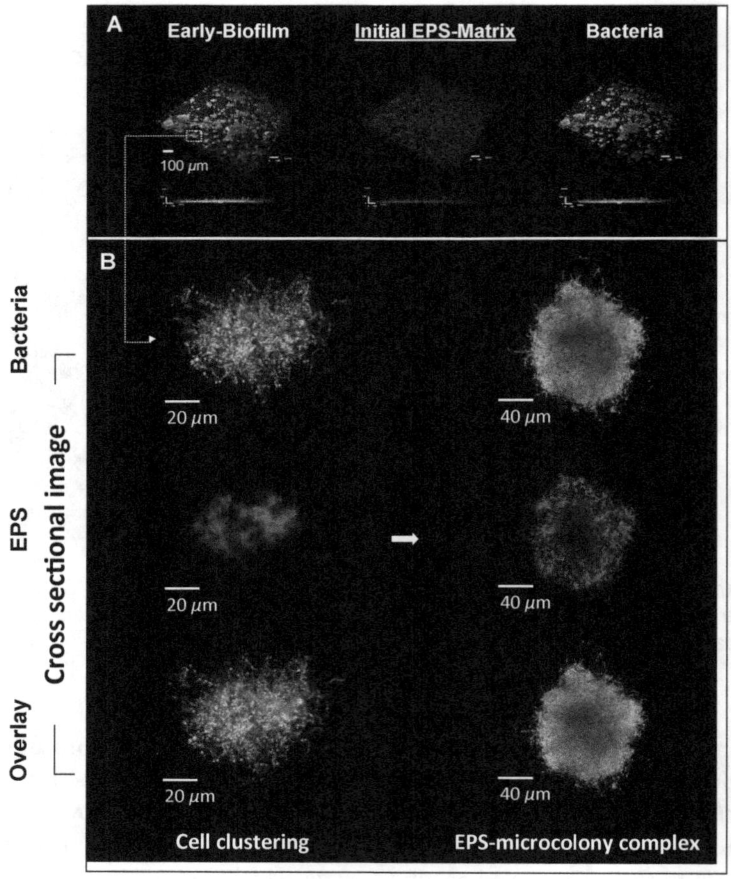


**Figure 5.3** The initial EPS-matrix and EPS-microcolony complex. (A) 3D reconstruction of EPS-matrix (in red) and bacterial cells (in green) within early-formed biofilm. (B) EPS-mediated cell clustering and further development into localized EPS-microcolony complexes. (Adapted from Xiao *et al.*, 2012.) A colour version of this figure is available in the plate section at the back of the book.

contributes to the insolubility of the matrix because the enzyme hydrolyses α-1,6-glucosyl-linkages during glucan synthesis, resulting in an increase in the proportion, while also providing primers for the synthesis of insoluble glucans rich in α–1,3-linkages (Hayacibara *et al.*, 2004). Likewise, degradation of soluble glucans (and also fructans) increases the availability of fermentable sugars, which are metabolized into acids by *S. mutans* and other acidogenic organisms (Paes Leme *et al.*, 2006; Bowen and Koo, 2011). Collectively, it appears that the continued and concerted actions of GtfB and GtfC with glucanohydrolases lead to the formation of a highly insoluble, cohesive, and web-like 3D matrix of cross-linked glucans (Fig. 5.4) as the biofilm grows and matures (Schilling and Bowen, 1992; Kopec *et al.*, 1997; Hayacibara *et al.*, 2004; Koo *et al.*, 2009; Xiao and Koo, 2010; Bowen and Koo, 2011).

It is noteworthy that neither the EPS-matrix nor microcolony complexes are observed when the biofilms are grown in glucose and/or fructose, which are not substrates for EPS synthesis (Koo *et al.*, 2009, 2010) and are less cariogenic than sucrose *in vivo* (Cury *et al.*, 2000; Zero, 2004). Moreover, *S. mutans* strains defective in *gtfB* and/or in *gtfC* do not form an EPS matrix, and the formation of microcolonies is completely abolished in the absence of both GtfB and GtfC activity (Fig. 5.5) (Koo *et al.*, 2010). These observations demonstrate the importance of the matrix in the development of structured 3D microbial

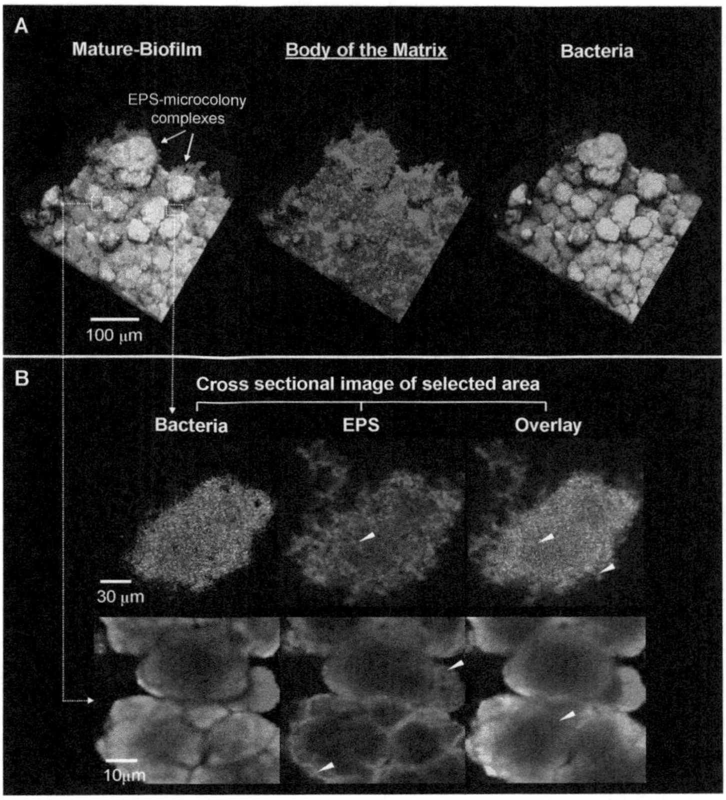


**Figure 5.4** The body of the EPS-matrix. (A) 3D reconstruction of EPS-matrix (in red) and bacterial cells (in green) within mature biofilm. (B) Close-up view of cell-matrix structural organization. Arrowheads show the intricate architecture of the matrix providing a 3D web-like meshwork surrounding and enmeshing the bacterial cells, forming compartmentalized microcolony complexes. (Adapted from Bowen and Koo, 2011. Image reproduced with the permission of S. Karger AG, Basle.) A colour version of this figure is available in the plate section at the back of the book.

communities within biofilms through the formation of EPS–microcolony complexes, which may explain why strains defective in *gtfB* and/or in *gtfC* are less virulent than the parental strain in animal models of dental caries (Tanzer, 1985; Munro *et al.*, 1991; Yamashita *et al.*, 1993).

Clearly, the sequential assembly of EPS-matrix leads to a dense concentration of microorganisms clustered at specific sites: (i) across the biofilm and (ii) on the tooth surface. In parallel, the surrounding milieu becomes highly acidic due to the abundance of metabolizable sugars (e.g. soluble glucans, fructans and intracellular polysaccharides) and acidogenic/aciduric organisms (Paes Leme *et al.*, 2006; Klein *et al.*, 2009, 2010; Koo *et al.*, 2010; Xiao and Koo, 2010; Bowen and Koo, 2011), which are protected from the neutralizing effects of the surrounding saliva by the EPS-matrix (see later). The matrix of plaque is neither static nor homogeneous; rather it undergoes constant restructuring, which can dramatically influence the architectural and biological transformations within biofilms to modulate the expression of virulence.

## The matrix structure influences the microenvironment within biofilms

The presence of highly adherent and insoluble glucans *in situ* provides a mesh framework that holds the bacterial cells together forming structurally cohesive 3D cell-clusters, which are bound to the apatite surface and tethered within the biofilm matrix (Schilling and Bowen, 1992; Cross *et al.*, 2007; Kreth *et al.*, 2008; Koo *et al.*, 2010;

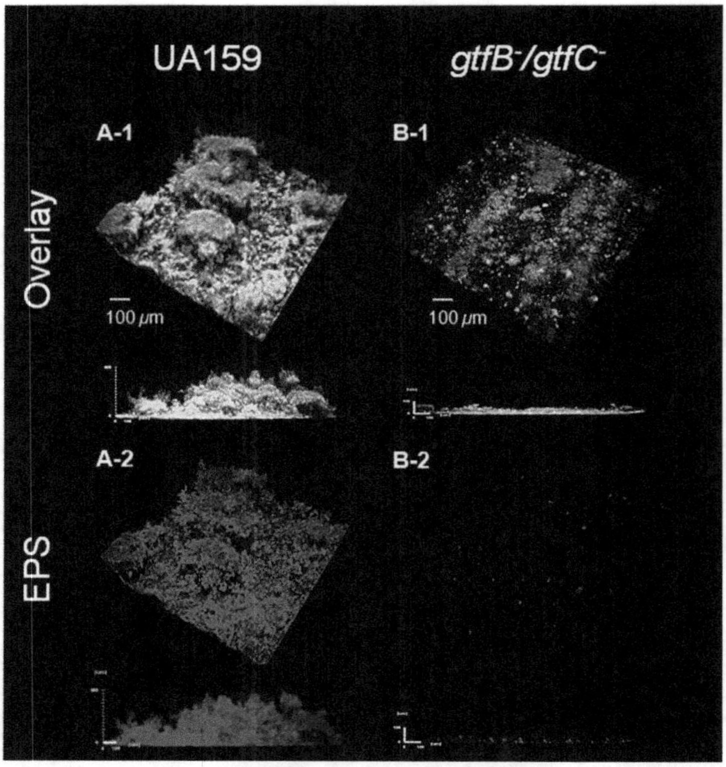


**Figure 5.5** Mixed-species biofilm formed by parental strain of *S. mutans* (UA159) and *gtfB-/gtfC-* mutant strain (Δ*gtfBC*::kan). (A-1, B-1) Representative 3D reconstruction of the biofilms. (A-2, B-2) 3D reconstruction of EPS-matrix in the same biofilms. (Adapted from Xiao *et al.*, 2012.) A colour version of this figure is available in the plate section at the back of the book.

Xiao and Koo, 2010). The insoluble exopolysaccharides provide structural stability, rigidity, and a supporting core for the formation of highly compartmentalized EPS–microcolony complexes (Fig. 5.4). These unique 3D structures may create localized and acidic microenvironments across the biofilm because EPS and other matrix materials may affect the diffusion of substances into and out of the biofilm, limiting the access of saliva and its neutralizing effects. Therefore, the diffusion limitation caused by the EPS-rich matrix may be a key trait for expression of virulence in biofilms (Xiao *et al.*, 2012). In the context of biofilm virulence, acid production alone may not be the key factor, but rather how and where acidic microenvironments are formed and maintained within the complex biofilm milieu.

The role of EPS in the formation of pH-delineated microenvironments within oral biofilms is poorly understood and remains controversial because the polysaccharides may not be evenly distributed in intact plaque-biofilms (Saxton and Kolendo, 1967; Reese and Guggenheim, 2007; Bowen and Koo, 2011). Studies that have attempted to characterize the unique microenvironments within biofilms have relied heavily on the use of invasive techniques (e.g. using probes or microsensors) (Werner *et al.*, 2004; Rani *et al.*, 2007) or extra oral modelling, which can alter the physical properties of the matrix, complicating data interpretation. We have devised a non-invasive technique that successfully determined the spatial distribution of pH within intact biofilm 3D architecture (Xiao *et al.*, 2012).

The formation of compartmentalized structures (EPS–microcolony complexes) may protect particular areas of the biofilm from neutralization. pH Mapping (Fig. 5.6) reveals the presence of several acidic regions heterogeneously distributed across the EPS–microcolony complex as follows:

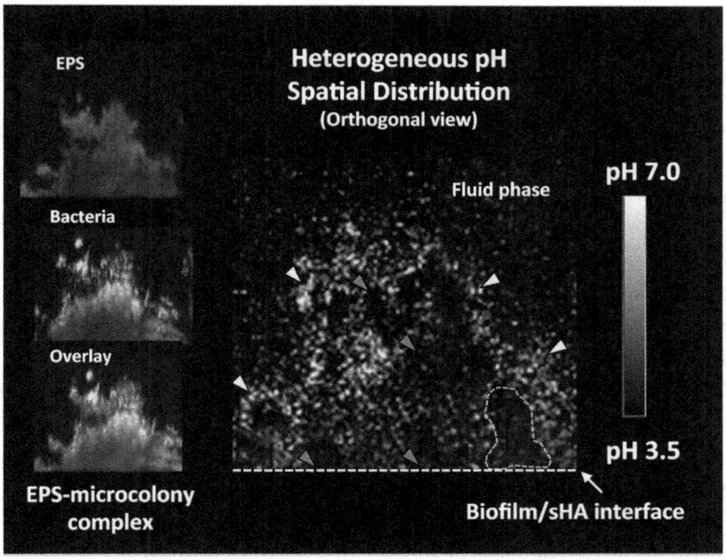


**Figure 5.6** Mapping of *in situ* pH of intact mixed-species biofilm. Orthogonal view of a representative image of the spatial distribution of pH within a EPS-microcolony complex. Red arrows indicate acidic pH regions (i) across the structure and (ii) at the microcolony/sHA interface (also highlighted with orange lines). Yellow arrows indicate pH close to neutral at the microcolony/fluid phase. (Adapted from Xiao *et al.*, 2012.) A colour version of this figure is available in the plate section at the back of the book.

(1) regions that display pH values that are close to neutral at the microcolony/fluid phase, (2) a range of acidic 'pockets' across the structure (from the sHA surface to the fluid phase interface), and (3) acidic pH at the microcolony/sHA interface (see arrows in Fig. 5.6). The presence of discrete regions containing low pH (3.5 to 5.5) indicates that the acids accumulated and confined in these particular areas are not rapidly neutralized when incubated in neutral pH (7.0) buffer; more than 120 minutes of exposure are required before neutralization occurs. In contrast, regions of the biofilm without EPS–microcolony complexes are immediately neutralized upon exposure to (pH 7.0) buffer, and no acidic pockets are formed (Xiao *et al.*, 2012). These observations suggest that diffusion limitation occurs within specific areas of the biofilm due to the presence of the EPS-mediated 3D structures.

The environmental changes associated with the architectural modification of the biofilms caused by the formation of EPS-matrix, may have direct relevance to the pathogenesis of dental caries. The presence of localized and elevated concentrations of acid trapped at the surface of biofilm attachment would increase local demineralization of the hydroxyapatite mineral, the main constituent of the tooth enamel. Such distribution of acidic pools may explain the localized/punctate appearance of regions of demineralization clinically known as 'white spots'. These white spots eventually coalesce into broader lesions, rather than a diffuse and homogeneous layer of erosion.

The creation of a heterogeneous acidic environment is biologically relevant, as pH changes could trigger dynamic shifts in the microbial population within the body of the biofilm (Marsh, 2003). Niches with low pH values would activate the acid stress adaptive responses of particular organisms clustered in these EPS-sheltered acidic pockets, favouring the survival/dominance of aciduric and glucan-binding pathogens, such as *S. mutans* (Burne *et al.*, 1996; Quivey *et al.*, 2000; Marsh, 2003; Bowen and Koo, 2011). Furthermore, bacterial cells within the EPS–microcolony complexes are less susceptible to killing by antimicrobials, such as chlorhexidine, than those residing outside the structured communities (Xiao *et al.*, 2012).

In summary, the EPS formed *in situ* can: (1) provide avid binding sites for colonization and accumulation of *S. mutans* (and other acidogenic and aciduric organisms) on the apatitic surface, (2) enhance microbial adhesion to surfaces, (3) promote cellular and biofilm coherence/cohesion, (4) serve as reserve source of energy, (5) protect microorganisms from antimicrobials, (6) affect inorganic and protein (both mammalian and microbial-derived) composition in plaque, and (7) may affect diffusion creating localized and confined acidic milieus (Marsh, 2003; Paes Leme *et al.*, 2006, 2008; Reese and Guggenheim, 2007; Koo *et al.*, 2010; Bowen and Koo, 2011; Xiao *et al.*, 2012).

Clearly, the assembly of the EPS-matrix is a major driving force associated with the evolution of virulent-cariogenic biofilms. The formation of 3D structures within biofilm, particularly close to the tooth surface, generates a pathological cycle, i.e. acidic environment selects only aciduric-acidogenic microorganisms, which in turn ensures continued localized acid-production. If the biofilm remains undisturbed, the more acidogenic it becomes leading to localized hard-tissue surface demineralization. Although *S. mutans* is frequently not the dominant organism numerically in plaque, it exercises influences far beyond what mere numbers would appear to warrant, as the Gtfs released even in early stages mediate the construction of the EPS-matrices on surfaces.

## Disrupting EPS-matrix as a therapeutic approach against formation of virulent biofilms

EPS and *S. mutans* Gtfs have been shown to be critical virulence factors in the pathogenesis of dental caries *in vivo* using rodent models of the disease (Tanzer *et al.*, 1974; Tanzer, 1985; Yamashita *et al.*, 1993). Furthermore, results from clinical studies have linked the presence of water insoluble glucans in the plaque with the cariogenicity of the biofilm in humans (Mattos-Graner *et al.*, 1998; Mattos-Graner *et al.*, 2000; Nobre dos Santos *et al.*, 2002; Ribeiro *et al.*, 2005; Aires *et al.*, 2006). Also, the level of GtfB in saliva correlates with the presence of carious lesions in young children (Vacca Smith *et al.*, 2007). Clearly, inhibition of glucan synthesis and/or its early removal may be an attractive and highly selective approach aimed at disrupting EPS-matrix assembly in virulent oral biofilms to prevent plaque-related diseases. It has been shown that inhibitors, either natural or synthetic, can partially inhibit GtfB and GtfC in solution, but only a few can inhibit the activity of the surface-adsorbed enzymes (see detailed review in Bowen and Koo, 2011; and in Jeon *et al.*, 2011). When surface-adsorbed, Gtfs display higher resistance to the most common inhibiting agents (including the anti-plaque mouth rinses) than when in solution (Wunder and Bowen, 1999; Koo *et al.*, 2002). The reason for this phenomenon is unclear, although it may be related to conformational changes that these enzymes undergo during adsorption, which can alter the enzyme kinetics and physicochemical properties (Schilling and Bowen, 1988, 1992; Venkitaraman *et al.*, 1995; Bowen and Koo, 2011). Nevertheless, it appears that agents that bind to the active site are equally effective when the enzyme is bound to a surface. Effective and potent inhibitors of *in situ* glucans synthesis by surface Gtfs (displaying at least > 90% inhibition) are yet to be discovered. However, recent advances in medicinal, biosynthetic, and combinatorial chemistry may offer enhanced opportunities to find inhibitors that could effectively target surface-adsorbed enzymes and in *situ* glucan synthesis (Jeon *et al.*, 2011).

## Conclusion and future directions

The 3D architecture of the EPS-matrix changes over time as a result of environmental changes in the mouth, such as exposure to dietary sucrose and starch as well as through the action of mutanase and dextranase found within plaque. As the EPS content and 3D structures mature in the matrix, localized acidic and protective microenvironments are created and the surviving populations of aciduric, acidogenic, and glucan-binding pathogens (such as *S. mutans*) and other less pathogenic or non-pathogenic co-habitants are dynamically altered, favouring the dominance of potentially virulent organisms.

The sequential assembly and spatial organization of the matrix provides the *in situ* foundation for formation of complex 3D bacterial islets through highly localized cell-to-matrix interactions (structures termed EPS–microcolony complexes) (Fig.

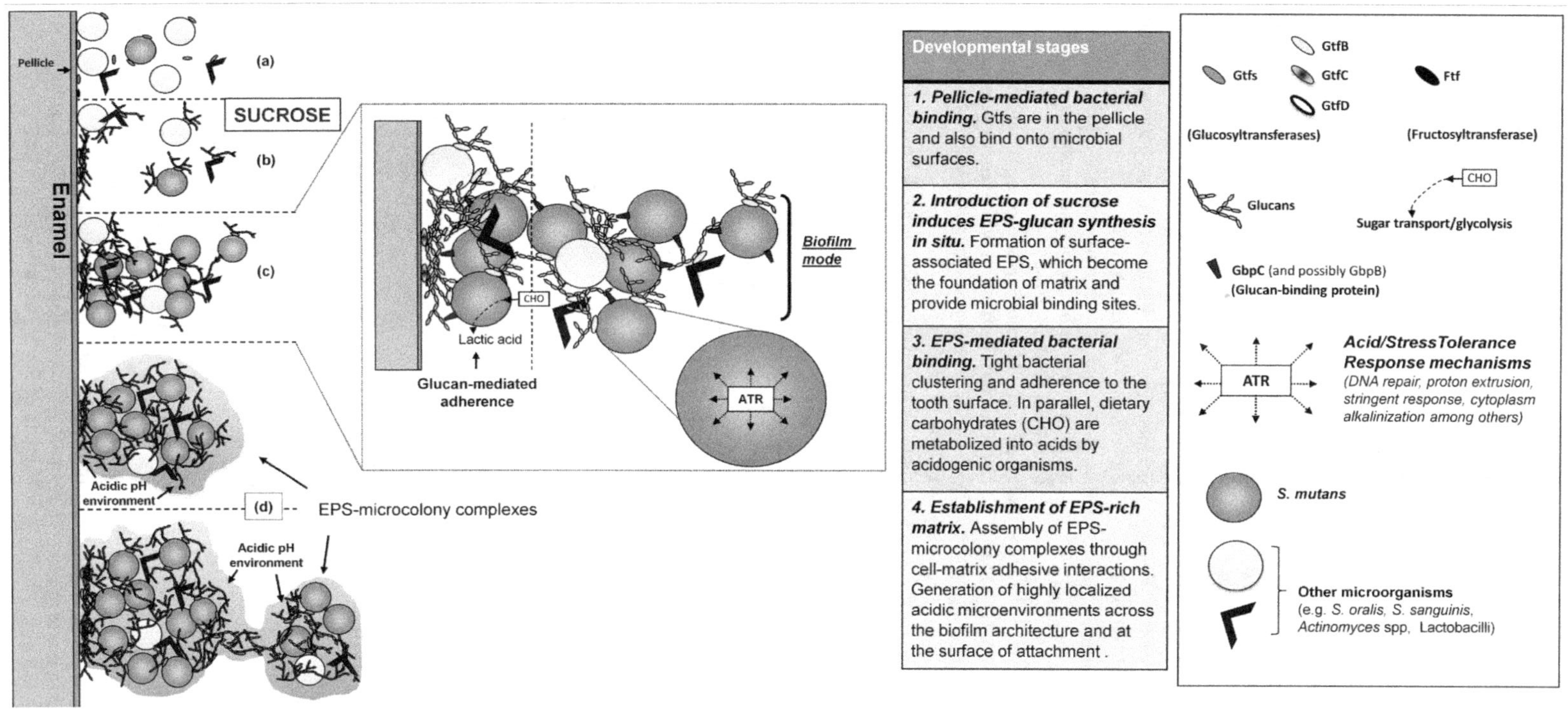


**Figure 5.7** Virulent-cariogenic biofilm formation. (a) The Gtfs and Ftf secreted by *S. mutans* are incorporated into the pellicle (particularly GtfC) and are adsorbed to bacterial surfaces (mainly GtfB), including microorganisms that do not produce Gtfs (e.g. Actinomyces spp.). (b) Surface-adsorbed GtfB and GtfC rapidly utilize dietary sucrose (and starch hydrolysates) to synthesize insoluble and soluble glucans *in situ*; the soluble glucans formed by GtfD could serve as primers for GtfB, enhancing overall synthesis of EPS. The Gtfs adsorbed onto surfaces provide an insoluble matrix for dental plaque-biofilm *in situ*. Concomitantly, dietary carbohydrates are metabolized into acids by acidogenic/aciduric organisms (e.g. *S. mutans*). (c) The glucan molecules provide avid binding sites for a variety of oral microorganisms, particularly *S. mutans*, mediating tight bacterial clustering and adherence to the tooth enamel (through glucan–glucan and glucan–Gbp interactions). Furthermore, Gtf-adsorbed bacteria become de facto glucan producers, binding to tooth and microbial surfaces by the same mechanisms. This model could explain the rapid formation and accumulation of highly cohesive-adherent plaque in the presence of sucrose (and starch hydrolysates), even if the starting number of *S. mutans* is relatively low. After the establishment of a glucan-rich biofilm matrix, ecological pressure (e.g. pH) will determine which bacteria may survive and dominate within plaque under frequent sucrose (or other fermentable carbohydrate) exposure. (d) If biofilm remains on tooth surfaces with frequent consumption of high carbohydrate diet (especially sucrose), the amounts of EPS and extent of acidification of the matrix will be increased (formation of EPS–microcolony complexes). Such conditions cause biochemical, ecological and structural changes favouring the survival and dominance of highly acid-stress tolerant organisms in cohesive and firmly attached biofilms. The low-pH environment at tooth–biofilm interface results in enhanced demineralization of enamel.

5.7). Formation of these structures is essential for the creation of localized acidic microenvironments. It is clear that Gtfs bound to surfaces, and acting in concert, play a critical role in the development of the EPS-matrix and the pathogenesis of dental caries. It is equally apparent that surface-adsorbed Gtf could be a major and selective target for chemotherapeutic agents chosen to prevent the formation of virulent-cariogenic plaque-biofilms. However, most of the agents that have been identified to inhibit Gtf in solution have either reduced efficacy or no effect on surface-adsorbed enzymes. Resistance to inhibitory agents may be related to conformational changes associated with adsorption of enzymes to surfaces, including microbial cells. Clearly, the reactions of Gtfs present in the pellicle, adsorbed on the microbial surfaces, and within plaque are diverse and complex, and they may dictate how bacterial cells interact with the extracellular matrix and the resultant microenvironments.

Elucidation of the exact conformational changes undergone by the enzymes during the adsorption to surfaces, and how these changes affect the production of EPS *in situ*, provision of binding sites, and diffusion within the matrix, will advance our understanding about the mechanisms of biofilm construction and the expression of virulence. Furthermore, exploration of the dynamics of matrix-to-cell interactions and the creation of microenvironments within the matrix using real time *in situ* models could enhance our knowledge about the biological as well as the physical changes occurring within dental plaque during its development. This valuable information will usher the way towards the development of novel and effective methods to prevent virulent biofilm formation and/or promote its disruption. The results of such investigations may have importance beyond the mouth and may be applied to any biofilm system where a polymeric matrix is formed and microenvironments are created.

## References


Abo, H., Matsumura, T., Kodama, T., Ohta, H., Fukui, K., Kato, K., and Kagawa, H. (1991). Peptide sequences for sucrose splitting and glucan binding within *Streptococcus sobrinus* glucosyltransferase (water-insoluble glucan synthetase). J. Bacteriol. *173*, 989–996.

Aires, C.P., Tabchoury, C.P., Del Bel Cury, A.A., Koo, H., and Cury, J.A. (2006). Effect of sucrose concentration on dental biofilm formed *in situ* and on enamel demineralization. Caries Res. *40*, 28–32.

Allen, P.Z., and Bowen, W.H. (1990). Immunochemical studies on levans from several strains of *Actinomyces viscosus*. Arch. Oral Biol. *35*, 55–62.

Aoki, H., Shiroza, T., Hayakawa, M., Sato, S., and Kuramitsu, H.K. (1986). Cloning of a *Streptococcus mutans* glucosyltransferase gene coding for insoluble glucan synthesis. Infect. Immun. *53*, 587–594.

Arndt, A., and Davey, M. (2009). *Porphyromonas gingivalis*: surface polysaccharides as virulence determinants. In Interface Oral Health Science 2009, Sasano, T., and Suzuki, O., eds. (Springer, Tokyo, Berlin, Heidelberg, New York), pp. 382–387.

Badel, S., Bernardi, T., and Michaud, P. (2011). New perspectives for Lactobacilli exopolysaccharides. Biotechnol. Adv. *29*, 54–66.

Banas, J.A., and Vickerman, M.M. (2003). Glucan-binding proteins of the oral streptococci. Crit. Rev. Oral Biol. Med. *14*, 89–99.

Bergeron, L.J., Morou-Bermudez, E., and Burne, R.A. (2000). Characterization of the fructosyltransferase gene of *Actinomyces naeslundii* WVU45. J. Bacteriol. *182*, 3649–3654.

Birkhed, D., and Rosell, K.G. (1975). Structural studies on polysaccharides synthesized from sucrose by soluble enzymes in human dental plaque material. Odontol. Revy. *26*, 281–290.

Bowen, W.H., and Koo, H. (2011). Biology of *Streptococcus mutans*-derived glucosyltransferases: Role in extracellular matrix formation of cariogenic biofilms. Caries Res. *45*, 69–86.

Branda, S.S., Vik, S., Friedman, L., and Kolter, R. (2005). Biofilms: the matrix revisited. Trends Microbiol. *13*, 20–26.

Burne, R.A. (1991). Oral ecological disasters: the role of short-term extracellular storage polysaccharides. In Cariology for the Nineties, Bowen, W.H., and Tabak, L.A., eds. (University of Rochester Press, Rochester), pp. 351–364.

Burne, R.A., and Penders, J.E. (1992). Characterization of the *Streptococcus mutans* GS-5 *fruA* gene encoding exo-beta-D-fructosidase. Infect. Immun. *60*, 4621–4632.

Burne, R.A., Chen, Y.Y., Wexler, D.L., Kuramitsu, H., and Bowen, W.H. (1996). Cariogenicity of *Streptococcus mutans* strains with defects in fructan metabolism assessed in a program-fed specific-pathogen-free rat model. J. Dent. Res. *75*, 1572–1577.

Cole, M.F., and Bowen, W.H. (1974). The effect of calcium beta glycerophosphate on phosphatase activity in plaque from monkeys (*Macaca fascicularis*). Arch. Oral Biol. *19*, 335–337.

Comstock, L.E., and Kasper, D.L. (2006). Bacterial glycans: key mediators of diverse host immune responses. Cell *126*, 847–850.

Corrigan, A.J., and Robyt, J.F. (1979). Nature of the fructan of *Streptococcus mutans* OMZ 176. Infect. Immun. *26*, 387–389.

Critchley, P., Wood, J.M., Saxton, C.A., and Leach, S.A. (1967). The polymerisation of dietary sugars by dental plaque. Caries Res. *1*, 112–129.

Cross, S.E., Kreth, J., Zhu, L., Sullivan, R., Shi, W., Qi, F., and Gimzewski, J.K. (2007). Nanomechanical properties of glucans and associated cell-surface adhesion of *Streptococcus mutans* probed by atomic force microscopy under in situ conditions. Microbiology *153*, 3124–3132.

Cury, J.A., Rebelo, M.A., Del Bel Cury, A.A., Derbyshire, M.T., and Tabchoury, C.P. (2000). Biochemical composition and cariogenicity of dental plaque formed in the presence of sucrose or glucose and fructose. Caries Res. *34*, 491–497.

de Carvalho, F.G., Silva, D.S., Hebling, J., Spolidorio, L.C., and Spolidorio, D.M. (2006). Presence of mutans streptococci and *Candida* spp. in dental plaque/dentine of carious teeth and early childhood caries. Arch. Oral Biol. *51*, 1024–1028.

Dewhirst, F.E., Chen, T., Izard, J., Paster, B.J., Tanner, A.C., Yu, W.H., Lakshmanan, A., and Wade, W.G. (2010). The human oral microbiome. J. Bacteriol. *192*, 5002–5017.

Dyer, J.K., and Bolton, R.W. (1985). Purification and chemical characterization of an exopolysaccharide isolated from *Capnocytophaga ochracea*. Can. J. Microbiol. *31*, 1–5.

Dye, B.A., Tan, S., Smith, V., Lewis, B.G., Barker, L.K., Thornton-Evans, G., Eke, P.I., Beltran-Aguilar, E.D., Horowitz, A.M., and Li, C.H. (2007). Trends in oral health status: United States, 1988–1994 and 1999–2004. Vital Health Stat *11*, 1–92.

Ebisu, S., Kato, K., Kotani, S., and Misaki, A. (1975). Structural differences in fructans elaborated by *Streptococcus mutans* and *Strep. salivarius*. J. Biochem. *78*, 879–887.

Eifuku, H., Yoshimitsu-Narita, A., Sato, S., Yakushiji, T., and Inoue, M. (1989). Production and partial characterization of the extracellular polysaccharides from oral *Streptococcus salivarius*. Carbohydr. Res. *194*, 247–260.

Emilson, C.G., Nilsson, B., and Bowen, W.H. (1984). Carbohydrate composition of dental plaque from primates with irradiation caries. J. Oral Pathol. *13*, 213–220.

Faber, E.J., Kamerling, J.P., and Vliegenthart, J.F. (2001). Structure of the extracellular polysaccharide produced by *Lactobacillus delbrueckii* subsp. *bulgaricus* 291. Carbohydr. Res. *331*, 183–194.

Farquharson, S.I., Germaine, G.R., and Gray, G.R. (2000). Isolation and characterization of the cell-surface polysaccharides of *Porphyromonas gingivalis* ATCC 53978. Oral Microbiol. Immunol. *15*, 151–157.

Ferretti, J.J., Gilpin, M.L., and Russell, R.R. (1987). Nucleotide sequence of a glucosyltransferase gene from *Streptococcus sobrinus* MFe28. J. Bacteriol. *169*, 4271–4278.

Flemming, H.C., and Wingender, J. (2010). The biofilm matrix. Nat. Rev. Microbiol. *8*, 623–633.

Foster, J.S., Palmer, R.J., Jr., and Kolenbrander, P.E. (2003). Human oral cavity as a model for the study of genome–genome interactions. Biol. Bull. *204*, 200–204.

Frengova, G.I., Simova, E.D., Beshkova, D.M., and Simov, Z.I. (2000). Production and monomer composition of exopolysaccharides by yogurt starter cultures. Can. J. Microbiol. *46*, 1123–1127.

Fu, D.T., and Robyt, J.F. (1991). Maltodextrin acceptor reactions of *Streptococcus mutans* 6715 glucosyltransferases. Carbohydr. Res. *217*, 201–211.

Fujiwara, T., Hoshino, T., Ooshima, T., Sobue, S., and Hamada, S. (2000). Purification, characterization, and molecular analysis of the gene encoding glucosyltransferase from *Streptococcus oralis*. Infect. Immun. *68*, 2475–2483.

Fukui, K., and Moriyama, T. (1983). Effect of maltose on glucan synthesis by glucosyltransferases of *Streptococcus mutans*. Microbiol. Immunol. *27*, 917–927.

Gibbons, R.J. (1968). Formation and significance of bacterial polysaccharides in caries etiology. Caries Res. *2*, 164–171.

Giffard, P.M., Simpson, C.L., Milward, C.P., and Jacques, N.A. (1991). Molecular characterization of a cluster of at least two glucosyltransferase genes in *Streptococcus salivarius* ATCC 25975. J. Gen. Microbiol. *137*, 2577–2593.

Gilmore, K.S., Russell, R.R., and Ferretti, J.J. (1990). Analysis of the *Streptococcus downei gtfS* gene, which specifies a glucosyltransferase that synthesizes soluble glucans. Infect. Immun. *58*, 2452–2458.

Gilpin, M.L., Russell, R.R., and Morrissey, P. (1985). Cloning and expression of two *Streptococcus mutans* glucosyltransferases in *Escherichia coli* K-12. Infect. Immun. *49*, 414–416.

Gregoire, S., Xiao, J., Silva, B.B., Gonzalez, I., Agidi, P.S., Klein, M.I., Ambatipudi, K.S., Rosalen, P.L., Bauserman, R., Waugh, R.E., *et al.* (2011). Role of glucosyltransferase B in the interactions of *Candida albicans* with *Streptococcus mutans* and experimental pellicle formed on hydroxyapatite surface. Appl. Environ. Microbiol. *77*, 6357–6367.

Guggenheim, B., and Burckhardt, J.J. (1974). Isolation and properties of a dextranase from *Streptococcus mutans* OMZ 176. Helv. Odontol. Acta *18*, 101–113.

Haisman, R.J., and Jenkinson, H.F. (1991). Mutants of *Streptococcus gordonii* Challis over-producing glucosyltransferase. J. Gen. Microbiol. *137*, 483–489.

Hamada, S., Tai, S., and Slade, H.D. (1978). Binding of glucosyltransferase and glucan synthesis by *Streptococcus mutans* and other bacteria. Infect. Immun. *21*, 213–220.

Hamada, S., Koga, T., and Ooshima, T. (1984). Virulence factors of *Streptococcus mutans* and dental caries prevention. J. Dent. Res. *63*, 407–411.

Hammond, B.F. (1969). Dextran production by a human oral strain of *Lactobacillus casei*. Arch. Oral Biol. *14*, 879–890.

Hanada, N., and Kuramitsu, H.K. (1989). Isolation and characterization of the *Streptococcus mutans gtfD* gene, coding for primer-dependent soluble glucan synthesis. Infect. Immun. *57*, 2079–2085.

Hanada, N., and Takehara, T. (1987). (1----3)-alpha-D-glucan synthase from *Streptococcus mutans* AHT (serotype g) does not synthesise glucan without primer. Carbohydr. Res. *168*, 120–124.

Hanada, N., Isobe, Y., Aizawa, Y., Katayama, T., Sato, S., and Inoue, M. (1993). Nucleotide sequence analysis of the *gtfT* gene from *Streptococcus sobrinus* OMZ176. Infect. Immun. *61*, 2096–2103.

Hanada, N., Fukushima, K., Nomura, Y., Senpuku, H., Hayakawa, M., Mukasa, H., Shiroza, T., and Abiko, Y. (2002). Cloning and nucleotide sequence analysis of the *Streptococcus sobrinus gtfU* gene that produces a highly branched water-soluble glucan. Biochim. Biophys. Acta *1570*, 75–79.

Hannig, C., Ruggeri, A., Al-Khayer, B., Schmitz, P., Spitzmuller, B., Deimling, D., Huber, K., Hoth-Hannig, W., Bowen, W.H., and Hannig, M. (2008). Electron microscopic detection and activity of glucosyltransferase B, C, and D in the in situ formed pellicle. Arch. Oral Biol. 53, 1003–1010.

Harding, L.P., Marshall, V.M., Hernandez, Y., Gu, Y., Maqsood, M., McLay, N., and Laws, A.P. (2005). Structural characterisation of a highly branched exopolysaccharide produced by *Lactobacillus delbrueckii* subsp. *bulgaricus* NCFB2074. Carbohydr. Res. *340*, 1107–1111.

Hayacibara, M.F., Koo, H., Vacca-Smith, A.M., Kopec, L.K., Scott-Anne, K., Cury, J.A., and Bowen, W.H. (2004). The influence of mutanase and dextranase on the production and structure of glucans synthesized by streptococcal glucosyltransferases. Carbohydr. Res. *339*, 2127–2137.

Hayakawa, M., Fukushima, K., Abiko, Y., Ikeda, T., and Takiguchi, H. (1993). Cloning of a *Streptococcus sobrinus* oligo-isomaltosaccharide synthase gene and characterization of its product. Biochem. Mol. Biol. Int. *31*, 1167–1175.

van Hijum, S.A., Kralj, S., Ozimek, L.K., Dijkhuizen, L., and van Geel-Schutten, I.G. (2006). Structure–function relationships of glucansucrase and fructansucrase enzymes from lactic acid bacteria. Microbiol. Mol. Biol. Rev. *70*, 157–176.

Honda, O., Kato, C., and Kuramitsu, H.K. (1990). Nucleotide sequence of the *Streptococcus mutans gtfD* gene encoding the glucosyltransferase-S enzyme. J. Gen. Microbiol. *136*, 2099–2105.

Honma, K., Inagaki, S., Okuda, K., Kuramitsu, H.K., and Sharma, A. (2007). Role of a *Tannerella forsythia* exopolysaccharide synthesis operon in biofilm development. Microb. Pathog. *42*, 156–166.

Hoshino, T., Kawaguchi, M., Shimizu, N., Hoshino, N., Ooshima, T., and Fujiwara, T. (2004). PCR detection and identification of oral streptococci in saliva samples using *gtf* genes. Diagn. Microbiol. Infect. Dis. *48*, 195–199.

Hotz, P., Guggenheim, B., and Schmid, R. (1972). Carbohydrates in pooled dental plaque. Caries Res. *6*, 103–121.

Izano, E.A., Sadovskaya, I., Wang, H., Vinogradov, E., Ragunath, C., Ramasubbu, N., Jabbouri, S., Perry, M.B., and Kaplan, J.B. (2008). Poly-*N*-acetylglucosamine mediates biofilm formation and detergent resistance in *Aggregatibacter actinomycetemcomitans*. Microb. Pathog. *44*, 52–60.

Jacques, N.A. (1984). Calcium dependence of the cell-associated fructosyltransferase of *Streptococcus salivarius*. Carbohydr. Res. *127*, 349–355.

Jeon, J.G., Rosalen, P.L., Falsetta, M.L., and Koo, H. (2011). Natural products in caries research: current (limited) knowledge, challenges and future perspective. Caries Res. *45*, 243–263.

Kaplan, J.B., Velliyagounder, K., Ragunath, C., Rohde, H., Mack, D., Knobloch, J.K., and Ramasubbu, N. (2004). Genes involved in the synthesis and degradation of matrix polysaccharide in *Actinobacillus actinomycetemcomitans* and *Actinobacillus pleuropneumoniae* biofilms. J. Bacteriol. *186*, 8213–8220.

Karatan, E., and Watnick, P. (2009). Signals, regulatory networks, and materials that build and break bacterial biofilms. Microbiol. Mol. Biol. Rev. *73*, 310–347.

Klein, M.I., Duarte, S., Xiao, J., Mitra, S., Foster, T.H., and Koo, H. (2009). Structural and molecular basis of the role of starch and sucrose in *Streptococcus mutans* biofilm development. Appl. Environ. Microbiol. *75*, 837–841.

Klein, M.I., DeBaz, L., Agidi, S., Lee, H., Xie, G., Lin, A.H., Hamaker, B.R., Lemos, J.A., and Koo, H. (2010). Dynamics of *Streptococcus mutans* transcriptome in response to starch and sucrose during biofilm development. PLoS One *5*, e13478.

Koga, T., Horikoshi, T., Fujiwara, T., and Hamada, S. (1988). Effects of panose on glucan synthesis and cellular adherence by *Streptococcus mutans*. Microbiol. Immunol. *32*, 25–31.

Kojic, M., Vujcic, M., Banina, A., Cocconcelli, P., Cerning, J., and Topisirovic, L. (1992). Analysis of exopolysaccharide production by *Lactobacillus casei* CG11, isolated from cheese. Appl. Environ. Microbiol. *58*, 4086–4088.

Koo, H., Rosalen, P.L., Cury, J.A., Park, Y.K., and Bowen, W.H. (2002). Effects of compounds found in propolis on *Streptococcus mutans* growth and on glucosyltransferase activity. Antimicrob. Agents Chemother. *46*, 1302–1309.

Koo, H., Xiao, J., and Klein, M.I. (2009). Extracellular polysaccharides matrix – an often forgotten virulence factor in oral biofilm research. Int. J. Oral Sci. *1*, 229–234.

Koo, H., Xiao, J., Klein, M.I., and Jeon, J.G. (2010). Exopolysaccharides produced by *Streptococcus mutans* glucosyltransferases modulate the establishment of microcolonies within multispecies biofilms. J. Bacteriol. *192*, 3024–3032.

Kopec, L.K., Vacca-Smith, A.M., and Bowen, W.H. (1997). Structural aspects of glucans formed in solution and on the surface of hydroxyapatite. Glycobiology *7*, 929–934.

Kralj, S., van Geel-Schutten, G.H., Dondorff, M.M., Kirsanovs, S., van der Maarel, M.J., and Dijkhuizen, L. (2004). Glucan synthesis in the genus *Lactobacillus*: isolation and characterization of glucansucrase genes, enzymes and glucan products from six different strains. Microbiology *150*, 3681–3690.

Kralj, S., Stripling, E., Sanders, P., van Geel-Schutten, G.H., and Dijkhuizen, L. (2005). Highly hydrolytic reuteransucrase from probiotic *Lactobacillus reuteri* strain ATCC 55730. Appl. Environ. Microbiol. *71*, 3942–3950.

Kreth, J., Zhu, L., Merritt, J., Shi, W., and Qi, F. (2008). Role of sucrose in the fitness of *Streptococcus mutans*. Oral Microbiol. Immunol. 23, 213–219.

Lal, P., Sharma, D., Pruthi, P., and Pruthi, V. (2010). Exopolysaccharide analysis of biofilm-forming *Candida albicans*. J. Appl. Microbiol. *109*, 128–136.

Lemos, J.A., and Burne, R.A. (2008). A model of efficiency: stress tolerance by *Streptococcus mutans*. Microbiology *154*, 3247–3255.

Lewis, K. (2001). Riddle of biofilm resistance. Antimicrob. Agents Chemother. *45*, 999–1007.

Li, Y., and Burne, R.A. (2001). Regulation of the *gtfBC* and *ftf* genes of *Streptococcus mutans* in biofilms in response to pH and carbohydrate. Microbiology *147*, 2841–2848.

Loesche, W.J. (1986). Role of *Streptococcus mutans* in human dental decay. Microbiol. Rev. *50*, 353–380.

Ma, L., Conover, M., Lu, H., Parsek, M.R., Bayles, K., and Wozniak, D.J. (2009). Assembly and development of the *Pseudomonas aeruginosa* biofilm matrix. PLoS Pathog. *5*, e1000354.

McCabe, R.M., and Donkersloot, J.A. (1977). Adherence of *Veillonella* species mediated by extracellular glucosyltransferase from *Streptococcus salivarius*. Infect. Immun. *18*, 726–734.

McDougall, W.A. (1963). Studies on the dental plaque. I. The histology of the dental plaque and its attachment. Aust. Dent. J. *8*, 261–273.

Mann, E.E., Rice, K.C., Boles, B.R., Endres, J.L., Ranjit, D., Chandramohan, L., Tsang, L.H., Smeltzer, M.S., Horswill, A.R., and Bayles, K.W. (2009). Modulation of eDNA release and degradation affects *Staphylococcus aureus* biofilm maturation. PLoS One *4*, e5822.

Marsh, P.D. (2003). Are dental diseases examples of ecological catastrophes? Microbiology *149*, 279–294.

Marshall, V.M., Laws, A.P., Gu, Y., Levander, F., Radstrom, P., De Vuyst, L., Degeest, B., Vaningelgem, F., Dunn, H., and Elvin, M. (2001). Exopolysaccharide-producing strains of thermophilic lactic acid bacteria cluster into groups according to their EPS structure. Lett. Appl. Microbiol. *32*, 433–437.

Mattos-Graner, R.O., Zelante, F., Line, R.C., and Mayer, M.P. (1998). Association between caries prevalence and clinical, microbiological and dietary variables in 1.0 to 2.5-year-old Brazilian children. Caries Res. *32*, 319–323.

Mattos-Graner, R.O., Smith, D.J., King, W.F., and Mayer, M.P. (2000). Water-insoluble glucan synthesis by mutans streptococcal strains correlates with caries incidence in 12- to 30-month-old children. J. Dent. Res. *79*, 1371–1377.

Mayhall, C.W., and Butler, W.T. (1976). The carbohydrate composition of experimental salivary pellicles. J. Oral Pathol. *5*, 358–370.

Maze, A., Boel, G., Zuniga, M., Bourand, A., Loux, V., Yebra, M.J., Monedero, V., Correia, K., Jacques, N., Beaufils, S., *et al.* (2010). Complete genome sequence of the probiotic *Lactobacillus casei* strain BL23. J. Bacteriol. *192*, 2647–2648.

Mozzi, F., Savoy de Giori, G., and Font de Valdez, G. (2003). UDP-galactose 4-epimerase: a key enzyme in exopolysaccharide formation by *Lactobacillus casei* CRL 87 in controlled pH batch cultures. J. Appl. Microbiol. *94*, 175–183.

Munro, C., Michalek, S.M., and Macrina, F.L. (1991). Cariogenicity of *Streptococcus mutans* V403 glucosyltransferase and fructosyltransferase mutants constructed by allelic exchange. Infect. Immun. *59*, 2316–2323.

Munro, C.L., Michalek, S.M., and Macrina, F.L. (1995). Sucrose-derived exopolymers have site-dependent roles in *Streptococcus mutans*-promoted dental decay. FEMS Microbiol. Lett. *128*, 327–332.

Nett, J.E., Sanchez, H., Cain, M.T., and Andes, D.R. (2010). Genetic basis of *Candida* biofilm resistance due to drug-sequestering matrix glucan. J. Infect. Dis. *202*, 171–175.

Nobbs, A.H., Lamont, R.J., and Jenkinson, H.F. (2009). *Streptococcus* adherence and colonization. Microbiol. Mol. Biol. Rev. *73*, 407–450.

Nobre dos Santos, M., Melo dos Santos, L., Francisco, S.B., and Cury, J.A. (2002). Relationship among dental plaque composition, daily sugar exposure and caries in the primary dentition. Caries Res. *36*, 347–352.

Ozimek, L.K., Kralj, S., van der Maarel, M.J., and Dijkhuizen, L. (2006). The levansucrase and inulosucrase enzymes of *Lactobacillus reuteri* 121 catalyse processive and non-processive transglycosylation reactions. Microbiology *152*, 1187–1196.

Paes Leme, A.F., Koo, H., Bellato, C.M., Bedi, G., and Cury, J.A. (2006). The role of sucrose in cariogenic dental biofilm formation – new insight. J. Dent. Res. *85*, 878–887.

Paes Leme, A.F., Bellato, C.M., Bedi, G., Cury, A.A., Koo, H., and Cury, J.A. (2008). Effects of sucrose on the extracellular matrix of plaque-like biofilm formed in vivo, studied by proteomic analysis. Caries Res. *42*, 435–443.

Pessan, J.P., Silva, S.M., Lauris, J.R., Sampaio, F.C., Whitford, G.M., and Buzalaf, M.A. (2008). Fluoride uptake by plaque from water and from dentifrice. J. Dent. Res. *87*, 461–465.

Pitty, L.J., and Jacques, N.A. (1989). Dissimilar effects of Na+ and K+ on the promotion of glucosyltransferase secretion in *Streptococcus salivarius*. J. Gen. Microbiol. *135*, 1431–1439.

Quivey, R.G., Jr., Kuhnert, W.L., and Hahn, K. (2000). Adaptation of oral streptococci to low pH. Adv. Microb. Physiol. *42*, 239–274.

Rani, S.A., Pitts, B., Beyenal, H., Veluchamy, R.A., Lewandowski, Z., Davison, W.M., Buckingham-Meyer, K., and Stewart, P.S. (2007). Spatial patterns of DNA replication, protein synthesis, and oxygen concentration within bacterial biofilms reveal diverse physiological states. J. Bacteriol. *189*, 4223–4233.

Reese, S., and Guggenheim, B. (2007). A novel TEM contrasting technique for extracellular polysaccharides in in vitro biofilms. Microsc. Res. Tech. *70*, 816–822.

Ribeiro, C.C., Tabchoury, C.P., Del Bel Cury, A.A., Tenuta, L.M., Rosalen, P.L., and Cury, J.A. (2005). Effect of starch on the cariogenic potential of sucrose. Br. J. Nutr. *94*, 44–50.

Rolla, G., Oppermann, R.V., Bowen, W.H., Ciardi, J.E., and Knox, K.W. (1980). High amounts of lipoteichoic acid in sucrose-induced plaque *in vivo*. Caries Res. *14*, 235–238.

Rolla, G., Ciardi, J.E., and Schultz, S.A. (1983). Adsorption of glucosyltransferase to saliva coated hydroxyapatite. Possible mechanism for sucrose dependent bacterial colonization of teeth. Scand. J. Dent. Res. *91*, 112–117.

Rozen, R., Steinberg, D., and Bachrach, G. (2004). *Streptococcus mutans* fructosyltransferase interactions with glucans. FEMS Microbiol. Lett. *232*, 39–43.

Russell, R.R., Gilpin, M.L., Hanada, N., Yamashita, Y., Shibata, Y., and Takehara, T. (1990). Characterization of the product of the *gtfS* gene of *Streptococcus downei*, a primer-independent enzyme synthesizing oligo-isomaltosaccharides. J. Gen. Microbiol. *136*, 1631–1637.

Saxton, C.A., and Kolendo, A.B. (1967). Light and electron microscopical studies of the degradation of salivary glycoproteins and their contribution in plaque formation. Arch. Oral Biol. *12*, 1541–1559.

Schilling, K.M., and Bowen, W.H. (1988). The activity of glucosyltransferase adsorbed onto saliva-coated hydroxyapatite. J. Dent. Res. *67*, 2–8.

Schilling, K.M., and Bowen, W.H. (1992). Glucans synthesized in situ in experimental salivary pellicle function as specific binding sites for *Streptococcus mutans*. Infect. Immun. *60*, 284–295.

Schooling, S.R., Hubley, A., and Beveridge, T.J. (2009). Interactions of DNA with biofilm-derived membrane vesicles. J. Bacteriol. *191*, 4097–4102.

Shemesh, M., and Steinberg, D. (2006). *In vitro* binding interactions of oral bacteria with immobilized fructosyltransferase. J. Appl. Microbiol. *100*, 871–877.

Shimamura, A., Nakano, Y.J., Mukasa, H., and Kuramitsu, H.K. (1994). Identification of amino acid residues in *Streptococcus mutans* glucosyltransferases influencing the structure of the glucan product. J. Bacteriol. *176*, 4845–4850.

Shiroza, T., and Kuramitsu, H.K. (1988). Sequence analysis of the *Streptococcus mutans* fructosyltransferase gene and flanking regions. J. Bacteriol. *170*, 810–816.

Shiroza, T., Ueda, S., and Kuramitsu, H.K. (1987). Sequence analysis of the *gtfB* gene from *Streptococcus mutans*. J. Bacteriol. *169*, 4263–4270.

Simms, P.J., Boyko, W.J., and Edwards, J.R. (1990). The structural analysis of a levan produced by *Streptococcus salivarius* SS2. Carbohydr. Res. *208*, 193–198.

Simpson, C.L., Cheetham, N.W., Giffard, P.M., and Jacques, N.A. (1995). Four glucosyltransferases, GtfJ, GtfK, GtfL and GtfM, from *Streptococcus salivarius* ATCC 25975. Microbiology *141*, 1451–1460.

Song, D.D., and Jacques, N.A. (1999a). Mutation of aspartic acid residues in the fructosyltransferase of *Streptococcus salivarius* ATCC 25975. Biochem. J. *344*, 259–264.

Song, D.D., and Jacques, N.A. (1999b). Purification and enzymic properties of the fructosyltransferase of *Streptococcus salivarius* ATCC 25975. Biochem. J. *341*, 285–291.

Steichen, C.T., Shao, J.Q., Ketterer, M.R., and Apicella, M.A. (2008). Gonococcal cervicitis: a role for biofilm in pathogenesis. J. Infect. Dis. *198*, 1856–1861.

Stephen, K.W., and Jenkins, G.N. (1971). Studies on some factors affecting plaque pH. Caries Res. *5*, 28.

Stewart, P.S., and Franklin, M.J. (2008). Physiological heterogeneity in biofilms. Nat. Rev. Microbiol. *6*, 199–210.

Tamesada, M., Kawabata, S., Fujiwara, T., and Hamada, S. (2004). Synergistic effects of streptococcal glucosyltransferases on adhesive biofilm formation. J. Dent. Res. *83*, 874–879.

Tanzer, J.M., Freedman, M.L., Fitzgerald, R.J., and Larson, R.H. (1974). Diminished virulence of glucan synthesis-defective mutants of *Streptococcus mutans*. Infect. Immun. *10*, 197–203.

Tanzer, J.M., Freedman, M.L., and Fitzgerald, R.J. (1985). Virulence of mutants defective in glucosyltransferase, dextran-mediated aggregation, or dextranase activity. In Molecular Basis of Oral Microbial Adhesion, Mergenhagen, S.E., and Rosan, B., eds. (ASM Press, Washington, DC) pp. 204–211.

Thomas, D.P., Pitarch, A., Monteoliva, L., Gil, C., and Lopez-Ribot, J.L. (2006). Proteomics to study *Candida albicans* biology and pathogenicity. Infect. Disord. Drug Targets *6*, 335–341.

Tsumuraya, Y., and Misaki, A. (1979). Structure of the water-insoluble alpha-D-glucan of *Streptococcus salivarius* HHT. Carbohydr. Res. *74*, 217–225.

Ueda, S., Shiroza, T., and Kuramitsu, H.K. (1988). Sequence analysis of the *gtfC* gene from *Streptococcus mutans* GS-5. Gene *69*, 101–109.

Vacca-Smith, A.M., and Bowen, W.H. (1998). Binding properties of streptococcal glucosyltransferases for hydroxyapatite, saliva-coated hydroxyapatite, and bacterial surfaces. Arch. Oral Biol. *43*, 103–110.

Vacca Smith, A.M., and Bowen, W.H. (2000). *In situ* studies of pellicle formation on hydroxyapatite discs. Arch. Oral Biol. *45*, 277–291.

Vacca-Smith, A.M., Venkitaraman, A.R., Quivey, R.G., Jr., and Bowen, W.H. (1996a). Interactions of streptococcal glucosyltransferases with alpha-amylase and starch on the surface of saliva-coated hydroxyapatite. Arch. Oral Biol. *41*, 291–298.

Vacca-Smith, A.M., Venkitaraman, A.R., Schilling, K.M., and Bowen, W.H. (1996b). Characterization of glucosyltransferase of human saliva adsorbed onto hydroxyapatite surfaces. Caries Res. *30*, 354–360.

Vacca Smith, A.M., Scott-Anne, K.M., Whelehan, M.T., Berkowitz, R.J., Feng, C., and Bowen, W.H. (2007). Salivary glucosyltransferase B as a possible marker for caries activity. Caries Res. *41*, 445–450.

Valm, A.M., Welch, J.L., Rieken, C.W., Hasegawa, Y., Sogin, M.L., Oldenbourg, R., Dewhirst, F.E., and Borisy, G.G. (2011). Systems-level analysis of microbial community organization through combinatorial labeling and spectral imaging. Proc. Natl. Acad. Sci. U.S.A. *108*, 4152–4157.

Vanhaverbeke, C., Bosso, C., Colin-Morel, P., Gey, C., Gamar-Nourani, L., Blondeau, K., Simonet, J.M., and

Heyraud, A. (1998). Structure of an extracellular polysaccharide produced by *Lactobacillus rhamnosus* strain C83. Carbohydr. Res. *314*, 211–220.

Venkitaraman, A.R., Vacca-Smith, A.M., Kopec, L.K., and Bowen, W.H. (1995). Characterization of glucosyltransferaseB, GtfC, and GtfD in solution and on the surface of hydroxyapatite. J. Dent. Res. *74*, 1695–1701.

Vickerman, M.M., Sulavik, M.C., Minick, P.E., and Clewell, D.B. (1996). Changes in the carboxyl-terminal repeat region affect extracellular activity and glucan products of *Streptococcus gordonii* glucosyltransferase. Infect. Immun. *64*, 5117–5128.

Vickerman, M.M., Sulavik, M.C., Nowak, J.D., Gardner, N.M., Jones, G.W., and Clewell, D.B. (1997). Nucleotide sequence analysis of the *Streptococcus gordonii* glucosyltransferase gene, *gtfG*. DNA Seq. *7*, 83–95.

Vickerman, M.M., Iobst, S., Jesionowski, A.M., and Gill, S.R. (2007). Genome-wide transcriptional changes in *Streptococcus gordonii* in response to competence signalling peptide. J. Bacteriol. *189*, 7799–7807.

Walker, G.J., Pulkownik, A., and Morrey-Jones, J.G. (1981). Metabolism of the polysaccharides of human dental plaque: release of dextranase in batch cultures of *Streptococcus mutans*. J. Gen. Microbiol. *127*, 201–208.

Werner, E., Roe, F., Bugnicourt, A., Franklin, M.J., Heydorn, A., Molin, S., Pitts, B., and Stewart, P.S. (2004). Stratified growth in *Pseudomonas aeruginosa* biofilms. Appl. Environ. Microbiol. *70*, 6188–6196.

Wilson, R.F., and Ashley, F.P. (1990). Relationships between the biochemical composition of both free smooth surface and approximal plaque and salivary composition and a 24-hour retrospective dietary history of sugar intake in adolescents. Caries Res. *24*, 203–210.

Wood, J.M. (1969). The state of hexose sugar in human dental plaque and its metabolism by the plaque bacteria. Arch. Oral Biol. *14*, 161–168.

Wunder, D., and Bowen, W.H. (1999). Action of agents on glucosyltransferases from *Streptococcus mutans* in solution and adsorbed to experimental pellicle. Arch. Oral Biol. *44*, 203–214.

Xiao, J., and Koo, H. (2010). Structural organization and dynamics of exopolysaccharide matrix and microcolonies formation by *Streptococcus mutans* in biofilms. J. Appl. Microbiol. *108*, 2103–2113.

Xiao, J., Klein, M.I., Falsetta, M.L., Lu, B., Delahunty, C.M., Yates III, J.R., Heydorn, A., and Koo, H. (2012). The exopolysaccharide matrix modulates the interaction between 3D architecture and virulence of a mixed-species oral biofilm. PLoS Pathog. *8*, e1002623.

Xu, P., Alves, J.M., Kitten, T., Brown, A., Chen, Z., Ozaki, L.S., Manque, P., Ge, X., Serrano, M.G., Puiu, D., *et al.* (2007). Genome of the opportunistic pathogen *Streptococcus sanguinis*. J. Bacteriol. *189*, 3166–3175.

Yamanaka, T., Furukawa, T., Matsumoto-Mashimo, C., Yamane, K., Sugimori, C., Nambu, T., Mori, N., Nishikawa, H., Walker, C.B., Leung, K.P., *et al.* (2009). Gene expression profile and pathogenicity of biofilm-forming *Prevotella intermedia* strain 17. BMC Microbiol. *9*, 11.

Yamane, K., Yamanaka, T., Yamamoto, N., Furukawa, T., Fukushima, H., Walker, C.B., and Leung, K.P. (2005). A novel exopolysaccharide from a clinical isolate of *Prevotella nigrescens*: purification, chemical characterization and possible role in modifying human leukocyte phagocytosis. Oral Microbiol. Immunol. *20*, 1–9.

Yamashita, Y., Hanada, N., and Takehara, T. (1988). A novel glucosyltransferase from *Streptococcus mutans* produces oligo-isomaltosaccharides. Biochem. Biophys. Res. Commun. *150*, 687–693.

Yamashita, Y., Bowen, W.H., Burne, R.A., and Kuramitsu, H.K. (1993). Role of the *Streptococcus mutans* gtf genes in caries induction in the specific-pathogen-free rat model. Infect. Immun. *61*, 3811–3817.

Zero, D.T. (2004). Sugars – the arch criminal? Caries Res. *38*, 277–285.

# Extracellular Proteins and DNA in the Matrix of Oral Biofilms

# 6

Nicholas S. Jakubovics

## Abstract

A core component of microbial biofilms is the extracellular matrix, which binds together the assembled micro-organisms and regulates the passage of small molecules to and from cells. The scaffolding of the matrix is composed of biological macromolecules including carbohydrates, nucleic acids and proteins. The production and function of extracellular polysaccharides in oral biofilms has been the subject of a great deal of research over many decades, and is considered in a separate chapter. More recently, it has become clear that proteins and extracellular DNA (eDNA) play key roles in maintaining the structure of many different biofilms, including oral biofilms. This chapter reviews the recent research on proteins and eDNA in the biofilm matrix, and discusses the evidence that oral micro-organisms utilize these macromolecules for specific functions within mixed-species biofilms.

## Introduction

Dental plaque has an important place in the microbiological history books, as one of the first surface-associated microbial communities to have been studied in detail. Indeed, the first observations of microorganisms in the mouth were recorded over 400 years ago by Anton van Leeuwenhoek, who described his own dental plaque as 'a little white matter, about as thick as batter' (Bardell, 1983). The introduction of the transmission electron microscope (TEM) enabled the visualization of dental plaque without dislodging the constituent microorganisms from the surface. The images produced by Listgarten and others through the 1960s and 1970s showed dental plaque as a collection of many different types of bacteria organized into a structured community (Fig. 6.1A) (Listgarten *et al.*, 1975; Listgarten, 1976). The dehydration steps required for TEM imaging caused the biofilm to collapse somewhat and to appear as a densely packed assemblage of cells. Nevertheless, intercellular material was clearly visible in the biofilm, and in some cases had a 'fibrillar' appearance (Fig. 6.1B). This was an early demonstration of a key component of biofilms: the extracellular matrix. Subsequent imaging with the confocal scanning laser microscope has revealed that, like many other biofilm systems, dental plaque has a relatively open architecture with channels and voids running between the cells (Wood *et al.*, 2000). The extracellular matrix plays a key role in maintaining this structure and controlling the flow of molecules through the biofilm. The matrix also promotes adhesion of microbial cells to surfaces and to other cells, and acts as a barrier against external insults.

Most of the work on biofilm matrices in the latter part of the 20th century focussed on the carbohydrate component of the structure. The term 'glycocalyx' was appropriated from eukaryotic biology to describe the biofilm matrix, and reflected the belief that exopolysaccharides are primarily responsible for the functions of biofilm matrices. In supragingival dental plaque, the extracellular glucans and fructans produced by oral streptococci have been extensively analysed and shown to contribute to the bulky structure of biofilms formed in the presence of sucrose. Poly-*N*-acetyl-D-glucosamine is important for the attachment of *Aggregatibacter* (formerly

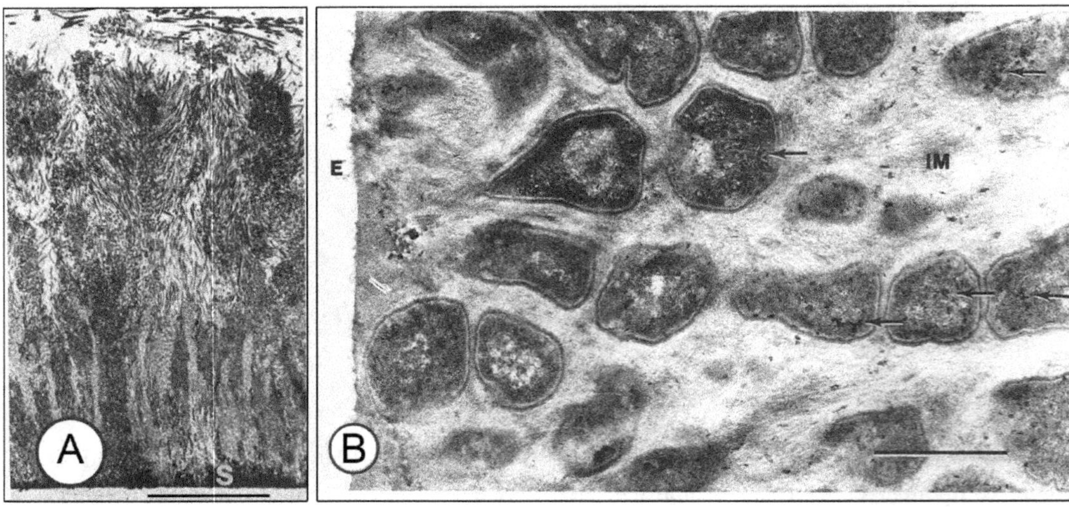


**Figure 6.1** Transmission electron microscopy images of dental plaque showing the biofilm architecture and matrix. (A) Structure of 1-week subgingival dental plaque formed on an epoxy resin crown substratum ('S'). The biofilm shows a clear organization. Despite the fixing and dehydration processes required for microscopy, channels between cells are evident. Scale bar = 20 µm. (B) High-resolution structure of subgingival dental plaque on a tooth extracted from an individual with gingivitis. Bacterial cells with Gram-positive type walls are evident. Some of these contain intracellular foci of mineralization (arrows). Between the cells is an intercellular matrix (IM) containing material with a fibrillar appearance. The enamel surface is marked (E). Scale bar = 1 µm. Images reproduced from Listgarten *et al.* (1975) and Listgarten *et al.* (1976) with permission from the American Academy of Periodontology.

*Actinobacillus*) *actinomycetemcomitans* to surfaces (Izano *et al.*, 2008), and may play a role in the structure of mixed-species subgingival biofilms. These carbohydrates are discussed in detail in Chapter 5.

Carbohydrates are also well known to contribute to the structure of biofilms formed by the extensively studied biofilm model organism *Pseudomonas aeruginosa*. However, a surprise came about in 2002 when it was found that a carbazole assay employed to measure acidic sugars in *P. aeruginosa* biofilms reacted primarily with extracellular DNA (eDNA) (Whitchurch *et al.*, 2002). Perhaps more importantly, the addition of DNase I, an endonuclease that digests single- or double-stranded DNA, inhibited the growth of *P. aeruginosa* biofilms. These data indicated that eDNA is not merely a passive bystander in the biofilm process, but is critical for developing stable biofilms. Of course, DNA is simply a chain of modified carbohydrates linked by phosphodiester bonds rather than by glycosidic linkages, and therefore it is perhaps not surprising that eDNA can perform similar functions to polysaccharides.

Are other biological macromolecules also involved in the biofilm matrix? To date, most of the work addressing this question has focused on proteins that are actively secreted and released from bacterial cells. Extracellular proteins are involved in adhesion of cells to surfaces, nutrient scavenging and in the catalysis of chemical reactions outside the cells. In addition, recent evidence indicates that protein fibrils can be found in some biofilms, and that these support the biofilm structure. This chapter reviews recent advances in understanding the extracellular protein and DNA components of biofilms, and explores the roles of these molecules in maintaining the structure of oral biofilms. An overview of the key biofilm matrix components discussed is presented in Fig. 6.2.

## Extracellular DNA

When the pioneering work of Avery, MacLeod and McCarty in the 1940s revealed DNA to be the coding material (Avery *et al.*, 1979), attention became firmly fixed on the structure and function of DNA inside the cell. However, DNA is a

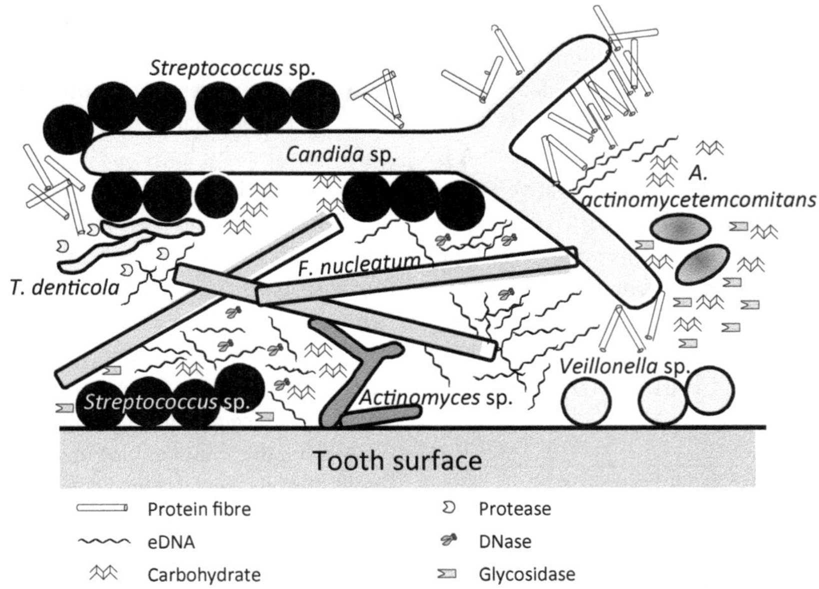


**Figure 6.2** Model to illustrate the nature of the biofilm matrix in subgingival dental plaque. Micro-organisms in dental plaque have been shown to release proteins, DNA and carbohydrates into the extracellular environment, and these macromolecules form the scaffold of the matrix. It is predicted that matrix components are unevenly distributed. For example, amyloid fibrils are shown concentrated around the cells that produce them (*Candida* sp.). Many species release DNA and extracellular DNA (eDNA) is shown throughout the biofilm. Polysaccharides are also produced by different bacteria, for example poly-*N*-acetyl-D-glucosamine of *A. actinomycetemcomitans* and capsular polysaccharides, glucans and fructans of *Streptococcus* spp. The matrix contains degradative enzymes including proteases, DNases and glycosidases. It is likely that these enzymes continuously remodel the matrix.

relatively stable molecule and does not disappear when cells die. Thus, it was soon found that DNA exists in relatively large quantities outside cell membranes (Catlin, 1956). More recently, it has been shown that extracellular DNA (eDNA) is a major source of nutrients for living organisms. For example, in the top layers of the world's ocean sediments, eDNA provides micro-organisms with approximately 50% of their total organic phosphorous (Dell'Anno and Corinaldesi, 2004). In fact, Avery, MacLeod and McCarty's early experiments had already hinted at the importance of DNA in the extracellular milieu. By demonstrating that streptococci could uptake and incorporate DNA, these workers indirectly showed that bacterial cells possess the machinery for actively transporting DNA across the cell membrane. Bacterial traits that have no physiological importance are quickly lost through natural selection, and therefore these studies indicated that streptococci almost certainly encounter eDNA within their natural environments. There is now increasing evidence to suggest that eDNA is not only important for nutrition and natural transformation, but also plays a major structural role as a core component of the biofilm matrix.

## Release of DNA from cells

When cells die and lyse, DNA is released naturally into the surrounding environment. Under some circumstances, the lysis of cells within a microbial population is enhanced by fratricidal mechanisms, which involve the release of molecules by a portion of the cells that bring about the death of neighbouring bacteria. For example, within *S. mutans* populations, cell–cell signalling results in the production of a bacteriocin that acts intracellularly to kill the cells that make it (Perry *et al.*, 2009). The signalling pathways that govern bacteriocin production and cell lysis are complex and appear to be linked to the development of competence within the remaining cells (these mechanisms are described in more detail in Chapter 8). It is not yet clear how the population as a whole benefits from microbial fratricide, but one possibility is that the community is aided by the release of eDNA into the biofilm matrix. DNA is also released actively from Gram-negative bacteria within vesicles that bud from the cell surface or, in the case of *Neisseria gonorrhoea*, by a type IV secretion system, again suggesting that the release of DNA from cells is an important process within microbial communities (Yaron *et al.*, 2000; Alvarez-Martinez and Christie, 2009).

## Functions of eDNA in biofilms

### Gene transfer

Streptococci are well known to enter a competent state where they can take up free DNA from the environment and incorporate it into their own genetic material. Recently, *Porphyromonas gingivalis* has also been shown to undergo natural transformation, and this appears to be the major mechanism for gene exchange in this organism (Tribble *et al.*, 2012). The trapping of eDNA within the biofilm matrix provides a local source of genetic material as a substrate for natural transformation. Thus, it is unsurprising that the expression of competence genes is enhanced within the biofilm environment (Eaton and Jacques, 2010). It has been shown experimentally that the conjugative transposon Tn*916* can be transferred from *Veillonella dispar* eDNA to *Streptococcus* spp. in a biofilm model (Hannan *et al.*, 2010).

### Nutrient uptake

Extracellular DNA provides a source of nutrients, including carbon and phosphorous, for the cells that can use it. Many bacteria produce extracellular DNase enzymes (see below) that digest eDNA into nucleotides, which can be more easily taken up by the cells. In the case of *P. aeruginosa*, it has been shown that extracellular nuclease production enables cells to grow in the presence of eDNA under phosphate-limiting conditions (Mulcahy *et al.*, 2010). The role of eDNA in the acquisition of trace metals by bacteria is less clear. Extracellular DNA is a relatively strong metal ion chelator and, at very high concentrations, eDNA can lyse bacterial cells by stripping them of metals (Mulcahy *et al.*, 2008). However, eDNA could also serve as a metal ion store: the degradation of eDNA potentially could release metal ions for utilization by bacteria. In keeping with this, $Mg^{2+}$ limitation promotes biofilm formation by *Pseudomonas aeruginosa*, indicating that the biofilm lifestyle can compensate for low metal ion availability (Mulcahy and Lewenza, 2011). Saliva is relatively low in trace metal ions, and it has been shown, for example, that *Streptococcus gordonii* expresses manganese-scavenging systems in the presence of saliva (Jakubovics *et al.*, 2000). Therefore, the ability of biofilms to modulate local metal ion concentrations is likely to be of great importance for bacterial growth in the oral cavity.

### Interactions with the host

Microbial eDNA is sensed by the host at the cell surface. Epithelial cells recognize unmethylated CpG sequences through Toll-like receptor 9, resulting in a signalling cascade that ultimately releases pro-inflammatory cytokines such as IL-8 (Fuxman Bass *et al.*, 2010). Oral bacteria can also trigger these signalling pathways, although their relevance for oral inflammatory diseases such as periodontitis is currently unclear (Kim *et al.*, 2012). Host neutrophils also produce eDNA, in the form of neutrophil extracellular traps (NETs), apparently to capture bacteria and prevent their spread (Brinkmann *et al.*, 2004). Some oral bacteria, including strains of *Prevotella* spp., *Streptococcus* spp., *Aggregatibacter actinomycetemcomitans* and *Fusobacterium nucleatum*, secrete DNase enzymes (Palmer *et al.*, 2012). In the case of *Prevotella* spp., it has been shown that these enzymes can degrade NETs. In periodontitis, neutrophils form an important protective barrier between periodontal bacteria and the junctional epithelium. Therefore, any bacterial mechanisms that help micro-organisms to evade neutrophil-mediated killing may have an important bearing on periodontal disease pathogenesis. Methods for experimentally inducing NET formation have been neatly presented in a video article (Brinkmann *et al.*, 2010), and it is anticipated that such approaches will rapidly advance our understanding of the interactions between host eDNA and bacteria in subgingival biofilms.

## Biofilm structure

The accumulation of microbial biofilms occurs in several stages, beginning with the initial adhesion of free-living micro-organisms to a surface. Treating *S. mutans* cells with DNase I to remove naturally occurring surface-bound eDNA reduced the rate of bacterial accumulation on hydrophilic or hydrophobic surfaces, indicating that eDNA may act as an adhesin (Das *et al.*, 2010). However, in the oral cavity, tissues are coated with salivary pellicle components and bacterial binding often involves the recognition of specific receptors in the pellicle by adhesin proteins or glycoproteins on the bacterial cell surface (Nobbs *et al.*, 2011). It has been shown that eDNA can retain proteins on the surface of bacterial cells (Ohta *et al.*, 1991), and it is therefore possible that eDNA acts primarily through stabilizing surface proteins rather than by interacting with the substratum directly. In addition to cell-surface adhesion, eDNA has been shown to play a role in bacterial aggregation. Thus, early serological techniques for the identification of *Neisseria* spp. employed nucleases to disrupt clumps of cells (Arko *et al.*, 1979). Aggregation may contribute to the accumulation of cells on surfaces either through promoting the formation of stable microcolonies, or by enabling the attachment of planktonic bacteria to surface-bound cells. Overall, therefore, there is strong evidence that eDNA is important in the attachment of bacteria to surfaces.

There have been numerous investigations into the role of eDNA in the development of biofilm communities on surfaces (Montanaro *et al.*, 2011). Many of these studies have focussed on monospecies systems and have assessed DNase sensitivity as a measure of the importance of eDNA for biofilm structure. For example, Whitchurch *et al.* (2002) demonstrated that the formation of *P. aeruginosa* biofilms could be inhibited by including DNase I in the biofilm growth medium. In addition, DNase I disrupted early biofilms, formed within 60 hours in flow-chambers. Once the biofilms had become well-established, after 84 hours in the system, DNase I had little effect on the biofilm structure, suggesting that other factors had become more important for biofilm stability. Using similar approaches, eDNA was found to be important for biofilm formation by *Candida albicans* (Martins *et al.*, 2010). *Candida* spp. are estimated to be present as oral commensals in over 50% of adults (Ghannoum *et al.*, 2010), and are associated with a variety of superficial and invasive infections. Significant quantities of eDNA could be extracted from *C. albicans* biofilms (Martins *et al.*, 2010). Treatment of biofilms with DNase I resulted in significant decreases in the biofilm biomass. Further, the addition of exogenous eDNA to biofilms increased the total biofilm present by up to 63% (Martins *et al.*, 2010).

Additional evidence that eDNA may contribute to the structural integrity of oral biofilms has come from recent work investigating links between cell lysis and biofilm formation in oral streptococci. Thus, the disruption of the *atlS* gene, encoding the major *S. gordonii* autolysin, resulted in reduced eDNA accumulation and a reduced capacity for biofilm formation (Liu and Burne, 2011). The exogenous addition of eDNA to the *atlS* mutant partially restored biofilm formation, indicating that eDNA contributes somewhat to the structure of biofilms in this organism. In *S. mutans*, treatment of wild-type biofilms with DNase I reduced the biomass by > 20%, indicating that eDNA contributes to biofilm stability. Disruption of the *cipI* gene encoding the bacteriocin immunity protein CipI resulted in increased cell lysis and the formation of thicker biofilms than the isogenic wild-type strain. However, treatment of *cipI* mutant biofilms with DNase I reduced the biomass to wild-type levels, indicating that eDNA was responsible for increased biomass in the mutant. Taken together, these data demonstrate that oral streptococci utilize eDNA for the stabilization of biofilm matrices, at least when grown in laboratory monocultures.

Oral biofilms are complex microbial networks consisting of many different species of bacteria, along with fungi and viruses. There is now clear evidence that eDNA is an important component of the biofilm matrix in many different single-culture systems. It remains to be determined how eDNA influences the structure of mixed-species oral biofilms, where different micro-organisms produce eDNA at different rates, and where some species produce extracellular DNase enzymes to digest the eDNA in the local environment (Fig. 6.2).

### Extracellular proteins

A proportion of the proteins produced by bacteria are secreted across the cell envelope and act outside the cells, either tethered to the outer cell surface or released into the external environment. Cell surface-associated proteins are critical for the initial phases of oral biofilm development since they mediate the adhesion of microbial cells to salivary pellicle-coated surfaces and to other bacteria (Nobbs *et al.*, 2011). Proteins that are released from the outer layers of cells enter the surrounding milieu, where they can interact with the biofilm substratum, microbial cells in the biofilm or with the host. The proteomic analysis of *Streptococcus pneumoniae* culture supernatants has identified > 200 secreted proteins, many of which would not have been predicted to have signal sequences for secretion across the cell membrane on the basis of genome analysis (Choi *et al.*, 2012). There was little similarity between the ratios of proteins in secreted versus cytosolic fractions, indicating that secreted proteins were not simply released by cell lysis, but were actively enriched by some currently unknown mechanism. Clearly, it will take a great deal of work to unravel the role of each of these proteins within biofilms. The discussion below focuses on classes of proteins that have been shown to be released from oral microbial cells, and therefore potentially play key roles as components of the biofilm matrix, either in supporting the biofilm structure or in providing enzyme activity within the extracellular biofilm environment.

## Structural proteins

### Amyloids

The amyloid plaques associated with neurodegenerative disorders including Alzheimer's, Parkinson's and prion diseases are formed by the aggregation of β-sheet domains of proteins into fibres that are highly resistant to heat and proteases. These fibres are considered to be misfolded proteins and are not known to have any beneficial function. In contrast, a variety of microorganisms produce extracellular amyloid fibres that have key functions such as promoting the formation of microbial communities, modulating cell surface properties or acting as toxins to increase virulence (Blanco *et al.*, 2012). For example, curli fibrils produced by Gram-negative Enterobacteriaceae including *E. coli* and *Salmonella* Typhimurium promote adhesion to host cells and biofilm formation (Kikuchi *et al.*, 2005; Jonas *et al.*, 2007). Curli act in concert with cellulose to produce a robust biofilm matrix (Jonas *et al.*, 2007; Saldana *et al.*, 2009). In *Pseudomonas* spp., extracellular amyloid fibres formed by FapC protein mediate the aggregation of cells and biofilm formation (Dueholm *et al.*, 2010). Amyloid fibres are also produced by Gram-positive bacteria. For example, *Bacillus subtilis* produces TasA fibres that are linked to the cell wall peptidoglycan by the action of TapA protein (Driks, 2011; Ostrowski *et al.*, 2011; Romero *et al.*, 2011). The addition of purified TasA fibres to a *tasA* mutant of *B. subtilis* rescued a biofilm-deficient phenotype, indicating that TasA is an important component of the extracellular matrix (Romero *et al.*, 2010). Pili resembling curli amyloid fibres are expressed on the cell surface of *Mycobacterium tuberculosis* and mediate adhesion to laminin (Blanco *et al.*, 2012). It is not yet known whether these structures can function as part of the extracellular matrix or whether their function is dependent upon cell surface expression.

The bacterial amyloid proteins identified to date have little primary sequence homology with each other or with proteins from different bacteria, including the species commonly found in the mouth. Nevertheless, the ability to form amyloid fibres under appropriate conditions may be a widespread property of proteins that can adopt β-sheet conformations. Thus, proteins that form inclusion bodies within bacterial cells adopt amyloid-like structures (Wang *et al.*, 2008). A variety of computer algorithms have been published that aim to predict amyloid-forming capacity from primary or secondary protein structures (Belli *et al.*, 2011). Using one of these algorithms, named TANGO, it was found that the vast majority of known cell surface proteins of *C. albicans* have the potential to form amyloids (Otoo *et al.*, 2008; Ramsook *et al.*, 2010). The production of amyloid fibres by several Als family proteins and by the adhesin Eap1p was confirmed by analysis of their capacity to bind to Congo Red and thiamine T, and by the detection of fibres under the transmission electron microscope (Otoo *et al.*, 2008; Ramsook *et al.*, 2010). Further, it was shown that Als family adhesins Als1p and Als5p formed amyloid structures when expressed on the cell surface of *Saccharomyces cerevisiae*, and that these proteins promoted cell flocculation, providing clear evidence that they play functional roles *in vivo* (Ramsook *et al.*, 2010). A point mutation of Als5p that abolishes amyloid formation abrogated the formation of cell–cell aggregates and the development of biofilms on plastic surfaces (Garcia *et al.*, 2011). Overall, these data demonstrate that amyloid structures form on the outer layers of at least one common oral micro-organism. It remains to be determined whether these are a critical component of the biofilm matrix in mixed-species oral microbial communities.

### Glucan-binding proteins (GBPs)

The mutans streptococci (*S. mutans* and *S. sobrinus*) produce insoluble glucans which add bulk to the extracellular matrix, and provide a substrate for biofilm adhesion by glucan binding proteins (GBPs). *S. mutans* produces three different GBPs: GbpA (59 kDa), GbpC (64 kDa) and GbpD (76 kDa). Of these, GbpC is retained on the cell surface, whilst GbpA and GbpD are released from cells. The contribution of GBPs to biofilm architecture has been assessed by comparing the formation of biofilms by isogenic mutants knocked out in each of the GBP-encoding genes (Banas *et al.*, 2007). In the presence of sucrose, each of the mutants formed thinner biofilms than the wild-type. The most pronounced differences occurred with the *gbpC* mutant, where the gene disruption resulted in a major change in the biofilm architecture, measured by the surface area to biovolume ratio, in addition to a reduction in the total biomass. This presumably reflects the key role of GbpC as a cell surface-associated biofilm adhesin. Nevertheless, mutants lacking GbpA and GbpD also showed clear defects in biofilm formation. The *gbpA* mutant formed thin, spread out biofilms, whilst the *gbpD* mutant formed thin biofilms with significantly reduced biomass. These observations suggest that GbpA and GbpD play important roles in strengthening the biofilm by interacting with glucan in the matrix.

### Nucleoprotein

Recent evidence suggests that proteins can stabilize biofilms by binding to eDNA in the matrix to form nucleoprotein complexes. Thus, staphylococcal beta toxin, a neutral sphingomyelinase, binds to single- or double-stranded eDNA and forms stable fibres (Huseby *et al.*, 2010). Beta toxin exhibits secondary structure homology with DNase I. However, unlike DNase I, it does not digest DNA, but instead covalently binds to DNA and forms a precipitate. Mutants that were disrupted in the *hlb* gene encoding beta toxin formed significantly thinner biofilms than the wild-type, indicating that beta toxin is important for biofilm structure (Huseby *et al.*, 2010). This report represents the first description of stable nucleoprotein matrices in bacterial biofilms. However, it is very likely that these structures are common to a variety of different micro-organisms and it will be of great interest to investigate whether they are present in oral biofilms.

### Enzymes

Enzymes that are secreted from bacterial cells do not have access to bacterial cell energy pools, and are therefore limited in the reactions that they can catalyse. Many extracellular enzyme-catalysed reactions involve degradation of substrates, releasing energy from the breaking of bonds. In some cases, the energy released is coupled to synthetic reactions. For example, the glucosyltransferases and fructosyltransferases of oral streptococci utilize the energy from breaking the glycosidic linkage in sucrose to synthesize long-chain glucans and fructans. The contribution of these polysaccharides to biofilm architecture is discussed in Chapter 5 and will not be considered further here. Instead, the following section explores the role of secreted enzymes in degrading and remodelling the biofilm matrix. If we can understand how biofilms are remodelled naturally, then it is hoped that in the future it will be possible to manipulate this process to control biofilm formation and to promote oral health.

### Polysaccharide lyases

The biofilm matrix plays an important role in retaining bacteria at surfaces. However, as the microbial community grows, it becomes more difficult for the constituent micro-organisms to compete for space and nutrients. In order to find new, less crowded, surfaces for colonization bacteria require mechanisms for escaping from the biofilm and dispersing back into the planktonic phase. One such mechanism has been described in the periodontal pathogen *Aggregatibacter actinomycetemcomitans*. Strains of *A. actinomycetemcomitans* bind avidly to surfaces by synthesizing the extracellular polysaccharide poly-*N*-acetyl-D-glucosamine (PNAG) (Kaplan *et al*., 2004). Cells also produce a soluble enzyme, Dispersin B, which degrades PNAG and promotes the release of cells from the biofilm (Ramasubbu *et al*., 2005). The PNAG produced by *A. actinomycetemcomitans* is structurally and functionally related to the polysaccharide intercellular adhesin of *Staphylococcus* spp., as well as to extracellular polysaccharides produced by several Gram-negative bacteria including *E. coli* and *Yersinia* spp. (Branda *et al*., 2005). Attempts are ongoing to develop Dispersin B as a commercial biofilm control enzyme (Kaplan, 2009). However, the limited specificity of this enzyme is likely to be a significant barrier to its widespread use against complex mixed-species oral biofilms.

Microbial glycosidases also act on carbohydrates produced by the host, including receptors in the salivary pellicle, and in some cases these reactions can influence biofilm structure. For example, isogenic mutants of *Tannerella forsythia* lacking sialidase (NanH) or β-hexaminidase activity are impaired in their ability to form biofilms on saliva- or serum-coated surfaces compared with wild-type cells (Roy *et al*., 2011, 2012). Enzyme activity was shown to be present in whole cell suspensions, but it is not clear whether these enzymes also escape into the biofilm matrix where they could act remotely from the cells that produce them. However, it has been demonstrated that *Streptococcus sanguinis* produces an extracellular sialidase (Varki and Diaz, 1983). Therefore, in mixed-species biofilms it is quite possible that glycosidases in the biofilm matrix have a profound influence on the structure of biofilms.

## Proteases

Proteins are a key nutrient, particularly for subgingival bacteria, and many oral micro-organisms secrete protein-degrading enzymes. The major secreted protease of the oral spirochaete *Treponema denticola*, dentilisin, has been examined for its role in biofilm formation. A mutant defective in dentilisin production formed thicker and more stable biofilms than the wild-type progenitor strain on surfaces coated with either fibronectin or *Porphyromonas gingivalis* cells, indicating that dentilisin hinders colonization by degrading receptors for *T. denticola* binding (Vesey and Kuramitsu, 2004). Extracellular proteases also have the potential to interfere with cell–cell communication processes that are critical for biofilm formation. For example, supernatants from *S. gordonii* cell cultures contain the protease challisin, which can degrade bacteriocins produced by *S. mutans* (Wang and Kuramitsu, 2005). Moreover, challisin degrades *S. mutans* competence stimulating peptide and hinders colonization by *S. mutans* (Wang *et al*., 2011).

## Nucleases

The production of extracellular nucleases by bacteria is well-documented. In fact, the analysis of nuclease production is an important test for distinguishing between *Staphylococcus aureus* and coagulase-negative staphylococci and for differentiating *Corynebacterium diphtheriae* from other members of the genus. *Streptococcus pyogenes* strains produce up to four extracellular nucleases that contribute to virulence in animal models of disease (Sumby *et al*., 2005). The principal extracellular DNase, Sda1, mediates escape from NETs and enables dissemination from the initial site of infection (Buchanan *et al*., 2006). Invasive isolates of *S. pyogenes* up-regulate the gene encoding Sda1 and select against the production of cysteine protease, SpeB, which is capable of degrading Sda1 at the cell surface (Walker *et al*., 2007). The expression of SpeB inhibits the binding and activation of plasminogen at the *S. pyogenes* cell surface, and loss of SpeB further increases the invasive capacity of *S. pyogenes*. Therefore in *S. pyogenes*, the production of extracellular nucleases plays several functions that promote tissue invasion and virulence. As described above, extracellular NET-degrading DNases produced by oral bacteria such as *Prevotella* spp. have recently been described (Palmer *et al*., 2012). However, their contribution to disease pathogenesis is not yet clear.

Extracellular nucleases have also been shown to influence biofilm architecture. In *Staph. aureus*, for example, the extracellular DNase Nuc negatively regulates biofilm formation and disruption of the *nuc* gene leads to thicker biofilms (Kiedrowski *et al.*, 2011). Similarly, mutants of *Neisseria gonorrhoeae, Vibrio cholerae* or *Shewanella oneidensis* lacking functional extracellular nuclease-encoding genes produce relatively thick biofilms containing abundant eDNA (Godeke *et al.*, 2011; Seper *et al.*, 2011; Steichen *et al.*, 2011). However, it is important to note that all of these studies have used simple *in vitro* assays on single-species biofilms. It is likely that the situation is far more complex *in vivo*, where extracellular nucleases play a number of different roles. Indeed, in a murine catheter infection model, mutants of *Staph. aureus* disrupted in two nuclease-encoding genes, *nuc1* and *nuc2*, produced relatively weak biofilms that were sensitive to daptomycin (Beenken *et al.*, 2012). By contrast, these mutants produced thick biofilms *in vitro*. Many oral bacteria produce extracellular nucleases (Palmer *et al.*, 2012), and it will be interesting to see how these enzymes influence the structure of mixed-species biofilms *in vitro* and *in vivo*.

## Future directions

The extracellular biofilm matrix is sometimes viewed as an inert structure, composed of a scaffold consisting almost entirely of polysaccharides. However, recent evidence points towards a different picture: a dynamic matrix of protein, nucleic acids and carbohydrates that is constantly restructuring to allow the movement of microbial cells into and out of the biofilm. A major challenge will be to understand how different species in mixed biofilms remodel the biofilm matrix, and how these processes affect the subsequent colonization by other micro-organisms. Central to this will be the development of new techniques to visualize specific components of the matrix under fully hydrated conditions. It is still far from clear whether ordered multimeric proteins such as amyloids play a role in the structure of mixed-species oral biofilms. Similarly, the function of eDNA in natural oral biofilms has not yet been established. Ultimately, the key goal is to develop new methods that target the biofilm matrix, in order to prevent the accumulation of pathogenic biofilms in the oral cavity.

## References


Alvarez-Martinez, C.E., and Christie, P.J. (2009). Biological diversity of prokaryotic type IV secretion systems. Microbiol. Mol. Biol. Rev. *73*, 775–808.

Arko, R.J., Wong, K.H., and Peacock, W.L. (1979). Nuclease enhancement of specific cell agglutination in a serodiagnostic test for *Neisseria gonorrhoeae*. J. Clin. Microbiol. *9*, 517–519.

Avery, O.T., MacLeod, C.M., and McCarty, M. (1979). Studies on the chemical nature of the substance inducing transformation of pneumococcal types. Inductions of transformation by a desoxyribonucleic acid fraction isolated from pneumococcus type III. J. Exp. Med. *149*, 297–326.

Banas, J.A., Fountain, T.L., Mazurkiewicz, J.E., Sun, K., and Vickerman, M. (2007). *Streptococcus mutans* glucan-binding protein-A affects *Streptococcus gordonii* biofilm architecture. FEMS Microbiol. Lett. *267*, 80–88.

Bardell, D. (1983). The roles of the sense of taste and clean teeth in the discovery of bacteria by Antoni van Leeuwenhoek. Microbiol. Rev. *47*, 121–126.

Beenken, K.E., Spencer, H., Griffin, L.M., and Smeltzer, M.S. (2012). Impact of extracellular nuclease production on the biofilm phenotype of *Staphylococcus aureus* under *in vitro* and *in vivo* conditions. Infect. Immun. *80*, 1634–1638.

Belli, M., Ramazzotti, M., and Chiti, F. (2011). Prediction of amyloid aggregation *in vivo*. EMBO Rep. *12*, 657–663.

Blanco, L.P., Evans, M.L., Smith, D.R., Badtke, M.P., and Chapman, M.R. (2012). Diversity, biogenesis and function of microbial amyloids. Trends Microbiol. *20*, 66–73.

Branda, S.S., Vik, S., Friedman, L., and Kolter, R. (2005). Biofilms: the matrix revisited. Trends Microbiol. *13*, 20–26.

Brinkmann, V., Reichard, U., Goosmann, C., Fauler, B., Uhlemann, Y., Weiss, D.S., Weinrauch, Y., and Zychlinsky, A. (2004). Neutrophil extracellular traps kill bacteria. Science *303*, 1532–1535.

Brinkmann, V., Laube, B., Abu Abed, U., Goosmann, C., and Zychlinsky, A. (2010). Neutrophil extracellular traps: how to generate and visualize them. J. Vis. Exp. *36*, 1724.

Buchanan, J.T., Simpson, A.J., Aziz, R.K., Liu, G.Y., Kristian, S.A., Kotb, M., Feramisco, J., and Nizet, V. (2006). DNase expression allows the pathogen group A *Streptococcus* to escape killing in neutrophil extracellular traps. Curr. Biol. *16*, 396–400.

Catlin, B.W. (1956). Extracellular deoxyribonucleic acid of bacteria and a deoxyribonuclease inhibitor. Science *124*, 441–442.

Choi, C.W., Lee, Y.G., Kwon, S.O., Kim, H.Y., Lee, J.C., Chung, Y.H., Yun, C.Y., and Kim, S.I. (2012). Analysis of *Streptococcus pneumoniae* secreted antigens by immuno-proteomic approach. Diagn. Microbiol. Infect. Dis. *72,* 318–327.

Das, T., Sharma, P.K., Busscher, H.J., van der Mei, H.C., and Krom, B.P. (2010). Role of extracellular DNA in initial bacterial adhesion and surface aggregation. Appl. Environ. Microbiol. *76,* 3405–3408.

Dell'Anno, A., and Corinaldesi, C. (2004). Degradation and turnover of extracellular DNA in marine sediments: ecological and methodological considerations. Appl. Environ. Microbiol. *70,* 4384–4386.

Driks, A. (2011). Tapping into the biofilm: insights into assembly and disassembly of a novel amyloid fibre in *Bacillus subtilis*. Mol. Microbiol. *80,* 1133–1136.

Dueholm, M.S., Petersen, S.V., Sonderkaer, M., Larsen, P., Christiansen, G., Hein, K.L., Enghild, J.J., Nielsen, J.L., Nielsen, K.L., Nielsen, P.H., *et al.* (2010). Functional amyloid in *Pseudomonas*. Mol. Microbiol. *77,* 1009–1020.

Eaton, R.E., and Jacques, N.A. (2010). Deletion of competence-induced genes over-expressed in biofilms caused transformation deficiencies in *Streptococcus mutans*. Mol. Oral Microbiol. *25,* 406–417.

Fuxman Bass, J.I., Russo, D.M., Gabelloni, M.L., Geffner, J.R., Giordano, M., Catalano, M., Zorreguieta, A., and Trevani, A.S. (2010). Extracellular DNA: a major proinflammatory component of *Pseudomonas aeruginosa* biofilms. J. Immunol. *184,* 6386–6395.

Garcia, M.C., Lee, J.T., Ramsook, C.B., Alsteens, D., Dufrene, Y.F., and Lipke, P.N. (2011). A role for amyloid in cell aggregation and biofilm formation. PLoS One *6,* e17632.

Ghannoum, M.A., Jurevic, R.J., Mukherjee, P.K., Cui, F., Sikaroodi, M., Naqvi, A., and Gillevet, P.M. (2010). Characterization of the oral fungal microbiome (mycobiome) in healthy individuals. PLoS Pathog. *6,* e1000713.

Godeke, J., Heun, M., Bubendorfer, S., Paul, K., and Thormann, K.M. (2011). Roles of two *Shewanella oneidensis* MR-1 extracellular endonucleases. Appl. Environ. Microbiol. *77,* 5342–5351.

Hannan, S., Ready, D., Jasni, A.S., Rogers, M., Pratten, J., and Roberts, A.P. (2010). Transfer of antibiotic resistance by transformation with eDNA within oral biofilms. FEMS Immunol. Med. Microbiol. *59,* 345–349.

Huseby, M.J., Kruse, A.C., Digre, J., Kohler, P.L., Vocke, J.A., Mann, E.E., Bayles, K.W., Bohach, G.A., Schlievert, P.M., Ohlendorf, D.H., *et al.* (2010). Beta toxin catalyzes formation of nucleoprotein matrix in staphylococcal biofilms. Proc. Natl. Acad. Sci. U.S.A. *107,* 14407–14412.

Izano, E.A., Sadovskaya, I., Wang, H., Vinogradov, E., Ragunath, C., Ramasubbu, N., Jabbouri, S., Perry, M.B., and Kaplan, J.B. (2008). Poly-*N*-acetylglucosamine mediates biofilm formation and detergent resistance in *Aggregatibacter actinomycetemcomitans*. Microb. Pathog. *44,* 52–60.

Jakubovics, N.S., Smith, A.W., and Jenkinson, H.F. (2000). Expression of the virulence-related Sca ($Mn^{2+}$) permease in *Streptococcus gordonii* is regulated by a diphtheria toxin metallorepressor-like protein ScaR. Mol. Microbiol. *38,* 140–153.

Jonas, K., Tomenius, H., Kader, A., Normark, S., Romling, U., Belova, L.M., and Melefors, O. (2007). Roles of curli, cellulose and BapA in *Salmonella* biofilm morphology studied by atomic force microscopy. BMC Microbiol. *7,* 70.

Kaplan, J.B. (2009). Therapeutic potential of biofilm-dispersing enzymes. Int. J. Artif. Organs *32,* 545–554.

Kaplan, J.B., Velliyagounder, K., Ragunath, C., Rohde, H., Mack, D., Knobloch, J.K., and Ramasubbu, N. (2004). Genes involved in the synthesis and degradation of matrix polysaccharide in *Actinobacillus actinomycetemcomitans* and *Actinobacillus pleuropneumoniae* biofilms. J. Bacteriol. *186,* 8213–8220.

Kiedrowski, M.R., Kavanaugh, J.S., Malone, C.L., Mootz, J.M., Voyich, J.M., Smeltzer, M.S., Bayles, K.W., and Horswill, A.R. (2011). Nuclease modulates biofilm formation in community-associated methicillin-resistant *Staphylococcus aureus*. PLoS One *6,* e26714.

Kikuchi, T., Mizunoe, Y., Takade, A., Naito, S., and Yoshida, S. (2005). Curli fibers are required for development of biofilm architecture in *Escherichia coli* K-12 and enhance bacterial adherence to human uroepithelial cells. Microbiol. Immunol. *49,* 875–884.

Kim, Y., Jo, A.R., Jang, D.H., Cho, Y.J., Chun, J., Min, B.M., and Choi, Y. (2012). Toll-like receptor 9 mediates oral bacteria-induced IL-8 expression in gingival epithelial cells. Immunol. Cell Biol. *90,* 655–663.

Listgarten, M.A. (1976). Structure of the microbial flora associated with periodontal health and disease in man. A light and electron microscopic study. J. Periodontol. *47,* 1–18.

Listgarten, M.A., Mayo, H.E., and Tremblay, R. (1975). Development of dental plaque on epoxy resin crowns in man. A light and electron microscopic study. J. Periodontol. *46,* 10–26.

Liu, Y., and Burne, R.A. (2011). The major autolysin of *Streptococcus gordonii* is subject to complex regulation and modulates stress tolerance, biofilm formation, and extracellular-DNA release. J. Bacteriol. *193,* 2826–2837.

Martins, M., Uppuluri, P., Thomas, D.P., Cleary, I.A., Henriques, M., Lopez-Ribot, J.L., and Oliveira, R. (2010). Presence of extracellular DNA in the *Candida albicans* biofilm matrix and its contribution to biofilms. Mycopathologia *169,* 323–331.

Montanaro, L., Poggi, A., Visai, L., Ravaioli, S., Campoccia, D., Speziale, P., and Arciola, C.R. (2011). Extracellular DNA in biofilms. Int. J. Artif. Organs *34,* 824–831.

Mulcahy, H., and Lewenza, S. (2011). Magnesium limitation is an environmental trigger of the *Pseudomonas aeruginosa* biofilm lifestyle. PLoS One *6,* e23307.

Mulcahy, H., Charron-Mazenod, L., and Lewenza, S. (2008). Extracellular DNA chelates cations and induces antibiotic resistance in *Pseudomonas aeruginosa* biofilms. PLoS Pathog. *4,* e1000213.

Mulcahy, H., Charron-Mazenod, L., and Lewenza, S. (2010). *Pseudomonas aeruginosa* produces an extracellular deoxyribonuclease that is required for utilization of DNA as a nutrient source. Environ. Microbiol. *12,* 1621–1629.

Nobbs, A.H., Jenkinson, H.F., and Jakubovics, N.S. (2011). Stick to your gums: mechanisms of oral microbial adherence. J. Dent. Res. *90*, 1271–1278.

Ohta, H., Kato, K., Kokeguchi, S., Hara, H., Fukui, K., and Murayama, Y. (1991). Nuclease-sensitive binding of an *Actinobacillus actinomycetemcomitans* leukotoxin to the bacterial cell surface. Infect. Immun. *59*, 4599–4605.

Ostrowski, A., Mehert, A., Prescott, A., Kiley, T.B., and Stanley-Wall, N.R. (2011). YuaB functions synergistically with the exopolysaccharide and TasA amyloid fibers to allow biofilm formation by *Bacillus subtilis*. J. Bacteriol. *193*, 4821–4831.

Otoo, H.N., Lee, K.G., Qiu, W., and Lipke, P.N. (2008). *Candida albicans* Als adhesins have conserved amyloid-forming sequences. Eukaryot. Cell *7*, 776–782.

Palmer, L.J., Chapple, I.L., Wright, H.J., Roberts, A., and Cooper, P.R. (2012). Extracellular deoxyribonuclease production by periodontal bacteria. J. Periodontal. Res. *47*, 439–445.

Perry, J.A., Jones, M.B., Peterson, S.N., Cvitkovitch, D.G., and Levesque, C.M. (2009). Peptide alarmone signalling triggers an auto-active bacteriocin necessary for genetic competence. Mol. Microbiol. *72*, 905–917.

Ramasubbu, N., Thomas, L.M., Ragunath, C., and Kaplan, J.B. (2005). Structural analysis of dispersin B, a biofilm-releasing glycoside hydrolase from the periodontopathogen *Actinobacillus actinomycetemcomitans*. J. Mol. Biol. *349*, 475–486.

Ramsook, C.B., Tan, C., Garcia, M.C., Fung, R., Soybelman, G., Henry, R., Litewka, A., O'Meally, S., Otoo, H.N., Khalaf, R.A., *et al.* (2010). Yeast cell adhesion molecules have functional amyloid-forming sequences. Eukaryot. Cell *9*, 393–404.

Romero, D., Aguilar, C., Losick, R., and Kolter, R. (2010). Amyloid fibers provide structural integrity to *Bacillus subtilis* biofilms. Proc. Natl. Acad. Sci. U.S.A. *107*, 2230–2234.

Romero, D., Vlamakis, H., Losick, R., and Kolter, R. (2011). An accessory protein required for anchoring and assembly of amyloid fibres in *B. subtilis* biofilms. Mol. Microbiol. *80*, 1155–1168.

Roy, S., Honma, K., Douglas, C.W., Sharma, A., and Stafford, G.P. (2011). Role of sialidase in glycoprotein utilization by *Tannerella forsythia*. Microbiology *157*, 3195–3202.

Roy, S., Phansopa, C., Stafford, P., Honma, K., Douglas, C.W., Sharma, A., and Stafford, G.P. (2012). Beta-hexosaminidase activity of the oral pathogen *Tannerella forsythia* influences biofilm formation on glycoprotein substrates. FEMS Immunol. Med. Microbiol. *65*, 116–120.

Saldana, Z., Xicohtencatl-Cortes, J., Avelino, F., Phillips, A.D., Kaper, J.B., Puente, J.L., and Giron, J.A. (2009). Synergistic role of curli and cellulose in cell adherence and biofilm formation of attaching and effacing *Escherichia coli* and identification of Fis as a negative regulator of curli. Environ. Microbiol. *11*, 992–1006.

Seper, A., Fengler, V.H., Roier, S., Wolinski, H., Kohlwein, S.D., Bishop, A.L., Camilli, A., Reidl, J., and Schild, S. (2011). Extracellular nucleases and extracellular DNA play important roles in *Vibrio cholerae* biofilm formation. Mol. Microbiol. *82*, 1015–1037.

Steichen, C.T., Cho, C., Shao, J.Q., and Apicella, M.A. (2011). The *Neisseria gonorrhoeae* biofilm matrix contains DNA, and an endogenous nuclease controls its incorporation. Infect. Immun. *79*, 1504–1511.

Sumby, P., Barbian, K.D., Gardner, D.J., Whitney, A.R., Welty, D.M., Long, R.D., Bailey, J.R., Parnell, M.J., Hoe, N.P., Adams, G.G., *et al.* (2005). Extracellular deoxyribonuclease made by group A *Streptococcus* assists pathogenesis by enhancing evasion of the innate immune response. Proc. Natl. Acad. Sci. U.S.A. *102*, 1679–1684.

Tribble, G.D., Rigney, T.W., Dao, D.H., Wong, C.T., Kerr, J.E., Taylor, B.E., Pacha, S., and Kaplan, H.B. (2012). Natural competence is a major mechanism for horizontal DNA transfer in the oral pathogen *Porphyromonas gingivalis*. MBio 3, e00231–11.

Varki, A., and Diaz, S. (1983). A neuraminidase from *Streptococcus sanguis* that can release O-acetylated sialic acids. J. Biol. Chem. *258*, 12465–12471.

Vesey, P.M., and Kuramitsu, H.K. (2004). Genetic analysis of *Treponema denticola* ATCC 35405 biofilm formation. Microbiology *150*, 2401–2407.

Walker, M.J., Hollands, A., Sanderson-Smith, M.L., Cole, J.N., Kirk, J.K., Henningham, A., McArthur, J.D., Dinkla, K., Aziz, R.K., Kansal, R.G., *et al.* (2007). DNase Sda1 provides selection pressure for a switch to invasive group A streptococcal infection. Nat. Med. *13*, 981–985.

Wang, B.Y., and Kuramitsu, H.K. (2005). Interactions between oral bacteria: inhibition of *Streptococcus mutans* bacteriocin production by *Streptococcus gordonii*. Appl. Environ. Microbiol. *71*, 354–362.

Wang, B.Y., Deutch, A., Hong, J., and Kuramitsu, H.K. (2011). Proteases of an early colonizer can hinder *Streptococcus mutans* colonization *in vitro*. J. Dent. Res. *90*, 501–505.

Wang, L., Maji, S.K., Sawaya, M.R., Eisenberg, D., and Riek, R. (2008). Bacterial inclusion bodies contain amyloid-like structure. PLoS Biol. *6*, e195.

Whitchurch, C.B., Tolker-Nielsen, T., Ragas, P.C., and Mattick, J.S. (2002). Extracellular DNA required for bacterial biofilm formation. Science *295*, 1487.

Wood, S.R., Kirkham, J., Marsh, P.D., Shore, R.C., Nattress, B., and Robinson, C. (2000). Architecture of intact natural human plaque biofilms studied by confocal laser scanning microscopy. J. Dent. Res. *79*, 21–27.

Yaron, S., Kolling, G.L., Simon, L., and Matthews, K.R. (2000). Vesicle-mediated transfer of virulence genes from *Escherichia coli* O157:H7 to other enteric bacteria. Appl. Environ. Microbiol. *66*, 4414–4420.

# A Holistic View of Interspecies Bacterial Interactions Within Human Dental Plaque

7

Alexander H. Rickard, Adam J. Underwood and William Nance


## Abstract

Mature dental plaque biofilm communities contain hundreds of bacterial species. The potential for these communities to cause caries or periodontal disease relates to bacterial spatiotemporal biofilm development and species composition. At least three forms of interspecies interactions can conceivably mediate altered biofilm development and species composition: coaggregation, metabolic interactions and cell–cell signalling. Coaggregation is the specific recognition and adhesion of different species of bacteria and likely contributes to the ordered (sequential) integration of species into biofilms as well as improving species retention in a flowing environment. 'Metabolic interactions' is an umbrella term that describes the exchange of metabolites or environmental protection afforded between adjacent species within dental plaque. Cell–cell signalling is a phenomenon that has gained increasing research interest over the past decade. One broad interspecies signalling molecule system consists of a collection of inter-convertible cell–cell signal molecules that are collectively called autoinducer-2 (AI-2). Evidence indicates that AI-2 can alter bacterial phenotypes when present in saliva at concentrations as low as the nanomolar range. It is the aim of this chapter to describe each of these interspecies phenomena, with case-examples, and extrapolate singular and combined roles in the spatio-temporal development of dental plaque. The potential for these phenomena to create shifts in community species composition has implications for the development of polymicrobial diseases.


## Introduction

Human oral bacteria are not solitary units of life; rather they are highly interactive and arguably social entities. Reports of their proclivity to interact with other species can be found up to 40 years ago (Gibbons and Nygaard, 1970; Listgarten *et al.*, 1973). Early electron microscopy studies demonstrated complex cellular arrangements in dental plaque, and planktonic investigations revealed that isolated dental plaque bacteria can specifically recognize and adhere to other oral species (Bourgeau and McBride, 1976; Cisar *et al.*, 1979; Mouton *et al.*, 1980). Interestingly, some of the dental plaque bacteria shown to interact with one another have also been indicated to benefit through metabolic interactions (Chalmers *et al.*, 2008; Delwiche *et al.*, 1985; Lancy *et al.*, 1983). In some cases, both partners benefit mutualistically. Alternatively, only one partner benefits or interactions can lead to a direct or indirect competitive effect. Until the last decade, however, the ability of a broad range of oral bacteria to produce and detect interspecies cell–cell signal molecules had not been demonstrated (Frias *et al.*, 2001) and roles for cell–cell signalling in biofilm development not been described (Novak *et al.*, 2010; Rickard *et al.*, 2008a; Whitmore and Lamont, 2011). Thus, it is the aim of this chapter to address and discuss three distinct interspecies phenomena and their potential for altering the ecology of dental plaque biofilms: bacterial coaggregation, bacterial metabolic interactions, and bacterial cell–cell signalling.

In this chapter, information from the last four decades will be collated and a broad holistic summary of key research in interspecies interactions will be discussed. From an ecological perspective, interspecies interaction studies, such as the study of coaggregation, are essential to understanding how dental plaque develops. Furthermore, interaction studies may allow us to predict shifts in the community species composition and changes from oral health towards oral disease. Thus, particular emphasis will be placed on how seemingly discrete, interspecies interactions between oral bacteria may be intertwined to exert a combined effect in mediating changes in dental plaque biofilms.

## Dental plaque: a taxonomically diverse and functionally interactive multispecies biofilm

Most bacteria exist within surface-attached assemblages that are structurally organized and where component bacteria are functionally interactive. These assembled bacterial communities are called biofilms – a term coined in the late 1970s by Costerton *et al.* (1978). Biofilms are typically composed of multiple species and, in the case of dental plaque, develop from the sequential adhesion (often also described as successional adhesion) and integration of species. A common attribute of biofilm communities is their persistent nature and recalcitrance to physical and chemical treatment regimens; a trait shared by dental plaque. Indeed, bacteria within biofilms are typically 100–1000 times more resistant to antimicrobials than their planktonic counterparts (Mah and O'Toole, 2001). Such resilience is afforded by a number of factors unique to biofilms. These include: (i) strong cell–surface and cell–cell physicochemical interactions, promoting rapid biofilm spatio-temporal growth/re-growth (Bos *et al.*, 1999; Daep *et al.*, 2006; Hannig and Hannig, 2009); (ii) the production of biofilm-specific, extra-cellular polymeric material that act as a bacterial scaffolding that can reduce the dispersive effect of antimicrobials and can (depending upon the antimicrobial being used) also retard the penetration of such agents (Sutherland, 2001; Takenaka *et al.*, 2008); (iii) proximity-driven, cross-species protection from the environment (Bradshaw *et al.*, 1998; Gilbert *et al.*, 2002); (iv) altered growth rates due to metabolic interactions, nutrient and/or oxygen heterogeneity within biofilms (Fux *et al.*, 2005; Stewart and Franklin, 2008); and finally, (v) cell–cell signalling (Jakubovics, 2010; Mitchell *et al.*, 2011). Thus, biofilms represent a bacterial state that promotes species retention, especially in flowing conditions such as those found in the human oral cavity, and a niche that confers protection from a potentially hostile environment. If left uncontrolled, such communities are open to the integration of species that may exploit the community and cause periodontal disease. In dental plaque biofilms, two well-studied species associated with periodontal disease are *Porphyromonas gingivalis* and *Aggregatibacter actinomycetemcomitans* (Jakubovics and Kolenbrander, 2010; Nonnenmacher *et al.*, 2001; Socransky *et al.*, 1998). The likelihood that these species will be isolated from the community increases with uncontrolled biofilm development and the establishment of mature dental plaque (Kuramitsu *et al.*, 2007).

The processes and mechanisms that mediate the development of dental plaque have received a great deal of research interest over the last thirty years. Based upon classic microbiological techniques (Ritz, 1967, 1969; Rosan *et al.*, 1976; Socransky *et al.*, 1977), more recent molecular techniques (Aas *et al.*, 2005; Nobbs *et al.*, 2011), and advanced microscopy (Diaz *et al.*, 2006; Guggenheim *et al.*, 2001; Zijnge *et al.*, 2010), it has become increasingly clear that the colonization of clean tooth surfaces occurs in an orchestrated and reproducible manner. A number of models have been used to describe this process (Busscher and van der Mei, 2000; Marsh, 1994; Rickard *et al.*, 2003; Scheie, 1994; Wimpenny, 2000). One recent model that has gained popularity describes the development of dental plaque biofilms in the context of highly specific adhesion and multiplication events (Kolenbrander *et al.*, 2005; Kolenbrander and London, 1993; Rickard *et al.*, 2003). This model relies upon the observation that dental plaque biofilms develop through an ordered sequence or 'succession' of adhesion events, and that many of these component bacteria also coaggregate with one another (Kolenbrander *et al.*, 1999; Palmer *et al.*, 2006; Palmer *et al.*, 2001) (Fig. 7.1). In this model, the first species to adhere to tooth surfaces are called

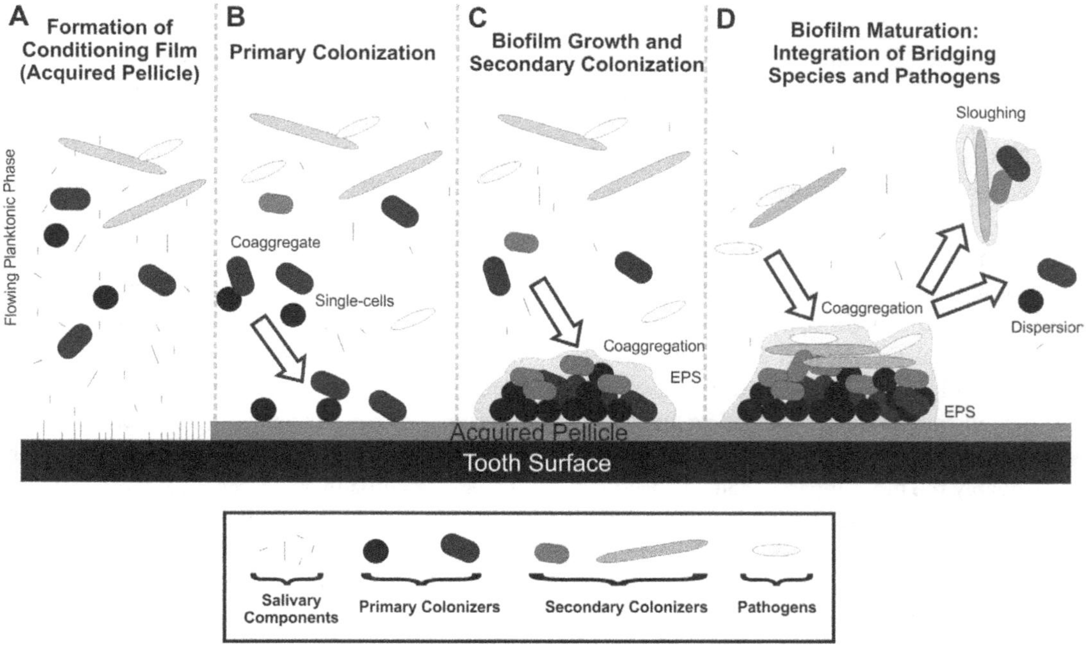


**Figure 7.1** A diagrammatic representation of multispecies biofilm development on a tooth surface. (A) A clean tooth surface is covered by salivary components to create a conditioning film (also known as the acquired pellicle). (B) Primary colonizing bacteria exclusively adhere to the acquired pellicle. Single cells or coaggregates adhere. (C) Cells within the immature biofilm divide and the biofilm grows. Extracellular polymeric substances (EPS) are produced and secondary colonizers begin to sequentially adhere. Secondary colonizing single cells or coaggregates adhere. (D) The biofilm continues to develop with the further integration of secondary colonizers and ultimately pathogenic species. Cells leave the biofilm via dispersion (single cells) or sloughing (aggregates/coaggregates of cells).

primary (or early) colonizers. Primary colonization occurs through both specific and non-specific physicochemical interactions with the salivary components of an adsorbed, organic conditioning film, often referred to as the acquired pellicle (Fig. 7.1A and B) (Hannig and Hannig, 2009; Hannig *et al.*, 2005; Hannig and Joiner, 2006; Ruhl *et al.*, 2004). Typical primary colonizers include members of the genera *Streptococcus* and *Actinomyces*. If conditions are amenable, the primary colonizers multiply on the substratum and their presence facilitates the adhesion of secondary colonizers (Fig. 7.1C). These secondary colonizers adhere to the primary colonizers through highly specific interactions (coaggregation, described in detail below) or through non-specific cell–cell adhesion (Bos *et al.*, 1996, 1999; Kolenbrander, 2011). Secondary colonizers include members of the genera *Veillonella, Eikenella* and *Haemophilus* (Kolenbrander, 2000). Later secondary colonizers (sometimes also called tertiary colonizers) can then integrate, including the bridging organism *Fusobacterium nucleatum* (Kolenbrander and London, 1993; Rickard *et al.*, 2003) (Fig. 7.1D). *F. nucleatum* not only coaggregates with commensal species (promoting biofilm integration) but also coaggregates with the oral pathogens *P. gingivalis* and *A. actinomycetemcomitans* (Kolenbrander and Andersen, 1989; Rosen *et al.*, 2003). *F. nucleatum* is often a dominant member of dental plaque biofilms within individuals with periodontal disease (Socransky *et al.*, 1998). Such a finding suggests a key role for *F. nucleatum* in the recruitment of oral pathogens into dental plaque biofilms through coaggregation. As the dental plaque biofilm reaches a climax community state, cells will conceivably be lost by sloughing of aggregates and coaggregates (van Loosdrecht *et al.*, 2002) or by the dispersion of single cells (Stoodley *et al.*, 2002).

## The role of coaggregation

Coaggregation, the specific recognition and adhesion of different species of bacteria, was first discovered by Gibbons and Nygaard in 1970 (Gibbons and Nygaard, 1970). It was not until further research and the mapping of coaggregation between representative dental plaque biofilm species, however, that the potential implications of such interactions were realized. Pioneering work by the Kolenbrander group (National Institute for Dental Craniofacial Research, National Institutes of Health, Bethesda, MD) in particular has demonstrated that coaggregation likely has a multifaceted role in dental plaque development.

Arguably the most apparent role of coaggregation was well detailed in a 1993 paper entitled 'Adhere today, here tomorrow: oral bacterial adherence' (Kolenbrander and London, 1993). This was the role of coaggregation in promoting adhesion between cells and the contribution of such interactions to biofilm development under flowing conditions (described in Fig. 7.1 and the above text). Indeed, coaggregation interactions conceivably contribute to the development of biofilms by two routes (Rickard *et al.*, 2003). The first is by promoting the adhesion of single planktonic cells to cells within the developing biofilm. The second is by the prior aggregation/coaggregation in the planktonic-phase followed by the subsequent adhesion of this aggregate/coaggregate to the developing biofilm. In both cases, planktonic cells specifically adhere to cells within the biofilm; this is often referred to as co-adhesion (Busscher and van der Mei, 2000; Jin and Yip, 2002). Interesting, supporting studies have also demonstrated that adhesive forces between cells are greater for coaggregating bacterial pairs than for non-coaggregating pairs (Postollec *et al.*, 2003, 2006). Thus, it is likely that developing (and developed) biofilms composed of coaggregating bacteria are more resilient to flow/shear forces than biofilms containing non-coaggregating bacteria. A variety of studies have demonstrated that coaggregation interactions are typically mediated by lectin (adhesin) and polysaccharide (receptor) interactions between coaggregating bacteria (although protein–protein interactions also occur between some coaggregating species). It is these interactions that likely contribute to the strength of the interactions (Doyle *et al.*, 1998; Kolenbrander *et al.*, 2006). The identities of some of the adhesins and receptors have been described using molecular techniques, although when considering the species diversity of dental plaque biofilms, this area of work is still in its infancy (Nobbs *et al.*, 2009; Rickard *et al.*, 2003; Yang *et al.*, 2009; Yoshida *et al.*, 2006).

While improved attachment/adhesion is clearly advantageous in flowing environments, such as those in the oral cavity, the ability to coaggregate may confer certain other subtle advantages. In particular, coaggregation is a *targeting* mechanism that allows bacteria to specifically recognize and adhere to genetically distinct species. The coaggregation of two genetically distinct species could conceivably enhance the growth of one or both species, although it is also possible that antagonistic interactions could prevent either partner species from benefiting from the union (Fig. 7.2). It is important to remember, however, that aggregation/accumulation will likely also occur through non-specific stochastic adhesion events between cells of different bacterial species; no doubt, these also occur in dental plaque biofilms. Coaggregation therefore increases the likelihood of specific species being in close proximity, which may allow a variety of other interactions to occur effectively. These interactions include the exchange of genetic material, environmental protection, contact-dependent gene expression, enhanced antimicrobial resistance, metabolic interactions, and cell–cell signalling (Fig. 7.2).

Although receiving less attention, the importance of genetic material exchange, contact-dependent gene expression, environmental protection, and antimicrobial resistance cannot be overstated. In particular, the ability of oral bacteria to exchange genetic information likely occurs between species of bacteria within dental plaque bacteria. This is especially likely as dental plaque biofilm species will be in close proximity, as compared to being in the planktonic phase, and the often numerically dominant biofilm streptococci may produce and respond to competence stimulating peptides to aid in DNA uptake and biofilm formation (Petersen *et al.*, 2004; Suntharalingam and Cvitkovitch, 2005). However, it is currently unclear whether competence stimulating peptides

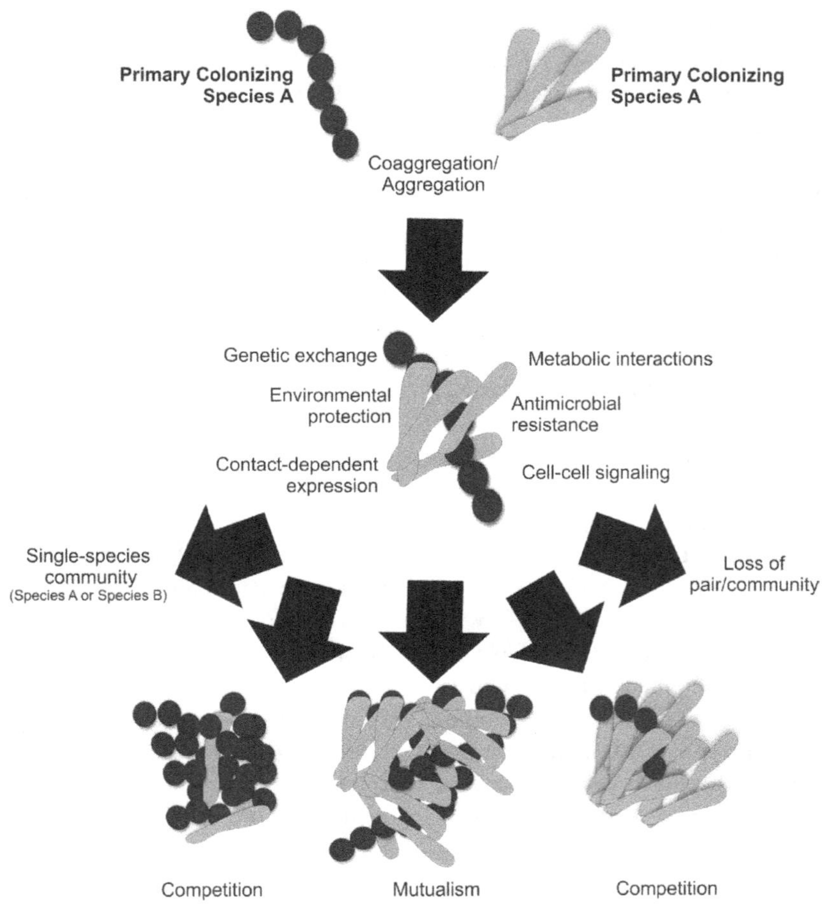


**Figure 7.2** Diagram demonstrating potential interactions and possible outcomes as a result of two primary colonizing species non-specifically aggregating or coaggregating in a flowing environment. This may occur in planktonic phase or on a surface (also called co-adhesion). Assuming interaction between the two species and a constant environment, at least six types of interactions could conceivably contribute towards mutualism/synergy or competition between the pair. Mutualism will be observed as interdigitated biofilm growth and competition will be observed as the dominance of one species over the other. The extreme outcome would be the loss of both species into the flowing environment or the development of single-species biofilms. Cells are not to scale and singular or combined interactions with the host or environment are not considered.

from a given species of streptococci can influence genetic exchange between a broad (coaggregating) community of streptococci (Jakubovics and Kolenbrander, 2010). Environmental protection is yet another consideration for possible interspecies interactions within dental plaque biofilms and may facilitate the expansion of one or more populations. For example, evidence suggests that the presence of *F. nucleatum* in dental plaque biofilms supports the growth and expansion of obligate anaerobic species (Bradshaw *et al.*, 1998). This is likely because of the bridging capability of *F. nucleatum* and a promiscuous ability to coaggregate with not only oxygen-tolerant/facultative anaerobes but also other obligate anaerobic species (Bradshaw *et al.*, 1998). Contact-dependent gene expression is a term given to the response of cells coming into contact with other cells (Kumamoto, 2002). A recent study by Inagaki *et al.* (2005) indicated that contact-dependent gene expression occurs in the oral pathogen *Tannerella forsythia* as a consequence of coaggregation with other oral species. This finding suggests that coaggregation (or even possibly non-specific aggregation) by species could alter their cellular properties as a consequence of physical contact within biofilms. The prospect of enhanced antimicrobial resistance as a consequence of interactions between juxtaposed species could occur through a variety of mechanisms. For example, one species will have a greater binding affinity to certain antimicrobials, such as quaternary ammonium compounds (used in mouthwashes), than the partner species and thus reduce lethal concentrations to sub-inhibitory concentrations (Gilbert *et al.*, 2002). Also, altered growth rates may occur as a consequence of changes in metabolism due to coaggregation. Growth rate is a well-established factor in altering the susceptibility of bacterial cells to antimicrobials (Brown *et al.*, 1988; Gilbert *et al.*, 1990; Sbordone and Bortolaia, 2003).

## Metabolic interactions

Taxonomically diverse, multispecies biofilms are proposed to possess a combined metabolic activity that is greater than that of the individual component species (Rickard *et al.*, 2003). Furthermore, it has been suggested that the ability of a microbial community to maintain an ecologically quasi-steady state increases with taxonomic breadth of communities (Alexander, 1971; Marsh, 1989). In reality, dental plaque biofilm communities are more likely to be stable with increased species diversity, functional diversity, and favourable community structure (Marsh, 1989, 1994; Vaun McArthur, 2006). Functional diversity allows for the development of food-chains between bacterial species and ultimately promotes complex metabolic interactions. Taxonomically diverse species often possess different nutritional requirements and, by virtue, produce different metabolic by-products.

In dental plaque biofilms, numerous examples of metabolic interactions between component species have been described (Wilson, 2005). For example, *Treponema denticola* and *P. gingivalis* have been shown to not only coaggregate but to also form a mutualistic relationship that facilitates the growth of both organisms in laboratory medium (Grenier, 1992a,b). The growth-stimulating factors involved in this interaction were indicated to be isobutyric acid (*P. gingivalis*) and succinic acid (*T. denticola*) and suggested to be of importance in the initiation and progression of periodontal disease (Grenier, 1992b). A later study by Nilus *et al.* (1993), which used a different type of culture media, also demonstrated that the *T. denticola* and *P. gingivalis* benefit when grown together, although the authors did not determine isobutyric acid to be involved and instead suggested that protein(s) with molecular weight >50 kDa facilitate this interaction. Vitamin K (menadione) is required by *P. gingivalis* for growth; this compound is also produced by *Veillonella* and *Bacteroides* species (Gibbons and Engle, 1964). Recently it was shown that oral *Bifidobacterium* species also utilize vitamin K as a growth factor (Hojo *et al.*, 2007). It is possible that *Bifidobacteria*, which are common to the human oral cavity and digestive tract (Fakhry *et al.*, 2009; Modesto *et al.*, 2006), may limit the expansion of *P. gingivalis* populations in dental plaque biofilms by competing for vitamin K (Hojo *et al.*, 2007). Of further interest, *Bifidobacterium* coaggregates with *F. nucleatum* and, thus, could be in close proximity to *P. gingivalis* because both coaggregate with *F. nucleatum* (Nagaoka *et al.*, 2008). Therefore, metabolic interactions could lead to competition and reduced numbers or possible loss of one of the competing species (Fig. 7.2). An extreme form of competition and/or antagonism is linked to the production of inhibitory metabolites such as bacteriocins. While beyond the scope of this chapter, bacteriocins have received attention elsewhere within this book (see Chapter 8) and some excellent reviews of the subject are available (Chatterjee *et al.*, 2005; Cotter *et al.*, 2005; Hojo *et al.*, 2009; Kirkup, 2006). In stark contrast, another example of metabolic interactions is the mutualistic interaction which occurs between streptococci and *Veillonella* species. *Veillonella* species are unable to use carbohydrates or amino acids for energy generation and instead ferment organic acids, such as lactic acid. Streptococci produce lactic acid from the fermentation of carbohydrates and can thus act as a significant energy provider to veillonellae. Of relevance, members of each genus have been reported to coaggregate with one another (Hughes *et al.*, 1988) and evidence presented by Chalmers *et al* (2008) showed the presence of the two genera within coaggregated micro-colonies in early dental plaque biofilms. Coaggregation likely aids juxtaposition by reducing the effective distance between partner cells to facilitate metabolic interactions (i.e. the utilization of lactic acid by *Veillonella* species) (Kolenbrander *et al.*, 2010). Another dimension to this metabolic interaction was demonstrated by Egland *et al.* (2004), who showed that signalling occurred between *Veillonella atypica* PK1910 and *Streptococcus gordonii* V288 and resulted in the *S. gordonii* alpha-amylase-encoding gene *amyB* being up-regulated. This was demonstrated by constructing an amylase B-*gfp* transcriptional fusion (P*amyB-gfp*) and monitoring green fluorescence. Of direct relevance to this chapter, confocal scanning laser microscopy of *S. gordonii-V. atypica* biofilms grown in flowing human saliva showed that *V. atypica* PK1910 only caused *S. gordonii* V288 P*amyB-gfp* up-expression when *V. atypica* PK1910 were in close proximity within the biofilm. *S. gordonii* V288 P*amyB-gfp* biofilms that lacked *V. atypica* PK1910 did not express GFP. Such a finding is intriguing as it demonstrates that *Veillonella* and *S. gordonii* can be juxtaposed by coaggregation and, potentially using cell–cell signal molecules, are able to coordinate metabolic interactions.

## Autoinducer-2 mediated interspecies interactions

Cell–cell signalling (also called quorum sensing) between bacteria was first described approximately 40 years ago (Nealson *et al.*, 1970; Eberhard, 1972; Ruby and Nealson, 1976). Notable examples, which initiated decades of research and are now the mainstay of *sociomicrobiology* (Parsek and Greenberg, 2005), include cell–cell signalling-mediated bioluminescence by the marine bacteria *Vibrio harveyi* and *Vibrio fischeri* (Ng and Bassler, 2009). The cell–cell signalling molecules that mediate bioluminescence are called autoinducers and allow the bacteria to monitor their own population density (Bassler, 1999). The mechanism by which cell–cell signalling operates relies upon the accumulation of autoinducers in the surrounding environment during growth. When a threshold concentration of autoinducer is achieved, the genes required for bioluminescence are activated (Mok *et al.*, 2003). An interesting development came with the discovery that *Vibrio harveyi* uses not one but two autoinducers for cell–cell signalling (Bassler *et al.*, 1994). These are called autoinducer-1 (AI-1) and autoinducer-2 (AI-2). Unlike AI-1, which is restricted to Gram-negative bacteria, AI-2 was shown to be produced from taxonomically diverse species including the Gram-negative genera *Escherichia* and *Salmonella* (Bassler *et al.*, 1997; Surette *et al.*, 1999) and the Gram-positive genera *Staphylococcus* and *Bacillus* (Jones and Blaser, 2003; Li *et al.*, 2008). AI-2 is generated as a consequence of the bacterial activated methyl cycle and is produced as a by-product of the enzyme LuxS (S-ribosyl-L-homocysteinase). LuxS converts S-ribosylhomocysteine into homocysteine and the by-product 4,5-dihydroxy-2,3-pentanedione (DPD). DPD undergoes spontaneous structural rearrangements, depending upon environmental conditions, and the inter-convertible forms are collectively called AI-2 (Semmelhack *et al.*, 2005). Different forms are recognized by various bacterial species. For example, *Vibrio harveyi* recognizes a boronated form of DPD while *Salmonella enterica* serovar Typhimurium recognizes a non-boronated form (Miller *et al.*, 2004; Semmelhack *et al.*, 2004). Controversy surrounds AI-2, as evidence from the study of some species suggests that AI-2/LuxS does not mediate cell–cell signalling (Winzer *et al.*, 2002; Turovskiy *et al.*, 2007; Heurlier *et al.*, 2009; Holmes *et al.*, 2009). Studies of other bacterial species, however, have shown that AI-2 mediates cell–cell signalling (Federle, 2009; Gospodarek *et al.*, 2009; Kendall and Sperandio, 2007; Van Houdt *et al.*, 2007). An intriguing finding is that even species that do not produce AI-2 have been indicated to respond to AI-2 (Duan *et al.*, 2003; Pereira *et al.*, 2008; Armbruster *et al.*, 2010)

In 2001, Frias *et al.* (2001) demonstrated that many human oral bacteria including members of the genera *Actinomyces, Fusobacterium, Porphyromonas,* and *Streptococcus* produce AI-2. As inferred from a *Vibrio harveyi* reporter system (Bassler *et al.*, 1997), that indicates the amount of AI-2 produced by bacteria, the periodontal pathogens generally produced more (10- to 100-fold greater signal) than commensal bacteria. Since then, numerous reports have been published describing the production, and more recently, the mechanism of detection of AI-2 by dental plaque bacteria. Considerable focus on the role and mechanism of AI-2 mediated cell–cell signalling has been applied to the cariogenic organism *Streptococcus mutans* and the periodontal pathogens *A. actinomycetemcomitans* and *P. gingivalis*.

Recent work by Merritt and coworkers (2003) showed that the *luxS* mutant of *S. mutans* UA159 formed biofilms that possessed a granular structure, as opposed to a relatively smooth, confluent structure seen in the wild type. The effect of the addition of AI-2, however, was not investigated. Subsequent microarray studies on the effect of AI-2 on *Streptococcus mutans* U159 demonstrated that AI-2 altered the expression of 59 genes in an isogenic *luxS* mutant (Sztajer *et al.*, 2008). These genes are involved in cell division, metabolism, nucleotide synthesis, protein synthesis, and stress. In addition, three membrane transport proteins were up-expressed and three transcription factors were identified. The delta subunit of the RNA polymerase (*rpoE*), a global regulatory protein, was also highly up-regulated by the addition of AI-2 (Sztajer *et al.*, 2008). When considering the importance of AI-2 in mediating biofilm-specific interspecies interactions involving the wild-type or *luxS* mutant of *S. mutans*, the effect of the co-inoculation of other oral species (wild-type or *luxS* mutant) was recently investigated (Yoshida *et al.*, 2005). This was an interesting and arguably ground-breaking study as it indicated that other oral species could provide AI-2 to biologically complement the *luxS* mutant of *S. mutans*. Specifically, using a novel two-compartment assay system, biofilm defects of a *luxS* mutant of *S. mutans* GS-5 were shown to be complemented by *S. anginosus* FW73, *S. gordonii* DL1, and *S. sobrinus* MT8145, but not *S. salivarius* HT9R, *S. sanguinis* ATCC10556, or *S. oralis* ATCC10557. The periodontal pathogens *P. gingivalis* 381 and *A. actinomycetemcomitans* Y4 restored the biofilm properties of the *S. mutans* UA159 mutant but, interestingly, not by a *luxS* mutant of *P. gingivalis*. Relating these findings to the work of Frias *et al.* (2001) and others working in the AI-2 research field, it is possible that either the strains that did not complement the *luxS* biofilm do not produce the correct concentration of AI-2 or they produce AI-2 that is in the incorrect structural form. The possibility that AI-2 concentration plays a role was addressed by Rickard *et al.* (2006, 2008b), although the focus was not *S. mutans* but another *Streptococcus* species, *S. oralis* 34. It was demonstrated that nanomolar concentrations of AI-2 (0.8–8 nM) mediated mutualistic biofilm growth with *Actinomyces oris* T14V when flowing saliva was used as the sole nutrient source. Taken together, the importance of AI-2 mediated interactions between primary colonizing streptococci in the crafting of multispecies biofilms is potentially significant. For instance, if AI-2 mediates biofilm mass and architecture of primary streptococcal biofilms, then specific pioneering streptococcal communities could form on tooth surfaces and recruit/support a unique cohort of secondary colonizing species. Such a possibility has begun to be addressed, although not in the context of AI-2 mediated signalling, by Periasamy and Kolenbrander (2009) who demonstrated that in flowing saliva, *P. gingivalis* ATCC 33277 formed mutualistic biofilms with *S. gordonii* DL1 but not with *S. oralis* 34.

When considering periodontal pathogens such as *P. gingivalis* and *A. actinomycetemcomitans*, work on the importance of AI-2 in cell–cell signalling has been pioneered by the Demuth and Lamont labs (University of Louisville Dental School, Louisville). The Lamont lab has focused primarily on *P. gingivalis* 33277 and interactions between this species and *S. gordonii* DL1. Some interesting aspects of *luxS* expression in *P. gingivalis* have been described. In particular, *P. gingivalis* is one of only a few species in which transcriptional control by AI-2/*luxS* has been demonstrated, and it is involved in the regulation of haemin and iron acquisition pathways (James *et al.*, 2006). The *P. gingivalis* GppX hybrid two-component system, a putative DNA-binding response regulator/sensor histidine kinase, was also linked to *luxS* transcription (James *et al.*, 2006). Additional studies have shown that a low molecular weight tyrosine phosphatase (designated as Ltp1) suppresses transcription of *luxS* and would therefore also moderate AI-2-mediated cell–cell signalling in multispecies biofilm communities (Maeda *et al.*, 2008). Increased *ltp1* transcription also increases hemin uptake and, conversely, hemin uptake is diminished in an *ltp1* mutant. Seeking to unravel what controls *luxS* and AI-2 interaction, it was shown that when *P. gingivalis* was grown together with *S. gordonii* in biofilms, the *S. gordonii* cells generate an Ltp1-dependent intra-cellular signal within *P. gingivalis* that results in the reduced transcription of adhesin and signalling genes. A *P. gingivalis* LuxR family transcription factor (described as a community development and hemin regulator, CdhR) was also determined to be involved in the complex regulatory process which ensues from contact between *P. gingivalis* and *S. gordonii*. It was shown that CdhR bound to the upstream regulatory regions of *mfa* (which encodes structural subunits of fimbriae that recognize streptococcal multifunctional SspA/B adhesins) and *luxS*, resulting in reduced gene expression and thus contributing to restricted community development (Chawla *et al.*, 2010; Whitmore and Lamont, 2011). As a result of these findings, the current model posits that through *S. gordonii* SspA/B interaction with Mfa, *lpt1* expression is directly up-regulated while CdhR is indirectly up-regulated. CdhR consequently down-regulates *mfa* and *luxS*, but AI-2 (either from *P. gingivalis* or presumably *S. gordonii* DL1) has the ability to down-regulate *cdhR* (Whitmore and Lamont, 2011).

Studies of the role of AI-2 in cell–cell signalling by *A. actinomycetemcomitans* JP2 have also revealed a complex and intricate recognition and response mechanism. Work by the Demuth lab has demonstrated that AI-2 is required for the development of *A. actinomycetemcomitans* JP2 single-species biofilms (Shao *et al.*, 2007b). Furthermore, an *lsrB* mutant and an *rbsB* mutant which have been shown to be involved in AI-2 sensing by *A. actinomycetemcomitans* JP2 (Shao *et al.*, 2007a) formed biofilms that were similar to the *luxS* mutant (Shao *et al.*, 2007b). Presumably, this was because the cells could produce but not detect AI-2. Continuing studies of AI-2 mediated communication in *A. actinomycetemcomitans* JP2 have also shown that inactivation of *qseC*, which was originally described to be involved in the AI-2-dependent response pathway for *E. coli* biofilm formation (Sperandio *et al.*, 2002), resulted in reduced biofilm development (Novak *et al.*, 2010). A particularly relevant finding from this work was that animals infected with the *qseC* mutant exhibited much less alveolar bone loss than those infected with the wild-type strain. Interestingly, the *luxS* mutant behaved like the wild-type and caused similar levels of alveolar bone resorption (Novak *et al.*, 2010). The researchers suggest that this was probably due to cross-species biological complementation: AI-2 from indigenous species provided AI-2 to the *luxS* mutant of *A. actinoycetemcomitans* JP2.

## The future: studying interspecies interactions in multispecies communities

Beighton (2009) asked the question 'Can the ecology of the dental biofilm be beneficially altered?'. In short, the answer was that, while dental plaque biofilm ecology can be altered by excluding fermentable carbohydrates from the human diet, there does not seem to be any current strategy to modify the ecology of the oral biofilm beneficially. The author did, however, highlight possibly using novel antimicrobials, probiotics, and manipulating cell–cell signalling systems. Here we describe a variety of systems employed by bacteria to interact with one another at the interspecies level. Of course, intra-species systems that have not been discussed (such as competence stimulating peptide signalling systems and autoaggregation ability) will also play a role (Merritt *et al.*, 2009; Liang *et al.*, 2011). Although, considering that dental plaque biofilms consist of numerous interactive species, an improved knowledge of the mechanisms mediating interspecies interactions and the roles they play may provide novel strategies to prevent periodontal disease. These strategies could require the manipulation of a single type of interspecies interaction (such as the inhibition of coaggregation) or require a multifaceted approach (such as the inhibition of coaggregation and certain metabolic interactions).

A key component to the development of dental plaque biofilms is the interaction of the host with the microbial community. This was not discussed as it both adds another dimension beyond the remit of this paper and is an area research of investigation still in its infancy. A paper of note by Foxman and coworkers, however, does introduce how to possibly begin studying the interplay between multispecies community members, the community as a whole, and the host (Foxman *et al.*, 2008). The authors ultimately focus on developing predictive models, methodologies, and research strategies to study a 'dynamic ecological community consisting of multiple taxa each potentially interacting with each other, the host, and the environment' (Foxman *et al.*, 2008). Such approaches would inevitably be needed in order to effectively study how multiple species interact with each other in dental plaque biofilms. Only then can single or combined approaches describe how inter-species interactions occur and the impact it has on the community be determined. For example, the potential use of prebiotics and probiotics in preventing disease has been gaining support in recent years (Meurman, 2005; Devine and Marsh, 2009; He *et al.*, 2009). New probiotic strains and chemical treatment strategies need to be evaluated in multispecies dental plaque biofilms. With the advent of new and/or improved microscopic technologies, molecular approaches, and modelling techniques such investigations are possible.

## References

Aas, J.A., Paster, B.J., Stokes, L.N., Olsen, I., and Dewhirst, F.E. (2005). Defining the normal bacterial flora of the oral cavity. J. Clin. Microbiol. *43*, 5721–5732.

Alexander, M. (1971). Microbial Ecology (John Wiley, New York).

Armbruster, C.E., Hong, W., Pang, B., Weimer, K.E., Juneau, R.A., Turner, J., and Swords, W.E. (2010). Indirect pathogenicity of Haemophilus influenzae and Moraxella catarrhalis in polymicrobial otitis media occurs via interspecies quorum signalling. MBio *1*.

Bassler, B.L. (1999). How bacteria talk to each other: regulation of gene expression by quorum sensing. Curr. Opin. Microbiol. *2*, 582–587.

Bassler, B.L., Wright, M., and Silverman, M.R. (1994). Multiple signalling systems controlling expression of luminescence in *Vibrio harveyi*: sequence and function of genes encoding a second sensory pathway. Mol. Microbiol. *13*, 273–286.

Bassler, B.L., Greenberg, E.P., and Stevens, A.M. (1997). Cross-species induction of luminescence in the quorum-sensing bacterium *Vibrio harveyi*. J. Bacteriol. *179*, 4043–4045.

Beighton, D. (2009). Can the ecology of the dental biofilm be beneficially altered? Adv. Dent. Res. *21*, 69–73.

Bos, R., van der Mei, H.C., and Busscher, H.J. (1996). Co-adhesion of oral microbial pairs under flow in the presence of saliva and lactose. J. Dent. Res. *75*, 809–815.

Bos, R., van der Mei, H.C., and Busscher, H.J. (1999). Physico-chemistry of initial microbial adhesive interactions – its mechanisms and methods for study. FEMS Microbiol. Rev. *23*, 179–230.

Bourgeau, G., and McBride, B.C. (1976). Dextran-mediated interbacterial aggregation between dextran-synthesizing streptococci and *Actinomyces viscosus*. Infect. Immun. *13*, 1228–1234.

Bradshaw, D.J., Marsh, P.D., Watson, G.K., and Allison, C. (1998). Role of *Fusobacterium nucleatum* and coaggregation in anaerobe survival in planktonic and biofilm oral microbial communities during aeration. Infect. Immun. *66*, 4729–4732.

Brown, M.R., Allison, D.G., and Gilbert, P. (1988). Resistance of bacterial biofilms to antibiotics: a growth-rate related effect? J. Antimicrob. Chemother. *22*, 777–780.

Busscher, H.J., and van der Mei, H.C. (2000). Initial microbial adhesion events: mechanisms and implications. In Community Structure and Cooperation in Biofilms, Allison, D.G., Gilbert, P., Lappin-Scott, H.M., and Wilson, M., eds. (Cambridge University Press, Cambridge), pp. 25–36.

Chalmers, N.I., Palmer, R.J., Jr., Cisar, J.O., and Kolenbrander, P.E. (2008). Characterization of a *Streptococcus* sp.-*Veillonella* sp. community micromanipulated from dental plaque. J. Bacteriol. *190*, 8145–8154.

Chatterjee, C., Paul, M., Xie, L., and van der Donk, W.A. (2005). Biosynthesis and mode of action of lantibiotics. Chem. Rev. *105*, 633–684.

Chawla, A., Hirano, T., Bainbridge, B.W., Demuth, D.R., Xie, H., and Lamont, R.J. (2010). Community signalling between *Streptococcus gordonii* and *Porphyromonas gingivalis* is controlled by the transcriptional regulator CdhR. Mol. Microbiol. *78*, 1510–1522.

Cisar, J.O., Kolenbrander, P.E., and McIntire, F.C. (1979). Specificity of coaggregation reactions between human oral streptococci and strains of *Actinomyces viscosus* or *Actinomyces naeslundii*. Infect. Immun. *24*, 742–752.

Costerton, J.W., Geesey, G.G., and Cheng, K.J. (1978). How bacteria stick. Sci. Am. *238*, 86–95.

Cotter, P.D., Hill, C., and Ross, R.P. (2005). Bacterial lantibiotics: strategies to improve therapeutic potential. Curr. Protein Pept. Sci. *6*, 61–75.

Daep, C.A., James, D.M., Lamont, R.J., and Demuth, D.R. (2006). Structural characterization of peptide-mediated inhibition of *Porphyromonas gingivalis* biofilm formation. Infect. Immun. *74*, 5756–5762.

Delwiche, E.A., Pestka, J.J., and Tortorello, M.L. (1985). The veillonellae: Gram-negative cocci with a unique physiology. Annu. Rev. Microbiol. *39*, 175–193.

Devine, D.A., and Marsh, P.D. (2009). Prospects for the development of probiotics and prebiotics for oral applications. J. Oral Microbiol. *1*.

Diaz, P.I., Chalmers, N.I., Rickard, A.H., Kong, C., Milburn, C.L., Palmer, R.J., Jr., and Kolenbrander, P.E. (2006). Molecular characterization of subject-specific oral microflora during initial colonization of enamel. Appl. Environ. Microbiol. *72*, 2837–2848.

Doyle, R.J., Calderon de La Barca, A.M., and Buck, G. (1998). Lectin interactions in oral microbial adhesion. In Oral Biofilms and Plaque Control, Busscher,

H.J., and Evans, L.V., eds. (Harwood Academic, Amsterdam), pp. 111–124.

Duan, K., Dammel, C., Stein, J., Rabin, H., and Surette, M.G. (2003). Modulation of *Pseudomonas aeruginosa* gene expression by host microflora through interspecies communication. Mol. Microbiol. *50,* 1477–1491.

Eberhard, A. (1972). Inhibition and activation of bacterial luciferase synthesis. J. Bacteriol. *109,* 1101–1105.

Egland, P.G., Palmer, R.J., Jr., and Kolenbrander, P.E. (2004). Interspecies communication in *Streptococcus gordonii-Veillonella atypica* biofilms: signalling in flow conditions requires juxtaposition. Proc. Natl. Acad. Sci. U.S.A. *101,* 16917–16922.

Fakhry, S., Manzo, N., D'Apuzzo, E., Pietrini, L., Sorrentini, I., Ricca, E., De Felice, M., and Baccigalupi, L. (2009). Characterization of intestinal bacteria tightly bound to the human ileal epithelium. Res. Microbiol. *160,* 817–823.

Federle, M.J. (2009). Autoinducer-2-based chemical communication in bacteria: complexities of interspecies signalling. Contrib. Microbiol. *16,* 18–32.

Foxman, B., Goldberg, D., Murdock, C., Xi, C., and Gilsdorf, J.R. (2008). Conceptualizing human microbiota: from multicelled organ to ecological community. Interdiscip. Perspect. Infect. Dis. *2008,* 613979.

Frias, J., Olle, E., and Alsina, M. (2001). Periodontal pathogens produce quorum sensing signal molecules. Infect. Immun. *69,* 3431–3434.

Fux, C.A., Costerton, J.W., Stewart, P.S., and Stoodley, P. (2005). Survival strategies of infectious biofilms. Trends Microbiol. *13,* 34–40.

Gibbons, R.J., and Engle, L.P. (1964). Vitamin K compounds in bacteria that are obligate anaerobes. Science *146,* 1307–1309.

Gibbons, R.J., and Nygaard, M. (1970). Interbacterial aggregation of plaque bacteria. Arch. Oral Biol. *15,* 1397–1400.

Gilbert, P., Collier, P.J., and Brown, M.R. (1990). Influence of growth rate on susceptibility to antimicrobial agents: biofilms, cell cycle, dormancy, and stringent response. Antimicrob. Agents Chemother. *34,* 1865–1868.

Gilbert, P., Maira-Litran, T., McBain, A.J., Rickard, A.H., and Whyte, F.W. (2002). The physiology and collective recalcitrance of microbial biofilm communities. Adv. Microb. Physiol. *46,* 202–256.

Gospodarek, E., Bogiel, T., and Zalas-Wiecek, P. (2009). Communication between microorganisms as a basis for production of virulence factors. Pol. J. Microbiol. *58,* 191–198.

Grenier, D. (1992a). Demonstration of a bimodal coaggregation reaction between *Porphyromonas gingivalis* and *Treponema denticola*. Oral Microbiol. Immunol. *7,* 280–284.

Grenier, D. (1992b). Nutritional interactions between two suspected periodontopathogens, *Treponema denticola* and *Porphyromonas gingivalis*. Infect. Immun. *60,* 5298–5301.

Guggenheim, M., Shapiro, S., Gmur, R., and Guggenheim, B. (2001). Spatial arrangements and associative behavior of species in an in vitro oral biofilm model. Appl. Environ. Microbiol. *67,* 1343–1350.

Hannig, C., and Hannig, M. (2009). The oral cavity – a key system to understand substratum-dependent bioadhesion on solid surfaces in man. Clin. Oral Investig. *13,* 123–139.

Hannig, C., Hannig, M., and Attin, T. (2005). Enzymes in the acquired enamel pellicle. Eur. J. Oral Sci. *113,* 2–13.

Hannig, M., and Joiner, A. (2006). The structure, function and properties of the acquired pellicle. Monogr. Oral Sci. *19,* 29–64.

He, X., Lux, R., Kuramitsu, H.K., Anderson, M.H., and Shi, W. (2009). Achieving probiotic effects via modulating oral microbial ecology. Adv. Dent. Res. *21,* 53–56.

Heurlier, K., Vendeville, A., Halliday, N., Green, A., Winzer, K., Tang, C.M., and Hardie, K.R. (2009). Growth deficiencies of *Neisseria* meningitidis pfs and luxS mutants are not due to inactivation of quorum sensing. J. Bacteriol. *191,* 1293–1302.

Hojo, K., Nagaoka, S., Murata, S., Taketomo, N., Ohshima, T., and Maeda, N. (2007). Reduction of vitamin K concentration by salivary Bifidobacterium strains and their possible nutritional competition with *Porphyromonas gingivalis*. J. Appl. Microbiol. *103,* 1969–1974.

Hojo, K., Nagaoka, S., Ohshima, T., and Maeda, N. (2009). Bacterial interactions in dental biofilm development. J. Dent. Res. *88,* 982–990.

Holmes, K., Tavender, T.J., Winzer, K., Wells, J.M., and Hardie, K.R. (2009). AI-2 does not function as a quorum sensing molecule in *Campylobacter jejuni* during exponential growth in vitro. BMC Microbiol. *9,* 214.

Van Houdt, R., Givskov, M., and Michiels, C.W. (2007). Quorum sensing in Serratia. FEMS Microbiol. Rev. *31,* 407–424.

Hughes, C.V., Kolenbrander, P.E., Andersen, R.N., and Moore, L.V. (1988). Coaggregation properties of human oral *Veillonella* spp.: relationship to colonization site and oral ecology. Appl. Environ. Microbiol. *54,* 1957–1963.

Inagaki, S., Kuramitsu, H.K., and Sharma, A. (2005). Contact-dependent regulation of a *Tannerella forsythia* virulence factor, BspA, in biofilms. FEMS Microbiol. Lett. *249,* 291–296.

Jakubovics, N.S. (2010). Talk of the town: interspecies communication in oral biofilms. Mol. Oral Microbiol. *25,* 4–14.

Jakubovics, N.S., and Kolenbrander, P.E. (2010). The road to ruin: the formation of disease-associated oral biofilms. Oral Dis. *16,* 729–739.

James, C.E., Hasegawa, Y., Park, Y., Yeung, V., Tribble, G.D., Kuboniwa, M., Demuth, D.R., and Lamont, R.J. (2006). LuxS involvement in the regulation of genes coding for hemin and iron acquisition systems in *Porphyromonas gingivalis*. Infect. Immun. *74,* 3834–3844.

Jin, Y., and Yip, H.K. (2002). Supragingival calculus: formation and control. Crit. Rev. Oral Biol. Med. *13*, 426–441.

Jones, M.B., and Blaser, M.J. (2003). Detection of a luxS-signalling molecule in *Bacillus anthracis*. Infect. Immun. *71*, 3914–3919.

Kendall, M.M., and Sperandio, V. (2007). Quorum sensing by enteric pathogens. Curr. Opin. Gastroenterol. *23*, 10–15.

Kirkup, B.C., Jr. (2006). Bacteriocins as oral and gastrointestinal antibiotics: theoretical considerations, applied research, and practical applications. Curr. Med. Chem. *13*, 3335–3350.

Kolenbrander, P.E. (2000). Oral microbial communities: biofilms, interactions, and genetic systems. Annu. Rev. Microbiol. *54*, 413–437.

Kolenbrander, P.E. (2011). Multispecies communities: interspecies interactions influence growth on saliva as sole nutritional source. Int. J. Oral. Sci. *3*, 49–54.

Kolenbrander, P.E., and Andersen, R.N. (1989). Inhibition of coaggregation between *Fusobacterium nucleatum* and *Porphyromonas* (Bacteroides) *gingivalis* by lactose and related sugars. Infect. Immun. *57*, 3204–3209.

Kolenbrander, P.E., and London, J. (1993). Adhere today, here tomorrow: oral bacterial adherence. J. Bacteriol. *175*, 3247–3252.

Kolenbrander, P.E., Andersen, R.N., Clemans, D.L., Whittaker, C.J., and Klier, C. (1999). Potential role of functionally similar coaggregation mediators in bacterial sucession. In Dental Plaque Revisited: Oral Biofilms in Health and Disease, Newman, H.N., and Wilson, M., eds. (Bioline, Cardiff, Wales), pp. 171–186.

Kolenbrander, P.E., Egland, P.G., Diaz, P.I., and Palmer, R.J., Jr. (2005). Genome–genome interactions: bacterial communities in initial dental plaque. Trends Microbiol. *13*, 11–15.

Kolenbrander, P.E., Palmer, R.J., Jr., Periasamy, S., and Jakubovics, N.S. (2010). Oral multispecies biofilm development and the key role of cell–cell distance. Nat. Rev. Microbiol. *8*, 471–480.

Kolenbrander, P.E., Palmer, R.J., Jr., Rickard, A.H., Jakubovics, N.S., Chalmers, N.I., and Diaz, P.I. (2006). Bacterial interactions and successions during plaque development. Periodontol. 2000 *42*, 47–79.

Kumamoto, C.A. (2002). Candida biofilms. Curr. Opin. Microbiol. *5*, 608–611.

Kuramitsu, H.K., He, X., Lux, R., Anderson, M.H., and Shi, W. (2007). Interspecies interactions within oral microbial communities. Microbiol. Mol. Biol. Rev. *71*, 653–670.

Lancy, P., Jr., Dirienzo, J.M., Appelbaum, B., Rosan, B., and Holt, S.C. (1983). Corncob formation between *Fusobacterium nucleatum* and *Streptococcus* sanguis. Infect. Immun. *40*, 303–309.

Li, M., Villaruz, A.E., Vadyvaloo, V., Sturdevant, D.E., and Otto, M. (2008). AI-2-dependent gene regulation in *Staphylococcus epidermidis*. BMC Microbiol. *8*, 4.

Liang, X., Chen, Y.Y., Ruiz, T., and Wu, H. (2011). New cell surface protein involved in biofilm formation by *Streptococcus parasanguinis*. Infect. Immun. *79*, 3239–3248.

Listgarten, M.A., Mayo, H., and Amsterdam, M. (1973). Ultrastructure of the attachment device between coccal and filamentous microorganisms in 'corn cob' formations of dental plaque. Arch. Oral Biol. *18*, 651–656.

van Loosdrecht, M.C., Heijnen, J.J., Eberl, H., Kreft, J., and Picioreanu, C. (2002). Mathematical modelling of biofilm structures. Antonie Van Leeuwenhoek *81*, 245–256.

Maeda, K., Tribble, G.D., Tucker, C.M., Anaya, C., Shizukuishi, S., Lewis, J.P., Demuth, D.R., and Lamont, R.J. (2008). A *Porphyromonas gingivalis* tyrosine phosphatase is a multifunctional regulator of virulence attributes. Mol. Microbiol. *69*, 1153–1164.

Mah, T.F., and O'Toole, G.A. (2001). Mechanisms of biofilm resistance to antimicrobial agents. Trends Microbiol. *9*, 34–39.

Marsh, P.D. (1989). Host defenses and microbial homeostasis: role of microbial interactions. J. Dent. Res. *68*, 1567–1575.

Marsh, P.D. (1994). Microbial ecology of dental plaque and its significance in health and disease. Adv. Dent. Res. *8*, 263–271.

Merritt, J., Qi, F., Goodman, S.D., Anderson, M.H., and Shi, W. (2003). Mutation of luxS affects biofilm formation in *Streptococcus mutans*. Infect. Immun. *71*, 1972–1979.

Merritt, J., Niu, G., Okinaga, T., and Qi, F. (2009). Autoaggregation response of *Fusobacterium nucleatum*. Appl. Environ. Microbiol. *75*, 7725–7733.

Meurman, J.H. (2005). Probiotics: do they have a role in oral medicine and dentistry? Eur. J. Oral Sci. *113*, 188–196.

Miller, S.T., Xavier, K.B., Campagna, S.R., Taga, M.E., Semmelhack, M.F., Bassler, B.L., and Hughson, F.M. (2004). *Salmonella* typhimurium recognizes a chemically distinct form of the bacterial quorum-sensing signal AI-2. Mol. Cell *15*, 677–687.

Mitchell, R.J., Lee, S.K., Kim, T., and Ghim, C.M. (2011). Microbial linguistics: perspectives and applications of microbial cell-to-cell communication. BMB Rep. *44*, 1–10.

Modesto, M., Biavati, B., and Mattarelli, P. (2006). Occurrence of the family bifidobacteriaceae in human dental caries and plaque. Caries Res. *40*, 271–276.

Mok, K.C., Wingreen, N.S., and Bassler, B.L. (2003). *Vibrio harveyi* quorum sensing: a coincidence detector for two autoinducers controls gene expression. EMBO J. *22*, 870–881.

Mouton, C., Reynolds, H.S., and Genco, R.J. (1980). Characterization of tufted streptococci isolated from the 'corn cob' configuration of human dental plaque. Infect. Immun. *27*, 235–245.

Nagaoka, S., Hojo, K., Murata, S., Mori, T., Ohshima, T., and Maeda, N. (2008). Interactions between salivary Bifidobacterium adolescentis and other oral bacteria: in vitro coaggregation and coadhesion assays. FEMS Microbiol. Lett. *281*, 183–189.

Nealson, K.H., Platt, T., and Hastings, J.W. (1970). Cellular control of the synthesis and activity of the bacterial luminescent system. J. Bacteriol. *104*, 313–322.

Ng, W.L., and Bassler, B.L. (2009). Bacterial quorum-sensing network architectures. Annu. Rev. Genet. *43*, 197–222.

Nilius, A.M., Spencer, S.C., and Simonson, L.G. (1993). Stimulation of in vitro growth of *Treponema denticola* by extracellular growth factors produced by *Porphyromonas gingivalis*. J. Dent. Res. *72*, 1027–1031.

Nobbs, A.H., Lamont, R.J., and Jenkinson, H.F. (2009). *Streptococcus* adherence and colonization. Microbiol. Mol. Biol. Rev. *73*, 407–450, Table of Contents.

Nobbs, A.H., Jenkinson, H.F., and Jakubovics, N.S. (2011). Stick to your gums: mechanisms of oral microbial adherence. J. Dent. Res. *90*, 1271–1278.

Nonnenmacher, C., Mutters, R., and de Jacoby, L.F. (2001). Microbiological characteristics of subgingival microbiota in adult periodontitis, localized juvenile periodontitis and rapidly progressive periodontitis subjects. Clin. Microbiol. Infect. *7*, 213–217.

Novak, E.A., Shao, H., Daep, C.A., and Demuth, D.R. (2010). Autoinducer-2 and QseC control biofilm formation and in vivo virulence of *Aggregatibacter actinomycetemcomitans*. Infect. Immun. *78*, 2919–2926.

Palmer, R.J., Jr., Diaz, P.I., and Kolenbrander, P.E. (2006). Rapid succession within the Veillonella population of a developing human oral biofilm in situ. J. Bacteriol. *188*, 4117–4124.

Palmer, R.J., Jr., Wu, R., Gordon, S., Bloomquist, C.G., Liljemark, W.F., Kilian, M., and Kolenbrander, P.E. (2001). Retrieval of biofilms from the oral cavity. Methods Enzymol. *337*, 393–403.

Parsek, M.R., and Greenberg, E.P. (2005). Sociomicrobiology: the connections between quorum sensing and biofilms. Trends Microbiol. *13*, 27–33.

Pereira, C.S., McAuley, J.R., Taga, M.E., Xavier, K.B., and Miller, S.T. (2008). Sino*Rhizobium meliloti*, a bacterium lacking the autoinducer-2 (AI-2) synthase, responds to AI-2 supplied by other bacteria. Mol. Microbiol. *70*, 1223–1235.

Periasamy, S., and Kolenbrander, P.E. (2009). Mutualistic biofilm communities develop with *Porphyromonas gingivalis* and initial, early, and late colonizers of enamel. J. Bacteriol. *191*, 6804–6811.

Petersen, F.C., Pecharki, D., and Scheie, A.A. (2004). Biofilm mode of growth of *Streptococcus* intermedius favored by a competence-stimulating signalling peptide. J. Bacteriol. *186*, 6327–6331.

Postollec, F., Norde, W., van der Mei, H.C., and Busscher, H.J. (2003). Enthalpy of interaction between coaggregating and non-coaggregating oral bacterial pairs – a microcalorimetric study. J. Microbiol. Methods *55*, 241–247.

Postollec, F., Norde, W., de Vries, J., Busscher, H.J., and van der Mei, H.C. (2006). Interactive forces between co-aggregating and non-co-aggregating oral bacterial pairs. J. Dent. Res. *85*, 231–234.

Rickard, A.H., Bachrach, G., and Davies, D.G. (2008a). Cell-cell communication in oral microbial communities. In Molecular Oral Microbiology, Rogers, A.H., ed. (Caister Academic Press, New York), pp. 87–108.

Rickard, A.H., Campagna, S.R., and Kolenbrander, P.E. (2008b). Autoinducer-2 is produced in saliva-fed flow conditions relevant to natural oral biofilms. J. Appl. Microbiol. *105*, 2096–2103.

Rickard, A.H., Gilbert, P., High, N.J., Kolenbrander, P.E., and Handley, P.S. (2003). Bacterial coaggregation: an integral process in the development of multi-species biofilms. Trends Microbiol. *11*, 94–100.

Rickard, A.H., Palmer, R.J., Jr., Blehert, D.S., Campagna, S.R., Semmelhack, M.F., Egland, P.G., Bassler, B.L., and Kolenbrander, P.E. (2006). Autoinducer 2: a concentration-dependent signal for mutualistic bacterial biofilm growth. Mol. Microbiol. *60*, 1446–1456.

Ritz, H.L. (1967). Microbial population shifts in developing human dental plaque. Arch. Oral Biol. *12*, 1561–1568.

Ritz, H.L. (1969). Fluorescent antibody staining of *Neisseria, Streptococcus* and *Veillonella* in frozen sections of human dental plaque. Arch. Oral Biol. *14*, 1073–1083.

Rosan, B., Lai, C.H., and Listgarten, M.A. (1976). *Streptococcus* sanguis: a model in the application in immunochemical analysis for the in situ localization of bacteria in dental plaque. J. Dent. Res. *55*, A124–141.

Rosen, G., Nisimov, I., Helcer, M., and Sela, M.N. (2003). *Actinobacillus actinomycetemcomitans* serotype b lipopolysaccharide mediates coaggregation with *Fusobacterium nucleatum*. Infect. Immun. *71*, 3652–3656.

Ruby, E.G., and Nealson, K.H. (1976). Symbiotic association of Photobacterium fischeri with the marine luminous fish *Monocentris japonica*; a model of symbiosis based on bacterial studies. Biol Bull *151*, 574–586.

Ruhl, S., Sandberg, A.L., and Cisar, J.O. (2004). Salivary receptors for the proline-rich protein-binding and lectin-like adhesins of oral actinomyces and streptococci. J. Dent. Res. *83*, 505–510.

Sbordone, L., and Bortolaia, C. (2003). Oral microbial biofilms and plaque-related diseases: microbial communities and their role in the shift from oral health to disease. Clin. Oral Investig. *7*, 181–188.

Scheie, A.A. (1994). Mechanisms of dental plaque formation. Adv. Dent. Res. *8*, 246–253.

Semmelhack, M.F., Campagna, S.R., Federle, M.J., and Bassler, B.L. (2005). An expeditious synthesis of DPD and boron binding studies. Org Lett *7*, 569–572.

Semmelhack, M.F., Campagna, S.R., Hwa, C., Federle, M.J., and Bassler, B.L. (2004). Boron binding with the quorum sensing signal AI-2 and analogues. Org Lett *6*, 2635–2637.

Shao, H., James, D., Lamont, R.J., and Demuth, D.R. (2007a). Differential interaction of Aggregatibacter (Actinobacillus) actinomycetemcomitans LsrB and RbsB proteins with autoinducer 2. J. Bacteriol. *189*, 5559–5565.

Shao, H., Lamont, R.J., and Demuth, D.R. (2007b). Autoinducer 2 is required for biofilm growth of Aggregatibacter (Actinobacillus) actinomycetemcomitans. Infect. Immun. *75*, 4211–4218.

Socransky, S.S., Haffajee, A.D., Cugini, M.A., Smith, C., and Kent, R.L., Jr. (1998). Microbial complexes in subgingival plaque. J. Clin. Periodontol. *25*, 134–144.

Socransky, S.S., Manganiello, A.D., Propas, D., Oram, V., and van Houte, J. (1977). Bacteriological studies of developing supragingival dental plaque. J. Periodontal. Res. *12*, 90–106.

Sperandio, V., Torres, A.G., and Kaper, J.B. (2002). Quorum sensing *Escherichia* coli regulators B and C (QseBC): a novel two-component regulatory system involved in the regulation of flagella and motility by quorum sensing in *E. coli*. Mol. Microbiol. *43*, 809–821.

Stewart, P.S., and Franklin, M.J. (2008). Physiological heterogeneity in biofilms. Nat. Rev. Microbiol. *6*, 199–210.

Stoodley, P., Sauer, K., Davies, D.G., and Costerton, J.W. (2002). Biofilms as complex differentiated communities. Annu. Rev. Microbiol. *56*, 187–209.

Suntharalingam, P., and Cvitkovitch, D.G. (2005). Quorum sensing in streptococcal biofilm formation. Trends Microbiol. *13*, 3–6.

Surette, M.G., Miller, M.B., and Bassler, B.L. (1999). Quorum sensing in *Escherichia* coli, *Salmonella* typhimurium, and *Vibrio harveyi*: a new family of genes responsible for autoinducer production. Proc. Natl. Acad. Sci. U.S.A. *96*, 1639–1644.

Sutherland, I. (2001). Biofilm exopolysaccharides: a strong and sticky framework. Microbiology *147*, 3–9.

Sztajer, H., Lemme, A., Vilchez, R., Schulz, S., Geffers, R., Yip, C.Y., Levesque, C.M., Cvitkovitch, D.G., and Wagner-Dobler, I. (2008). Autoinducer-2-regulated genes in *Streptococcus mutans* UA159 and global metabolic effect of the luxS mutation. J. Bacteriol. *190*, 401–415.

Takenaka, S., Trivedi, H.M., Corbin, A., Pitts, B., and Stewart, P.S. (2008). Direct visualization of spatial and temporal patterns of antimicrobial action within model oral biofilms. Appl. Environ. Microbiol. *74*, 1869–1875.

Turovskiy, Y., Kashtanov, D., Paskhover, B., and Chikindas, M.L. (2007). Quorum sensing: fact, fiction, and everything in between. Adv. Appl. Microbiol. *62*, 191–234.

Vaun McArthur, J. (2006). Microbial Ecology: an Evolutionary Approach (Academic Press, New York).

Whitmore, S.E., and Lamont, R.J. (2011). The pathogenic persona of community-associated oral streptococci. Mol. Microbiol. *81*, 305–314.

Wilson, M. (2005). Microbial Inhabitants of Humans: Their Ecology and Role in Health and Disease (Cambridge, Cambridge University Press).

Wimpenny, J.W.T. (2000). An overview of biofilms as functional communities. In Community Structure and Cooperation in Biofilms, Allison, D.G., Gilbert, P., Lappin-Scott, H.M., and Wilson, M., eds. (Cambridge University Press, Cambridge).

Winzer, K., Hardie, K.R., Burgess, N., Doherty, N., Kirke, D., Holden, M.T., Linforth, R., Cornell, K.A., Taylor, A.J., Hill, P.J., *et al.* (2002). LuxS: its role in central metabolism and the in vitro synthesis of 4-hydroxy-5-methyl-3(2H)-furanone. Microbiology *148*, 909–922.

Yang, J., Ritchey, M., Yoshida, Y., Bush, C.A., and Cisar, J.O. (2009). Comparative structural and molecular characterization of ribitol-5-phosphate-containing *Streptococcus oralis* coaggregation receptor polysaccharides. J. Bacteriol. *191*, 1891–1900.

Yoshida, A., Ansai, T., Takehara, T., and Kuramitsu, H.K. (2005). LuxS-based signalling affects *Streptococcus mutans* biofilm formation. Appl. Environ. Microbiol. *71*, 2372–2380.

Yoshida, Y., Palmer, R.J., Yang, J., Kolenbrander, P.E., and Cisar, J.O. (2006). Streptococcal receptor polysaccharides: recognition molecules for oral biofilm formation. BMC Oral Health *6(Suppl. 1)*, S12.

Zijnge, V., van Leeuwen, M.B., Degener, J.E., Abbas, F., Thurnheer, T., Gmur, R., and Harmsen, H.J. (2010). Oral biofilm architecture on natural teeth. PLoS One *5*, e9321.

# Environmental Sensory Perception by Oral Streptococci

# 8

Justin Merritt and Jens Kreth

## Abstract

Oral streptococci encounter an exceptionally wide range of environmental stresses and population densities. These stimuli are sensed by efficient detection systems that also coordinate the appropriate adaptive genetic responses. The majority of these detection systems utilize membrane bound sensory proteins that are directly or indirectly regulated by their sensed stimuli. Such systems play an intimate role in mitigating the potential damage caused by changes in redox potential, fluctuations in local pH, and toxicity from antimicrobial agents. In addition, the typical life cycle of oral streptococci includes a transition from growth in a relatively low cell density planktonic state to an extremely high cell density biofilm environment. Consequently, various sensory systems are dedicated to detecting this increase in population density and regulating the genetic pathways that are essential for persistence in a highly competitive multispecies biofilm environment. Recent studies have identified many of the targets of these sensory systems and have provided unprecedented insight into the intimate connection between the constantly changing oral environment and the genetic machinery of oral bacteria.

## Environmental sensing by oral streptococci

The ability of bacteria to sense and respond to environmental stimuli is a fundamental aspect of survival in a dynamic environment and critical for regulating the virulence of pathogens. Consequently, there is a tremendous interest to understand how seemingly simple single-celled organisms can respond to an exceptional diversity of environmental information. Some of the earliest research in this area was focused upon the ability of bacteria to detect and selectively import preferred carbohydrates from the environment for catabolism. Initially, it was observed that many bacterial species are able to reliably discriminate between carbon sources when fed a mixture of sugars. Rather than importing all sugars through a general transport mechanism, bacteria were found to encode multiple sensory/transport systems that selectively import particular carbohydrates in a distinct hierarchy of preference. These sensory systems, referred to as phosphoenolpyruvate:sugar phosphotransferase systems (PTS systems) were initially described in *E. coli* in 1964 (Kundig *et al.*, 1964) and a few years later were also found to exist in oral streptococci (Kanapka and Hamilton, 1971). Since then, numerous carbohydrate-specific PTS systems have been identified through physiological and genetic studies and more recently, through genome sequencing (Vadeboncoeur and Pelletier, 1997; Ajdic *et al.*, 2002; Vickerman *et al.*, 2007; Xu *et al.*, 2007). PTS systems encode permeases that exhibit a high affinity for a particular carbohydrate substrate, which is phosphorylated upon importation to the cell (Vadeboncoeur and Pelletier, 1997). If a preferred PTS carbohydrate is available in the environment, its transport into the cell will activate the transcription of the genes encoding the corresponding PTS. In this way, the cell is able to connect the expression of its substrate-specific carbohydrate transport systems to the availability of carbohydrate sources in the environment.

More recently, a large proportion of oral streptococcal environmental sensory research has been devoted to studies of stress adaptation. The early studies in this area were more focused upon the systems employed by streptococci to mitigate the damage caused by environmental stress, rather than the actual stress detection systems. However, this research has directly led to the current state-of-the-art, which is now heavily focused upon the connection between stress signal detection and the regulation of these stress resistance pathways. One of the most recent topics to emerge in environmental sensing research involves cell density perception. It has been known for some time that certain species of bacteria regulate gene expression in response to cell density. However, it was not until the 1990s that this phenomenon was recognized to be a ubiquitous ability of bacteria (Bassler and Losick, 2006). Now it is known that cell density plays a major role in regulating numerous genetic pathways, and particularly those pathways involved in the production of virulence factors and accessory genes. In the following sections, we will begin with a description of some of the general mechanisms used by oral streptococci to sense environmental changes. This will be followed by a detailed examination of the current knowledge regarding several model sensory systems that were chosen to illustrate some of the most recent advances in environmental sensory perception among oral streptococci.

## Sensing mechanisms

In general, prokaryotes are unicellular organisms that typically reside in densely populated biofilm communities. For most organisms, growth occurs under a multitude of different environmental conditions, which requires that the individual cells have the capacity to effectively respond to a constantly changing environment. The cellular response is usually rapid and typically utilizes either membrane-embedded or cytoplasmic sensors. Through the interplay of these sensors, the cell is able to respond effectively, by integrating extracellular environmental stimuli and intracellular metabolic information into the optimal adaptive response. It is therefore not surprising that signal perception systems are intimately integrated into multigene regulatory networks. The regulatory read-out is often a change in the transcriptional profile of the cell, which occurs via the coordinated effort of sensor proteins and their associated DNA-binding transcription regulators.

### Features and mechanisms of two-component signal transduction systems

The best-characterized signalling proteins in bacteria are grouped under the general term two-component signal transduction systems; often abbreviated as TCSTS or TCS. (For a comprehensive collection of bacterial TCS, see http://www.p2cs.org/.) The general TCS architecture is composed of two protein components, a membrane-bound sensor kinase for signal perception and a cytosolic response regulator responsible for influencing specific gene expression via defined DNA sequences on the chromosome. A bioinformatic analysis of the putative TCS systems in the genome of the cariogenic organism *Streptococcus mutans* UA159 revealed the presence of 14 TCS (Ajdic *et al.*, 2002). Each of these TCS has been subjected to genetic manipulation to examine their specific role in environmental sensing in *S. mutans* (Levesque *et al.*, 2007; Biswas *et al.*, 2008). The oral commensal *Streptococcus sanguinis* encodes 14 TCS as well (Xu *et al.*, 2007). Though, considerably less is known about the function of its TCS in environmental adaptation. Both the sensor kinases and response regulators of TCS have distinct domains that are essential for properly transducing sensory information. Sensor kinases are usually membrane bound and contain at least one transmembrane region; however, examples of cytosolic sensor kinases are also described in the literature. In *S. mutans* UA159 (for genome annotation information see http://www.ncbi.nlm.nih.gov/genome/856), the uncharacterized sensor kinase annotated as SMU.45 is the only histidine kinase that does not seem to encode a transmembrane region, indicating that it may be an example of a cytosolic sensor kinase (Fig. 8.1A). Sensor kinases have a variable assortment of potential receptor domains, which allow them to sense a highly diverse array of stimuli. These domains are connected to a conserved transmitter domain that is responsible for altering the

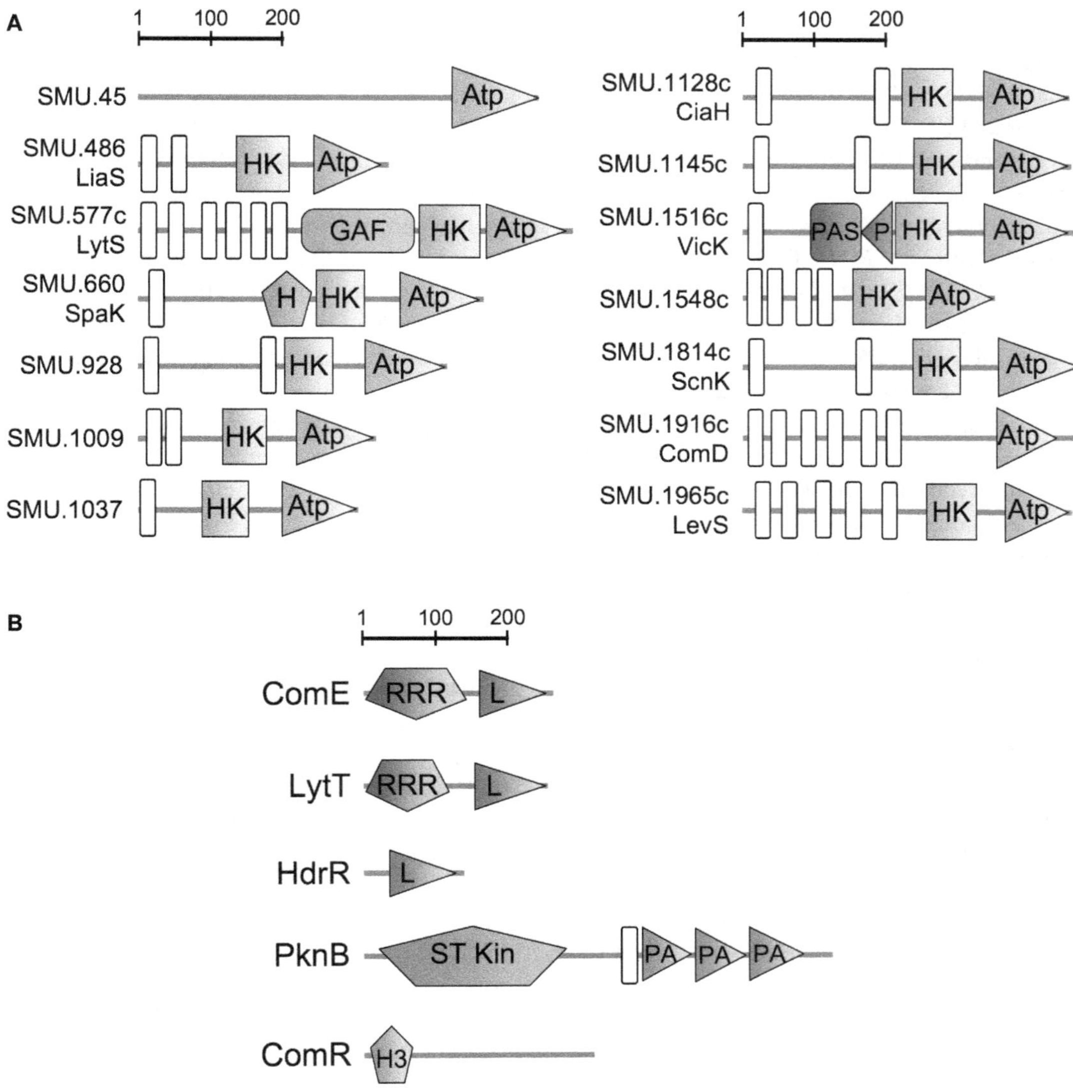


**Figure 8.1** Architecture of proteins used for environmental sensing. All proteins listed in this figure are adapted from the results of analyses performed using the SMART protein architecture tool available at http://smart.embl-heidelberg.de/. All proteins are drawn relative to the included scalebar, which indicates the number of amino acids. (A) Each of the confirmed and putative histidine kinases of *S. mutans* UA159 is shown with their predicted domains located in their approximate positions within the proteins. Proteins are drawn with their N- and C-termini oriented from left to right and white bars represent individual transmembrane segments. Domain abbreviations are listed as follows: Atp, ATPase domain; HK, histidine kinase domain; GAF, GAF domain; H, HAMP domain; PAS, PAS domain; P, PAC domain. All proteins are listed by their NCBI locus tag assignment and common protein names are also included if available. (B) Domain architecture of other proteins important for environmental sensing. Domain abbreviations are listed as follows: 'RRR' response regulator receiver domain; L, LytTR DNA binding domain; ST Kin, serine/threonine kinase domain; PA, PASTA domain; H3, HTH-3 DNA binding domain. Transmembrane segments are illustrated by white bars.

phosphorylation state of the kinase in response to a perceived stimulus. Many sensor kinases also encode structural domains that participate in signal perception. Examples found among *S. mutans* TCS include the PAS, PAC, HAMP and GAF domains (Fig. 8.1A).

PAS domains (named for Per, ARNT and Sim proteins) are a ubiquitous class of cytosolic signalling domains in proteins involved in monitoring changes in light, redox potential, oxygen, small ligands, and the energy status of the cells (Taylor and Zhulin, 1999). PAS domains are not only found in TCS, but also in other proteins involved in signalling. Several co-factors have been identified, including *p*-coumaric acid, flavine mononucleotide (FMN), flavine adenine dinucleotide (FAD), C3 and C4 carbohydrates, haem, and others (Moglich *et al.*, 2009).

Haem-binding PAS domains are frequently used as sensors for gaseous ligands like oxygen, carbon monoxide (Hefti *et al.*, 2004; Moglich *et al.*, 2009), and nitrous oxide (Rao *et al.*, 2011). In general, it seems that the sensing of light, oxygen, or redox potential requires co-factors, whereas signals such as voltage, xenobiotics, and nitrogen availability do not (Ashby, 2006). In *S. mutans*, the only TCS containing a PAS module is the sensor kinase VicK (Fig. 8.1A). By comparison, the Gram-positive model organism *B. subtilis* contains nine proteins with a single PAS domain and one protein with three PAS domains.

PAC domains (PAS-associated C-terminal motif) are usually associated with the C-terminal end of PAS domains and contribute to the PAS domain fold. This fold forms a specific cavity that was shown to be the signal-binding site. The PAS domain of the *S. mutans* sensor kinase VicK is followed by a PAC domain.

GAF domains (stands for cGMP-specific and -stimulated phosphodiesterases, *Anabaena* adenylate cyclases and *Escherichia coli* FhlA) are widely distributed small-molecule-binding-domains (Aravind and Ponting, 1997). Initially identified as a non-catalytic cGMP-binding domain in a bovine cone photoreceptor (Charbonneau *et al.*, 1990), structural analysis showed a close evolutionary relationship to PAS domains, as both share a similar three dimensional structure (Ho *et al.*, 2000). GAF domains have been shown to bind a variety of ligands, including cGMP, cAMP, haem and bilin (Cornilescu *et al.*, 2008; Ikeuchi and Ishizuka, 2008; Martinez *et al.*, 2008). The *S. mutans* TCS sensor kinase LytS contains a single GAF domain. However, *S. mutans* is not known to respond to any of the aforementioned GAF domain ligands.

HAMP domains are found in many bacterial signalling proteins. They mediate the input–output signalling of histidine kinases, adenyl cyclases, methyl-accepting chemotaxis proteins and phosphatases (hence, HAMP) (Aravind and Ponting, 1999; Williams and Stewart, 1999). Studies of HAMP domain function suggest that perceived stimuli induce conformational changes in the HAMP domain structure. Consequently, two alternative HAMP structures can be observed, a tightly packed HAMP(A) and a more loosely packed HAMP(B) arrangement of HAMP subunits. The different states are responsible for regulating the kinase activity associated with the C-terminal end of sensor kinases (Parkinson, 2010). A single HAMP domain is only found in the sensor kinase SpaK of *S. mutans* (Fig. 8.1A), whereas other oral streptococci such as *S. gordonii*, *S. mitis* and *S. sanguinis* have multiple sensor kinases with putative HAMP domains (J. Merritt and J. Kreth, unpublished).

The transfer of phosphoryl groups to response regulator proteins is essential for signal transduction by TCS. The transmitter domain of the sensor kinase is responsible for mediating *trans* auto-phosphorylation upon stimulus perception by the receptor domain. For auto-phosphorylation to occur, sensor kinases typically dimerize into homo-dimers with the γ-phosphoryl group of a bound ATP molecule being transferred from one sensor kinase monomer to a conserved histidine residue in the transmitter domain of the other. The phosphorylated transmitter domain then transfers the phosphoryl group to a conserved aspartyl residue within the response regulator receiver domain. Sensor kinases also have phosphatase activity, which enables dephosphorylation of the response regulator to reset the signalling circuit. Response regulators show differences in their phosphorylation half-lives, which results in TCS-specific response times for particular environmental stimuli. In addition, the sensor kinases themselves show individual variation in phosphorylation stability, which further increases the complexity and sensitivity of the signal adaptation mechanism (Krell *et al.*, 2010). Once a response regulator has been phosphorylated, the protein will undergo a significant change in its ability to interact with DNA. All of the TCS response regulator output domains of the sequenced oral streptococci contain conserved structural motifs required for DNA binding. Furthermore, activated response regulators frequently bind to multiple sites on the chromosome containing the specific consensus binding sequences for the respective response regulators. This enables a single environmental stimulus to be utilized for a multigene adaptive response.

## Response regulators with an unusual DNA binding mechanism – the LytTR domain

The response regulators LytT and ComE, from the TCS LytTS and ComDE belong to the family of AlgR/AgrA/LytR-like transcriptional regulators that possess the LytTR or 'litter' domain (Fig. 8.1B). These transcription regulators show an unusual mechanism of protein–DNA interaction that is distinct from the classic helix–turn–helix DNA-binding structure (Galperin, 2008). The LytTR domain of the *Staphylococcus aureus* AgrA transcription regulator has recently been cocrystallized with bound target DNA. Structural analysis revealed a 10-stranded beta fold interacting with two adjacent major grooves and the intervening minor groove of the target DNA sequence, leading to a significant bend within the DNA (Sidote *et al.*, 2008). Although mostly found as DNA interacting elements in TCS response regulators, the LytTR domain is also used by non-TCS transcription regulators. For example, as discussed later, LytTR domains can be found in the transcription regulators of the newly identified HdrRM (Fig. 8.1B) and BsrRM regulatory systems from *S. mutans* (Okinaga *et al.*, 2010a; Xie *et al.*, 2010).

## Eukaryote-like serine/threonine kinases (eSTKs)

TCS are probably the predominant signal transduction systems utilized by oral streptococci. However, in recent years it has become evident that prokaryotes also utilize serine/threonine phosphorylation to regulate gene expression. Previously, it was assumed that serine/threonine phosphorylation was exclusively a eukaryotic signal transduction mechanism. Now, eukaryote-like serine/threonine kinases (eSTKs) are known to be widely distributed among prokaryotes and are implicated in the regulation of important processes like bacterial development, secondary metabolism, cell division, cell wall synthesis, and virulence (Pereira *et al.*, 2011).

The molecular architecture of prokaryotic eSTKs is quite diverse, which indicates the development of specialized species-specific functions (Krupa and Srinivasan, 2005). The serine/threonine catalytic domain, however, is highly conserved and places the prokaryotic eSTKs within the eukaryotic protein kinase superfamily (Hanks and Hunter, 1995). The catalytic domain contains a characteristic two-lobed core structure with the catalytic site hidden deeply within the cleft formed by the two lobes. The phosphoryl group from ATP is transferred via the C-terminal lobe to the respective protein substrates (Pereira *et al.*, 2011). Although it is not uncommon for bacteria to encode multiple eSTKs, oral streptococci only possess one. The kinase domain at the N-terminus is located in the cytoplasm and is followed by a transmembrane segment connected to the extracellular portion of the protein. The extracellular portion contains the characteristic PASTA domains (for penicillin-binding protein and serine/threonine kinase associated) (Yeats *et al.*, 2002). In the case of oral streptococci, the eSTKs all contain three consecutive PASTA domain repeats (Fig. 8.1B). PASTA domains were first identified in the *Streptococcus pneumoniae* penicillin binding protein PBP2X and were suggested to bind non-cross-linked peptidoglycan molecules called muropeptides (Pares *et al.*, 1996). Recently, this ability was confirmed with *in vitro* assays (Maestro *et al.*, 2011). For *B. subtilis* spores, it seems that the PASTA domains might act as sensors for the presence of muropeptides to transmit environmental cues indicating active cell division and therefore favourable growth conditions (Shah *et al.*, 2008). However, it has also been suggested that the eSTK from *S. pneumoniae* might sense intracellular peptidoglycan, since it is activated continuously during growth and inhibited during growth arrest (Novakova *et al.*, 2010). This would be consistent with the observation that the eSTK from *S. pneumoniae* localizes with the cell division apparatus and can interact with one of the major cell division proteins FtsZ. Although, the same study used antibodies to demonstrate the surface exposure of the eSTK, which would also seem to indicate a role in extracellular signal detection (Giefing *et al.*, 2010). Unlike oral streptococci, the eSTK from *S. pneumoniae* contains four PASTA domains and is therefore about 60 aa longer than the predicted oral streptococcal eSTKs. Thus, there may be some species-specific differences in localization or function.

A search for substrates of eSTK-dependent phosphorylation has been performed in *S. pneumoniae* and several proteins have been identified. In addition to the aforementioned FtsZ, the cell division protein DivIVA, the pyrophosphatase PpaC, and the hypothetical protein Spr0334 were demonstrated to be phosphorylated *in vivo*. Phenotypic studies of eSTK mutants suggest that streptococcal eSTKs could be involved in the control or monitoring of cell division. Since cell division is directly linked to the availability of environmental nutrients and permissive growth conditions, eSTKs might provide a direct link to the decision making process used for cellular proliferation.

## Environmental stress detection

The oral environment imposes a multitude of stresses upon the resident flora that constantly challenge their basic survival as well as their ability to compete for ecological niches with other organisms. Unlike many other sites in the human body, the oral cavity is an open system that is constantly inoculated with foreign micro-organisms and is subject to extreme fluctuations in temperature, pH, osmolarity, redox potential, and numerous other potentially hostile environmental changes. This poses an interesting challenge for the oral flora because they need to remain more competitive than the numerous other species that constantly enter the oral cavity and do so under a wide variety of growth conditions. Not surprisingly, oral streptococci have evolved specific mechanisms to persist in such conditions and much is already known about the machinery required to mitigate the effects of cellular damage caused by environmental stress. However, until recently, much less was understood about how these cells are able to coordinate the complex genetic responses utilized to adapt to a suddenly hostile environment. Research in recent years has shown that oral streptococcal species have a variety of membrane bound sensory systems that are intimately involved in coordinating these functions. Most of the research in this area has focused upon sensory systems that are required to survive the effects of oxidation, pH, and toxins/antimicrobials. Each of these stresses is a constant threat for oral streptococci and originates from a variety of host- and/or flora-derived components in the oral cavity.

### Oxidative stress

As pioneering colonizers of the tooth surface, oral streptococci are exposed to a relatively high concentration of oxygen during the early stages of dental plaque formation, due to the aerobic environment of the oral cavity. As a consequence, streptococcal metabolism will continually yield $H_2O_2$ and various oxygen radicals as a byproduct of the reduction of $O_2$. In addition, certain oral streptococci like the mitis group species further augment their $H_2O_2$ production via pathways involving the pyruvate oxidase (Kreth *et al.*, 2009; Chen *et al.*, 2011; Zheng *et al.*, 2011), lactate oxidase (Tong *et al.*, 2007), and L-amino acid oxidase (Tong *et al.*, 2008). This increased production of $H_2O_2$ is moderately toxic to the producers and exquisitely lethal to more sensitive organisms like *S. mutans* (Kreth *et al.*, 2005b, 2008). In addition to the $H_2O_2$ produced from bacterial sources, multiple oxidases in the salivary glands ensure that saliva also contains a continuous supply of $H_2O_2$ (Geiszt *et al.*, 2003) and other oxygen radicals (Ihalin *et al.*, 2006; Ashby, 2008). In addition, oral neutrophils can further increase this pool of reactive oxygen species through the release of oxidative bursts (Sato *et al.*, 2008). Thus, there is a potentially wide range of oxidative stresses encountered by oral streptococci, depending upon the availability of oxygen in the community and the amount of exposure to saliva. As the biofilm grows, the surrounding environment will become increasingly more anaerobic and less affected by saliva (von Ohle *et al.*, 2010), but the whole process will be repeated once the biofilm has been removed and cells are once again exposed to the oral environment.

### pH stress

Saliva, like many other mucosal secretions, has a strong buffering capacity and a pH near to neutral (Tabak, 2006; Dawes, 2008). This buffering function is essential to prevent the acidic dissolution of the calcium phosphate-containing mineral components of the tooth (Tabak, 2006). Despite the presence of saliva, plaque biofilm pH values are subject to a dynamic range that can span three pH units or more, largely due to the acidic metabolic waste products excreted by lactic acid bacteria (Jensen *et al.*, 1982). The cycles of food ingestion from the host creates a situation where plaque pH values are constantly fluctuating between nearly neutral and highly acidic. This drop in pH can also be remarkably rapid depending upon the concentration of freely available fermentable carbohydrates (Englander, 1959). As lactic acid-producing bacteria, oral streptococci are well equipped to handle growth in acidic environments, but similar to oxidative stress resistance, acid tolerance abilities also vary widely between species. Multiple studies of oral streptococci have shown that acid tolerance is a largely inducible adaptation process, rather than an inherent resistance (Quivey *et al.*, 2000b, 2001). For example, cells grown at pH 5 will be much more resistant to acid killing than those maintained at a higher pH (Belli and Marquis, 1991). There are a variety of mechanisms used by oral streptococci to survive acid challenge, such as proton extrusion (Kuhnert *et al.*, 2004), changes in membrane structure (Quivey *et al.*, 2000a), production of alkaline molecules (Dong *et al.*, 2002; Griswold *et al.*, 2004), induction of DNA repair enzymes and protein chaperones (Jayaraman *et al.*, 1997; Hahn *et al.*, 1999; Hanna *et al.*, 2001), and others (Matsui and Cvitkovitch, 2010). If the cell is impaired in any of these abilities, there will be a measurable reduction in the survival rate at acidic pH values. Therefore, it is critical that oral streptococci have the ability to efficiently coordinate the regulation of a large number of different genetic pathways in response to sudden changes in plaque pH.

### Toxin/antimicrobial stress

Dental plaque and saliva are both particularly rich sources of antimicrobial molecules. These antimicrobials are typically involved in attacking various components of the bacterial cell envelope (cell membrane and cell wall). For biofilm-embedded streptococci, the principal antimicrobial threats come from neighbouring organisms that produce ribosomally synthesized peptide antibiotics called bacteriocins. In general, lactic acid bacteria are robust producers of a wide variety of both lantibiotic (extensive post-translational modifications) and non-lantibiotic bacteriocins (few, if any, post-translational modifications) (Nes *et al.*, 2007). Most bacteriocins are pore forming peptides that target the cell membrane, while some lantibiotics are also known to disrupt the cell wall by inhibiting peptidoglycan synthesis (Cotter *et al.*, 2005). In the densely populated biofilm environment, bacteriocins are particularly effective at inhibiting nearby competitor species to prevent their overgrowth and dominance of the population (Kreth *et al.*, 2005b). The saliva also contains a variety of antimicrobial substances, which are specifically produced by the host to control microbial growth in the oral cavity. Whole saliva contains an abundance of lysozyme (Klimiuk *et al.*, 2006), which is a cell wall degrading hydrolytic enzyme specific for the 1,4-beta-linkages in peptidoglycan (Callewaert and Michiels, 2010). In addition, saliva contains a variety of cationic antimicrobial peptides such as the α- and β-defensins (Mathews *et al.*, 1999; Puklo *et al.*, 2008), histatin (Johnson *et al.*, 2000), LL-37 (Tao *et al.*, 2005), and others (Gorr and Abdolhosseini, 2011). Like the bacteriocins, human antimicrobial peptides generally target the cell membrane and form pores that will result in cytoplasmic leakage and eventual lethality (De Smet and Contreras, 2005). Given the ubiquity of various cell envelope damaging substances in the oral cavity, it is essential that oral streptococci have the ability to detect when envelope damaging agents are present in order to mount the appropriate adaptive response.

## Stress adaptation via two-component signal transduction systems

Multiple studies suggest that TCS play a major role for mediating the cellular response to environmental stress in oral streptococci. As mentioned previously, the 14 histidine kinases of *S. mutans* have all been systematically mutated and tested phenotypically for various defects in stress tolerance (Levesque *et al.*, 2007; Biswas *et al.*, 2008). The LiaFSR, CiaXRH, and VicRKX TCS (Fig. 8.1A) were each found to participate in multiple stress resistance pathways (Biswas *et al.*, 2008), while the TCS encoded by SMU.1145 – 1146 and SMU.1964 – 1965 (Fig. 8.1A) seem to have more specific functions in acid resistance (Levesque *et al.*, 2007). The remaining *S. mutans* TCS either do not function in stress resistance or they are required for growth conditions that have yet to be tested experimentally. In the following sections, we will examine the two most thoroughly characterized TCS to illustrate how TCS are able to connect sensory perception with resistance to some of the previously described environmental stresses found in the oral cavity.

## The VicRKX redox sensory system

In *S. mutans*, the VicRKX operon encodes a two-component signal transduction system that plays a major role in stress resistance (Deng *et al.*, 2007; Senadheera *et al.*, 2007). In other oral streptococci, it has not been demonstrated whether similar VicRKX systems exist. However, highly homologous operons can be found in the genomes of *S. gordonii*, *S. mitis*, *S. sanguinis*, *S. pneumoniae*, and *S. pyogenes*, which suggests that most, if not all, streptococci encode VicRKX systems (Wagner *et al.*, 2002; Ng and Winkler, 2004; Liu *et al.*, 2006). In addition to sharing similar operon arrangements, each of the VicK sensor kinases from all of these organisms contains a PAS domain and a PAC motif (Wagner *et al.*, 2002; Deng *et al.*, 2007). Recent crystallographic and NMR studies suggest that the PAS domain and PAC motif are structurally linked and both are required to form the PAS fold (Hefti *et al.*, 2004). The PAS domain is highly conserved among numerous sensory proteins from all kingdoms of life. In particular, many PAS domains, such as the ones found in VicK, contain another conserved feature referred to as S boxes. S boxes have been shown to respond to changes in redox potential in a variety of prokaryotic and eukaryotic proteins (Zhulin *et al.*, 1997). Studies have yet to determine whether the PAS domain and S boxes are involved in redox sensing in streptococcal VicK kinases. However, the similarity to known oxidative sensors like FixL from *Rhizobium meliloti* (Gilles-Gonzalez *et al.*, 1991) and the oxidative stress phenotypes of *vicK* mutants (Deng *et al.*, 2007; Senadheera *et al.*, 2007) suggest that the VicK PAS domain could be a sensor of redox potential in oral streptococci. If so, VicK could mediate the direct sensing of $O_2$ or oxygen radicals like FixL, or it could indirectly sense the byproducts created by environmental stress. Given that many similar PAS domain kinases like FixL actually sense the redox potential within the cytoplasm (Gilles-Gonzalez and Gonzalez, 1993), it is also highly likely that the substrates for streptococcal VicK proteins are sensed intracellularly. Furthermore, *in silico* analyses of all of the aforementioned streptococcal VicK proteins predict that they possess only one transmembrane segment, which makes it unlikely that VicK sensors receive signals from outside the cell (Ng and Winkler, 2004). In addition, the PAS domains are all located in the cytoplasmic portions of the VicK proteins, where they are most likely involved with regulating the activity of the VicK histidine kinase domains via substrate binding (Fig. 8.1A) (Miyatake *et al.*, 1999, 2000; Gong *et al.*, 2000; Wagner *et al.*, 2002). Studies of VicX also implicate this protein as a modulator of VicK function, but the biochemical mechanism of this regulation has yet to be determined (Szurmant *et al.*, 2005).

## Function of VicRKX

The VicRK two-component system was first identified in *B. subtilis*, where it was given the name YycFG (Fabret and Hoch, 1998). In that study, it was discovered that mutations in either the response regulator (*yycF*) or the histidine kinase (*yycG*) resulted in lethality. A similar result was also found in subsequent studies of *S. aureus* (Martin *et al.*, 1999). However, in *S. pneumoniae* and *S. mutans*, only the *vicR* mutation is lethal (Wagner *et al.*, 2002; Senadheera *et al.*, 2005). Studies from all of these species suggest that the VicRKX system is involved in numerous regulatory pathways. However, the regulation of cell wall metabolism seems to be the most conserved feature among these organisms. In fact, multiple studies have demonstrated that this conserved role in cell wall metabolism likely accounts for the lethality associated with *vicK* and/or *vicR* mutations (Ng *et al.*, 2003; Senadheera *et al.*, 2005; Liu *et al.*, 2006; Ahn and Burne, 2007; Bisicchia *et al.*, 2007; Duque *et al.*, 2011). In addition to cell wall metabolism, the VicRKX system seems to function in the regulation of cell membrane structure as well (Martin *et al.*, 1999; Ng *et al.*, 2003; Mohedano *et al.*, 2005; Bisicchia *et al.*, 2007). For example, an *S. mutans vicK* mutant exhibits altered expression profiles for a large number of fatty acid and phospholipid metabolic genes (Senadheera *et al.*, 2009). It is conceivable that these changes in cell surface or cell membrane properties could play a direct role modulating cellular permeability to a variety of stressors from the environment. This may also partially explain the frequent occurrence of osmotic sensitivity phenotypes in *vicK* mutants (Martin *et al.*, 1999; Liu *et al.*, 2006) or their decreased resistance to acid and $H_2O_2$ stress (Deng *et al.*, 2007; Senadheera *et al.*, 2009). Consistent with this notion, it has been shown that reducing the expression of a VicRKX-controlled peptidoglycan hydrolase (*gbpB*) results in increased $H_2O_2$ sensitivity (Duque *et al.*, 2011). Likewise, modifications in cell membrane structure play an intimate role in the acid stress adaptation response (Fozo and Quivey, 2004a,b). Presumably, *S. mutans vicK* mutants are defective in acid adaptation, due to the membrane gene expression phenotypes described above (Senadheera *et al.*, 2009). From all of these data, a picture is emerging that suggests VicK likely coordinates the expression of a wide variety of genes that have the ability to influence the resistance to environmental stress in response to sensed changes in the redox status within the cytoplasm. Further research is still needed to determine the actual signal(s) sensed by VicK. This information would also help to address some of the significant remaining questions regarding the role of the VicRKX system in stress adaptation. For example, does the VicRKX system respond to particular toxic molecules directly or does this system respond to the byproducts of stress, such as accumulated metabolic intermediates or cofactors? What is the connection between the signal(s) sensed by VicK and the adaptive response triggered via VicR?

## The LiaFSR envelope stress sensory system

The *liaFSR* envelope stress sensory system is highly conserved within the phylum Firmicutes (low G+C Gram-positive bacteria) (Mascher, 2006). Studies from *Lactococcus, Staphylococcus* and *Streptococcus* suggest that the LiaFSR system is one of the principal cell envelope stress response systems in Firmicutes cocci (Jordan *et al.*, 2008). Not surprisingly, nearly identical *liaFSR* operons can be found in the genomes of the sequenced oral streptococci. All characterized LiaS proteins belong to a class of intramembrane-sensing histidine kinases (Mascher *et al.*, 2006). These sensor proteins encode two N-terminal transmembrane segments that have fewer than 25 amino acids separating them (Fig. 8.1A), which suggests that few, if any, amino acids are exposed to the outside environment (Mascher, 2006). In fact, it was this architecture that provided the initial evidence suggesting a role in intramembrane-sensing (Mascher *et al.*, 2003). A comparison of the putative protein sequences from oral streptococcal LiaS proteins suggests that they too share this same transmembrane architecture (J. Merritt and J. Kreth, unpublished). While it has yet to be determined in any species whether components of the LiaFSR system interact directly with particular substrates, all characterized LiaFSR sensory systems have been shown to respond to various agents that affect cell wall and cell membrane integrity (Mascher, 2006; Mascher *et al.*, 2006; Jordan *et al.*, 2008). For example, in *S. mutans,* the addition of peptidoglycan inhibiting antibiotics, such as bacitracin and vancomycin or the addition of cell membrane perturbing agents like chlorhexidine, all cause a rapid increase in *liaFSR* expression (Suntharalingam *et al.*, 2009). Mutations within the *liaFRS* operon also alter the susceptibility of *S. mutans* to these same agents and many other cell envelope damaging compounds (Suntharalingam *et al.*, 2009; Zhang and Biswas, 2009). As previously mentioned, human saliva contains a variety of cell envelope disrupting antimicrobial peptides like the defensins. Though these peptides have yet to be tested as *liaFSR* inducers in oral streptococci, several of the same antimicrobial peptides found in saliva have been shown to cause a rapid induction of *liaFSR* expression in *B. subtilis* (Pietiainen *et al.*, 2005). It is not yet known how envelope stress results in the activation of LiaS, but studies in both *B. subtilis* and *S. mutans* implicate LiaF as a negative regulator of LiaS function (Jordan *et al.*, 2006; Suntharalingam *et al.*, 2009). For example, a *liaF* mutation in *S. mutans* results in the increased expression of genes within the LiaR regulon in a stress-independent manner (Suntharalingam *et al.*, 2009). Further studies are still required to determine whether LiaF is the actual stress sensor.

### LiaFSR and stress adaptation

In the presence of cell envelope stress, all characterized LiaFSR systems have been shown to be autoregulatory for their own operons (Kuroda *et al.*, 2003; Martinez *et al.*, 2007; Suntharalingam *et al.*, 2009; Eldholm *et al.*, 2010; Wolf *et al.*, 2010). However, with the exception of this autoregulatory function, there appears to be a considerable divergence in the genes regulated by LiaR in Firmicutes bacilli versus cocci. For the bacilli, LiaR has a much smaller set of highly inducible genes under its control and a unique LiaR binding site (Jordan *et al.*, 2006; Wolf *et al.*, 2010). One of these highly responsive genes of *B. subtilis* called *liaH* appears to function in the adaptive response to oxidative and cell envelope stresses (Wolf *et al.*, 2010). It is not yet known how the LiaH protein offers protection to the cell, but based upon its similarity to the phage shock protein PspA in *E. coli*, it has been speculated that this protein could interact with the cell membrane in response to various stresses (Wolf *et al.*, 2010). Interestingly, there are no obvious *liaH* homologues in the genomes of Firmicutes cocci. Therefore, it appears as if these organisms utilize a distinct adaptive mechanism. In *L. lactis*, *S. aureus*, *S. pneumoniae*, and *S. mutans* the LiaR regulons consist of a diverse array of genes, but at least two of these genes appear to be regulated similarly among all of the species (Kuroda *et al.*, 2003; Martinez *et al.*, 2007; Suntharalingam *et al.*, 2009; Eldholm *et al.*, 2010). In *S. mutans*, these two genes are annotated as SMU.753 and SMU.2084c (Suntharalingam *et al.*, 2009). Both genes encode putative transcription regulators. Interestingly, SMU.753 is predicted to encode a protein with a PspC domain, which in *E. coli* is used to coordinate the phage shock response with PspA (Brissette *et al.*, 1991). This similarity to components of the *E. coli* phage shock response is highly reminiscent of the role of LiaH in *B. subtilis*. SMU.2084c encodes the transcription regulator SpxA, which plays a major role in regulating genes required for oxidative stress resistance in *S. mutans* (Kajfasz *et al.*, 2010). Given the potential role of both SMU.753 and SMU.2084c in stress adaptation and their conserved regulation by LiaR, it seems likely that these genes are central mediators of the LiaFSR stress adaptation response. The genomes of *S. gordonii*, *S. mitis*, and *S. sanguinis* all contain highly homologous genes to SMU.753 and SMU.2084c and in most cases the genomic organization surrounding these genes is conserved as well (J. Merritt and J. Kreth, unpublished). Presumably, these genes are also regulated similarly by their respective LiaFSR systems and may play similar roles in adapting to environmental stress.

## Stress adaptation via serine/threonine phosphorylation

### The PknB/PppL muropeptide sensory system

The serine/threonine protein kinase PknB and its cognate phosphatase PppL are the only known members of this protein family in oral streptococci (Hussain *et al.*, 2006; Jones and Dyson, 2006). Both proteins are highly conserved in Gram-positive bacteria and are involved in regulating a diverse array of cellular functions (Jones and Dyson, 2006). In *S. mutans*, the PknB/PppL system is required for proper biofilm formation, bacteriocin production, natural competence development, and resistance to a variety of stresses (Hussain *et al.*, 2006; Banu *et al.*, 2010; Zhu and Kreth, 2010). In other oral streptococci, the functional role of PknB/PppL has not yet been investigated. However, from the genomes of *S. gordonii*, *S. mitis*, and *S. sanguinis*, it is apparent that the *pknB* and *pppL* genes are present in nearly identical genetic loci to *S. mutans* with both genes located adjacent to multiple highly conserved housekeeping genes (J. Merritt and J. Kreth, unpublished). In addition, the PknB and PppL proteins from *S. mutans* are > 55% identical to the putative corresponding proteins from

*S. gordonii, S. mitis,* and *S. sanguinis* and all of these PknB proteins likely encode three PASTA domains (Fig. 8.1B), which is typical of Firmicute serine/threonine protein kinases (Paracuellos *et al.*, 2010). A variety of studies have indicated that the PASTA domains are likely exposed to the outside environment, where they are able to bind fragments of peptidoglycan or peptidoglycan analogues (Jones and Dyson, 2006; Maestro *et al.*, 2011). This ability is thought to be directly connected to the ability of PknB to phosphorylate its protein substrates in the cytoplasm (Barthe *et al.*, 2010; Paracuellos *et al.*, 2010).

### Function of PknB

Actively dividing bacteria are constantly releasing peptidoglycan fragments called muropeptides into their surrounding environments. For Gram-positive organisms, their cell walls typically contain more than 10 times the number of peptidoglycan layers as those found in Gram-negative organisms and consist of a much more complex and less uniform structure (Schleifer and Kandler, 1972). As such, murein remodelling plays a central role in numerous Gram-positive cellular processes beyond cell division, like stress adaptation (Severin *et al.*, 2004), competence development (Kausmally *et al.*, 2005; Okinaga *et al.*, 2010b), biofilm formation (Shibata *et al.*, 2005), etc. Not surprisingly, Gram-positive species are also copious producers of muropeptides (Doyle *et al.*, 1988), which in recent years have been shown to function as important intercellular signals sensed by PknB to reactivate cells from dormancy after episodes of starvation or extreme environmental stress (Shah *et al.*, 2008; Barthe *et al.*, 2010; Kana and Mizrahi, 2010). Consistent with this role, many of the genes/and or proteins affected by PknB in streptococci are involved in basic metabolic processes, cell division, and cell wall remodelling (Saskova *et al.*, 2007; Banu *et al.*, 2010; Novakova *et al.*, 2010). However, it is currently unknown whether streptococcal muropeptides sensed by PknB function in cellular reactivation or if they trigger an altogether unique response. Likewise, studies have yet to determine any of the direct protein targets of PknB in oral streptococci. Though, it is likely that muropeptide concentration in the oral biofilm serves as an important source of information used to adapt to environmental changes.

## Sensory systems for cell density perception

As biofilm-dwelling organisms, the typical life cycle of oral streptococci consists of growth within an exceptionally wide range of cell densities. Under laboratory conditions, cells that are growing in the free-floating planktonic state typically reach a maximum cell density near to $10^9$ CFU/ml. In the oral cavity, the actual cell density of planktonic streptococcal species in saliva is likely to be considerably lower than this value. For example, in extreme cases, such as in children afflicted by severe early childhood caries, the levels of *S. mutans* in saliva are still typically $< 10^6$ CFU/ml (Berkowitz *et al.*, 2009; Begzati *et al.*, 2010). Even the total bacterial population within unstimulated saliva fails to reach the cell density of a typical laboratory culture (Cousido *et al.*, 2008; Tomas *et al.*, 2008). However, once the planktonic cells attach to the tooth surface and form a biofilm, they can easily experience an increase in cell density that spans several orders of magnitude or greater. For supragingival biofilms, cell density values range between $10^{10}$ and $10^{11}$ CFU/ml, while subgingival biofilms range up to $10^{12}$ CFU/ml (Evaldson *et al.*, 1982). As with any population of organisms, growth in a sparsely populated environment requires vastly different abilities for survival compared to growth in a crowded environment. For example, intercellular communication for oral bacteria is much more difficult in a planktonic salivary population compared to a biofilm, due to the problem of dilution (Egland *et al.*, 2004). Therefore, genes regulated by intercellular signalling systems are less likely to be affected during growth in saliva as compared to a biofilm. Planktonic bacteria are also much less likely to require competitive behaviours, such as bacteriocin excretion, in order to inhibit other organisms, since growth competition is less problematic in the absence of crowding. Furthermore, in saliva, bacteriocins would be essentially useless anyway because they would be quickly diluted upon excretion from the cell. For bacteriocins to function as inhibitors, they must be present above a minimum threshold concentration, which is typically in the nanomolar range (Sang and Blecha, 2008). In the densely packed biofilm environment, it is much easier to increase the local concentration of bacteriocins above this threshold, due to the high numbers of closely associated producer cells. The biofilm matrix may also help to further concentrate the bacteriocins by slowing their diffusion away from the producers (Chan *et al.*, 2004). Thus, growth as a planktonic organism in saliva requires many distinct genetic responses that may not be appropriate for the biofilm lifestyle and vice versa. For this reason, oral streptococci utilize a variety of sensory systems to constantly monitor the population density and optimize gene expression patterns accordingly. The role of autoinducer 2 in interspecies communication has been discussed in detail in Chapter 7. In addition, oral streptococci produce competence stimulating peptides (discussed below), which comprise an essential component of the most thoroughly characterized cell density sensory systems among these species.

## The ComCDE system

Oral streptococci such as *S. mutans*, *S. sanguinis* and *S. gordonii* all have the remarkable ability to incorporate DNA from the environment in a process referred to as natural competence (Havarstein *et al.*, 1997). Natural competence seems to be a process adapted for the extremely high cell density found in the oral biofilm. Consequently, *S. mutans* develops natural competence with a substantially higher efficiency when grown in a biofilm. For example, biofilm cells exhibit a 600-fold increase in transformation frequency as compared to planktonic cells (Li *et al.*, 2001). This preference for the biofilm setting could be related to the high abundance of free DNA in the biofilm environment. Certainly, the energy investment in developing competence would only be worthwhile if there is a reasonable chance of contacting free DNA. The biofilm matrix may also serve to help concentrate the secreted signalling peptides required for triggering competence development (discussed below) (Vlassov *et al.*, 2007). In ecological terms, the release and uptake of DNA seems to promote species diversity and facilitates the process of evolution through the acquisition of new genetic traits encoded within the vast store of genetic information contained in the multispecies community. Over the long term, this ensures that organisms like oral streptococci can remain competitive in a mixed community of numerous other evolving species that vie for the same niches. In the short term, competence can help to survive episodes of environmental stress by providing templates for DNA repair via homologous recombination. Free DNA is surprisingly abundant in biofilms and it has been demonstrated that oral streptococci commonly release DNA into the surrounding environment to continually add to this pool (Kreth *et al.*, 2009; Flemming and Wingender, 2010; Liu and Burne, 2011). To help circumvent the restriction barrier imposed upon foreign DNA, the natural competence machinery will degrade one strand of the DNA upon entry into the cell (Dubnau, 1991; Claverys *et al.*, 2009). Therefore, the only cellular limitation is in the ability to incorporate the DNA into the chromosome via homologous recombination. This is a rare event even with highly homologous DNA, but under the right selective pressure, natural competence could easily result in a new population of cells containing a novel genotype.

Competence development is a peptide-based cell–cell communication event that has been best studied in the pneumococci (Claverys *et al.*, 2009; Johnsborg and Havarstein, 2009). An integral part of competence development in *S. pneumoniae* is the expression of the *comCDE* operon, encoding the competence stimulating peptide (referred to as CSP or ComC) as well as the ComDE TCS, which serves as a detection system for excreted extracellular CSP (Havarstein *et al.*, 1996; Pestova *et al.*, 1996). Once the cell population and thus, CSP concentration has reached a critical concentration in the environment, extracellular CSP molecules can bind to the membrane spanning ComD sensor kinase and induce a conformational change that activates its kinase domain and results in autophosphorylation (Havarstein *et al.*, 1995). This is followed by the transfer of the phosphoryl group to the response regulator ComE, which results in a highly increased DNA binding affinity for a 9-bp imperfect direct repeat consensus sequence (ComE-box). This sequence is found in the promoter regions of several genes referred to as the early competence genes. The early genes include the *comCDE* operon itself as well as the *comAB* operon encoding the CSP-specific transporter ComAB, responsible for the export and processing of the 41 amino acid CSP pre-peptide. CSP activation therefore leads to an increase of its own expression, production, secretion, and detection apparatuses (Johnsborg and Havarstein, 2009). In addition, the gene encoding the RNA polymerase alternative sigma factor ComX is transcribed as part of the early competence genes. ComX by itself is essential for competence development and regulates about 60 genes, including the genes required for the production of the DNA uptake system and recombination machinery (termed the late competence genes) (Johnsborg and Havarstein, 2009). ComX recognizes a unique DNA consensus sequence (TACGAATA) named the *cin*-box, which is typically located about ten bases upstream from the start of transcription of the late competence genes and is essential for their expression (Campbell *et al.*, 1998; Okinaga *et al.*, 2010b).

Competence development in *S. sanguinis* and *S. gordonii* seems to follow a similar blueprint as *S. pneumoniae*. *S. gordonii* develops a transient competent state peaking 10–20 minutes after the addition of its specific CSP to non-competent cell cultures. The initial set of genes responding after 5 minutes of exposure are *comCDE*, *comAB*, the *comX* orthologue *comR*, and a variety of newly identified genes. This is followed by the late competence genes regulated by the ComR sigma factor and includes genes encoding proteins required for DNA uptake and recombination. Competence declines about 30–40 minutes after CSP addition (Vickerman *et al.*, 2007).

A comprehensive study of competence development was recently performed with *S. sanguinis* (Rodriguez *et al.*, 2011). One major distinction from the other oral streptococcal competence systems is that *S. sanguinis* does not encode a ComAB transporter for the excretion and processing of CSP (Xu *et al.*, 2007). In agreement with this observation, the *S. sanguinis* CSP pre-peptide lacks the double glycine processing site found in the N-terminal leader sequence of all other ComAB-processed peptides (Zhu *et al.*, 2011). In addition, the usage of DNA microarray technology has enabled the identification of the competence-specific transcriptome. The addition of CSP to *S. sanguinis* increased the expression of 122 genes and decreased 83, a number comparable to *S. gordonii* (162 increased and 89 decreased). However, the number of genes belonging to the early competence gene-set was substantially lower in *S. sanguinis* with only five genes compared with 28 for *S. pneumoniae* and 35 for *S. gordonii*. The early competence genes of *S. sanguinis* are only comprised of *comCDE*, *comX*, and two uncharacterized open reading frames, suggesting that this constitutes the minimal set of early competence genes (Rodriguez *et al.*, 2011; Zhu *et al.*, 2011).

Competence development in *S. pneumoniae* and *S. sanguinis* is abolished in *comC* deletion mutants, indicating the importance of CSP in the competence regulation cascade (Cheng *et al.*, 1997; Rodriguez *et al.*, 2011). Interestingly, the deletion of *comC* in *S. mutans* reduces its ability to take up and incorporate DNA, but does not abolish it (Li *et al.*, 2001). Despite this difference, it was thought that competence development in *S. mutans* follows the same scheme described for *S. pneumoniae*. The *S. mutans* ComCDE locus was first described in 2001 and exhibited an obvious distinction in its organization. Rather than being encoded in one operon with *comDE*, *comC* is transcribed independently in the opposite orientation to *comDE* (Li *et al.*, 2001). The addition of CSP nonetheless induces the *S. mutans* competence machinery, but in addition, it also increases the expression of multiple bacteriocin genes (van der Ploeg, 2005; Yonezawa and Kuramitsu, 2005; Kreth *et al.*, 2006; Perry *et al.*, 2009). This includes the bacteriocins mutacin IV, which has been shown to induce lysis of competing oral streptococci and mutacin V, which lyses competitor species and functions in self-lysis as well (Kreth *et al.*, 2005a; Perry *et al.*, 2009). An analysis of the genomic organization of the *comCDE* locus suggests a closer relationship to the *blpABCHR* bacteriocin regulatory system of *S. pneumoniae*, rather than the pneumococcal competence system (Martin *et al.*, 2006). *S. mutans* seems to have evolved a unique mechanism for coordinating the control of competence and bacteriocin gene expression via the response regulator ComE. Several bacteriocin and bacteriocin-like genes, including mutacin IV and V, have a conserved ComE-binding site in their gene promoters and expression is abolished in a ComE mutant (Kreth *et al.*, 2006). In contrast, *S. mutans* competence development does not have a stringent requirement for ComE, similar to what is observed with ComC mutants (Li *et al.*, 2001). Therefore a basal level of CSP- and ComDE-independent competence exists (Ahn *et al.*, 2006). The control point for this unusual observation seems to be the ComX sigma factor. Without a functional ComX, no competence develops even in the presence of intact early competence genes. The promoter region of *comX* in *S. mutans* lacks the typical ComE binding site found in other oral streptococci, suggesting the presence of other regulatory elements controlling competence independently of ComE (Lemme *et al.*, 2011).

The ecological advantage for a combined control of competence development and bacteriocin production is an increased competitiveness of *S. mutans* in the multispecies environment of the dental biofilm. The accumulation of CSP first leads to an increase in the expression of bacteriocin genes. Consequently, the bacteriocins are able to liberate DNA from physically proximal species, which in turn can be internalized after a delayed activation of competence (Kreth *et al.*, 2005a). *S. mutans* seems to ensure that DNA is present before developing competence. Accordingly, the CSP-dependent bacteriocins, such as mutacin IV, lyse only closely related species, thereby increasing the access to a gene pool with a high chance of recombination with the *S. mutans* chromosome (Kreth *et al.*, 2005a). From an ecological standpoint, the coordination of competence development and bacteriocin production would only be practical in the high cell density biofilm environment, where there is an ample supply of closely associated competitor organisms. First, this ensures that competent cells are in an environment where there is a high likelihood of encountering liberated DNA for uptake. Second, the production of bacteriocins is only effective over a limited range, due to the problem of diffusion. In the biofilm, competitors are present in juxtaposition, where they constantly compete for space and resources. It is this physical proximity that makes bacteriocins effective weapons. Thus, both competence and bacteriocin production are ideally suited for the oral biofilm environment and are strongly regulated as such via cell density perception.

## The ComRS system

As mentioned previously, the main regulator for the late competence genes in *S. mutans* is the sigma factor ComX. The direct activation of *comX* expression is ComE-independent, unlike the other oral streptococci. The source of this unusual regulation was finally identified through studies of competence regulation in *Streptococcus thermophilus* (Gardan *et al.*, 2009; Fontaine *et al.*, 2010). Like the oral streptococci, *S. thermophilus* can develop natural competence and encodes all of the necessary late competence genes, as well as *comX*. However, this species does not encode the typical set of early competence genes, such as the ComDE TCS or the ComAB CSP transporter. In *S. thermophilus*, competence is regulated with a TCS-independent peptide based signalling system. A small precursor peptide, named ComS seems to be excreted from the cell, probably through the general secretion system. Like CSP, the mature ComS peptide accumulates in the environment and serves to signal population density. ComS is recognized by an oligopeptide transporter, Opp and is transported into the cell where it directly interacts with a transcriptional activator named ComR (no relation to the ComR from *S. gordonii*). ComR can bind to a specific binding site called a ComR box. A ComR box is located in the promoters of the *comS* gene to trigger autoinduction and the *comX* gene to initiate late competence gene expression (Fontaine *et al.*, 2010). ComR belongs to a group of stand-alone transcription regulators of the Rgg family, which has several members in the genus *Streptococcus* (Qi *et al.*, 1999; Chaussee *et al.*, 2004; Declerck *et al.*, 2007). A recent study identified paralogous Rgg regulators in the genome of several streptococci, including *S. mutans* by screening for conserved ComR consensus sequences (Mashburn-Warren *et al.*, 2010). In *S. mutans*, this pattern is only found in the promoter regions of *comX* and an uncharacterized small peptide-encoding open reading frame adjacent to the Rgg-type regulator SMU.61. These genes encoding the Rgg regulator and peptide were subsequently renamed as *comR* and *comS*, respectively. Mutagenesis studies of *comR*, *comS*, and the putative ComS transporter gene *oppD* indicate that each of these genes is indispensable for competence development, similar to what was found in *S. thermophilus*. Furthermore, the addition of the ComS mature peptide can induce competence in a non-competent wild type and an isogenic ComE mutant (Mashburn-Warren *et al.*, 2010). Thus, the evidence suggests that the ComRS system is the proximal regulator of competence in *S. mutans*.

### HdrRM

Similar to the ComDE TCS, the HdrRM regulatory system is a sensory/regulatory system involved in the regulation of competence, bacteriocin production, and other cell density-dependent functions in *S. mutans* (Merritt *et al.*, 2007; Okinaga *et al.*, 2010a,b). Initially identified in a screen for genes induced under high cell density conditions, HdrRM consists of a membrane protein, HdrM and a transcriptional regulator HdrR (Merritt *et al.*, 2007). Interestingly, although the *hdrRM* genes are encoded in a two-gene operon similar to other bacterial TCS, they represent a novel class of two-gene regulatory systems now referred to as LytTR Regulatory Systems. Currently, the only other characterized LytTR Regulatory System in bacteria is the BrsRM system also in *S. mutans* (Xie *et al.*, 2010). Though, *in silico* analyses of *S. gordonii*, *S. mitis*, and *S. sanguinis* suggest that these organisms each encode similar LytTR Regulatory Systems as well (J. Merritt, and J. Kreth, unpublished). The initial phenotypic characterization of HdrRM mutants revealed that HdrM is a strong negative regulator of genetic competence, whereas all other identified competence regulators are involved in activation. Furthermore, the HdrRM system is also able to directly induce the expression of *nlmA*, the gene encoding the bacteriocin mutacin IV. Epistatic analysis determined that the regulation of competence development and bacteriocin production by HdrRM is independent of the ComCDE system and therefore constitutes a parallel regulatory system possibly integrating an unknown environmental signal found in a high cell density environment like that of the oral biofilm (Okinaga *et al.*, 2010a,b). As mentioned earlier, HdrR and ComE both belong to the class of AlgR/AgrA/LytR transcriptional regulators containing a LytTR or 'litter' domain (Fig. 8.1B). Consequently, both proteins regulate their target genes by binding to direct repeat promoter elements matching the consensus pattern for the LytTR family of transcription regulators (Kreth *et al.*, 2007; Okinaga *et al.*, 2010a). However, it is currently not known what connects these two parallel pathways in competence control, since the *comX* promoter does not contain a binding site for either ComE or HdrR. It has been suggested that the ComRS system could function as an integrator of the various signal detection systems, since it seems to be the master regulator of *S. mutans* competence (Mashburn-Warren *et al.*, 2010; Lemme *et al.*, 2011). Though, detailed mechanistic studies have yet to corroborate this hypothesis.

## Future perspectives

Studies of environmental sensing are just beginning to offer significant insight into the mechanisms utilized by oral streptococci to adapt to the dynamic environment in the oral cavity. Already, studies have detailed many of the gene responses controlled by sensory systems in response to environmental changes or genetic mutations. Currently, work is under way to clarify which genes are the direct targets of sensory system regulation. The recent successes in characterizing the output responses of these sensory systems are largely due to the explosion of publicly available genome sequences, microarray technology, and the widespread incorporation of molecular biology and biochemistry techniques into microbiology. As this downstream portion of the genetic adaptation response becomes increasingly characterized, it will be necessary to focus upon some of the other significant areas of environmental sensory research that have yet to receive much attention. For example, most environmental sensory systems respond to unknown signals. Does a particular molecule actually bind to the sensor protein to modulate its activity or does some other factor affect it? Which technology would be useful for identifying the stimulus? At the present, these questions are still difficult to address. The difficulty of these studies is further compounded by the added complexity of working with membrane proteins, which are notoriously challenging to stably produce and purify using the currently available technologies. Consequently, only a small number of sensory systems are presently suitable for thorough mechanistic analyses.

Recent progress in imaging technologies and single cell analytical techniques may offer new options to characterize additional sensory systems in the near future. With a more thorough understanding of how specific signals from the environment modulate the function of sensor proteins, it is anticipated that entirely new approaches for the study and treatment of oral diseases would emerge. For example, as more sensory circuits are characterized from the input signal to the final output genetic response, increasingly sophisticated mathematical models could be developed to predict bacterial behaviour and theoretical outcomes *in silico*. From a therapeutic perspective, sensory systems offer tremendous potential for the development of entirely new classes of drugs that modify bacterial gene responses by acting as agonists or inhibitors of important sensor proteins. For example, a recent study demonstrated the potential utility of the histidine kinase inhibitor Walkmycin C as an anti-*S. mutans* therapeutic drug (Eguchi *et al.*, 2011). Ideally, these next generation therapeutics would also specifically target particular organisms within mixed species populations, rather than affecting the entire population like most antibiotics. In this way, it may be possible to reduce the virulence of specific pathogenic species by interfering with their ability to sense critical changes in the environment or changes in cell density. This is a particularly attractive approach for treating oral diseases because it would prevent further disruption of the oral microbial ecology. These targeted anti-virulence approaches are also less likely to promote resistance, since the drug would not be directly involved in killing the organism. Growth competition with the other members of the oral flora should be all that is required to reduce the prevalence of the targeted pathogens. In the fiercely competitive multispecies environment of the oral biofilm, slight increases or decreases in fitness can have a profound influence upon the ability of organisms to persist, proliferate, and ultimately cause disease.

## References


Ahn, S.J., and Burne, R.A. (2007). Effects of oxygen on biofilm formation and the AtlA autolysin of *Streptococcus mutans*. J. Bacteriol. *189*, 6293–6302.

Ahn, S.J., Wen, Z.T., and Burne, R.A. (2006). Multilevel control of competence development and stress tolerance in *Streptococcus mutans* UA159. Infect. Immun. *74*, 1631–1642.

Ajdic, D., McShan, W.M., McLaughlin, R.E., Savic, G., Chang, J., Carson, M.B., Primeaux, C., Tian, R., Kenton, S., Jia, H., *et al.* (2002). Genome sequence of *Streptococcus mutans* UA159, a cariogenic dental pathogen. Proc. Natl. Acad. Sci. U.S.A. *99*, 14434–14439.

Aravind, L., and Ponting, C.P. (1997). The GAF domain: an evolutionary link between diverse phototransducing proteins. Trends Biochem. Sci. *22*, 458–459.

Aravind, L., and Ponting, C.P. (1999). The cytoplasmic helical linker domain of receptor histidine kinase and methyl-accepting proteins is common to many prokaryotic signalling proteins. FEMS Microbiol. Lett. *176*, 111–116.

Ashby, M.K. (2006). Distribution, structure and diversity of 'bacterial' genes encoding two-component proteins in the Euryarchaeota. Archaea *2*, 11–30.

Ashby, M.T. (2008). Inorganic chemistry of defensive peroxidases in the human oral cavity. J. Dent. Res. *87*, 900–914.

Banu, L.D., Conrads, G., Rehrauer, H., Hussain, H., Allan, E., and van der Ploeg, J.R. (2010). The *Streptococcus mutans* serine/threonine kinase, PknB, regulates competence development, bacteriocin production, and cell wall metabolism. Infect. Immun. *78*, 2209–2220.

Barthe, P., Mukamolova, G.V., Roumestand, C., and Cohen-Gonsaud, M. (2010). The structure of PknB extracellular PASTA domain from mycobacterium tuberculosis suggests a ligand-dependent kinase activation. Structure *18*, 606–615.

Bassler, B.L., and Losick, R. (2006). Bacterially speaking. Cell *125*, 237–246.

Begzati, A., Berisha, M., and Meqa, K. (2010). Early childhood caries in preschool children of Kosovo – a serious public health problem. BMC public health *10*, 788.

Belli, W.A., and Marquis, R.E. (1991). Adaptation of *Streptococcus mutans* and *Enterococcus* hirae to acid stress in continuous culture. Appl. Environ. Microbiol. *57*, 1134–1138.

Berkowitz, R.J., Koo, H., McDermott, M.P., Whelehan, M.T., Ragusa, P., Kopycka-Kedzierawski, D.T., Karp, J.M., and Billings, R. (2009). Adjunctive chemotherapeutic suppression of mutans streptococci in the setting of severe early childhood caries: an exploratory study. J. Public Health Dent. *69*, 163–167.

Bisicchia, P., Noone, D., Lioliou, E., Howell, A., Quigley, S., Jensen, T., Jarmer, H., and Devine, K.M. (2007). The essential YycFG two-component system controls cell wall metabolism in *Bacillus subtilis*. Mol. Microbiol. *65*, 180–200.

Biswas, I., Drake, L., Erkina, D., and Biswas, S. (2008). Involvement of sensor kinases in the stress tolerance response of *Streptococcus mutans*. J. Bacteriol. *190*, 68–77.

Brissette, J.L., Weiner, L., Ripmaster, T.L., and Model, P. (1991). Characterization and sequence of the *Escherichia* coli stress-induced psp operon. J. Mol. Biol. *220*, 35–48.

Callewaert, L., and Michiels, C.W. (2010). Lysozymes in the animal kingdom. J. Biosci. *35*, 127–160.

Campbell, E.A., Choi, S.Y., and Masure, H.R. (1998). A competence regulon in *Streptococcus pneumoniae* revealed by genomic analysis. Mol. Microbiol. *27*, 929–939.

Chan, C., Burrows, L.L., and Deber, C.M. (2004). Helix induction in antimicrobial peptides by alginate in biofilms. J. Biol. Chem. *279*, 38749–38754.

Charbonneau, H., Prusti, R.K., LeTrong, H., Sonnenburg, W.K., Mullaney, P.J., Walsh, K.A., and Beavo, J.A. (1990). Identification of a noncatalytic cGMP-binding domain conserved in both the cGMP-stimulated and photoreceptor cyclic nucleotide phosphodiesterases. Proc. Natl. Acad. Sci. U.S.A. *87*, 288–292.

Chaussee, M.A., Callegari, E.A., and Chaussee, M.S. (2004). Rgg regulates growth phase-dependent expression of proteins associated with secondary metabolism and stress in *Streptococcus pyogenes*. J. Bacteriol. *186*, 7091–7099.

Chen, L., Ge, X., Dou, Y., Wang, X., Patel, J.R., and Xu, P. (2011). Identification of hydrogen peroxide production-related genes in *Streptococcus sanguinis* and their functional relationship with pyruvate oxidase. Microbiology (Reading, England) *157*, 13–20.

Cheng, Q., Campbell, E.A., Naughton, A.M., Johnson, S., and Masure, H.R. (1997). The com locus controls genetic transformation in *Streptococcus pneumoniae*. Mol. Microbiol. *23*, 683–692.

Claverys, J.P., Martin, B., and Polard, P. (2009). The genetic transformation machinery: composition, localization, and mechanism. FEMS Microbiol. Rev. *33*, 643–656.

Cornilescu, G., Ulijasz, A.T., Cornilescu, C.C., Markley, J.L., and Vierstra, R.D. (2008). Solution structure of a cyanobacterial phytochrome GAF domain in the red-light-absorbing ground state. J. Mol. Biol. *383*, 403–413.

Cotter, P.D., Hill, C., and Ross, R.P. (2005). Bacteriocins: developing innate immunity for food. Nat. Rev. *3*, 777–788.

Cousido, M.C., Tomas, M., Tomas, I., Limeres, J., Garcia-Caballero, L., and Diz, P. (2008). Effect of a neutralising agent on the evaluation of the antimicrobial activity of chlorhexidine on the bacterial salivary flora. Arch. Oral Biol. *53*, 981–984.

Dawes, C. (2008). Salivary flow patterns and the health of hard and soft oral tissues. J. Am. Dent. Assoc. *139* (Suppl.), 18S–24S.

De Smet, K., and Contreras, R. (2005). Human antimicrobial peptides: defensins, cathelicidins and histatins. Biotechnol. Lett. *27*, 1337–1347.

Declerck, N., Bouillaut, L., Chaix, D., Rugani, N., Slamti, L., Hoh, F., Lereclus, D., and Arold, S.T. (2007). Structure of PlcR: Insights into virulence regulation and evolution of quorum sensing in Gram-positive bacteria. Proc. Natl. Acad. Sci. U.S.A. *104*, 18490–18495.

Deng, D.M., Liu, M.J., ten Cate, J.M., and Crielaard, W. (2007). The VicRK system of *Streptococcus mutans* responds to oxidative stress. J. Dent. Res. *86*, 606–610.

Dong, Y., Chen, Y.Y., Snyder, J.A., and Burne, R.A. (2002). Isolation and molecular analysis of the gene cluster for the arginine deiminase system from *Streptococcus gordonii* DL1. Appl. Environ. Microbiol. *68*, 5549–5553.

Doyle, R.J., Chaloupka, J., and Vinter, V. (1988). Turnover of cell walls in microorganisms. Microbiol. Rev. *52*, 554–567.

Dubnau, D. (1991). Genetic competence in *Bacillus subtilis*. Microbiol. Rev. *55*, 395–424.

Duque, C., Stipp, R.N., Wang, B., Smith, D.J., Hofling, J.F., Kuramitsu, H.K., Duncan, M.J., and Mattos-Graner, R.O. (2011). Downregulation of GbpB, a component of the VicRK regulon, affects biofilm formation and cell surface characteristics of *Streptococcus mutans*. Infect. Immun. *79*, 786–796.

Egland, P.G., Palmer, R.J., Jr., and Kolenbrander, P.E. (2004). Interspecies communication in *Streptococcus gordonii–Veillonella atypica* biofilms: signalling in flow conditions requires juxtaposition. Proc. Natl. Acad. Sci. U.S.A. *101*, 16917–16922.

Eguchi, Y., Kubo, N., Matsunaga, H., Igarashi, M., and Utsumi, R. (2011). Development of an antivirulence drug against *Streptococcus mutans*: repression of biofilm formation, acid tolerance, and competence by a histidine kinase inhibitor, Walkmycin C. Antimicrob. Agents Chemother. *55*, 1475–1484.

Eldholm, V., Gutt, B., Johnsborg, O., Bruckner, R., Maurer, P., Hakenbeck, R., Mascher, T., and Havarstein, L.S. (2010). The pneumococcal cell envelope stress-sensing system LiaFSR is activated by murein hydrolases and lipid II-interacting antibiotics. J. Bacteriol. *192*, 1761–1773.

Englander, H.R. (1959). SHKLAIR IL, FOSDICK LS: The effects of saliva on the pH and lactate concentration in dental plaques. I. Caries-rampant individuals. J. Dent. Res. *38*, 848–853.

Evaldson, G., Heimdahl, A., Kager, L., and Nord, C.E. (1982). The normal human anaerobic microflora. Scand. J. Infect. Dis. Suppl. *35*, 9–15.

Fabret, C., and Hoch, J.A. (1998). A two-component signal transduction system essential for growth of *Bacillus subtilis*: implications for anti-infective therapy. J. Bacteriol. *180*, 6375–6383.

Flemming, H.C., and Wingender, J. (2010). The biofilm matrix. Nat. Rev. Microbiol. *8*, 623–633.

Fontaine, L., Boutry, C., de Frahan, M.H., Delplace, B., Fremaux, C., Horvath, P., Boyaval, P., and Hols, P. (2010). A novel pheromone quorum-sensing system controls the development of natural competence in *Streptococcus thermophilus* and *Streptococcus salivarius*. J. Bacteriol. *192*, 1444–1454.

Fozo, E.M., and Quivey, R.G., Jr. (2004a). The fabM gene product of *Streptococcus mutans* is responsible for the synthesis of monounsaturated fatty acids and is necessary for survival at low pH. J. Bacteriol. *186*, 4152–4158.

Fozo, E.M., and Quivey, R.G., Jr. (2004b). Shifts in the membrane fatty acid profile of *Streptococcus mutans* enhance survival in acidic environments. Appl. Environ. Microbiol. *70*, 929–936.

Galperin, M.Y. (2008). Telling bacteria: do not LytTR. Structure *16*, 657–659.

Gardan, R., Besset, C., Guillot, A., Gitton, C., and Monnet, V. (2009). The oligopeptide transport system is essential for the development of natural competence in *Streptococcus thermophilus* strain LMD-9. J. Bacteriol. *191*, 4647–4655.

Geiszt, M., Witta, J., Baffi, J., Lekstrom, K., and Leto, T.L. (2003). Dual oxidases represent novel hydrogen peroxide sources supporting mucosal surface host defense. FASEB J. *17*, 1502–1504.

Giefing, C., Jelencsics, K.E., Gelbmann, D., Senn, B.M., and Nagy, E. (2010). The pneumococcal eukaryotic-type serine/threonine protein kinase StkP co-localizes with the cell division apparatus and interacts with FtsZ in vitro. Microbiology *156*, 1697–1707.

Gilles-Gonzalez, M.A., and Gonzalez, G. (1993). Regulation of the kinase activity of heme protein FixL from the two-component system FixL/FixJ of *Rhizobium meliloti*. J. Biol. Chem. *268*, 16293–16297.

Gilles-Gonzalez, M.A., Ditta, G.S., and Helinski, D.R. (1991). A haemoprotein with kinase activity encoded by the oxygen sensor of *Rhizobium meliloti*. Nature *350*, 170–172.

Gong, W., Hao, B., and Chan, M.K. (2000). New mechanistic insights from structural studies of the oxygen-sensing domain of *Bradyrhizobium japonicum* FixL. Biochemistry *39*, 3955–3962.

Gorr, S.U., and Abdolhosseini, M. (2011). Antimicrobial peptides and periodontal disease. J. Clin. Periodontol. *38(Suppl. 11)*, 126–141.

Griswold, A.R., Chen, Y.Y., and Burne, R.A. (2004). Analysis of an agmatine deiminase gene cluster in *Streptococcus mutans* UA159. J. Bacteriol. *186*, 1902–1904.

Hahn, K., Faustoferri, R.C., and Quivey, R.G., Jr. (1999). Induction of an AP endonuclease activity in *Streptococcus mutans* during growth at low pH. Mol. Microbiol. *31*, 1489–1498.

Hanks, S.K., and Hunter, T. (1995). Protein kinases 6. The eukaryotic protein kinase superfamily: kinase (catalytic) domain structure and classification. FASEB J *9*, 576–596.

Hanna, M.N., Ferguson, R.J., Li, Y.H., and Cvitkovitch, D.G. (2001). uvrA is an acid-inducible gene involved in the adaptive response to low pH in *Streptococcus mutans*. J. Bacteriol. *183*, 5964–5973.

Havarstein, L.S., Coomaraswamy, G., and Morrison, D.A. (1995). An unmodified heptadecapeptide pheromone induces competence for genetic transformation in *Streptococcus pneumoniae*. Proc. Natl. Acad. Sci. U.S.A. *92*, 11140–11144.

Havarstein, L.S., Gaustad, P., Nes, I.F., and Morrison, D.A. (1996). Identification of the streptococcal competence-pheromone receptor. Mol. Microbiol. *21*, 863–869.

Havarstein, L.S., Hakenbeck, R., and Gaustad, P. (1997). Natural competence in the genus *Streptococcus*: evidence that streptococci can change pherotype by interspecies recombinational exchanges. J. Bacteriol. *179*, 6589–6594.

Hefti, M.H., Francoijs, K.J., de Vries, S.C., Dixon, R., and Vervoort, J. (2004). The PAS fold. A redefinition of the PAS domain based upon structural prediction. Eur. J. Biochem. *271*, 1198–1208.

Ho, Y.S., Burden, L.M., and Hurley, J.H. (2000). Structure of the GAF domain, a ubiquitous signalling motif and a new class of cyclic GMP receptor. EMBO J. *19*, 5288–5299.

Hussain, H., Branny, P., and Allan, E. (2006). A eukaryotic-type serine/threonine protein kinase is required for biofilm formation, genetic competence, and acid resistance in *Streptococcus mutans*. J. Bacteriol. *188*, 1628–1632.

Ihalin, R., Loimaranta, V., and Tenovuo, J. (2006). Origin, structure, and biological activities of peroxidases in human saliva. Arch. Biochem. Biophys. *445*, 261–268.

Ikeuchi, M., and Ishizuka, T. (2008). Cyanobacteriochromes: a new superfamily of tetrapyrrole-binding photoreceptors in cyanobacteria. Photochem. Photobiol. Sci. *7*, 1159–1167.

Jayaraman, G.C., Penders, J.E., and Burne, R.A. (1997). Transcriptional analysis of the *Streptococcus mutans* hrcA, grpE and dnaK genes and regulation of expression in response to heat shock and environmental acidification. Mol. Microbiol. *25*, 329–341.

Jensen, M.E., Polansky, P.J., and Schachtele, C.F. (1982). Plaque sampling and telemetry for monitoring acid production on human buccal tooth surfaces. Arch. Oral Biol. *27*, 21–31.

Johnsborg, O., and Havarstein, L.S. (2009). Regulation of natural genetic transformation and acquisition of transforming DNA in *Streptococcus pneumoniae*. FEMS Microbiol. Rev. *33*, 627–642.

Johnson, D.A., Yeh, C.K., and Dodds, M.W. (2000). Effect of donor age on the concentrations of histatins in human parotid and submandibular/sublingual saliva. Arch. Oral Biol. *45*, 731–740.

Jones, G., and Dyson, P. (2006). Evolution of transmembrane protein kinases implicated in coordinating remodeling of Gram-positive peptidoglycan: inside versus outside. J. Bacteriol. *188*, 7470–7476.

Jordan, S., Junker, A., Helmann, J.D., and Mascher, T. (2006). Regulation of LiaRS-dependent gene expression in *Bacillus subtilis*: identification of inhibitor proteins, regulator binding sites, and target genes of a conserved cell envelope stress-sensing two-component system. J. Bacteriol. *188*, 5153–5166.

Jordan, S., Hutchings, M.I., and Mascher, T. (2008). Cell envelope stress response in Gram-positive bacteria. FEMS Microbiol. Rev. *32*, 107–146.

Kajfasz, J.K., Rivera-Ramos, I., Abranches, J., Martinez, A.R., Rosalen, P.L., Derr, A.M., Quivey, R.G., and Lemos, J.A. (2010). Two Spx proteins modulate stress tolerance, survival, and virulence in *Streptococcus mutans*. J. Bacteriol. *192*, 2546–2556.

Kana, B.D., and Mizrahi, V. (2010). Resuscitation-promoting factors as lytic enzymes for bacterial growth and signalling. FEMS Immunol. Med. Microbiol. *58*, 39–50.

Kanapka, J.A., and Hamilton, I.R. (1971). Fluoride inhibition of enolase activity in vivo and its relationship to the inhibition of glucose-6-P formation in *Streptococcus salivarius*. Arch. Biochem. Biophys. *146*, 167–174.

Kausmally, L., Johnsborg, O., Lunde, M., Knutsen, E., and Havarstein, L.S. (2005). Choline-binding protein D (CbpD) in *Streptococcus pneumoniae* is essential for competence-induced cell lysis. J. Bacteriol. *187*, 4338–4345.

Klimiuk, A., Waszkiel, D., Jankowska, A., Zelazowska-Rutkowska, B., and Choromanska, M. (2006). The evaluation of lysozyme concentration and peroxidase activity in non-stimulated saliva of patients infected with HIV. Adv. Med. Sci. *51(Suppl. 1)*, 49–51.

Krell, T., Lacal, J., Busch, A., Silva-Jimenez, H., Guazzaroni, M.E., and Ramos, J.L. (2010). Bacterial sensor kinases: diversity in the recognition of environmental signals. Annu. Rev. Microbiol. *64*, 539–559.

Kreth, J., Merritt, J., Shi, W., and Qi, F. (2005a). Co-ordinated bacteriocin production and competence development: a possible mechanism for taking up DNA from neighbouring species. Mol. Microbiol. *57*, 392–404.

Kreth, J., Merritt, J., Shi, W., and Qi, F. (2005b). Competition and coexistence between *Streptococcus mutans* and *Streptococcus sanguinis* in the dental biofilm. J. Bacteriol. *187*, 7193–7203.

Kreth, J., Merritt, J., Zhu, L., Shi, W., and Qi, F. (2006). Cell density- and ComE-dependent expression of a group of mutacin and mutacin-like genes in *Streptococcus mutans*. FEMS Microbiol. Lett. *265*, 11–17.

Kreth, J., Hung, D.C., Merritt, J., Perry, J., Zhu, L., Goodman, S.D., Cvitkovitch, D.G., Shi, W., and Qi, F. (2007). The response regulator ComE in *Streptococcus mutans* functions both as a transcription activator of mutacin production and repressor of CSP biosynthesis. Microbiology (Reading, England) *153*, 1799–1807.

Kreth, J., Zhang, Y., and Herzberg, M.C. (2008). Streptococcal antagonism in oral biofilms: *Streptococcus sanguinis* and *Streptococcus gordonii* interference with *Streptococcus mutans*. J. Bacteriol. *190*, 4632–4640.

Kreth, J., Vu, H., Zhang, Y., and Herzberg, M.C. (2009). Characterization of hydrogen peroxide-induced DNA release by *Streptococcus sanguinis* and *Streptococcus gordonii*. J. Bacteriol. *191*, 6281–6291.

Krupa, A., and Srinivasan, N. (2005). Diversity in domain architectures of Ser/Thr kinases and their homologues in prokaryotes. BMC Genomics *6*, 129.

Kuhnert, W.L., Zheng, G., Faustoferri, R.C., and Quivey, R.G., Jr. (2004). The F-ATPase operon promoter of *Streptococcus mutans* is transcriptionally regulated in response to external pH. J. Bacteriol. *186*, 8524–8528.

Kundig, W., Ghosh, S., and Roseman, S. (1964). Phosphate bound to histidine in a protein as an intermediate in a novel phospho-transferase system. Proc. Natl. Acad. Sci. U.S.A. *52*, 1067–1074.

Kuroda, M., Kuroda, H., Oshima, T., Takeuchi, F., Mori, H., and Hiramatsu, K. (2003). Two-component system VraSR positively modulates the regulation of cell-wall biosynthesis pathway in *Staphylococcus aureus*. Mol. Microbiol. *49*, 807–821.

Lemme, A., Grobe, L., Reck, M., Tomasch, J., and Wagner-Dobler, I. (2011). Subpopulation-specific transcriptome analysis of competence-stimulating-peptide-induced *Streptococcus mutans*. J. Bacteriol. *193*, 1863–1877.

Levesque, C.M., Mair, R.W., Perry, J.A., Lau, P.C., Li, Y.H., and Cvitkovitch, D.G. (2007). Systemic inactivation and phenotypic characterization of two-component systems in expression of *Streptococcus mutans* virulence properties. Lett Appl. Microbiol. *45*, 398–404.

Li, Y.H., Lau, P.C., Lee, J.H., Ellen, R.P., and Cvitkovitch, D.G. (2001). Natural genetic transformation of *Streptococcus mutans* growing in biofilms. J. Bacteriol. *183*, 897–908.

Liu, M., Hanks, T.S., Zhang, J., McClure, M.J., Siemsen, D.W., Elser, J.L., Quinn, M.T., and Lei, B. (2006). Defects in ex vivo and in vivo growth and sensitivity to osmotic stress of group A *Streptococcus* caused by interruption of response regulator gene vicR. Microbiology (Reading, England) *152*, 967–978.

Liu, Y., and Burne, R.A. (2011). The major autolysin of *Streptococcus gordonii* is subject to complex regulation and modulates stress tolerance, biofilm formation and eDNA release. J. Bacteriol. *193*, 2826–2837.

Maestro, B., Novakova, L., Hesek, D., Lee, M., Leyva, E., Mobashery, S., Sanz, J.M., and Branny, P. (2011). Recognition of peptidoglycan and beta-lactam antibiotics by the extracellular domain of the Ser/Thr protein kinase StkP from *Streptococcus pneumoniae*. FEBS Lett. *585*, 357–363.

Martin, B., Quentin, Y., Fichant, G., and Claverys, J.P. (2006). Independent evolution of competence regulatory cascades in streptococci? Trends Microbiol. *14*, 339–345.

Martin, P.K., Li, T., Sun, D., Biek, D.P., and Schmid, M.B. (1999). Role in cell permeability of an essential two-component system in *Staphylococcus aureus*. J. Bacteriol. *181*, 3666–3673.

Martinez, B., Zomer, A.L., Rodriguez, A., Kok, J., and Kuipers, O.P. (2007). Cell envelope stress induced by the bacteriocin Lcn972 is sensed by the Lactococcal two-component system CesSR. Mol. Microbiol. *64*, 473–486.

Martinez, S.E., Heikaus, C.C., Klevit, R.E., and Beavo, J.A. (2008). The structure of the GAF A domain from phosphodiesterase 6C reveals determinants of cGMP binding, a conserved binding surface, and a large cGMP-dependent conformational change. J. Biol. Chem. *283*, 25913–25919.

Mascher, T. (2006). Intramembrane-sensing histidine kinases: a new family of cell envelope stress sensors in Firmicutes bacteria. FEMS Microbiol. Lett. *264*, 133–144.

Mascher, T., Margulis, N.G., Wang, T., Ye, R.W., and Helmann, J.D. (2003). Cell wall stress responses in *Bacillus subtilis*: the regulatory network of the bacitracin stimulon. Mol. Microbiol. *50*, 1591–1604.

Mascher, T., Helmann, J.D., and Unden, G. (2006). Stimulus perception in bacterial signal-transducing histidine kinases. Microbiol. Mol. Biol. Rev. *70*, 910–938.

Mashburn-Warren, L., Morrison, D.A., and Federle, M.J. (2010). A novel double-tryptophan peptide pheromone controls competence in *Streptococcus* spp. via an Rgg regulator. Mol. Microbiol. 78, 589–606.

Mathews, M., Jia, H.P., Guthmiller, J.M., Losh, G., Graham, S., Johnson, G.K., Tack, B.F., and McCray, P.B., Jr. (1999). Production of beta-defensin antimicrobial peptides by the oral mucosa and salivary glands. Infect. Immun. *67*, 2740–2745.

Matsui, R., and Cvitkovitch, D. (2010). Acid tolerance mechanisms utilized by *Streptococcus mutans*. Future Microbiol. 5, 403–417.

Merritt, J., Zheng, L., Shi, W., and Qi, F. (2007). Genetic characterization of the hdrRM operon: a novel high-cell-density-responsive regulator in *Streptococcus mutans*. Microbiology *153*, 2765–2773.

Miyatake, H., Mukai, M., Adachi, S., Nakamura, H., Tamura, K., Iizuka, T., Shiro, Y., Strange, R.W., and Hasnain, S.S. (1999). Iron coordination structures of oxygen sensor FixL characterized by Fe K-edge extended x-ray absorption fine structure and resonance raman spectroscopy. J. Biol. Chem. *274*, 23176–23184.

Miyatake, H., Mukai, M., Park, S.Y., Adachi, S., Tamura, K., Nakamura, H., Nakamura, K., Tsuchiya, T., Iizuka, T., and Shiro, Y. (2000). Sensory mechanism of oxygen sensor FixL from *Rhizobium meliloti*: crystallographic, mutagenesis and resonance Raman spectroscopic studies. J. Mol. Biol. *301*, 415–431.

Moglich, A., Ayers, R.A., and Moffat, K. (2009). Structure and signalling mechanism of Per-ARNT-Sim domains. Structure *17*, 1282–1294.

Mohedano, M.L., Overweg, K., de la Fuente, A., Reuter, M., Altabe, S., Mulholland, F., de Mendoza, D., Lopez, P., and Wells, J.M. (2005). Evidence that the essential response regulator YycF in *Streptococcus pneumoniae* modulates expression of fatty acid biosynthesis genes and alters membrane composition. J. Bacteriol. *187*, 2357–2367.

Nes, I.F., Diep, D.B., and Holo, H. (2007). Bacteriocin diversity in *Streptococcus* and *Enterococcus*. J. Bacteriol. *189*, 1189–1198.

Ng, W.L., and Winkler, M.E. (2004). Singular structures and operon organizations of essential two-component systems in species of *Streptococcus*. Microbiology (Reading, England) *150*, 3096–3098.

Ng, W.L., Robertson, G.T., Kazmierczak, K.M., Zhao, J., Gilmour, R., and Winkler, M.E. (2003). Constitutive expression of PcsB suppresses the requirement for the essential VicR (YycF) response regulator in *Streptococcus pneumoniae* R6. Mol. Microbiol. *50*, 1647–1663.

Novakova, L., Bezouskova, S., Pompach, P., Spidlova, P., Saskova, L., Weiser, J., and Branny, P. (2010). Identification of multiple substrates of the StkP Ser/Thr protein kinase in *Streptococcus pneumoniae*. J. Bacteriol. *192*, 3629–3638.

von Ohle, C., Gieseke, A., Nistico, L., Decker, E.M., DeBeer, D., and Stoodley, P. (2010). Real-time microsensor measurement of local metabolic activities in ex vivo dental biofilms exposed to sucrose and treated with chlorhexidine. Appl. Environ. Microbiol. *76*, 2326–2334.

Okinaga, T., Niu, G., Xie, Z., Qi, F., and Merritt, J. (2010a). The hdrRM operon of *Streptococcus mutans* encodes a novel regulatory system for coordinated competence development and bacteriocin production. J. Bacteriol. *192*, 1844–1852.

Okinaga, T., Xie, Z., Niu, G., Qi, F., and Merritt, J. (2010b). Examination of the hdrRM regulon yields insight into the competence system of *Streptococcus mutans*. Molecular Oral Microbiol. *25*, 165–177.

Paracuellos, P., Ballandras, A., Robert, X., Kahn, R., Herve, M., Mengin-Lecreulx, D., Cozzone, A.J., Duclos, B., and Gouet, P. (2010). The extended conformation of the 2.9-A crystal structure of the three-PASTA domain of a Ser/Thr kinase from the human pathogen *Staphylococcus aureus*. J. Mol. Biol. *404*, 847–858.

Pares, S., Mouz, N., Petillot, Y., Hakenbeck, R., and Dideberg, O. (1996). X-ray structure of *Streptococcus pneumoniae* PBP2x, a primary penicillin target enzyme. Nat. Struct. Biol. 3, 284–289.

Parkinson, J.S. (2010). Signalling mechanisms of HAMP domains in chemoreceptors and sensor kinases. Annu. Rev. Microbiol. *64*, 101–122.

Pereira, S.F., Goss, L., and Dworkin, J. (2011). Eukaryote-like serine/threonine kinases and phosphatases in bacteria. Microbiol. Mol. Biol. Rev. *75*, 192–212.

Perry, J.A., Jones, M.B., Peterson, S.N., Cvitkovitch, D.G., and Levesque, C.M. (2009). Peptide alarmone signalling triggers an auto-active bacteriocin necessary for genetic competence. Mol. Microbiol. *72*, 905–917.

Pestova, E.V., Havarstein, L.S., and Morrison, D.A. (1996). Regulation of competence for genetic transformation in *Streptococcus pneumoniae* by an auto-induced peptide pheromone and a two-component regulatory system. Mol. Microbiol. *21*, 853–862.

Pietiainen, M., Gardemeister, M., Mecklin, M., Leskela, S., Sarvas, M., and Kontinen, V.P. (2005). Cationic antimicrobial peptides elicit a complex stress response in *Bacillus subtilis* that involves ECF-type sigma factors and two-component signal transduction systems. Microbiology (Reading, England) *151*, 1577–1592.

van der Ploeg, J.R. (2005). Regulation of bacteriocin production in *Streptococcus mutans* by the quorum-sensing system required for development of genetic competence. J. Bacteriol. *187*, 3980–3989.

Puklo, M., Guentsch, A., Hiemstra, P.S., Eick, S., and Potempa, J. (2008). Analysis of neutrophil-derived antimicrobial peptides in gingival crevicular fluid suggests importance of cathelicidin LL-37 in the innate immune response against periodontogenic bacteria. Oral Microbiol. Immunol. *23*, 328–335.

Qi, F., Chen, P., and Caufield, P.W. (1999). Functional analyses of the promoters in the lantibiotic mutacin

II biosynthetic locus in *Streptococcus mutans*. Appl. Environ. Microbiol. *65*, 652–658.

Quivey, R.G., Jr., Faustoferri, R., Monahan, K., and Marquis, R. (2000a). Shifts in membrane fatty acid profiles associated with acid adaptation of *Streptococcus mutans*. FEMS Microbiol. Lett. *189*, 89–92.

Quivey, R.G., Jr., Kuhnert, W.L., and Hahn, K. (2000b). Adaptation of oral streptococci to low pH. Adv. Microb. Physiol. *42*, 239–274.

Quivey, R.G., Kuhnert, W.L., and Hahn, K. (2001). Genetics of acid adaptation in oral streptococci. Crit. Rev. Oral Biol. Med. *12*, 301–314.

Rao, F., Ji, Q., Soehano, I., and Liang, Z.X. (2011). Unusual heme-binding PAS domain from YybT family proteins. J. Bacteriol. *193*, 1543–1551.

Rodriguez, A.M., Callahan, J.E., Fawcett, P., Ge, X., Xu, P., and Kitten, T. (2011). Physiological and molecular characterization of genetic competence in *Streptococcus sanguinis*. Mol. Oral Microbiol. *26*, 99–116.

Sang, Y., and Blecha, F. (2008). Antimicrobial peptides and bacteriocins: alternatives to traditional antibiotics. Anim. Health Res. Rev. *9*, 227–235.

Saskova, L., Novakova, L., Basler, M., and Branny, P. (2007). Eukaryotic-type serine/threonine protein kinase StkP is a global regulator of gene expression in *Streptococcus pneumoniae*. J. Bacteriol. *189*, 4168–4179.

Sato, E.F., Choudhury, T., Nishikawa, T., and Inoue, M. (2008). Dynamic aspect of reactive oxygen and nitric oxide in oral cavity. J. Clin. Biochem. Nutr. *42*, 8–13.

Schleifer, K.H., and Kandler, O. (1972). Peptidoglycan types of bacterial cell walls and their taxonomic implications. Bacteriol. Rev. *36*, 407–477.

Senadheera, M.D., Guggenheim, B., Spatafora, G.A., Huang, Y.C., Choi, J., Hung, D.C., Treglown, J.S., Goodman, S.D., Ellen, R.P., and Cvitkovitch, D.G. (2005). A VicRK signal transduction system in *Streptococcus mutans* affects gtfBCD, gbpB, and ftf expression, biofilm formation, and genetic competence development. J. Bacteriol. *187*, 4064–4076.

Senadheera, M.D., Lee, A.W., Hung, D.C., Spatafora, G.A., Goodman, S.D., and Cvitkovitch, D.G. (2007). The *Streptococcus mutans* vicX gene product modulates gtfB/C expression, biofilm formation, genetic competence, and oxidative stress tolerance. J. Bacteriol. *189*, 1451–1458.

Senadheera, D., Krastel, K., Mair, R., Persadmehr, A., Abranches, J., Burne, R.A., and Cvitkovitch, D.G. (2009). Inactivation of VicK affects acid production and acid survival of *Streptococcus mutans*. J. Bacteriol. *191*, 6415–6424.

Severin, A., Tabei, K., Tenover, F., Chung, M., Clarke, N., and Tomasz, A. (2004). High level oxacillin and vancomycin resistance and altered cell wall composition in *Staphylococcus aureus* carrying the staphylococcal mecA and the enterococcal vanA gene complex. J. Biol. Chem. *279*, 3398–3407.

Shah, I.M., Laaberki, M.H., Popham, D.L., and Dworkin, J. (2008). A eukaryotic-like Ser/Thr kinase signals bacteria to exit dormancy in response to peptidoglycan fragments. Cell *135*, 486–496.

Shibata, Y., Kawada, M., Nakano, Y., Toyoshima, K., and Yamashita, Y. (2005). Identification and characterization of an autolysin-encoding gene of *Streptococcus mutans*. Infect. Immun. *73*, 3512–3520.

Sidote, D.J., Barbieri, C.M., Wu, T., and Stock, A.M. (2008). Structure of the *Staphylococcus aureus* AgrA LytTR domain bound to DNA reveals a beta fold with an unusual mode of binding. Structure *16*, 727–735.

Suntharalingam, P., Senadheera, M.D., Mair, R.W., Levesque, C.M., and Cvitkovitch, D.G. (2009). The LiaFSR system regulates the cell envelope stress response in *Streptococcus mutans*. J. Bacteriol. *191*, 2973–2984.

Szurmant, H., Nelson, K., Kim, E.J., Perego, M., and Hoch, J.A. (2005). YycH regulates the activity of the essential YycFG two-component system in *Bacillus subtilis*. J. Bacteriol. *187*, 5419–5426.

Tabak, L.A. (2006). In defense of the oral cavity: the protective role of the salivary secretions. Pediatr. Dent. *28*, 110–117; discussion 192–118.

Tao, R., Jurevic, R.J., Coulton, K.K., Tsutsui, M.T., Roberts, M.C., Kimball, J.R., Wells, N., Berndt, J., and Dale, B.A. (2005). Salivary antimicrobial peptide expression and dental caries experience in children. Antimicrob. Agents Chemother. *49*, 3883–3888.

Taylor, B.L., and Zhulin, I.B. (1999). PAS domains: internal sensors of oxygen, redox potential, and light. Microbiol. Mol. Biol. Rev. *63*, 479–506.

Tomas, I., Cousido, M.C., Tomas, M., Limeres, J., Garcia-Caballero, L., and Diz, P. (2008). In vivo bactericidal effect of 0.2% chlorhexidine but not 0.12% on salivary obligate anaerobes. Arch. Oral Biol. *53*, 1186–1191.

Tong, H., Chen, W., Merritt, J., Qi, F., Shi, W., and Dong, X. (2007). *Streptococcus* oligofermentans inhibits *Streptococcus mutans* through conversion of lactic acid into inhibitory H2O2: a possible counteroffensive strategy for interspecies competition. Mol. Microbiol. *63*, 872–880.

Tong, H., Chen, W., Shi, W., Qi, F., and Dong, X. (2008). SO-LAAO, a novel L-amino acid oxidase that enables *Streptococcus oligofermentans* to outcompete *Streptococcus mutans* by generating H2O2 from peptone. J. Bacteriol. *190*, 4716–4721.

Vadeboncoeur, C., and Pelletier, M. (1997). The phosphoenolpyruvate:sugar phosphotransferase system of oral streptococci and its role in the control of sugar metabolism. FEMS Microbiol. Rev. *19*, 187–207.

Vickerman, M.M., Iobst, S., Jesionowski, A.M., and Gill, S.R. (2007). Genome-wide transcriptional changes in *Streptococcus gordonii* in response to competence signalling peptide. J. Bacteriol. *189*, 7799–7807.

Vlassov, V.V., Laktionov, P.P., and Rykova, E.Y. (2007). Extracellular nucleic acids. Bioessays *29*, 654–667.

Wagner, C., Saizieu Ad, A., Schonfeld, H.J., Kamber, M., Lange, R., Thompson, C.J., and Page, M.G. (2002). Genetic analysis and functional characterization of the *Streptococcus pneumoniae* vic operon. Infect. Immun. *70*, 6121–6128.

Williams, S.B., and Stewart, V. (1999). Functional similarities among two-component sensors and methyl-accepting chemotaxis proteins suggest a role for linker region amphipathic helices in transmembrane signal transduction. Mol. Microbiol. *33*, 1093–1102.

Wolf, D., Kalamorz, F., Wecke, T., Juszczak, A., Mader, U., Homuth, G., Jordan, S., Kirstein, J., Hoppert, M., Voigt, B., *et al.* (2010). In-depth profiling of the LiaR response of *Bacillus subtilis*. J. Bacteriol. *192*, 4680–4693.

Xie, Z., Okinaga, T., Niu, G., Qi, F., and Merritt, J. (2010). Identification of a novel bacteriocin regulatory system in *Streptococcus mutans*. Mol. Microbiol. *78*, 1431–1447.

Xu, P., Alves, J.M., Kitten, T., Brown, A., Chen, Z., Ozaki, L.S., Manque, P., Ge, X., Serrano, M.G., Puiu, D., *et al.* (2007). Genome of the opportunistic pathogen *Streptococcus sanguinis*. J. Bacteriol. *189*, 3166–3175.

Yeats, C., Finn, R.D., and Bateman, A. (2002). The PASTA domain: a beta-lactam-binding domain. Trends Biochem. Sci. *27*, 438.

Yonezawa, H., and Kuramitsu, H.K. (2005). Genetic analysis of a unique bacteriocin, Smb, produced by *Streptococcus mutans* GS5. Antimicrob. Agents Chemother. *49*, 541–548.

Zhang, J., and Biswas, I. (2009). A phenotypic microarray analysis of a *Streptococcus mutans* liaS mutant. Microbiology (Reading, England) *155*, 61–68.

Zheng, L., Chen, Z., Itzek, A., Ashby, M., and Kreth, J. (2011). Catabolite control protein A controls hydrogen peroxide production and cell death in *Streptococcus sanguinis*. J. Bacteriol. *193*, 516–526.

Zhu, L., and Kreth, J. (2010). Role of *Streptococcus mutans* eukaryotic-type serine/threonine protein kinase in interspecies interactions with *Streptococcus sanguinis*. Arch. Oral Biol. *55*, 385–390.

Zhu, L., Zhang, Y., Fan, J., Herzberg, M.C., and Kreth, J. (2011). Characterization of competence and biofilm development of a *Streptococcus sanguinis* endocarditis isolate. Mol. Oral Microbiol. *26*, 117–126.

Zhulin, I.B., Taylor, B.L., and Dixon, R. (1997). PAS domain S-boxes in Archaea, Bacteria and sensors for oxygen and redox. Trends Biochem. Sci. *22*, 331–333.

# Microbial Community Interactions of the Cariogenic Organism *Streptococcus mutans*

9

Saswat Sourav Mohapatra and Indranil Biswas


## Abstract

The human oral cavity is estimated to host more than 700 bacterial species formed into distinct biofilm communities, of which more than 50% are yet to be cultured in the laboratory. Though oral streptococci constitute two-thirds of the total commensals, only a fraction, known as mutans streptococci, are involved in producing dental caries. The oral streptococci are the primary colonizers of the tooth and other mucosal surfaces in the oral cavity and initiate plaque biofilm formation. Mutual interaction in the form of cooperation and competition shapes the constitution of the oral microflora. Colonization by *Streptococcus mutans*, having significant acidogenicity and aciduricity properties, is primarily responsible for dental caries formation. Recent advances in nucleotide sequencing and other high throughput methods have provided significant clues to the biology and gene regulation of *S. mutans* in a community structure. Many therapeutic methods are being devised to specifically target the *S. mutans* in the biofilm, without disturbing other bacterial species. Significant among them are targeting the interbacterial signalling, replacing cariogenic flora with non-cariogenic flora, and specifically targeted antimicrobial peptides (STAMPs). As the research progresses in this field, better therapeutic methods are on the horizon.


## Introduction

In the postgenomic era, it has been established that human health or disease is the outcome of many intricate interactions among billions of microorganisms associated with the human body. With more and more bacterial pathogens and commensals being sequenced, the concept of the human microbiome has been developed to address the effect of these organisms on human health and disease. According to Koch's postulates, a single biological agent must be responsible for a particular transmissible disease; this agent must be isolated and cultured and should cause the disease if introduced into a healthy host. The discovery of various bacterial communities in the form of biofilms associated with different diseases has revolutionized the present understanding of infectious disease processes. Therefore, the reductionist thinking of a single organism as the causative agent of a particular disease has to be reviewed, and a more unifying view of a consortium of microorganisms controlling the state of our health should be considered.

From the time of birth, humans are densely populated with diverse bacterial species colonizing various mucosal surfaces. These bacteria are known as the human commensal microflora. Present estimates indicate that the number of bacterial cells in the human body may be more than ten times that of the body's own cells (100 trillion vs. 10 trillion) (Knights *et al.*, 2011). Therefore, it is important to understand the role of such a huge number of bacteria in human health. Commensal microorganisms have been observed to become opportunistic pathogens depending on various host derived and environmental factors. Therefore, instead of targeting a single bacterial species for a particular disease, one should look into manipulating the community balance in favour of a commensal setting (Jenkinson and Lamont, 2005). The oral microflora represents a wonderfully developed community responsible for maintaining the overall state of our oral health.

## Oral microbiome

The human oral cavity includes several anatomical sites such as teeth, gingival sulcus with attached gingiva, tongue, cheek, hard and soft palate, all of which serve as distinct habitats for a large number of bacterial species (Dewhirst *et al.*, 2010). The oral microbiome consists of all microorganisms that are found on or in the human oral cavity and the associated anatomical sites such as the tonsils, pharynx, oesophagus, Eustachian tube, middle ear, trachea, lungs, nasal passages and sinuses (Dewhirst *et al.*, 2010). Approximately 700 bacterial species are estimated to reside in the oral cavity, making it a vibrant ecological niche, and it is interesting to note that about 100–200 bacterial species are present in the mouth of any healthy individual (Paster *et al.*, 2006). Recently, more advanced sequencing technologies indicate that the number of bacterial species in a healthy mouth could be around 500 (Keijser *et al.*, 2008; Zaura *et al.*, 2009). Earlier studies have shown that distinct microbial communities colonize different oral structures and tissues (Mager *et al.*, 2003; Aas *et al.*, 2005). The Human Oral Microbiome Database (www.homd.org) has currently catalogued 619 taxa belonging to 13 phyla, based upon 16S rRNA gene sequences. These phyla are *Actinobacteria, Bacteroidetes, Chlamydiae, Chloroflexi, Euryarchaeota, Firmicutes, Fusobacteria, Proteobacteria, Spirochaetes,* SR1, *Synergistetes, Tenericutes,* and TM7 (Dewhirst *et al.*, 2010). To date, less than 50% of the total bacterial species have been isolated in culture and identified. Cultivation-independent molecular methods using 16S rRNA gene sequences have validated these estimates by identifying approximately 600 species or phylotypes (Dewhirst *et al.*, 2010). The presence of such a significant number of bacterial species in the oral cavity alone indicates the importance of their mutual interactions affecting the overall oral health.

Oral microorganisms have been associated with a number of infectious diseases such as dental caries, periodontitis, endodontic infections, alveolar osteitis and tonsillitis (Dewhirst *et al.*, 2010). Interestingly, the bacterial constituent progressively changes from relatively thin biofilms dominated by Gram-positive found in healthy individuals to thicker biofilms containing increased proportions of Gram-negative anaerobic rods in patients with periodontitis (Hojo *et al.*, 2009). Traditionally, it has been considered that the mutans group of streptococci are the aetiological agent of dental caries, and the anaerobic bacteria *Porphyromonas gingivalis* and *Tannerella forsythia* are responsible for periodontitis (Jenkinson and Lamont, 2005), though the role of other groups of pathogens can not be ruled out. These mutans streptococci are found to be associated with not only the oral diseases, but also a number of systemic diseases including cardiovascular diseases (Joshipura *et al.*, 1996, 2003; Beck and Offenbacher, 2005; Nakano *et al.*, 2010), preterm birth (Offenbacher *et al.*, 1998) diabetes (Genco *et al.*, 2005), intrauterine infection (Fardini *et al.*, 2010) and pneumonia (Awano *et al.*, 2008). In this chapter, we will discuss the bacterial species that are present as distinct biofilm communities in the healthy mouth and their mutual interactions in dental caries progression.

## Colonization of the oral cavity

The oral microbiota are present as distinct biofilms formed by polymicrobial communities on the surfaces of the teeth, gums and tongue. These biofilms are three-dimensional structured assemblages of microorganisms that are embedded in a matrix of extracellular polymeric substances (Zijnge *et al.*, 2010). Studies over the last two to three decades have shown that many of the constituents of these biofilms cannot survive as monocultures, indicating that their mutual interaction is a key factor for their survival in a polymicrobial community structure (Jenkinson and Lamont, 2005). These observations also explain why more than half of the bacterial species present in oral biofilms have not yet been cultured in the laboratory. The biofilm thrives because of the mutual cooperation among the members; moreover, the community structure provides increased security for them, as shown by recent studies indicating that biofilms can be hundreds of times more resistant to antibiotics than their planktonic counterparts

(Sedlacek and Walker, 2007; Zhang and Mah, 2008). In the biofilm community, the bacterial species maintain a state of homeostasis; however, in the oral cavity, where the microenvironment is in a state of perpetual change, the bacteria also manipulate the balance in order to favour their own growth and survival. When this delicate balance is compromised, more pathogenic species become predominant, leading to caries or periodontitis. The host's oral environment provides conditions that support multispecies microbial community structures. Minute changes in the host environment shift the balance from a commensal community towards a biofilm community consisting of cariogenic bacterial species (Kreth *et al.*, 2009; Marsh and Devine, 2011).

## The early colonizers

The freshly cleaned tooth surface is quickly coated with a salivary pellicle, which provides a variety of components that are used as receptors for pioneer bacteria to attach. Oral biofilm development is initiated by adhesion of the primary microorganisms to the saliva-coated surface (Fig. 9.1). Streptococci are the pioneers in the oral biofilm communities, with about 60–80% of the bacteria found in early communities on tooth enamel belonging to this group (Nyvad and Kilian, 1987, 1990; Diaz *et al.*, 2006; Dige *et al.*, 2009). Oral streptococci are in general referred to as viridans streptococci and are divided into five different groups: (1) Mutans group (e.g. *Streptococcus mutans* and *Streptococcus sobrinus*), (2) Salivarius group (*Streptococcus salivarius*), (3) Anginosus group (e.g. *Streptococcus anginosus* and *Streptococcus intermedius*), (4) Sanguinis group (e.g. *Streptococcus sanguinis* and *Streptococcus gordonii*), and (5) Mitis group (e.g. *Streptococcus mitis* and *Streptococcus oralis*) (Facklam, 2002). It is interesting to note that not all streptococci belong to the early colonizers. Among them commensals belonging to the mitis group of streptococci are the predominant early colonizers. Sometimes *S. gordonii* and *S. sanguinis* have also been reported to be present in the early stages of dental biofilm (Kolenbrander *et al.*, 2006). Apart from the streptococci, other early colonizers include *Actinomyces* (Palmer *et al.*, 2003), *Veillonella* and *Neisseria* (Li

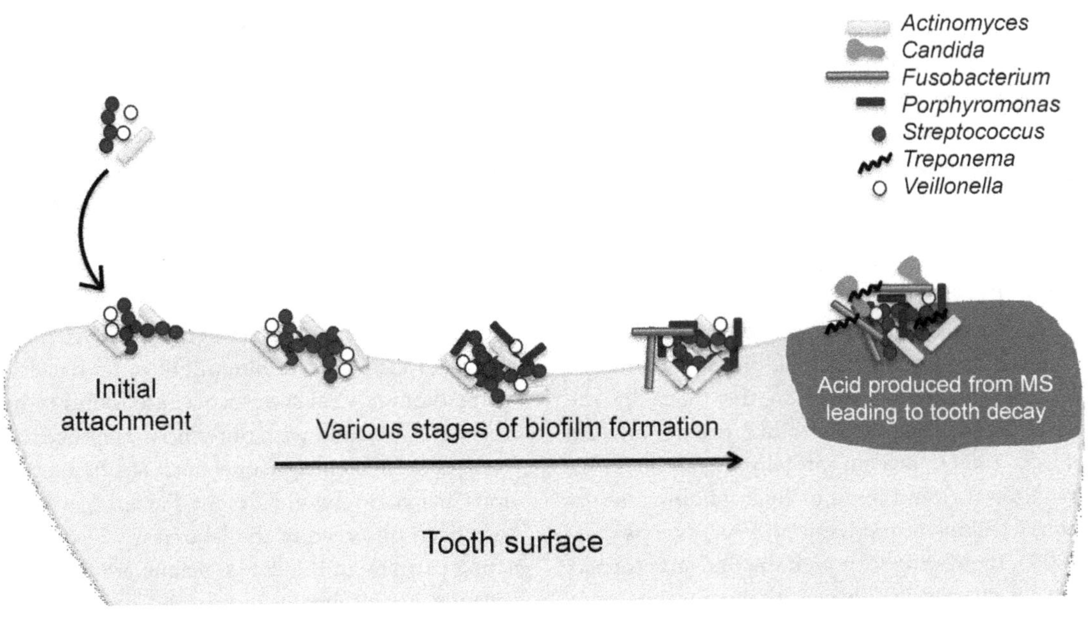


**Figure 9.1** Formation of microbial community on tooth surface. Biofilm formation is initiated by the attachment of the early colonizers on mucosal surface, co-aggregation and co-adhesion helps in bringing more bacterial diversity in the community. Biofilm maturation favours the growth of various pathogenic strains of Mutans Streptococci, *Porphyromonas* and *Fusobacterium* in an increasingly anaerobic condition followed by colonization by *Treponema* and *Candida* species belonging to the late colonizers. Communication among the bacterial species and various interaction shapes the final outcome of the community composition.

*et al.*, 2004). Initial adhesion involves binding of bacteria to saliva components that are adsorbed to the surfaces in the oral cavity. The components also show specificity towards certain oral surfaces, and therefore different receptors can be found at different locations. Interestingly, there are some salivary receptors, termed cryptitopes, which attain an active configuration only after binding to a surface (Gibbons, 1989; Jenkinson and Lamont, 2005).

Streptococci also recognize the proline-rich proteins and statherin components of saliva, which are tightly bound by enamel. Salivary mucins and agglutinins, such as gp340 glycoprotein, also adsorb to enamel and the epithelial surfaces, providing receptors for streptococci and other early colonizing bacteria (Loimaranta *et al.*, 2005). Apart from the above components, human saliva contains an array of proteins such as a-amylase, immunoglobulins, fibronectin, lactoferrin and $a_2$-macroglobulin, which can also act as receptors for the streptococci (Scannapieco, 1994). The oral streptococci also express numerous protein and lipoprotein adhesins on their cell surface. The *S. sanguinis* genome encodes at least 93 putative surface anchored polypeptides, which have a role in adhesion (Xu *et al.*, 2007). The presence of such diverse cell-surface exposed molecules enables the streptococci to bind to a vast number of human and bacterial receptors, also providing selective advantage over other bacteria that have limited receptor recognition variability. Streptococcal adhesion mechanisms have been described in several recent reviews (Nobbs *et al.*, 2009, 2011; Moschioni *et al.*, 2010).

The initial bacterial adhesion mechanism includes the recognition of oligosaccharide or protein receptors present in the saliva by the bacterial adhesins, in lectin-like reactions (Ruhl *et al.*, 2004), protein–protein interactions (Li *et al.*, 2001a) and ionic or hydrophobic associations (Jenkinson and Lamont, 1997; Nobbs *et al.*, 2009). Interestingly, though the oral streptococci use a vast array of adhesins, there are some principal adhesin molecules that have a functional homology across many different species of streptococci. For example, the antigen I/II (AgI/II) family of proteins are expressed by commensal oral streptococci such as *S. gordonii*, *S. oralis* and *S. sanguinis*, by cariogenic mutans streptococci such as *S. mutans* and *S. sobrinus*, and also by the group A *Streptococcus S. pyogenes* (Jakubovics *et al.*, 2005b; Zhang *et al.*, 2006). AgI/II family polypeptides are 160–180 kDa and have an N-terminal region containing a signal for secretion across the cell membrane and a C-terminal domain containing the LPXTG motif required for cross-linking to the cell wall. Salivary glycoprotein gp340 is incorporated into the enamel pellicle and promotes the adhesion of streptococci by binding to the AgI/II polypeptides (Jakubovics *et al.*, 2005a,b). Interestingly, in the fluid phase of the saliva, gp340 interacts with the AgI/II proteins, resulting in the aggregation of bacterial cells. Most of the time, the bacterial aggregates do not adhere well to surfaces, and as a result are removed from the mouth by swallowing (Jakubovics and Kolenbrander, 2010). The primary colonizer *S. gordonii* produces two AgI/II polypeptides, SspA and SspB, which are shown to interact with extracellular matrix proteins such as collagen and fibronectin (Heddle *et al.*, 2003; Jakubovics *et al.*, 2009). As a result of such interactions between the adhesins and the salivary receptors, the primary colonizers adhere tenaciously to surfaces, even under high shear stress forces (Ding *et al.*, 2010). These properties can be attributed to the adhesins that respond positively to fluid flow stresses and that can discriminate between immobilized or fluid phase isoforms of the same receptor (Brady *et al.*, 1992; Isberg and Barnes, 2002; Jakubovics *et al.*, 2005b).

## The late colonizers

Non-Mutans streptococci, which are the primary colonizing bacterial species in the oral cavity, are shown to facilitate the binding of other bacteria to the biofilm. Oral streptococci are found to be in close association with other non-streptococcal bacteria in the dental plaque. Such *in situ* interactions between these different bacterial species have been observed in the laboratory. The interacting partners in the dental plaque are found to coaggregate in the culture media. Coaggregation occurs via adhesin-receptor mediated interaction between compatible partner organisms. A similar phenomenon that facilitates the binding of planktonic bacteria to immobilized cells on the tooth surface is termed coadhesion. Streptococcal

AgI/II proteins are the most important group of bacterial cell-surface molecules that have been associated with the phenomenon of coaggregation and coadhesion. *S. gordonii* coaggregates with *Actinomyces oris* strain T14V, and interestingly, this interaction is mediated by the *S. gordonii* SspB protein and not by SspA (Egland *et al.*, 2001; Jakubovics *et al.*, 2005b). However, SspA and a number of other AgI/II proteins from other streptococci mediate coaggregation with *A. oris* strain PK606. Therefore, the presence of different types of AgI/II proteins among the oral streptococci provides increased opportunities to establish partnership with different groups of bacteria (Jakubovics and Kolenbrander, 2010). *S. gordonii* AgI/II proteins also bind to the periodontal pathogen *Porphyromonas gingivalis*, establishing the subgingival biofilms (Lamont *et al.*, 2002). In addition, the *S. gordonii* AgI/II polypeptide also recognizes the receptors on the surface of *Candida albicans* hyphae (Bamford *et al.*, 2009).

Since coaggregation between the oral bacteria establishes the foundation for the biofilms, thousands of their pairwise interactions have been studied (Kolenbrander *et al.*, 2006). These investigations have identified *Fusobacterium nucleatum* as the most promiscuous coaggregating partner, as strains of *F. nucleatum* coaggregate with almost all other oral bacteria. *F. nucleatum* is an obligate anaerobe and is not generally associated in the newly formed supragingival biofilms. As the plaque ages, the level of *F. nucleatum* increases, making up almost 20% or more of the bacteria in subgingival dental plaque (Suzuki *et al.*, 2004) (Fig. 9.1). Another important aspect of *F. nucleatum* in the dental biofilm community is its ability to interact with both early and late colonizers. Therefore, it acts as a bridge in the progression of the dental biofilm from a superficial to a matured plaque. Recently, a 350-kDa cell-surface protein RadD of *F. nucleatum* has been described which mediates arginine-inhibitable coaggregation with *S. sanguinis* (Kaplan *et al.*, 2009). The *radD* knockout mutant could not form a structured biofilm with *S. sanguinis*, indicating that coaggregation influences the spatial distribution of the partner organisms in biofilms. During the initial stages of biofilm formation, adhesive interactions between partner bacterial cells coordinate the spatial distribution of cells in the biofilm. Further evidence for the role of *F. nucleatum* as a structural bridge between early and late colonizers is presented in Chapter 4. Subsequently, growth and coadhesion among the bacteria helps in the development of the microenvironment within the biofilm. In the biofilm, close interaction among the partners in the form of metabolic cooperation and interspecies competition further modifies the microenvironment. Sometimes these mutual interactions are detrimental to human oral health, producing dental caries and periodontitis (Fig. 9.1).

## Cariogenic oral microflora or mutans streptococci (MS)

The human oral cavity provides a habitat for more than 700 bacterial species, though not all of them are involved in dental caries. In the late 19th century, it was believed that the oral microorganisms produce various enzymes, which break down the dietary carbohydrates, creating acids responsible for tooth demineralization and caries progression. This hypothesis, the 'non-specific plaque hypothesis' considered all oral microorganisms to be potentially cariogenic. However, as microbiology advanced in the 20th century, it became apparent that a group of oral streptococci are the primary etiologic agent of dental caries. From human carious lesions, Clarke isolated these organisms with distinct characteristics and named them *Streptococcus mutans* (Balakrishnan *et al.*, 2000). Later, studies by various groups established that *S. mutans* is in fact one of the most important cariogenic agents in the human oral cavity. Further extensive studies have indicated that the bacteria originally classified as '*S. mutans*' actually form a collection of seven distinct species (Coykendall, 1989). These species are collectively referred to as the mutans streptococci (MS), as they have a lot of common properties involved in caries induction. MS have been divided into eight serotypes 'a' through 'h' and the seven different MS species are *S. mutans* (serotypes c, e and f), *S. rattus* (serotype b), *S. cricetus* (serotype a), *S. sobrinus* (serotype d and g), *S. ferus* (serotype c), *S. macacae* (serotype c) and *S. downei* (serotype h) (Balakrishnan *et al.*, 2000). Though all of the above species are potentially cariogenic, *S. mutans* and *S. sobrinus* are by far the most common in humans, and between them *S. mutans* is more frequently isolated from the carious lesions. *S. sobrinus* is generally found associated with *S. mutans* and is involved primarily in smooth surface caries formation (Lindquist and Emilson, 1991). Though the number of MS in patients with caries is generally higher than that in healthy individuals, sometimes, MS has been shown to colonize the healthy teeth too. This finding indicates that the caries progression in an individual depends on a number of factors including food intake, as a high carbohydrate diet favours the growth of MS.

Mutans streptococci first colonize the infant's teeth at 19–31 months of age, which is known as the 'window of infectivity' (Caufield *et al.*, 1993). If the MS fail to colonize the tooth surface during this time, the child remains free of them until 6 years of age. However, MS start colonizing the permanent molar teeth, which begin to erupt at this time. The colonization process continues with age and depends on oral hygiene practices. The number of MS in the oral microflora also depends on the presence of a large number of other bacterial species, the interaction with which shapes the final oral microbial community. Apart from the natural colonization of the tooth with MS, saliva acts as a vehicle to transfer the bacteria from mother to infant. Studies have demonstrated that if the number of MS in mother's saliva is low, the chances of MS acquisition by the infant is reduced greatly (Kohler *et al.*, 1983).

As the principal cariogenic bacterial species, the mutans streptococci have a set of distinct properties such as extracellular polysaccharide synthesis, acid production (acidogenicity), survival in an acidic environment (aciduricity), intracellular polysaccharide synthesis and endodextranase synthesis. Mutans streptococci, and specifically *S. mutans*, colonize the tooth surface using cell wall associated adhesins (Fig. 9.2). This initial attachment is generally reversible, and is followed by a sucrose dependent irreversible attachment. In the presence of sucrose, *S. mutans* synthesizes both water-soluble (α1-6 linked) and water-insoluble (α1-3 linked) polyglucose molecules known as glucans. Various glucosyltransferases (GTFs) encoded by *gtfB*, *gtfC* and *gtfD* genes have been described in *S. mutans* and are responsible for the production of glucans (Fig. 9.2). Water-insoluble glucans are a significant constituent of plaque biofilms as they facilitate the accumulation and adherence of *S. mutans* forming stable biofilms (Mattos-Graner *et al.*, 2004; Paes Leme *et al.*, 2006; Koo *et al.*, 2010; Bowen and Koo, 2011). *S. mutans* also produces a group of proteins called glucan binding proteins (GBPs) on their cell surface, facilitating strong adhesion to the tooth surface (Banas and Vickerman, 2003). Sugar fermentation and production of a large amount of lactic acid is the major attribute of *S. mutans*, creating a localized drop of pH causing enamel demineralization (Fig. 9.2). In addition, *S. mutans* can also withstand very low pH because of its strong acid tolerance mechanisms (Matsui and Cvitkovitch, 2010). By producing acids, *S. mutans* makes the oral microenvironment inhospitable for other species. Therefore, *S. mutans*' acidogenicity compounded with its remarkable aciduricity makes it a formidable cariogenic pathogen.

Though the presence of a large number of *S. mutans* in the oral cavity increases the risk of dental caries formation, a strong strain-to-strain variation with respect to their caries causing potential has also been observed. However, at the genetic level, the strains are relatively homogeneous with respect to their core virulence genes, though considerable diversity has been observed for the horizontally acquired genetic clusters (Saxena *et al.*, 2005; Caufield *et al.*, 2007; Nakano *et al.*, 2007; Waterhouse *et al.*, 2007; Saxena *et al.*, 2008; Maruyama *et al.*, 2009; Argimon and Caufield, 2011). Recent research indicates that subtle variation in the gene expression level of the *S. mutans* determines their virulence potential. Also, biofilm formation in *S. mutans* has been suggested to be associated with a higher prevalence of caries in individuals with low counts of the cariogenic microflora such as mutans streptococci (Giacaman *et al.*, 2010). A number of genes have been shown to be differentially expressed when *S. mutans* is grown in planktonic and biofilm forms, indicating that virulence gene expression is modulated by the growth phase of the bacteria (Motegi *et al.*, 2006; Shemesh *et al.*, 2007, 2008).

Two-component signal transduction systems (TCS) in bacteria are known to be associated with

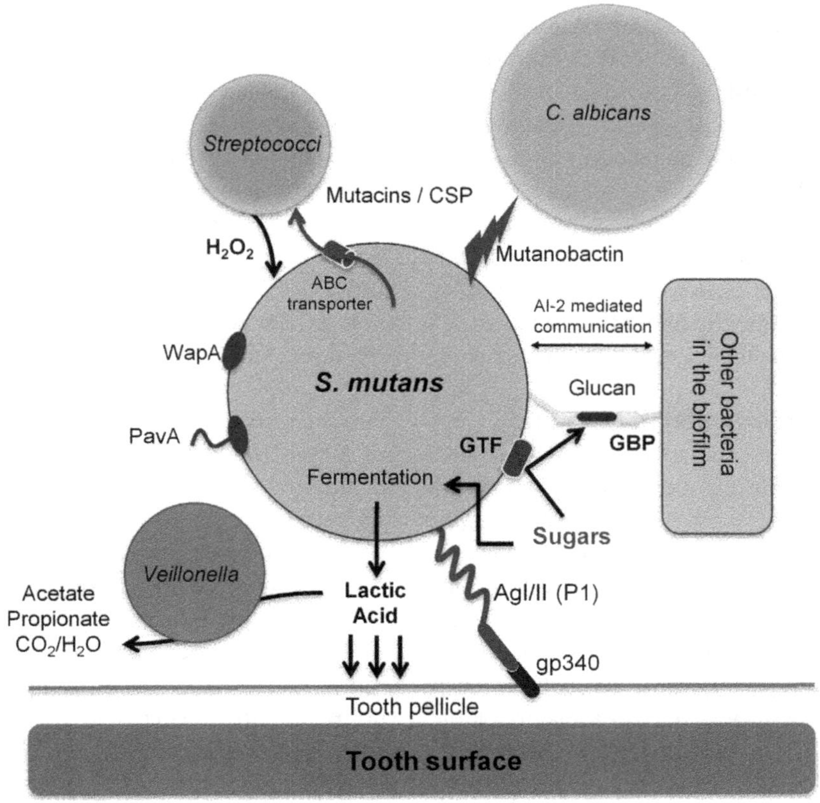


**Figure 9.2** Interaction of cariogenic pathogen *Streptococcus mutans* with other bacteria and *Candida* in the oral biofilm community. The relative size of *S. mutans* cell is enhanced for clarity.

bacteria's ability to sense and respond to their ever-changing surroundings. Genome sequencing of *S. mutans* UA159 identified 13 distinct TCSs (Ajdic *et al.*, 2002), of which a few are involved in the regulation of virulence gene expression. An orphan response regulator CovR has been demonstrated to regulate the expression of the virulence genes *gtfBC* and *gbpC* (Idone *et al.*, 2003; Biswas and Biswas, 2006; Biswas *et al.*, 2007). CovR generally acts as a negative regulator of the virulence genes. Recent work from our group identified a number of genes positively regulated by CovR (Chong *et al.*, 2010; Dmitriev, 2011). However, in the absence of a cognate sensor kinase, the pathway of gene regulation mediated by CovR remains to be established. In addition to its involvement with virulence genes, the TCSs are also involved in stress response and mutacin production in *S. mutans* (Biswas *et al.*, 2008; Chong *et al.*, 2008). All these studies indicate that the cariogenic agent *S. mutans* modulates its virulence gene expression by carefully sensing its environment. The sensory response becomes very important for the bacteria given the ever-changing environment where they reside (see Chapter 8).

## Interactions among the bacterial species of the oral communities

Traditionally, bacteria are studied in isolation. However, in the dental biofilm communities, the participating bacteria interact very closely, and in the process modify the microenvironment in such a way that creating such a condition *in vitro* is difficult. This is one of the reasons why more than 50% of oral bacteria are not yet cultured in the laboratory. The recent developments in high throughput nucleotide sequencing and microarray technologies have revolutionized our understanding of the lifestyle of the closely interacting bacterial species in the oral microflora. In ideal situations, the bacteria closely collaborate to obtain nutrients. It is interesting to note that sometimes the byproducts of one bacterium's metabolism are used up by another bacterium. Also, there are situations when the mutually cooperating partners become competitors and produce potent peptides called bacteriocins to kill others. We will discuss these interactions shaping the life of the bacterial species in the oral biofilms in the following sections.

## Mutualistic interaction among oral microflora

During the establishment of the oral community, mutualism is the most fundamental principle involved, as the partners need to cooperate in order to colonize the saliva-coated surface successfully. Saliva is the primary source of nutrients for the bacterial species in the oral cavity. The oral bacteria also metabolize the carbohydrates taken by the host, which plays a major role in caries progression. As carbohydrate metabolism produces acids, the pH of the biofilm microenvironment drops, causing demineralization of teeth and caries. Interestingly, most oral bacteria cannot utilize saliva as the sole source of nutrients in monoculture; however, in a dental plaque they can degrade the mucin and other components of saliva (Wickstrom and Svensater, 2008; Wickstrom *et al.*, 2009). Studies have indicated that consortia of microorganisms in the oral cavity can grow better than the individual species on saliva as a nutrient source (Periasamy and Kolenbrander, 2009a,b, 2010; Periasamy *et al.*, 2009; Jakubovics and Kolenbrander, 2010).

Another aspect of mutual cooperation discovered among the plaque community is the specific interaction between partners for metabolic purposes. Periasamy and Kolenbrander (2009b) described the interaction of *P. gingivalis*, which could be found within four hours of tooth cleaning, with the initial, early, middle, and late colonizers growing exclusively on saliva. This study revealed that *P. gingivalis* could not grow as a single species or together with the initial colonizer *S. oralis* but showed mutualistic growth when grown together with other initial colonizers such as *S. gordonii* and *A. oris*, *Veillonella* sp., the secondary colonizer *F. nucleatum*, and the late colonizer *Aggregatibacter actinomycetemcomitans*. Although *P. gingivalis* did not grow when paired with *S. oralis*, it was able to grow in a three-species consortium containing *S. gordonii*, indicating that a specific mutualistic interaction among the oral bacteria plays a major role in establishing the community. Another study from the same group demonstrated that *S. oralis* and *F. nucleatum* coaggregate, but this interaction did not lead to the growth of *F. nucleatum*. However, when *A. oris* that coaggregates with both *S. oralis* and *F. nucleatum* was introduced into the system, *F. nucleatum* grew (Periasamy *et al.*, 2009). Though such studies have convincingly demonstrated the mutualistic interactions among the partners in the biofilm, it is always difficult to predict such interactions in a natural setting.

In the oral biofilm, nutritional interrelationship is an important factor determining the composition of the flora. This interrelationship can be exemplified by lactic acid production and utilization. Streptococci and lactococci produce lactic acid as a byproduct of their metabolism. However, lactic acid can be used as a source of energy by *A. actinomycetemcomitans* and *Veillonella* spp. (Fig. 9.2). Thus, when sucrose is the only source of carbon, *A. actinomycetemcomitans* cannot grow as a monoculture, but can grow when *S. gordonii* is also present, as *S. gordonii* converts sucrose to lactic acid (Brown and Whiteley, 2007). Periasamy and Kolenbrander (2010) studied the interaction of *Veillonella* sp. with lactate producing *S. oralis* in flowcell biofilm model using saliva as the sole source of nutrients. Their findings indicated that neither *S. oralis* nor *Veillonella* sp. could grow in monoculture. However, when inoculated together, both organisms increased approximately 8-fold, resulting in a confluent biofilm. Furthermore, *Veillonella* sp. neutralized the incompatibility of *P. gingivalis* and *S. oralis*. Biofilms containing the three species *Veillonella* sp., *P. gingivalis* and *S. oralis* grew well. Measurements of the lactate concentration in the biofilm effluent indicated that *Veillonella* sp. utilized the residual lactate in the saliva and also removed lactate produced by *S. oralis* (Periasamy and Kolenbrander, 2010). Therefore, lactate consumption is an important factor influencing the mutualistic associations between oral biofilm bacteria.

## Competition among oral microflora

Interspecies competition shapes the course of evolution. This principle holds true in the oral biofilm microenvironment. Oral streptococci such as *S. mutans* and *S. salivarius* produce antibacterial peptides known as bacteriocins, which are potent against a range of other oral bacteria (van der Ploeg, 2005; Hyink *et al.*, 2007; Hossain and Biswas, 2011). The production of bacteriocins in *S. mutans*, known as mutacins, are shown to be coordinated with the development of competence, a state in which the competent cells can take up extracellular DNA from the environment (van der Ploeg, 2005; Kreth *et al.*, 2006). Since competence development is quorum sensing-dependent in *S. mutans*, it follows that mutacin production at a particular cell density would enhance its effectiveness against other organisms. Bacteriocin production provides another significant benefit to the producing organisms. These bacteria can acquire DNA from the dying cells in their surroundings. This DNA may be incorporated into the host chromosome, increasing its genetic versatility, or else can be broken down for energy. It has also been observed that the extracellular DNA (eDNA) acts as a stabilizer in the biofilm matrix. *S. mutans* utilizes another mechanism to release DNA from a fraction of its own population in the community, and subsequently acquires the DNA when other bacteria are not available. This altruistic behaviour involves the induction of an intracellular bacteriocin, CipB (mutacin V) in approximately 1% of the population in response to the competence stimulating peptide (CSP) (Perry *et al.*, 2009). Disruption of the gene encoding CipB significantly reduced competence for genetic transformation, indicating that CipB-dependent autolysis is important for the transfer of genetic material between *S. mutans* cells.

In addition to bacteriocins, oral streptococci also produce hydrogen peroxide ($H_2O_2$), which can cross bacterial cell membranes and oxidize intracellular macromolecules such as lipids, DNA and proteins, causing stress or cell death. $H_2O_2$ produced by the oral streptococci has been shown to either kill or prevent the growth of other organisms in the oral biofilm (Holmberg and Hallander, 1973; Jakubovics *et al.*, 2008). However, the producers are relatively resistant to its effects. Microbial composition in the oral biofilm often changes from predominantly commensals to pathogens as a result of changes in the microenvironment. In the case of dental caries, the population diversity decreases significantly (Aas *et al.*, 2008). The diminished diversity can be attributed to the production of acids by the acidogenic bacteria, leading to the drop of pH in the microenvironment. This event acts as a selective factor favouring the growth and survival of the aciduric and acidogenic organisms such as mutans streptococci, lactobacilli and bifidobacteria. Although this phenomenon is more pronounced in the matured dental plaque associated with active caries lesions, the onset of competition probably arises much earlier. The early colonizing streptococci such as *S. sanguinis*, *S. gordonii*, *S. mitis* and *S. oralis*, produce acid from sugars, though to a lesser extent than *S. mutans* or *S. sobrinus*; however, the process is important in preparing the foundation for the incorporation of more acidogenic and aciduric organisms into the biofilm (Takahashi and Nyvad, 2008, 2011).

Apart from bacteria, fungal species such as *Candida albicans* also constitute a small fraction of the oral commensal microflora (Kulak *et al.*, 1997; Thein *et al.*, 2006). In immunocompromised individuals, it causes candidiasis, which is associated with a morphological transition from yeast to a hyphal structure. This transition is associated with virulence and biofilm formation (Mitchell, 1998; Douglas, 2003). Interestingly, co-culturing of *C. albicans* with *S. mutans* in a model oral biofilm increased the growth of *C. albicans*, though the hyphae formation was suppressed (Pereira-Cenci *et al.*, 2008), indicating that *S. mutans* interferes in the transition of *C. albicans* by some interkingdom signalling method. Further investigation indicated that *S. mutans* CSP was the interfering substance (Jarosz *et al.*, 2009). The genomic island TnSmu2 in *S. mutans* carries a cluster of genes coding for a putative hybrid polyketide synthase–non-ribosomal peptide synthase (PKS-NRPS) responsible for producing secondary metabolites involved in oxygen and $H_2O_2$ tolerance and biofilm formation (Wu *et al.*, 2010). The secondary metabolites known as mutanobactins are supposed to have functions in interspecies interaction in the biofilm community. Recently, Joyner *et al.* (2010) found that mutanobactin interferes in *C. albicans* hyphal transition (Fig. 9.2). Their experiment involving co-culture of wild type *S. mutans* with *C. albicans* resulted in maintenance of yeast morphology, whereas co-culturing *C. albicans* with a mutanobactin cluster deleted *S. mutans* strain produced the hyphal morphology involved in the virulent form.

## Communication among the oral microflora

Communication among the bacterial species in the oral microflora often coordinates the expression of virulence genes, various physiological processes and also the biofilm formation. Various secreted diffusible molecules mediate this communication. One such group of molecules are the autoinducers (AI), which are involved in the density-dependent or quorum sensing systems operating within the monospecies societies of bacteria and in multispecies communities (Jenkinson and Lamont, 2005). The species-specific signalling molecule N-acyl homoserine lactones (autoinducer 1) has not been detected in oral bacteria; however, many of the bacteria produce and sense the interspecies signalling molecule autoinducer 2 (AI-2) through the action of LuxS. Rickard *et al.* (2006) demonstrated that AI-2 is essential for mutualistic biofilm growth in the cocultures of *S. oralis* and *A. oris* with saliva as the sole source of nutrients. Neither of the two bacteria can grow in monoculture, indicating that they are dependent on each other for nutrients or some other factors. The *luxS* mutant of *S. oralis* was unable to grow in monoculture or with *A. oris* in coculture. Genetic complementation of the *luxS* mutation restored the growth pattern. Also, the exogenous addition of AI-2 showed the wild type growth pattern (Rickard *et al.*, 2006). This study also found that the exogenous AI-2 was effective in a limited concentration, as the optimum concentration for mutualistic growth was determined to be 0.8 nM.

Until now the molecular basis of AI-2 sensing by oral bacteria has not been well understood. However, in *A. actinomycetemcomitans*, two receptors LsrB and RbsB, for AI-2, have been identified (Shao *et al.*, 2007a). The growth of *A. actinomycetemcomitans* in a flowcell biofilm model was dependent upon the production and sensing of AI-2 (Shao *et al.*, 2007b). Periasamy and Kolenbrander (2009a) have shown that in a microtitre plate biofilm model, *A. actinomycetemcomitans* cells were able to grow and attach to the surface; however, the growth occurred after 36 hours. It is speculated that AI-2 gradually accumulates over time and growth is accelerated when a threshold level of AI-2 has been reached (Jakubovics and Kolenbrander, 2010). Previously, Liljemark *et al.* (1997) showed that crude preparations of culture-free supernatant from oral bacteria stimulated the growth of other strains. Moreover, the development of natural dental plaque involves a rapid burst of growth following the accumulation of a threshold concentration of bacteria, approximately 2–6 ′ $10^6$ cells/mm$^2$. These data provide conclusive evidence in support of bacterial growth in dental plaque being dependent on cell density sensing (Jakubovics and Kolenbrander, 2010).

Recently, it has been shown that bacteria can physically interact with one another using a specialized structure called nanotubes that bridge neighbouring cells (Dubey and Ben-Yehuda, 2011). These researchers have shown that these nanotubes, which range from 30 to 130 nm in diameter, could connect bacteria of the same and different species. Through these tubes, bacteria were able to exchange small molecules, proteins and even small genetic elements such as plasmids (Dubey and Ben-Yehuda, 2011). This new mechanism of communication can facilitate the acquisition of new features in nature, such as antibiotic resistance, and may help explain how bacteria of many different species coordinate community life in biofilms. Since dental plaque harbours many different species of bacteria, it is possible that some of these organisms are engaged in nanotubes mediated communication with other species.

## Competence development and horizontal gene transfer among oral microflora

In a multispecies community structure like the oral biofilm, the bacteria try to outcompete each other to gain maximum benefit from the available resources. Horizontal gene transfer (HGT) is a rampant event in such a condition, helping them to increase their genetic repertoire. The oral biofilm environment provides the proper setting for DNA exchange as the cells are in close proximity. HGT has been observed among oral streptococci (Roberts *et al.*, 2001; Warburton *et al.*, 2007), and many genera of oral bacteria, such as *Actinomyces, Bifidobacterium, Fusobacterium, Haemophilus, Peptostreptococcus, Streptococcus* and *Veillonella*, contain conjugative transposons that facilitate the DNA transfer between bacteria through conjugation (Rice, 1998). Sequencing of the whole genome of various oral streptococci has indicated the presence of many acquired genomic islands (Ajdic *et al.*, 2002; Vickerman *et al.*, 2007; Denapaite *et al.*, 2010; Reichmann *et al.*, 2011). Comparative genome hybridization experiment with the cariogenic *S. mutans* strains have indicated that approximately 20% of the open reading frames are dispensable, as they are absent in some stains (Waterhouse *et al.*, 2007). These open reading frames are found to be associated with various genomic islands and insertion sequences, providing increased genotypic variability to the strains. Genome analysis of the sequenced oral bacteria suggests that approximately 5–45% of genes in different species are acquired by HGT (Kolenbrander *et al.*, 2010). Oral streptococci are found to be naturally competent: cell–cell contact is not required for DNA transfer. However, the spatial arrangement of cells in the biofilm structure greatly influences the transfer of DNA to competent cells (Kolenbrander *et al.*, 2010). As mentioned earlier, the development of competence in streptococci occurs in response to sensing a secreted signal molecule, competence-stimulating peptide (CSP), which is encoded by *comC*. The sequence of *comC* varies between streptococci, which respond only to the type of CSP that they produce. Therefore, CSP-mediated communication is species or even strain dependent. Competence development in *S. mutans* is more efficient in monospecies biofilms than in planktonic cells, and it has been postulated that CSP acts as a quorum-sensing regulator in this organism (Li *et al.*, 2001b). Recently, an alternative pathway for competence development among the streptococci has been described (Mashburn-Warren *et al.*, 2010). This study, discovered that a transcription factor ComR of Rgg family of transcription factors regulate the expression of a small pheromone peptide XIP (*sigX* inducing peptide), responsible for inducing the alternative sigma factor SigX, which is involved in competence development and transformation. Deletion of *comR* and *comS* (coding for XIP) genes abolishes transformation in *S. mutans*. Fleuchot *et al.* (2011) have described a novel quorum sensing mechanism involving the Rgg family transcriptions factors along with the small hydrophobic peptide (SHP) in *S. thermophilus*, where the SHP pheromone production is regulated by Rgg, subsequently detected and reimported into the cell via ABC transporters. Though such proteins are universal among the streptococci and many other *Firmicutes*, the context of their expression and role in competence development remains to be established.

The extracellular DNA has another important function in stabilizing the biofilm matrix. Therefore, oral streptococci have developed means to release the much-required DNA from the competitor bacteria. This competence dependent phenomenon is well studied in *S. mutans* (Petersen *et al.*, 2005). In multispecies biofilms, other oral streptococci target the CSP signalling by *S. mutans*. *S. salivarius* produces an extracellular product that interferes with *S. mutans* CSP and inhibits biofilm formation (Tamura *et al.*, 2009). *S. mutans* CSP activity has been found to be inhibited by *S. sanguinis, S. mitis, S. oralis* and *S. gordonii*. Oral streptococci degrade the CSP to protect themselves as high concentrations of *S. mutans* CSP trigger the production and secretion of mutacins, which kill neighbouring cells and release their DNA (Kreth *et al.*, 2005a). However, *S. mutans* is immune to its own extracellular bacteriocins, but the accumulation of CSP triggers the synthesis of an intracellular bacteriocin and subsequent autolysis in approximately 1% of the *S. mutans* population (Perry *et al.*, 2009). Therefore, in the absence of other species, CSP-mediated quorum sensing enables *S. mutans* to obtain extracellular DNA from a subpopulation of *S. mutans* cells.

## New approaches to control the cariogenic community

Application of good oral hygiene practices together with fluoridation is generally considered to be primarily responsible for the declining incidence of dental caries in industrialized countries. In the last few years, studies using modern genetic and genomics technologies have significantly increased our knowledge about dental caries. Traditionally, caries treatment has been focused around removing the whole dental plaque. However, this approach has its limitations, as dental biofilm consists of mostly commensals and few pathogens, and removal of the plaque disturbs this balance. After cleaning the plaque, the possibility arises of cariogenic bacteria recolonizing the tooth surface. Therefore, modern methods are being devised to target key pathogens, and in particular Mutans Streptococci, rather than the whole biofilm structure (Kuramitsu *et al.*, 2007). These approaches are discussed in the following sections.

### Inhibition of bacterial adhesion and biofilm formation on the tooth surface

The AgI/II adhesin of the cariogenic pathogen *S. mutans* has been targeted for anti-caries approaches for many years. Early studies have shown promising results using the *S. mutans* AgI/II as vaccine candidates (Lehner *et al.*, 1980; Hajishengallis *et al.*, 2005; Li *et al.*, 2009). Glucosyltransferases (GTF), and glucan binding proteins (GBP) produced by *S. mutans* and utilized for biofilm formation, have also been used as vaccine candidates (Taubman and Nash, 2006). However, as dental caries is not a life threatening disease, it now seems unlikely that vaccine development will receive sufficient support from industry or from the public to reach large-scale clinical trials. An alternative approach to vaccine development is blocking the adhesin molecules by small peptides or other molecules. A synthetic peptide P1025 has been described corresponding to a fragment of *S. mutans* AgI/II that blocks its adhesion to gp340 glycoprotein and has been shown to be effective in clinical trials (Younson and Kelly, 2004). Recently, a peptide targeting the *S. gordonii* SspB protein has been found to inhibit the binding of *P. gingivalis*, which has potential application against periodontitis (Daep *et al.*, 2008, 2011). Liu *et al.* (2011) screened a small-molecule library of 506 compounds against the *S. mutans*. Eight of the small-molecules in the library showed selective inhibition of *S. mutans* biofilm formation at a concentration of 4 μM or less. Interestingly, these molecules did not affect the growth or biofilm formation of commensal bacteria, *S. sanguinis* and *S. gordonii*, in monospecies biofilms. The most active compound also showed selective activity against *S. mutans* cell survival and biofilm formation in multispecies biofilm model. Genetic studies indicated that the inhibitor down-regulated the expression of six biofilm-associated genes (*ftf, pac, relA, comDE, gbpB* and *gtfB*) in planktonic *S. mutans* cells, whereas it only down-regulated the expression of *ftf, pac* and *relA* in biofilm cells of *S. mutans* (Liu *et al.*, 2011). The most potent compound also inhibited the production of two key adhesins, AgI/II and GTF-B/C of *S. mutans*. However, the compound did not alter the expression of the corresponding genes in both *S. sanguinis* and *S. gordonii*, indicating that it possesses a selective inhibitory activity against *S. mutans* (Liu *et al.*, 2011). Peptides against the catalytic and glucan binding domain of the *S. mutans* GTFs have been shown to inhibit their function, and a single molecule containing two copies of each peptide was effective at reducing colonization by mutans streptococci in a rodent model of caries (Taubman *et al.*, 2001).

Bacteria produce certain enzymes to disperse from the biofilm communities, and such enzymes have been used to disrupt the natural biofilm. An enzyme known as dispersin B (glycosidase) produced by *Aggregatibacter actinomycetemcomitans* (Kaplan *et al.*, 2003) has been shown to effectively dissolve monospecies biofilms produced by *Escherichia coli, Staphylococcus aureus, Pseudomonas fluorescens* and *Yersinia pestis* (Itoh *et al.*, 2005). Though dispersin B is effective against such a diverse group of bacterial species, its activity against the oral microflora has not been tested. Since extracellular DNA has a very important function in the biofilm matrix, the use of DNase enzymes has been suggested to dissolve the biofilm structure (Jakubovics and Kolenbrander, 2010)

## Manipulating the oral biofilm community with replacement therapy

Replacement of the cariogenic *S. mutans* in the dental plaque with their non-cariogenic counterparts has been tested by Hillman (2002). This study involved the use of a genetically modified strain of *S. mutans* (BCS3-L1) that is deficient in producing lactic acid and also is designed to produce elevated amounts of a novel mutacin 1140 acting against most strains of cariogenic *S. mutans*. In laboratory and rodent model studies, BCS3-L1 was found to be genetically stable and to produce no apparent deleterious side effects during prolonged colonization. The strong colonization ability accompanied with a lack of cariogenic activity makes this strain a suitable candidate to replace cariogenic *S. mutans* in the dental biofilm. The strain BCS3-L1 has further been mutated to create stain A2JM which is deficient in producing alanine racemase and also having reduced transformation ability (Hillman *et al.*, 2007). This strain is thus dependent on exogenous D-alanine for its metabolic requirement and deficient in transformation, which makes it a suitable candidate for replacement therapy. However, rigorous clinical trials are required before using them in human subjects. Recently, a spontaneous lactate dehydrogenase mutant of *S. rattus* has been used to reduce the load of *S. mutans* in the animal models for dental caries (Hillman *et al.*, 2009). Another method for manipulating the biofilm community is to enhance the number of bacterial species having anti-*S. mutans* activity, such as *S. gordonii*, *S. sanguinis* and *S. salivarius* (Kreth *et al.*, 2005b, 2008; Kuramitsu *et al.*, 2007). It has been demonstrated that *S. gordonii* and *S. sanguinis* produce $H_2O_2$, which inhibits the growth of cariogenic *S. mutans in vitro*. The *S. mutans* CSP mediated competence development and biofilm formation has been shown to be inhibited by *S. salivarius* (Tamura *et al.*, 2009), and the biofilm inhibiting factor has been described to be *S. salivarius* fructosyl transferase (FTF) and exo-beta-D-fructosidase (FruA) (Ogawa *et al.*, 2011). Interestingly, inhibition of *S. mutans* biofilm formation by *S. salivarius* FTF and FruA has been observed in the presence of sucrose in a concentration dependent manner in the biofilm assay medium (Ogawa *et al.*, 2011). Commercially available FruA purified from *Aspergillus niger* has also been found to completely inhibit *S. mutans* biofilm formation on saliva-coated polystyrene and hydroxyapatite surfaces, indicating its potential importance in the sucrose-dependent multispecies biofilm formation in the oral cavity.

Apart from the commensal oral streptococci, many probiotic lactobacilli have been shown to inhibit the growth of *S. mutans*. Lang *et al.* (2010) tested a library of 624 lactobacilli strains for their ability to co-aggregate with *S. mutans* in the biofilm. The aim of the study was to find specific interaction partners for *S. mutans*, which can be used *in vivo* without disturbing the normal microflora and selectively removing the interacting flora. Finally, they identified two strains belonging to *Lactobacillus paracasei* and *Lactobacillus rhamnosus*, which can selectively co-aggregate with *S. mutans in vitro*. Later study using the *L. paracasei* strain DSMZ16671 in a caries rat model decreased the number of *S. mutans* (Tanzer *et al.*, 2010). Though these concepts are quite promising, their application in human subjects needs more rigorous experimentation.

## Inhibition of bacterial communication in the biofilm

Communication between and among the bacterial species in the biofilm structure substantially affects the community structure. Recently, disruption of the communication mechanism among the cariogenic species has been suggested to be an effective anti caries strategy. Quorum sensing mediated by the competence stimulating peptide CSP is used by *S. mutans* to regulate its virulence gene expression (Cvitkovitch *et al.*, 2003; van der Ploeg, 2005). Inhibition of CSP activity can be used as a strategy to inhibit *S. mutans* mediated caries production. It has been observed that *S. gordonii* produces a protease challisin, which inactivates the CSP (Wang and Kuramitsu, 2005). It has also been found that *in vitro*, *S. mutans* colonized less efficiently on biofilms produced by oral streptococci than on those produced by *A. naeslundii* (Wang *et al.*, 2011). The basis of this difference was demonstrated to be the cell extracts from the oral streptococci, which were found to interfere with *S. mutans* colonization. However, *S. mutans* could colonize on biofilm produced by the *S. gordonii sgc* (encodes challisin) knockout strain. Thus, increasing the number of *S. gordonii* in the oral microflora can be an important strategy to suppress the growth and virulence gene expression of *S. mutans*.

Autoinducer 2 (AI-2) mediated signalling modulates the virulence properties of many bacterial species including their biofilm formation. Natural and synthetic compounds have been developed to block the AI-2-mediated quorum sensing and to control infections by *P. aeruginosa* (Njoroge and Sperandio, 2009). However, the specific pathways targeted by these synthetic compounds are not yet known. Furthermore, AI2 is involved both in monospecies and multispecies biofilm formation, so the compounds targeting AI-2 may disturb the balance of the microbial composition in the biofilm.

### Targeted antimicrobial therapy

Conventional antimicrobial agents cannot distinguish between the commensals and pathogens. Therefore, the use of such agents in the oral biofilm is not very effective and disturbs the ecological balance of the oral microenvironment. Eckert *et al.* (2006) developed a novel technology using a new class of antimicrobials called specifically targeted antimicrobial peptides (STAMPs). This method uses a fusion peptide with two moieties, a killing moiety made up of a non-specific antimicrobial peptide and a targeting moiety made up of a species-specific binding peptide. In the study, they fused the *S. mutans* CSP as the STAMP targeting domain to mediate *S. mutans* specific delivery of an antimicrobial peptide domain. This was found to be effective against *S. mutans* grown both in liquid and biofilm states and did not affect the growth of other oral bacteria. This novel method provides a new avenue to develop more such molecules to target specific pathogens in a multispecies community structure, and a number of studies in the recent years have produced promising results in this direction (He *et al.*, 2007, 2010; Li *et al.*, 2010).

## Summary and future directions

The human oral cavity supports the growth of a vast number of microorganisms in the form of multispecies biofilm communities. Currently, more than 700 bacterial species are estimated to dwell in the oral microenvironment. Of this huge number of diverse bacteria, less than 50% have been grown successfully in the lab. Though most of the oral microflora are commensals, there are a few, such as the mutans streptococci and *P. gingivalis*, that have well-established roles in the pathogenicity of a wide range of diseases, including dental caries, periodontitis and bacterial endocarditis. Bacterial communities in the biofilm structure have been shown to mutually interact, sometimes in a beneficial way for the host, and sometimes to cause diseases. Though modern genetics and genomics methods have substantially improved our understanding of these organisms, the contribution of the large number of unknown flora to both health and disease remains to be determined. In the case of dental caries, many recent studies have focused on the regulation of virulence gene expression in *S. mutans* and understanding the interbacterial interaction producing caries. This research will, it is hoped, provide ways to target various pathways in *S. mutans* to reduce its pathogenicity.

Another interesting approach discussed above is the manipulation of the oral microflora composition in such a way that the commensals would outnumber and suppress the mutans streptococci. However, much research is required in this regard to understand thoroughly the interactions taking place in such multispecies biofilms. Oral biofilms can be studied as a novel community in this regard, and the outcomes of these studies can be extrapolated to other diseases caused by biofilm communities. Recently, it has been suggested that we should consider various infectious diseases as the outcome of the interactions among billions of bacteria inhabiting our body. Therefore, targeting a single pathogen to cure a disease may be less successful than targeting the community as a whole. Arming the commensals to fight pathogens in a natural way could be a fruitful area for future research. The human microbiome project has highlighted this aspect. New treatment methods need to be developed, keeping in mind not to disturb the natural balance, but at the same time manipulating the interbacterial interaction favouring the commensal lifestyle. The belief that we should live in harmony with our bacterial microflora for a healthy life seems to be the currently accepted mindset.

## References


Aas, J.A., Paster, B.J., Stokes, L.N., Olsen, I., and Dewhirst, F.E. (2005). Defining the normal bacterial flora of the oral cavity. J. Clin. Microbiol. *43*, 5721–5732.

Aas, J.A., Griffen, A.L., Dardis, S.R., Lee, A.M., Olsen, I., Dewhirst, F.E., Leys, E.J., and Paster, B.J. (2008). Bacteria of dental caries in primary and permanent teeth in children and young adults. J. Clin. Microbiol. *46*, 1407–1417.

Ajdic, D., McShan, W.M., McLaughlin, R.E., Savic, G., Chang, J., Carson, M.B., Primeaux, C., Tian, R., Kenton, S., Jia, H., *et al.* (2002). Genome sequence of *Streptococcus mutans* UA159, a cariogenic dental pathogen. Proc. Natl. Acad. Sci. U.S.A. *99*, 14434–14439.

Argimon, S., and Caufield, P.W. (2011). Distribution of putative virulence genes in *Streptococcus mutans* strains does not correlate with caries experience. J. Clin. Microbiol. *49*, 984–992.

Awano, S., Ansai, T., Takata, Y., Soh, I., Akifusa, S., Hamasaki, T., Yoshida, A., Sonoki, K., Fujisawa, K., and Takehara, T. (2008). Oral health and mortality risk from pneumonia in the elderly. J. Dent. Res. *87*, 334–339.

Balakrishnan, M., Simmonds, R.S., and Tagg, J.R. (2000). Dental caries is a preventable infectious disease. Aust. Dent. J. *45*, 235–245.

Bamford, C.V., d'Mello, A., Nobbs, A.H., Dutton, L.C., Vickerman, M.M., and Jenkinson, H.F. (2009). *Streptococcus gordonii* modulates *Candida albicans* biofilm formation through intergeneric communication. Infect. Immun. *77*, 3696–3704.

Banas, J.A., and Vickerman, M.M. (2003). Glucan-binding proteins of the oral streptococci. Crit. Rev. Oral Biol. Med. *14*, 89–99.

Beck, J.D., and Offenbacher, S. (2005). Systemic effects of periodontitis: epidemiology of periodontal disease and cardiovascular disease. J. Periodontol. *76*, 2089–2100.

Biswas, S., and Biswas, I. (2006). Regulation of the glucosyltransferase (gtfBC) operon by CovR in *Streptococcus mutans*. J. Bacteriol. *188*, 988–998.

Biswas, I., Drake, L., and Biswas, S. (2007). Regulation of gbpC expression in *Streptococcus mutans*. J. Bacteriol. *189*, 6521–6531.

Biswas, I., Drake, L., Erkina, D., and Biswas, S. (2008). Involvement of sensor kinases in the stress tolerance response of *Streptococcus mutans*. J. Bacteriol. *190*, 68–77.

Bowen, W.H., and Koo, H. (2011). Biology of *Streptococcus mutans*-derived glucosyltransferases: role in extracellular matrix formation of cariogenic biofilms. Caries Res. *45*, 69–86.

Brady, L.J., Piacentini, D.A., Crowley, P.J., Oyston, P.C., and Bleiweis, A.S. (1992). Differentiation of salivary agglutinin-mediated adherence and aggregation of mutans streptococci by use of monoclonal antibodies against the major surface adhesin P1. Infect. Immun. *60*, 1008–1017.

Brown, S.A., and Whiteley, M. (2007). A novel exclusion mechanism for carbon resource partitioning in *Aggregatibacter actinomycetemcomitans*. J. Bacteriol. *189*, 6407–6414.

Caufield, P.W., Cutter, G.R., and Dasanayake, A.P. (1993). Initial acquisition of mutans streptococci by infants: evidence for a discrete window of infectivity. J. Dent. Res. *72*, 37–45.

Caufield, P.W., Saxena, D., Fitch, D., and Li, Y. (2007). Population structure of plasmid-containing strains of *Streptococcus mutans*, a member of the human indigenous biota. J. Bacteriol. *189*, 1238–1243.

Chong, P., Drake, L., and Biswas, I. (2008). LiaS regulates virulence factor expression in *Streptococcus mutans*. Infect. Immun. *76*, 3093–3099.

Chong, P., Chattoraj, P., and Biswas, I. (2010). Activation of the SMU.1882 transcription by CovR in *Streptococcus mutans*. PLoS One *5*, e15528.

Coykendall, A.L. (1989). Classification and identification of the viridans streptococci. Clin. Microbiol. Rev. *2*, 315–328.

Cvitkovitch, D.G., Li, Y.H., and Ellen, R.P. (2003). Quorum sensing and biofilm formation in streptococcal infections. J. Clin. Invest. *112*, 1626–1632.

Daep, C.A., Lamont, R.J., and Demuth, D.R. (2008). Interaction of *Porphyromonas gingivalis* with oral streptococci requires a motif that resembles the eukaryotic nuclear receptor box protein–protein interaction domain. Infect. Immun. *76*, 3273–3280.

Daep, C.A., Novak, E.A., Lamont, R.J., and Demuth, D.R. (2011). Structural dissection and in vivo effectiveness of a peptide inhibitor of *Porphyromonas gingivalis* adherence to *Streptococcus gordonii*. Infect. Immun. *79*, 67–74.

Denapaite, D., Bruckner, R., Nuhn, M., Reichmann, P., Henrich, B., Maurer, P., Schahle, Y., Selbmann, P., Zimmermann, W., Wambutt, R., *et al.* (2010). The genome of *Streptococcus mitis* B6 – what is a commensal? PLoS One *5*, e9426.

Dewhirst, F.E., Chen, T., Izard, J., Paster, B.J., Tanner, A.C., Yu, W.H., Lakshmanan, A., and Wade, W.G. (2010). The human oral microbiome. J. Bacteriol. *192*, 5002–5017.

Diaz, P.I., Chalmers, N.I., Rickard, A.H., Kong, C., Milburn, C.L., Palmer, R.J., Jr., and Kolenbrander, P.E. (2006). Molecular characterization of subject-specific oral microflora during initial colonization of enamel. Appl. Environ. Microbiol. *72*, 2837–2848.

Dige, I., Nyengaard, J.R., Kilian, M., and Nyvad, B. (2009). Application of stereological principles for quantification of bacteria in intact dental biofilms. Oral Microbiol. Immunol. *24*, 69–75.

Ding, A.M., Palmer, R.J., Jr., Cisar, J.O., and Kolenbrander, P.E. (2010). Shear-enhanced oral microbial adhesion. Appl. Environ. Microbiol. *76*, 1294–1297.

Dmitriev, A., Mohapatra, S.S., Chong, P., Neely, M., Biswas, S., and Biswas, I. (2011). CovR-controlled global regulation of gene expression in *Streptococcus mutans* PLoS One *6*, e20127.

Douglas, L.J. (2003). Candida biofilms and their role in infection. Trends Microbiol. *11*, 30–36.

Dubey, G.P., and Ben-Yehuda, S. (2011). Intercellular nanotubes mediate bacterial communication. Cell *144*, 590–600.

Eckert, R., He, J., Yarbrough, D.K., Qi, F., Anderson, M.H., and Shi, W. (2006). Targeted killing of *Streptococcus mutans* by a pheromone-guided 'smart' antimicrobial peptide. Antimicrob. Agents Chemother. *50*, 3651–3657.

Egland, P.G., Du, L.D., and Kolenbrander, P.E. (2001). Identification of independent *Streptococcus gordonii* SspA and SspB functions in coaggregation with *Actinomyces naeslundii*. Infect. Immun. *69*, 7512–7516.

Facklam, R. (2002). What happened to the streptococci: overview of taxonomic and nomenclature changes. Clin. Microbiol. Rev. *15*, 613–630.

Fardini, Y., Chung, P., Dumm, R., Joshi, N., and Han, Y.W. (2010). Transmission of diverse oral bacteria to murine placenta: evidence for the oral microbiome as a potential source of intrauterine infection. Infect. Immun. *78*, 1789–1796.

Fleuchot, B., Gitton, C., Guillot, A., Vidic, J., Nicolas, P., Besset, C., Fontaine, L., Hols, P., Leblond-Bourget, N., Monnet, V., *et al.* (2011). Rgg proteins associated with internalized small hydrophobic peptides: a new quorum-sensing mechanism in streptococci. Mol. Microbiol. *80*, 1102–1119.

Genco, R.J., Grossi, S.G., Ho, A., Nishimura, F., and Murayama, Y. (2005). A proposed model linking inflammation to obesity, diabetes, and periodontal infections. J. Periodontol. *76*, 2075–2084.

Giacaman, R.A., Araneda, E., and Padilla, C. (2010). Association between biofilm-forming isolates of mutans streptococci and caries experience in adults. Arch. Oral Biol. *55*, 550–554.

Gibbons, R.J. (1989). Bacterial adhesion to oral tissues: a model for infectious diseases. J. Dent. Res. *68*, 750–760.

Hajishengallis, G., Arce, S., Gockel, C.M., Connell, T.D., and Russell, M.W. (2005). Immunomodulation with enterotoxins for the generation of secretory immunity or tolerance: applications for oral infections. J. Dent. Res. *84*, 1104–1116.

He, J., Eckert, R., Pharm, T., Simanian, M.D., Hu, C., Yarbrough, D.K., Qi, F., Anderson, M.H., and Shi, W. (2007). Novel synthetic antimicrobial peptides against *Streptococcus mutans*. Antimicrob. Agents Chemother. *51*, 1351–1358.

He, J., Yarbrough, D.K., Kreth, J., Anderson, M.H., Shi, W., and Eckert, R. (2010). Systematic approach to optimizing specifically targeted antimicrobial peptides against *Streptococcus mutans*. Antimicrob. Agents Chemother. *54*, 2143–2151.

Heddle, C., Nobbs, A.H., Jakubovics, N.S., Gal, M., Mansell, J.P., Dymock, D., and Jenkinson, H.F. (2003). Host collagen signal induces antigen I/II adhesin and invasin gene expression in oral *Streptococcus gordonii*. Mol. Microbiol. *50*, 597–607.

Hillman, J.D. (2002). Genetically modified *Streptococcus mutans* for the prevention of dental caries. Antonie van Leeuwenhoek *82*, 361–366.

Hillman, J.D., McDonell, E., Cramm, T., Hillman, C.H., and Zahradnik, R.T. (2009). A spontaneous lactate dehydrogenase deficient mutant of *Streptococcus rattus* for use as a probiotic in the prevention of dental caries. J. Appl. Microbiol. *107*, 1551–1558.

Hillman, J.D., Mo, J., McDonell, E., Cvitkovitch, D., and Hillman, C.H. (2007). Modification of an effector strain for replacement therapy of dental caries to enable clinical safety trials. J. Appl. Microbiol. *102*, 1209–1219.

Hojo, K., Nagaoka, S., Ohshima, T., and Maeda, N. (2009). Bacterial interactions in dental biofilm development. J. Dent. Res. *88*, 982–990.

Holmberg, K., and Hallander, H.O. (1973). Production of bactericidal concentrations of hydrogen peroxide by *Streptococcus sanguis*. Arch. Oral Biol. *18*, 423–434.

Hossain, M.S., and Biswas, I. (2011). Mutacins from *Streptococcus mutans* UA159 are active against multiple streptococcal species. Appl. Environ. Microbiol. *77*, 2428–2434.

Hyink, O., Wescombe, P.A., Upton, M., Ragland, N., Burton, J.P., and Tagg, J.R. (2007). Salivaricin A2 and the novel lantibiotic salivaricin B are encoded at adjacent loci on a 190-kilobase transmissible megaplasmid in the oral probiotic strain *Streptococcus salivarius* K12. Appl. Environ. Microbiol. *73*, 1107–1113.

Idone, V., Brendtro, S., Gillespie, R., Kocaj, S., Peterson, E., Rendi, M., Warren, W., Michalek, S., Krastel, K., Cvitkovitch, D., *et al.* (2003). Effect of an orphan response regulator on *Streptococcus mutans* sucrose-dependent adherence and cariogenesis. Infect. Immun. *71*, 4351–4360.

Isberg, R.R., and Barnes, P. (2002). Dancing with the host; flow-dependent bacterial adhesion. Cell *110*, 1–4.

Itoh, Y., Wang, X., Hinnebusch, B.J., Preston, J.F., 3rd, and Romeo, T. (2005). Depolymerization of beta-1,6-N-acetyl-D-glucosamine disrupts the integrity of diverse bacterial biofilms. J. Bacteriol. *187*, 382–387.

Jakubovics, N.S., and Kolenbrander, P.E. (2010). The road to ruin: the formation of disease-associated oral biofilms. Oral Dis. *16*, 729–739.

Jakubovics, N.S., Kerrigan, S.W., Nobbs, A.H., Stromberg, N., van Dolleweerd, C.J., Cox, D.M., Kelly, C.G., and Jenkinson, H.F. (2005a). Functions of cell surface-anchored antigen I/II family and Hsa polypeptides in interactions of *Streptococcus gordonii* with host receptors. Infect. Immun. *73*, 6629–6638.

Jakubovics, N.S., Stromberg, N., van Dolleweerd, C.J., Kelly, C.G., and Jenkinson, H.F. (2005b). Differential binding specificities of oral streptococcal antigen I/II family adhesins for human or bacterial ligands. Mol. Microbiol. *55*, 1591–1605.

Jakubovics, N.S., Gill, S.R., Vickerman, M.M., and Kolenbrander, P.E. (2008). Role of hydrogen peroxide in competition and cooperation between *Streptococcus gordonii* and *Actinomyces naeslundii*. FEMS Microbiol. Ecol. *66*, 637–644.

Jakubovics, N.S., Brittan, J.L., Dutton, L.C., and Jenkinson, H.F. (2009). Multiple adhesin proteins on the cell surface of *Streptococcus gordonii* are involved in adhesion to human fibronectin. Microbiology *155*, 3572–3580.

Jarosz, L.M., Deng, D.M., van der Mei, H.C., Crielaard, W., and Krom, B.P. (2009). *Streptococcus mutans* competence-stimulating peptide inhibits *Candida albicans* hypha formation. Eukaryot. Cell *8*, 1658–1664.

Jenkinson, H.F., and Lamont, R.J. (1997). Streptococcal adhesion and colonization. Crit. Rev. Oral Biol. Med. *8*, 175–200.

Jenkinson, H.F., and Lamont, R.J. (2005). Oral microbial communities in sickness and in health. Trends Microbiol. *13*, 589–595.

Joshipura, K.J., Rimm, E.B., Douglass, C.W., Trichopoulos, D., Ascherio, A., and Willett, W.C. (1996). Poor oral health and coronary heart disease. J. Dent. Res. *75*, 1631–1636.

Joshipura, K.J., Hung, H.C., Rimm, E.B., Willett, W.C., and Ascherio, A. (2003). Periodontal disease, tooth loss, and incidence of ischemic stroke. Stroke *34*, 47–52.

Joyner, P.M., Liu, J., Zhang, Z., Merritt, J., Qi, F., and Cichewicz, R.H. (2010). Mutanobactin A from the human oral pathogen *Streptococcus mutans* is a cross-kingdom regulator of the yeast-mycelium transition. Org. Biomol. Chem. *8*, 5486–5489.

Kaplan, J.B., Ragunath, C., Ramasubbu, N., and Fine, D.H. (2003). Detachment of *Actinobacillus actinomycetemcomitans* biofilm cells by an endogenous beta-hexosaminidase activity. J. Bacteriol. *185*, 4693–4698.

Keijser, B.J., Zaura, E., Huse, S.M., van der Vossen, J.M., Schuren, F.H., Montijn, R.C., ten Cate, J.M., and Crielaard, W. (2008). Pyrosequencing analysis of the oral microflora of healthy adults. J. Dent. Res. *87*, 1016–1020.

Knights, D., Costello, E.K., and Knight, R. (2011). Supervised classification of human microbiota. FEMS Microbiol. Rev. *35*, 343–359.

Kohler, B., Bratthall, D., and Krasse, B. (1983). Preventive measures in mothers influence the establishment of the bacterium *Streptococcus mutans* in their infants. Arch. Oral Biol. *28*, 225–231.

Kolenbrander, P.E., Palmer, R.J., Jr., Rickard, A.H., Jakubovics, N.S., Chalmers, N.I., and Diaz, P.I. (2006). Bacterial interactions and successions during plaque development. Periodontol. 2000 *42*, 47–79.

Kolenbrander, P.E., Palmer, R.J., Jr., Periasamy, S., and Jakubovics, N.S. (2010). Oral multispecies biofilm development and the key role of cell–cell distance. Nat. Rev. Microbiol. *8*, 471–480.

Koo, H., Xiao, J., Klein, M.I., and Jeon, J.G. (2010). Exopolysaccharides produced by *Streptococcus mutans* glucosyltransferases modulate the establishment of microcolonies within multispecies biofilms. J. Bacteriol. *192*, 3024–3032.

Kreth, J., Merritt, J., Shi, W., and Qi, F. (2005a). Co-ordinated bacteriocin production and competence development: a possible mechanism for taking up DNA from neighbouring species. Mol. Microbiol. *57*, 392–404.

Kreth, J., Merritt, J., Shi, W., and Qi, F. (2005b). Competition and coexistence between *Streptococcus mutans* and *Streptococcus sanguinis* in the dental biofilm. J. Bacteriol. *187*, 7193–7203.

Kreth, J., Merritt, J., Zhu, L., Shi, W., and Qi, F. (2006). Cell density- and ComE-dependent expression of a group of mutacin and mutacin-like genes in *Streptococcus mutans*. FEMS Microbiol. Lett. *265*, 11–17.

Kreth, J., Zhang, Y., and Herzberg, M.C. (2008). Streptococcal antagonism in oral biofilms: *Streptococcus sanguinis* and *Streptococcus gordonii* interference with *Streptococcus mutans*. J. Bacteriol. *190*, 4632–4640.

Kreth, J., Merritt, J., and Qi, F. (2009). Bacterial and host interactions of oral streptococci. DNA Cell Biol. *28*, 397–403.

Kulak, Y., Arikan, A., and Kazazoglu, E. (1997). Existence of *Candida albicans* and microorganisms in denture stomatitis patients. J. Oral Rehabil. *24*, 788–790.

Kuramitsu, H.K., He, X., Lux, R., Anderson, M.H., and Shi, W. (2007). Interspecies interactions within oral microbial communities. Microbiol. Mol. Biol. Rev. *71*, 653–670.

Lamont, R.J., El-Sabaeny, A., Park, Y., Cook, G.S., Costerton, J.W., and Demuth, D.R. (2002). Role of the *Streptococcus gordonii* SspB protein in the development of *Porphyromonas gingivalis* biofilms on streptococcal substrates. Microbiology *148*, 1627–1636.

Lang, C., Bottner, M., Holz, C., Veen, M., Ryser, M., Reindl, A., Pompejus, M., and Tanzer, J.M. (2010). Specific Lactobacillus/Mutans *Streptococcus* co-aggregation. J. Dent. Res. *89*, 175–179.

Lehner, T., Russell, M.W., and Caldwell, J. (1980). Immunisation with a purified protein from *Streptococcus mutans* against dental caries in rhesus monkeys. Lancet *1*, 995–996.

Li, J., Helmerhorst, E.J., Leone, C.W., Troxler, R.F., Yaskell, T., Haffajee, A.D., Socransky, S.S., and Oppenheim, F.G. (2004). Identification of early microbial colonizers in human dental biofilm. J. Appl. Microbiol. *97*, 1311–1318.

Li, L.N., Guo, L.H., Lux, R., Eckert, R., Yarbrough, D., He, J., Anderson, M., and Shi, W.Y. (2010). Targeted antimicrobial therapy against *Streptococcus mutans* establishes protective non-cariogenic oral biofilms and reduces subsequent infection. Int. J. Oral. Sci. *2*, 66–73.

Li, T., Khah, M.K., Slavnic, S., Johansson, I., and Stromberg, N. (2001a). Different type 1 fimbrial genes and tropisms of commensal and potentially pathogenic *Actinomyces* spp. with different salivary acidic proline-rich protein and statherin ligand specificities. Infect. Immun. *69*, 7224–7233.

Li, Y., Jin, J., Yang, Y., Bian, Z., Chen, Z., and Fan, M. (2009). Enhanced immunogenicity of an anti-caries vaccine encoding a cell-surface protein antigen of *Streptococcus mutans* by intranasal DNA prime-protein boost immunization. J. Gene Med. *11*, 1039–1047.

Li, Y.H., Lau, P.C., Lee, J.H., Ellen, R.P., and Cvitkovitch, D.G. (2001b). Natural genetic transformation of

*Streptococcus mutans* growing in biofilms. J. Bacteriol. *183*, 897–908.

Liljemark, W.F., Bloomquist, C.G., Reilly, B.E., Bernards, C.J., Townsend, D.W., Pennock, A.T., and LeMoine, J.L. (1997). Growth dynamics in a natural biofilm and its impact on oral disease management. Adv. Dent. Res. *11*, 14–23.

Lindquist, B., and Emilson, C.G. (1991). Dental location of *Streptococcus mutans* and *Streptococcus sobrinus* in humans harboring both species. Caries Res. *25*, 146–152.

Liu, C., Worthington, R.J., Melander, C., and Wu, H. (2011). A new small-molecule specifically inhibits the cariogenic *Streptococcus mutans* in multispecies biofilms. Antimicrob Agents Chemother. *55*, 2679–2687.

Loimaranta, V., Jakubovics, N.S., Hytonen, J., Finne, J., Jenkinson, H.F., and Stromberg, N. (2005). Fluid- or surface-phase human salivary scavenger protein gp340 exposes different bacterial recognition properties. Infect. Immun. *73*, 2245–2252.

Mager, D.L., Ximenez-Fyvie, L.A., Haffajee, A.D., and Socransky, S.S. (2003). Distribution of selected bacterial species on intraoral surfaces. J. Clin. Periodontol. *30*, 644–654.

Marsh, P.D., and Devine, D.A. (2011). How is the development of dental biofilms influenced by the host? J. Clin. Periodontol. *38(Suppl. 11)*, 28–35.

Maruyama, F., Kobata, M., Kurokawa, K., Nishida, K., Sakurai, A., Nakano, K., Nomura, R., Kawabata, S., Ooshima, T., Nakai, K., *et al.* (2009). Comparative genomic analyses of *Streptococcus mutans* provide insights into chromosomal shuffling and species-specific content. BMC Genomics *10*, 358.

Mashburn-Warren, L., Morrison, D.A., and Federle, M.J. (2010). A novel double-tryptophan peptide pheromone controls competence in *Streptococcus* spp. via an Rgg regulator. Mol. Microbiol. *78*, 589–606.

Matsui, R., and Cvitkovitch, D. (2010). Acid tolerance mechanisms utilized by *Streptococcus mutans*. Future Microbiol. *5*, 403–417.

Mattos-Graner, R.O., Napimoga, M.H., Fukushima, K., Duncan, M.J., and Smith, D.J. (2004). Comparative analysis of Gtf isozyme production and diversity in isolates of *Streptococcus mutans* with different biofilm growth phenotypes. J. Clin. Microbiol. *42*, 4586–4592.

Mitchell, A.P. (1998). Dimorphism and virulence in *Candida albicans*. Curr. Opin. Microbiol. *1*, 687–692.

Moschioni, M., Pansegrau, W., and Barocchi, M.A. (2010). Adhesion determinants of the *Streptococcus* species. Microb Biotechnol *3*, 370–388.

Motegi, M., Takagi, Y., Yonezawa, H., Hanada, N., Terajima, J., Watanabe, H., and Senpuku, H. (2006). Assessment of genes associated with *Streptococcus mutans* biofilm morphology. Appl. Environ. Microbiol. *72*, 6277–6287.

Nakano, K., Lapirattanakul, J., Nomura, R., Nemoto, H., Alaluusua, S., Gronroos, L., Vaara, M., Hamada, S., Ooshima, T., and Nakagawa, I. (2007). *Streptococcus mutans* clonal variation revealed by multilocus sequence typing. J. Clin. Microbiol. *45*, 2616–2625.

Nakano, K., Nomura, R., Matsumoto, M., and Ooshima, T. (2010). Roles of oral bacteria in cardiovascular diseases – from molecular mechanisms to clinical cases: Cell-surface structures of novel serotype k *Streptococcus mutans* strains and their correlation to virulence. J. Pharmacol. Sci. *113*, 120–125.

Njoroge, J., and Sperandio, V. (2009). Jamming bacterial communication: new approaches for the treatment of infectious diseases. EMBO Mol. Med. *1*, 201–210.

Nobbs, A.H., Lamont, R.J., and Jenkinson, H.F. (2009). *Streptococcus* adherence and colonization. Microbiol. Mol. Biol. Rev. *73*, 407–450, Table of Contents.

Nobbs, A.H., Jenkinson, H.F., and Jakubovics, N.S. (2011). Stick to your gums: mechanisms of oral microbial adherence. J. Dent. Res. *90*, 1271–1278.

Nyvad, B., and Kilian, M. (1987). Microbiology of the early colonization of human enamel and root surfaces in vivo. Scand. J. Dent. Res. *95*, 369–380.

Nyvad, B., and Kilian, M. (1990). Comparison of the initial streptococcal microflora on dental enamel in caries-active and in caries-inactive individuals. Caries Res. *24*, 267–272.

Offenbacher, S., Jared, H.L., O'Reilly, P.G., Wells, S.R., Salvi, G.E., Lawrence, H.P., Socransky, S.S., and Beck, J.D. (1998). Potential pathogenic mechanisms of periodontitis associated pregnancy complications. Ann. Periodontol. *3*, 233–250.

Ogawa, A., Furukawa, S., Fujita, S., Mitobe, J., Kawarai, T., Narisawa, N., Sekizuka, T., Kuroda, M., Ochiai, K., Ogihara, H., *et al.* (2011). Inhibition of *Streptococcus mutans* biofilm formation by *Streptococcus salivarius* FruA. Appl. Environ. Microbiol. *77*, 1572–1580.

Paes Leme, A.F., Koo, H., Bellato, C.M., Bedi, G., and Cury, J.A. (2006). The role of sucrose in cariogenic dental biofilm formation – new insight. J. Dent. Res. *85*, 878–887.

Palmer, R.J., Jr., Gordon, S.M., Cisar, J.O., and Kolenbrander, P.E. (2003). Coaggregation-mediated interactions of streptococci and actinomyces detected in initial human dental plaque. J. Bacteriol. *185*, 3400–3409.

Paster, B.J., Olsen, I., Aas, J.A., and Dewhirst, F.E. (2006). The breadth of bacterial diversity in the human periodontal pocket and other oral sites. Periodontol. 2000 *42*, 80–87.

Pereira-Cenci, T., Deng, D.M., Kraneveld, E.A., Manders, E.M., Del Bel Cury, A.A., Ten Cate, J.M., and Crielaard, W. (2008). The effect of *Streptococcus mutans* and *Candida glabrata* on *Candida albicans* biofilms formed on different surfaces. Arch. Oral Biol. *53*, 755–764.

Periasamy, S., and Kolenbrander, P.E. (2009a). *Aggregatibacter actinomycetemcomitans* builds mutualistic biofilm communities with *Fusobacterium nucleatum* and *Veillonella* species in saliva. Infect. Immun. *77*, 3542–3551.

Periasamy, S., and Kolenbrander, P.E. (2009b). Mutualistic biofilm communities develop with *Porphyromonas gingivalis* and initial, early, and late colonizers of enamel. J. Bacteriol. *191*, 6804–6811.

Periasamy, S., and Kolenbrander, P.E. (2010). Central role of the early colonizer *Veillonella* sp. in establishing multispecies biofilm communities with initial, middle, and late colonizers of enamel. J. Bacteriol. *192*, 2965–2972.

Periasamy, S., Chalmers, N.I., Du-Thumm, L., and Kolenbrander, P.E. (2009). *Fusobacterium nucleatum* ATCC 10953 requires *Actinomyces naeslundii* ATCC 43146 for growth on saliva in a three-species community that includes *Streptococcus oralis* 34. Appl. Environ. Microbiol. 75, 3250–3257.

Perry, J.A., Cvitkovitch, D.G., and Levesque, C.M. (2009). Cell death in *Streptococcus mutans* biofilms: a link between CSP and extracellular DNA. FEMS Microbiol. Lett. *299*, 261–266.

Petersen, F.C., Tao, L., and Scheie, A.A. (2005). DNA binding-uptake system: a link between cell-to-cell communication and biofilm formation. J. Bacteriol. *187*, 4392–4400.

van der Ploeg, J.R. (2005). Regulation of bacteriocin production in *Streptococcus mutans* by the quorum-sensing system required for development of genetic competence. J. Bacteriol. *187*, 3980–3989.

Reichmann, P., Nuhn, M., Denapaite, D., Bruckner, R., Henrich, B., Maurer, P., Rieger, M., Klages, S., Reinhard, R., and Hakenbeck, R. (2011). Genome of *Streptococcus oralis* Strain Uo5. J. Bacteriol. *193*, 2888–2889.

Rice, L.B. (1998). Tn916 family conjugative transposons and dissemination of antimicrobial resistance determinants. Antimicrob. Agents Chemother. *42*, 1871–1877.

Rickard, A.H., Palmer, R.J., Jr., Blehert, D.S., Campagna, S.R., Semmelhack, M.F., Egland, P.G., Bassler, B.L., and Kolenbrander, P.E. (2006). Autoinducer 2: a concentration-dependent signal for mutualistic bacterial biofilm growth. Mol. Microbiol. *60*, 1446–1456.

Roberts, A.P., Cheah, G., Ready, D., Pratten, J., Wilson, M., and Mullany, P. (2001). Transfer of TN916-like elements in microcosm dental plaques. Antimicrob. Agents Chemother. *45*, 2943–2946.

Ruhl, S., Sandberg, A.L., and Cisar, J.O. (2004). Salivary receptors for the proline-rich protein-binding and lectin-like adhesins of oral actinomyces and streptococci. J. Dent. Res. *83*, 505–510.

Saxena, D., Li, Y., and Caufield, P.W. (2005). Identification of unique bacterial gene segments from *Streptococcus mutans* with potential relevance to dental caries by subtraction DNA hybridization. J. Clin. Microbiol. *43*, 3508–3511.

Saxena, D., Caufield, P.W., Li, Y., Brown, S., Song, J., and Norman, R. (2008). Genetic classification of severe early childhood caries by use of subtracted DNA fragments from *Streptococcus mutans*. J. Clin. Microbiol. *46*, 2868–2873.

Scannapieco, F.A. (1994). Saliva–bacterium interactions in oral microbial ecology. Crit. Rev. Oral Biol. Med. *5*, 203–248.

Sedlacek, M.J., and Walker, C. (2007). Antibiotic resistance in an in vitro subgingival biofilm model. Oral Microbiol. Immunol. 22, 333–339.

Shao, H., James, D., Lamont, R.J., and Demuth, D.R. (2007a). Differential interaction of *Aggregatibacter* (Actinobacillus) *actinomycetemcomitans* LsrB and RbsB proteins with autoinducer 2. J. Bacteriol. *189*, 5559–5565.

Shao, H., Lamont, R.J., and Demuth, D.R. (2007b). Autoinducer 2 is required for biofilm growth of *Aggregatibacter* (Actinobacillus) *actinomycetemcomitans*. Infect. Immun. *75*, 4211–4218.

Shemesh, M., Tam, A., and Steinberg, D. (2007). Differential gene expression profiling of *Streptococcus mutans* cultured under biofilm and planktonic conditions. Microbiology *153*, 1307–1317.

Shemesh, M., Tam, A., Kott-Gutkowski, M., Feldman, M., and Steinberg, D. (2008). DNA-microarrays identification of *Streptococcus mutans* genes associated with biofilm thickness. BMC Microbiol. *8*, 236.

Suzuki, N., Yoshida, A., Saito, T., Kawada, M., and Nakano, Y. (2004). Quantitative microbiological study of subgingival plaque by real-time PCR shows correlation between levels of *Tannerella forsythensis* and Fusobacterium spp. J. Clin. Microbiol. *42*, 2255–2257.

Takahashi, N., and Nyvad, B. (2008). Caries ecology revisited: microbial dynamics and the caries process. Caries Res. *42*, 409–418.

Takahashi, N., and Nyvad, B. (2011). The role of bacteria in the caries process: ecological perspectives. J. Dent. Res. *90*, 294–303.

Tamura, S., Yonezawa, H., Motegi, M., Nakao, R., Yoneda, S., Watanabe, H., Yamazaki, T., and Senpuku, H. (2009). Inhibiting effects of *Streptococcus salivarius* on competence-stimulating peptide-dependent biofilm formation by *Streptococcus mutans*. Oral Microbiol. Immunol. *24*, 152–161.

Tanzer, J.M., Thompson, A., Lang, C., Cooper, B., Hareng, L., Gamer, A., Reindl, A., and Pompejus, M. (2010). Caries inhibition by and safety of *Lactobacillus paracasei* DSMZ16671. J. Dent. Res. *89*, 921–926.

Taubman, M.A., and Nash, D.A. (2006). The scientific and public-health imperative for a vaccine against dental caries. Nat. Rev. Immunol. *6*, 555–563.

Taubman, M.A., Holmberg, C.J., and Smith, D.J. (2001). Diepitopic construct of functionally and epitopically complementary peptides enhances immunogenicity, reactivity with glucosyltransferase, and protection from dental caries. Infect. Immun. *69*, 4210–4216.

Thein, Z.M., Samaranayake, Y.H., and Samaranayake, L.P. (2006). Effect of oral bacteria on growth and survival of *Candida albicans* biofilms. Arch. Oral Biol. *51*, 672–680.

Vickerman, M.M., Iobst, S., Jesionowski, A.M., and Gill, S.R. (2007). Genome-wide transcriptional changes in *Streptococcus gordonii* in response to competence signalling peptide. J. Bacteriol. *189*, 7799–7807.

Wang, B.Y., Deutch, A., Hong, J., and Kuramitsu, H.K. (2011). Proteases of an early colonizer can hinder *Streptococcus mutans* colonization in vitro. J. Dent. Res. *90*, 501–505.

Wang, B.Y., and Kuramitsu, H.K. (2005). Interactions between oral bacteria: inhibition of *Streptococcus mutans* bacteriocin production by *Streptococcus gordonii*. Appl. Environ. Microbiol. *71*, 354–362.

Warburton, P.J., Palmer, R.M., Munson, M.A., and Wade, W.G. (2007). Demonstration of in vivo transfer of doxycycline resistance mediated by a novel transposon. J. Antimicrob. Chemother. *60*, 973–980.

Waterhouse, J.C., Swan, D.C., and Russell, R.R. (2007). Comparative genome hybridization of *Streptococcus mutans* strains. Oral Microbiol. Immunol. *22*, 103–110.

Wickstrom, C., and Svensater, G. (2008). Salivary gel-forming mucin MUC5B – a nutrient for dental plaque bacteria. Oral Microbiol. Immunol. *23*, 177–182.

Wickstrom, C., Hamilton, I.R., and Svensater, G. (2009). Differential metabolic activity by dental plaque bacteria in association with two preparations of MUC5B mucins in solution and in biofilms. Microbiology *155*, 53–60.

Wu, C., Cichewicz, R., Li, Y., Liu, J., Roe, B., Ferretti, J., Merritt, J., and Qi, F. (2010). Genomic island TnSmu2 of *Streptococcus mutans* harbors a nonribosomal peptide synthetase-polyketide synthase gene cluster responsible for the biosynthesis of pigments involved in oxygen and H2O2 tolerance. Appl. Environ. Microbiol. *76*, 5815–5826.

Xu, P., Alves, J.M., Kitten, T., Brown, A., Chen, Z., Ozaki, L.S., Manque, P., Ge, X., Serrano, M.G., Puiu, D., *et al.* (2007). Genome of the opportunistic pathogen *Streptococcus sanguinis*. J. Bacteriol. *189*, 3166–3175.

Younson, J., and Kelly, C. (2004). The rational design of an anti-caries peptide against *Streptococcus mutans*. Mol. Divers. *8*, 121–126.

Zaura, E., Keijser, B.J., Huse, S.M., and Crielaard, W. (2009). Defining the healthy 'core microbiome' of oral microbial communities. BMC Microbiol. *9*, 259.

Zhang, L., and Mah, T.F. (2008). Involvement of a novel efflux system in biofilm-specific resistance to antibiotics. J. Bacteriol. *190*, 4447–4452.

Zhang, S., Green, N.M., Sitkiewicz, I., Lefebvre, R.B., and Musser, J.M. (2006). Identification and characterization of an antigen I/II family protein produced by group A Streptococcus. Infect. Immun. *74*, 4200–4213.

Zijnge, V., van Leeuwen, M.B., Degener, J.E., Abbas, F., Thurnheer, T., Gmur, R., and Harmsen, H.J. (2010). Oral biofilm architecture on natural teeth. PLoS One *5*, e9321.

# Biofilms in Periodontal Health and Disease

# 10

Purnima S. Kumar, Matthew R. Mason and Janel Yu


## Abstract

Dental plaque biofilm is composed of a diverse microbial community. Several decades of research have been focused on the role played by the subgingival biofilm in the aetiology of periodontal diseases. However, recent evidence from other ecosystems within the human body indicates that these biofilms also play an important role in maintaining health. The purpose of this chapter is to explore the health benefits of subgingival plaque, and to outline the development of the biofilm as well as to characterize the bacteria present in the plaque biofilm in health and disease. Additionally, the virulence mechanisms of health and disease-associated biofilms will be outlined.


## Periodontal diseases – general concepts

The periodontium is made up of four structures – the gingiva, cementum, periodontal ligament and bone. Diseases of the periodontium can be broadly classified into those that affect the gingiva only (gingivitis) and those that affect the bone and attachment apparatus (periodontitis) (Armitage, 1999). Plaque-induced gingivitis, aggressive periodontitis and chronic periodontitis are diseases caused by bacteria in the plaque biofilm. Plaque-induced gingivitis is one of the most common infections to affect humans and manifests as a reversible inflammation of the marginal tissues. The duration and severity of the disease can be modified by several systemic factors that modulate host–bacterial interactions, for example, pregnancy, diabetes, malnutrition, and certain medications. Existing evidence indicates that gingivitis precedes the onset of periodontitis, however, not all gingivitis cases proceed to develop into periodontitis (Lindhe *et al.*, 1973; Listgarten *et al.*, 1985; Löe *et al.*, 1986). The reason for this is that accumulation of plaque bacteria is a necessary but not sufficient condition for the development of periodontitis: a susceptible host is also required (Lindhe *et al.*, 1973; Löe *et al.*, 1986; Page, 1999).

While several different forms of periodontitis are recognized, the two most common forms to affect a systemically healthy individual are aggressive periodontitis and chronic periodontitis. Aggressive periodontitis is characterized by a rapid rate of bone and attachment loss and is characteristically seen in children and young adults. The disease encompasses several clinical entities that were previously described as localized juvenile periodontitis [currently termed localized aggressive periodontitis, (LAP)], generalized juvenile periodontitis and rapidly progressing periodontitis [now collectively classified as generalized aggressive periodontitis (GAP)]. Aggressive periodontitis is typically seen in circum-pubertal children and young adults who are otherwise systemically healthy. The disease is characterized by rapid loss of attachment and bone loss in either specific teeth (first permanent molars and incisors in the case of LAP) or throughout the dentition (GAP). Chronic periodontitis is a slowly progressing disease that occurs predominantly in adults but may be seen in children. An important diagnostic feature is that the amount of destruction is consistent with the amount of plaque and calculus present.

## Periodontal diseases and biofilms – the specific, non-specific and ecological plaque hypotheses

The presence of bacteria in dental plaque has been known since 1683, when Anton von Leeuwenhoek first described his observations of 'animalcules' in dental plaque in letters to the Royal Society of London. However, for several centuries plaque was thought of as an amorphous collection of bacteria. A brief history of the search for periodontal pathogens is given in Fig. 10.1.

In 1882, Koch published a treatise describing a specific bacterium as the aetiological agent of tuberculosis along with criteria for establishing an organism as the aetiological agent of a disease (Koch, 1932). This began the search for species involved in the aetiology of several other communicable diseases. The time period from 1880 to 1930 became known as the golden age of microbiology. During this period, many microbial pathogens were linked to specific infections in the body. This led oral health researchers to seek out specific pathogens for the etiologic agent of dental caries and periodontal disease and the age of the specific plaque hypothesis was born. Several possible agents were isolated, however, no specific bacterium could be identified as the causative agent. Further, no organism could be isolated only from diseased individuals, on the contrary, organisms found in disease were also found in health. Additional important evidence came from epidemiological data indicating that the prevalence of periodontal disease was higher in older individuals (Anon., 1996). While it was later realized that this was because of the cumulative destruction from the disease, at the time it was suggested that the presence of plaque, and not necessarily specific organisms, resulted in disease. As a result, the 1930s brought a new view – the non-specific plaque hypothesis (Loesche, 1979).

This new hypothesis evolved around the idea that all plaque was considered pathogenic and diseases associated with plaque arose from 'elaboration of noxious products by the entire plaque flora' (Loesche, 1979). Large amounts of plaque would lead to production of toxins that could overwhelm the host defences and result in disease. This hypothesis led to wholesale removal of plaque by surgical or non-surgical methods and rigorous homecare as therapeutic options for periodontal disease.

The 1960s brought another change with many advances in bacterial detection and characterization. Evidence suggested that plaque behaved differently when certain species were present (Socransky and Haffajee, 2005). Also, longitudinal studies revealed that many individuals with significant amounts of plaque accumulation never developed destructive disease (Löe *et al.*, 1986). Further, within any individual, the disease was seen to be site specific, that is, only certain teeth demonstrated large amounts of destruction and these sites were seen in close juxtaposition to normal sites. The development of better techniques for microbial characterization demonstrated significant differences in the microbial profiles of periodontal health and disease. Thus, the specific plaque hypothesis was again adopted, as it was believed specific microbes were responsible for disease and once identified, would allow for targeted treatment of the disease. This concept gained acceptance when *Aggregatibacter* (formerly *Actinobacillus*)

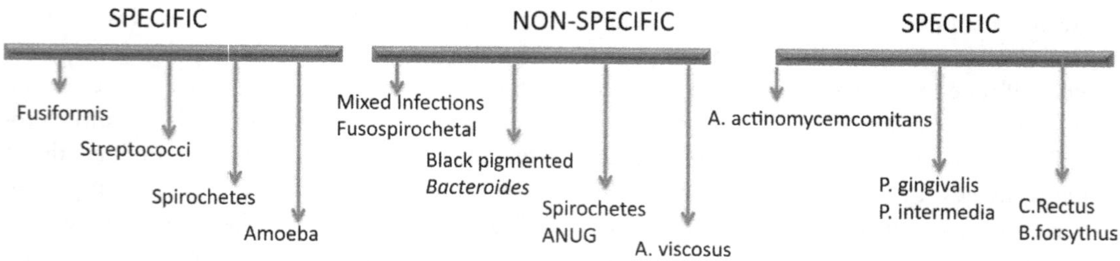


**Figure 10.1** Bacterial ecology of periodontal disease – historical perspective (adapted from 100 years of periodontal microbiology by Socransky and Haffajee, 1992).

*actinomycetemcomitans* was recognized as the predominant pathogen in LAP.

Investigations on the influence of the oral environment on the structure and composition of microbial communities led to the development of the ecological plaque hypothesis (Marsh, 2004). This hypothesis suggests that the resident microflora undergoes a transformation from a commensal to a pathogenic population due to environmental perturbations, for example, pH, oxygen tension, flow of gingival crevicular fluid and presence of blood and blood products. While the ecological hypothesis is similar to the specific plaque hypothesis in recognizing the varying pathogenic potentials of individual bacterial species, it maintains that perturbation in the microbial homeostasis is the result of environmental shifts. Thus, treatment of disease does not only require targeting specific species, it is also important to alter the environment from one that promotes pathogen enrichment to one that is compatible with commensal growth.

The identification of specific causative species, or periodontopathogens, has been hampered by some of the unique features of periodontal diseases. The foremost of these is that disease occurs in a site already colonized by a bacterial population. Thus, disease might be caused either by overgrowth of one or more species in the resident population, or by colonization by an exogenous pathogen. For example, it has been shown that *Capnocytophaga* spp. are seen in high levels prior to the onset of gingivitis, while *Prevotella* spp. are detected in areas with established gingivitis. Thus, *Capnocytophaga* is more likely an aetiological agent and *Prevotella* species are present as a consequence of the disease process (Mombelli *et al.*, 1990). Colonization by exogenous pathogens is thought to contribute to the episodic nature of disease progression (Goodson *et al.*, 1982), i.e. the fact that not all sites with baseline attachment loss demonstrate the same rate of disease progression or disease activity at the same time points.

## Technological advances in biofilm research

Our understanding of the structure and composition of dental plaque has grown in parallel with advances in bacterial identification and characterization methods. Microscopic identification and cultivation formed the basis of all early work on characterization of oral and gingival bacteria. Witzel (1882) and Miller (1890) identified and characterized different bacterial morphotypes and species found in caries and periodontitis (Gold, 1985). Over the next 80 years, cultivation and phenotypic characterization were routinely used for bacterial identification. Early microscopic studies by Listgarten (1976) described the ultrastructural characteristics of mature plaque on extracted healthy teeth. These studies provided the first evidence that bacteria in dental plaque aggregate in specific groups, as seen with the 'maize cob' and 'bristle brush' formations between different bacterial morphotypes. The health-associated biofilm consisted of a thin layer of adherent bacterial cells with the characteristics of Gram-positive cocci.

Direct visualization of the bacteria associated with health and disease using light and electron microscopy (Listgarten, 1976; Listgarten and Hellden, 1978) suggested a definite bacterial profile associated with health and a significant shift associated with alterations in health status. These studies showed that the healthy gingival crevice was colonized predominantly by Gram-positive bacteria, this profile changed to a predominantly Gram-negative one with the onset and progression of gingivitis.

During the 1980s, scanning electron microscopy was used to examine early colonizers of the biofilm. The cellular morphology at 4 hours was consistent with Gram-positive cocci, then rod shaped organisms were seen at 8 hours, and by 24–28 hours, thick deposits of cells with variable morphologies were noted: coccoid, cocco-bacillary, rod-shaped, and filamentous bacteria (Nyvad and Fejerskov, 1987). These studies showed that the health-associated biofilm was more diverse than was previously suspected.

Microscopic techniques provide valuable information on the general aspects of a bacterial community, since they can be used to detect shape, size and staining characteristics. However, lack of specificity is a major concern since morphotyping based on shape, size, motility and staining characteristics does not provide definitive identification. For example, several bacteria that were originally thought to be Gram-negative species (*Veillonella, Fusobacterium, Filifactor*) cluster with Gram-positive species based on their phylogenetic relationships (Tanner *et al.*, 1994). This type of phylogenetic clustering prompted a closer examination of cell wall characteristics and physiological behaviour of *Filifactor alocis* and revealed that it is in fact, a Gram-positive species (Downes *et al.*, 2001). It is possible that the traditional paradigm of a predominantly Gram-positive flora associated with health may undergo drastic changes when more sophisticated methods are used to characterize bacteria.

More recently, the emergence of newer technology has allowed for further characterization of the plaque biofilm. Studies beginning in the 1970s revealed that oral bacteria operate as a microbial community that is 'co-ordinated, spatially organized, and fully metabolically integrated' called the biofilm (Marsh, 2004). These biofilms are highly structured with channels traversing their depths, creating primary circulatory systems (Costerton *et al.*, 1994). Confocal laser scanning microscopy has enabled a three-dimensional examination of the ultrastructural characteristics of plaque. It is now seen that plaque has a highly structured architecture made up of bacterial microcolonies interspersed with highly permeable water channels (Costerton *et al.*, 1994). These studies have also shown that the bacteria within these microcolonies produce polysaccharides that concentrate densely around them and are less densely distributed in the spaces between microcolonies. Current evidence suggests that the plaque biofilm has a more open architecture than was suggested by early electron microscopy reports, with channels traversing from the outside of the biofilm to the enamel surface.

Evidence from culturing, microscopy and more recently, DNA-based molecular approaches have led to a paradigm shift on the nature of dental plaque. We are learning that it is not an amorphous collection of bacteria formed by random accumulation, but a structured community with specific patterns of aggregation and inter-bacterial interactions.

## Development of the biofilm

Research on plaque development has shown that oral bacteria colonize both the non-shedding hard surfaces as well as the shedding soft tissue surfaces. The physical and morphological characteristics of these surfaces create different ecosystems or 'niches' with distinct bacterial profiles. This is the result of a dynamic equilibrium that exists between the adhesion forces of microorganisms and forces such as swallowing; mastication, salivary and crevicular flow as well as oral hygiene measures.

Bacterial adhesion to oral soft tissue surfaces is a function both of bacterial surface characteristics and the susceptibility of the host epithelium. Studies on bacterial colonization of other niches in the body have shown that epithelial cells of individuals susceptible to infection harbour as many as five times more pathogenic bacteria than do resistant individuals (Danino *et al.*, 1998). Initial studies on animal models suggest that the same might be true of periodontal disease (Isogai *et al.*, 1986).

Development of dental biofilms can be classified into certain phases: adsorption of host and bacterial molecules to the tooth surface, passive transport of oral bacteria to the tooth surface, co-adhesion of later colonizers to already attached early colonizers, multiplication of attached microorganisms, and active detachment (Marsh, 2004).

Bacterial colonization of a tooth surface begins within two minutes following oral prophylaxis following adsorption of proteins from saliva on the tooth surface. The early colonizers belong to those species that have receptors for these salivary proteins. Two phenomena are important in the development of the biofilm- coaggregation and coadhesion. Coadhesion is the interaction between cells already adherent to a surface and cells in suspension. Coadhesion commonly occurs between the cells of a biofilm and bacterial aggregates in suspension.

Coaggregation is defined as the specific cell-to-cell recognition that occurs between genetically distinct cell types. In dental plaque, the succession of bacterial colonization follows basic principles of coaggregation: high specificity, bridging, competition, and functional similarity (Kolenbrander *et al.*, 2006). The specificity of coaggregation between oral bacteria allows for delineation of partnerships. For example, the coaggregation between *Streptococcus* and *Actinomyces* is highly specific and can be inhibited by lactose and not sucrose. One partner bears the receptor polysaccharide (carbohydrate) and the other partner bears the complementary adhesion (lectin). The adhesion-bearing partner is inactivated from participating in coaggregation when heated, but the partner bearing the complementary receptor polysaccharide is unaffected when heated. Bridging is a principle whereby a common partner bears two or more types of coaggregation mediators, which may be receptor polysaccharides, adhesins, or both. In general, bridging promotes a cooperative environment and symbiotic relationship so that two cell types do not seek the same binding site to the common partner, which acts as a coaggregation bridge to connect cell types. The principle of coaggregation competition occurs when multiple cell types recognize the same coaggregation mediator on the common coaggregation partner. It has been observed that *Actinomyces* spp. and *Prevotella* spp. can recognize the same receptor of *Streptococcus* and compete for binding. Functional similarity may be important in coaggregation in that strains from multiple species can bind to a common partner by recognizing and binding the same receptor on a common partner. The principle of functional similarity can be demonstrated by *Streptococcus oralis* 34, which serves as a common partner to multiple species of *Actinomyces, Streptococcus, Veillonella, and Prevotella*; however, a coaggregation-defective mutant of *S. oralis* cannot coaggregate with *Prevotella* or *Actinomyces,* but retains coaggregation with *Streptococcus.* It is believed that these principles of coaggregation play a role in succession of bacterial colonization in dental plaque.

With maturation, a biofilm has continued synthesis of exopolymers to form an extracellular matrix, which serves as a scaffold that determines the structure of biofilms. Additionally, its biological activity allows it to retain nutrients, water, and key enzymes (Marsh, 2005). A biofilm environment confers certain properties to bacteria that are not seen in the planktonic state, a fact that explains the importance of recognizing dental plaque as biofilm and not as bacteria in planktonic state.

### Cell–cell communication

An important characteristic seen only in biofilm-associated bacteria is quorum sensing, or cell-density mediated gene expression. This process is mediated by two groups of compounds, known as Autoinducer 1 and Autoinducer 2. Both Gram-positive and Gram-negative cells secrete molecules known as autoinducer-2, which is encoded by the *luxS* gene (Kolenbrander *et al.*, 2002). In *Streptococcus gordonii, luxS* up-regulates the expression of *sspA* and *sspB* genes, which encode for proteins that provide a binding site for the major fimbriae of *Porphyromonas gingivalis* (Chung *et al.*, 2001; Fong *et al.*, 2001; Lamont *et al.*, 2002; McNab *et al.*, 2003). Thus, *P. gingivalis,* a secondary colonizer, is capable of preferential binding in the presence of *S. gordonii,* an early colonizer.

### Gene transfer

Biofilm-associated bacteria communicate with each other via horizontal gene transfer. In *Streptococcus mutans,* quorum sensing is mediated by competence stimulating peptide (CSP). Genes of the CSP signalling system (*comC, comD, comE* and *comX*) are responsible for multiple functions- biofilm formation, competence (ability to accept foreign DNA) and acid tolerance (Li *et al.*, 2002a,b).

### Antimicrobial resistance

The biofilm provides a protected environment against antimicrobial agents. The biofilm acts as a barrier to diffusion, due to the presence of neutralizing enzymes (ß-lactamase, IgA protease) as well as a diffusion-resistant matrix (Stewart, 2003). Cells in the biofilm can develop antibiotic resistance from horizontal gene transfer, mutations and by expressing efflux pumps. Penicillin (Dowson *et al.*, 1990) and tetracycline (Roberts *et al.*, 2001) resistance occurs among bacteria by conjugative transfer of mobile genetic elements. Further, cells in a biofilm have a slower rate of division and hence might not be as susceptible to antibiotics as an actively dividing population. This resistance provided by biofilm living must be taken into account while selecting antibiotics. *In vitro* antibiotic susceptibility tests are typically conducted on planktonic cells, whose behaviour is significantly different from that of biofilm-associated cells. Thus it has been argued that 'biofilm inhibitory concentration' is a more realistic estimate of anti-microbial activity than the current measures (Anwar *et al.*, 1990).

## Regulation of gene expression

Biofilm living has been shown to regulate gene expression in certain bacteria. For example, exposure of *S. gordonii* to saliva results in induction of *sspA/B* genes that mediates host-surface binding as well as co-aggregation with *P. gingivalis* and *Actinomyces* spp. (Kolenbrander and Phucas, 1984; Du and Kolenbrander, 2000). Similarly, genes encoding glucan (*gtf*) and fructan (*ftf*) synthesis are differentially regulated in biofilm associated *S. mutans* (Li and Burne, 2001).

Biofilm living confers advantages not only to the co-existing bacteria but also to the human host since bacteria in a biofilm play an important role in preventing exogenous colonization. This colonization resistance may occur by means of effective competition for nutrients and attachment sites, the production of inhibitory factors, and creation of unfavourable growth conditions by resident microflora (Marsh, 2004). Thus, by using very specific mechanisms of aggregation and signalling, as well as by partitioning the available resources within the biofilm, a health-compatible community can prevent colonization by pathogens. This phenomenon, called niche saturation, has been observed in several biofilm systems in the human body- for example the gastro-intestinal tract, vagina etc. Recent evidence indicates that bacteria in dental plaque also play an important role in maintaining a healthy biofilm by preventing adhesion of pathogenic species (Van Hoogmoed *et al.*, 2008). There is also evidence to indicate that periodontitis is associated with loss of beneficial bacteria, for example, species belonging to *Veillonella* and *Streptococcus*, within the biofilm (Kumar *et al.*, 2005, 2006; Stingu *et al.*, 2008).

## Bacterial antigens and virulence factors in the biofilm

Studies on periodontal microbiology have focused on 'virulence factors' associated with putative periodontopathogens. A virulence factor is any moiety that is necessary for causing disease in the host. These factors may either be essential for host colonization or are 'true' antigens capable of causing an immune response by the host.

*In vitro* studies of biofilms have enabled us to study the antigenic potential of biofilms (Bradshaw and Marsh, 1999). *Aggregatibacter actinomycetemcomitans*, for example, produces an 116 kDa protein antigen, called leukotoxin, that is capable of immunomodulation (Ohta *et al.*, 1996). Leukotoxin causes apoptosis, or programmed cell death of leucocytes, the first line of defence against bacterial invasion. Since the organism grows optimally under pH of 7.0–8.0, it is suggested that the more virulent strains of *A. actinomycetemcomitans* are present on the outer surface of subgingival plaque (Ohta *et al.*, 1996). Similarly, *P. gingivalis* strain W50 shows maximal growth and proteolytic activity at a pH of 7.0–8.0 and haemin-excess suggesting it becomes more virulent in an inflamed gingival crevice (Marsh *et al.*, 1994). Among the specific virulence factors associated with putative periodontopathogens are the following molecules (Table 10.1).

## Lipopolysaccharide (LPS)

The outer membranes of most Gram-negative bacteria comprise repeating polysaccharide units attached to the O-side chains of a lipid component (LPS). LPS is highly antigenic and capable of inducing the release of inflammatory mediators and cytokines such as prostaglandin $E_2$ ($PGE_2$), interleukin 1a (IL-1α), IL-1β, tumour necrosis factor-a (TNF-α), and IL-8 through CD14 and Toll-like receptor (TLR) activation of several host cells. *A. actinomycetemcomitans* can be classified into six distinct serotypes of varying virulence based on LPS (Zambon *et al.*, 1983). Unlike other bacteria, LPS of *P. gingivalis* activates the cytokine cascade both through TLR2 and TLR4, perhaps because of its unique three-dimensional structure (Netea *et al.*, 2002; Darveau *et al.*, 2004).

## Heat shock proteins

When both eukaryotic and bacterial cells are exposed to environmental stress (e.g. temperature, pH, redox potential), they synthesize stress proteins such as heat shock proteins (HSPs) (Goulhen *et al.*, 2003). These proteins act as

**Table 10.1** Bacterial virulence factors

| Virulence factor | Effects on host |
|---|---|
| Lipopolysaccharide (LPS) | Increases cytokine release from PMNs, macrophages, fibroblasts<br>Induces nitric oxide secretion in macrophages<br>Increases differentiation of osteoclast precursors<br>Activates osteoclasts<br>Supports osteoclast survival<br>Stimulates T helper (Th) cell proliferation<br>Haemagglutination |
| Heat shock proteins | Molecular mimicry between bacterial and host HSPs leads to autoimmune response<br>Induce gingival epithelial cell proliferation in low doses<br>Osteolytic activity<br>PDL epithelial cell proliferation |
| Extracellular proteolytic enzymes | Degrades fibrinogen, fibronectin, albumin, laminin,<br>Hydrolyses collagen IV, IgG, IgA |
| Fimbriae | Important in colonization<br>Role in invasion of cells by receptor mediated endocytosis |
| Outer membrane proteins | Enhance innate immune response through cytokines<br>Suppress proliferation of T and B cells<br>Suppress proliferation of fibroblasts, monocytes and osteoblasts |
| Leukotoxin | Promotes apoptosis of T-cells, natural killer (NK) cells, neutrophils and PMNs in low doses<br>Causes cell death by necrosis in high concentrations |
| Flagellum | Motility, biofilm dispersal<br>Role in adherence due to fibronectin binding, thereby promotes biofilm formation |
| Capsule | Increases resistance to phagocytosis by PMNs<br>Inhibits fibroblast attachment to root surface |
| Membranous vesicles (blebs) | Mediate binding to red blood cells, hydroxyapatite surfaces and other bacteria<br>Promote targeted delivery of toxic material at site of invasion |
| Haemolysin | Destruction of red blood cells |
| Haemagglutinin | Promotes colonization by increasing iron acquisition |
| Exopolysaccharide (Slime layer) | Inhibition of phagocytosis |
| Cytotoxin | Cell distension, cell cycle arrest, apoptosis<br>Actin rearrangement |
| Fc-binding protein | Inhibition of phagocytosis<br>Inhibition of complement activation |

molecular chaperones in the assembly and folding of proteins and thus protect the cell from the damaging effects of environmental stress. HSP homologues such as GroEL, GroES, DnaK and HtpG have been studied in oral bacteria. HSPs from *A. actinomycetemcomitans* stimulate osteoclast activation and epithelial cell proliferation at low concentrations and are cytotoxic at high doses (Paju *et al.*, 2000). Microbial HSPs are highly immunogenic and have been shown to be associated with autoimmune diseases such as rheumatoid arthritis and atherosclerosis. It has been suggested that chronic infections such as periodontitis may result in prolonged exposure to HSPs, thereby promoting autoimmune disease. The molecular mimicry model suggests that the immune response to this prolonged exposure could be the mechanism underlying the progression of atherosclerosis. It is hypothesized that cross-reactivity between bacterial HSPs for example, GroEL, and human HSPs, for example, hHSP60, that is expressed on endothelial cells, may lead to endothelial dysfunction and the formation of atheromatous plaques (Seymour *et al.*, 2007). The evidence, however, is tenuous, with some studies demonstrating a definite link between the presence of *P. gingivalis* GroEL and atheromatous plaque and others unable to establish a correlation or connection (Ueki *et al.*, 2002; Yamazaki *et al.*, 2002).

## Fimbriae

Fimbriae are found in oral bacteria such as *A. actinomycetemcomitans* and *P. gingivalis*. These are long, single stranded filaments, present singly or in bundles on the surface of cells. The major component is fimbrillin, a highly antigenic protein encoded by *fimA* in *P. gingivalis* and *flp* in *A. actinomycetemcomitans*. In both these bacteria, fimbriae are thought to be important in colonization, fimbrial mutants showing reduced ability to bind and invade epithelial cells and fibroblasts (Njoroge *et al.*, 1997). Fimbriae-mediated epithelial cell invasion stimulates expression of host cell adhesion molecules such as intercellular adhesion molecule (ICAM-1), vascular cell adhesion molecule (VCAM-1), P- and E- selectins, thus inducing a massive leucocytic response at the site. *P. gingivalis* fimbriae also stimulate IL-1α, IL-1β, TNF-α and granulocyte–macrophage colony-stimulating factor (GM-CSF) leading to bone resorption.

## Extracellular proteolytic enzymes

*Tannerella forsythia*, *P. gingivalis*, *Treponema denticola* and other oral bacteria have proteolytic enzymes on their cell surfaces. Dentilisin (from *T. denticola*), PrtH (*T. forsythia*), RgpA, RgpB and Kgp (*P. gingivalis*) are the best-characterized proteins in this group (Curtis *et al.*, 1999). The RgpA/B and Kgp proteinases, major virulence factors of *P. gingivalis*, degrade fibrinogen, fibronectin, laminin, as well as adhesion molecules and several types of collagen (Kesavalu *et al.*, 1992, 1997; Chen *et al.*, 2001). *T. forsythia* with the *prtH* genotype is found to be more common in chronic periodontitis than in health. Dentilisin has been shown to hydrolyse fibrinogen, albumin, laminin, collagen type IV, IgG and IgA.

While there is compelling evidence that bacteria and bacterial products present a challenge to the host defence system, there is no single bacterial product that can lead to all the changes observed in periodontal disease. Henderson and coworkers (Henderson *et al.*, 1996) have proposed that bacterial modulins are responsible for immunomodulation of the host cells. Modulins are a group of bioactive bacterial compounds, which include exotoxins, endotoxins and metabolic end products. Both commensal organisms as well as pathogens are reported to produce modulins that induce host cytokine networks. What is evident from the preceding description of the various virulence factors is that bacteria can cause direct tissue damage. However, by triggering several types of host cells to release inflammatory cytokines and other mediators, bacteria also cause indirect tissue damage. Some of the mechanisms of causing damage to host tissue are listed in Table 10.2.

**Table 10.2** Mechanisms of host damage

| Pathogenic mechanisms |
|---|
| Coaggregation, colonization and quorum sensing |
| Adhesion to host proteins, epithelial cells, macrophages, erythrocytes, lymphocytes, extracellular matrix, fibroblasts |
| Epithelial cell invasion |
| Endothelial cell invasion |
| Proteolysis |
| Induction of pro-inflammatory response |
| Interference with host defence mechanisms, e.g. PMN chemotaxis, inhibition of epithelial growth |
| Immunosuppression |
| Apoptosis |
| Bone resorption, inhibition of bone formation |
| Haemolysis and haemagglutination |

## Specific microorganisms associated with periodontal health and disease

Different periodontal diseases have fairly unique profiles of associated bacteria. A partial list of disease-associated subgingival bacteria and their methods of detection are shown in Fig. 10.2. This, along with the fact that disease occurs in sporadic bursts in the mouth, strengthens the evidence for the role of specific microorganisms in disease causation and progression.

### Periodontal health

Bacteria that are associated with periodontal health include primary or early colonizers such as *Streptococcus sanguinis, Streptococcus mitis, Gemella* spp., *Atopobium* spp., *Fusobacterium nucleatum* and *Capnocytophaga* spp. (Socransky and Haffajee, 1992; Paster *et al.*, 2002; Kumar *et al.*, 2003). Species belonging to the genera *Veillonella, Streptococcus* and *Capnocytophaga* are thought to be beneficial to the host (Socransky and Haffajee, 1992). Molecular analysis has shown the presence of certain uncultivated species such as *Bacteroides* oral clone BU063 strongly associated with periodontal health (Leys *et al.*, 2002).

### Gingivitis

Gram-positive species, for example *Streptococcus* spp., *Actinomyces viscosus* and *Parvimonas micra* (formerly *Peptostreptococcus micros*), as well as Gram-negative species such as *Campylobacter gracilis, F. nucleatum, Prevotella intermedia* and *Veillonella* have been associated with gingivitis (Theilade *et al.*, 1966; Kremer *et al.*, 2000; Macuch and Tanner, 2000). Pregnancy-associated gingivitis, however, has a microflora predominated by *P. intermedia* (Kornman and Loesche, 1980).

### Chronic periodontitis

The bacterial profile of chronic periodontitis has been explored in cross-sectional and longitudinal studies. The effect of various treatment methods on changing the microbial ecology has also been investigated. *P. gingivalis, T. forsythia, P. intermedia, Campylobacter rectus, Eikenella corrodens, F. nucleatum, A. actinomycetemcomitans, P. micra, Treponema* spp. have been most commonly found. *P. gingivalis, T. forsythia, P. intermedia, C. rectus, F. nucleatum* are found in higher levels in sites with active disease or with progressing disease (Dzink *et al.*, 1985, 1988, 1989; Haffajee *et al.*, 1988). Clinical resolution of disease is also associated with a decrease in the levels of these species. More recent molecular approaches have found uncultivated bacterial species, such as *Desulfobulbus* sp. *oral clone R004, Deferribacteres* sp. *oral clones BH017* and *D084*, and *Bacteroides* sp. *oral clone AU126*, to be significantly associated with periodontitis (Paster *et al.*, 2001; Kumar *et al.*, 2003).

### Localized aggressive periodontitis (LAP)

Studies have implicated *A. actinomyecetemcomitans* as an important organism in the aetiology of LAP (Zambon, 1985, 1994; Haraszthy *et al.*, 2000). This species has been found as the predominant cultivable species in as many as 90% of sites with LAP. However, it should be noted that not all studies support the association of *A. actinomycetemcomitans* in aggressive periodontitis. This organism is not always found in disease sites, further, it has been found in healthy children, suggesting that it is a member of the healthy microbial flora (Gafan *et al.*, 2004). Other species such as *P. gingivalis, E. corrodens*, and *C. rectus* have also been found in high levels in certain cases of LAP (Kornman and Robertson, 1985). Viruses such as Epstein Barr virus (EBV-1) and human cytomegalovirus (HCMV) have also been associated with this disease (Michalowicz *et al.*, 2000).

### Generalized aggressive periodontitis (GAP)

The microbial aetiology of generalized aggressive periodontitis is not as well defined as other forms of periodontal diseases owing to multiple transformations in disease nomenclature. The disease now encompasses entities such as periodontosis, prepubertal periodontitis and rapidly progressing periodontitis. Nevertheless, available evidence suggests that the bacterial profile of generalized aggressive periodontitis is not significantly different from that of chronic periodontitis (Lee *et al.*, 2003; Takeuchi *et al.*, 2003; Kamma *et al.*, 2004).

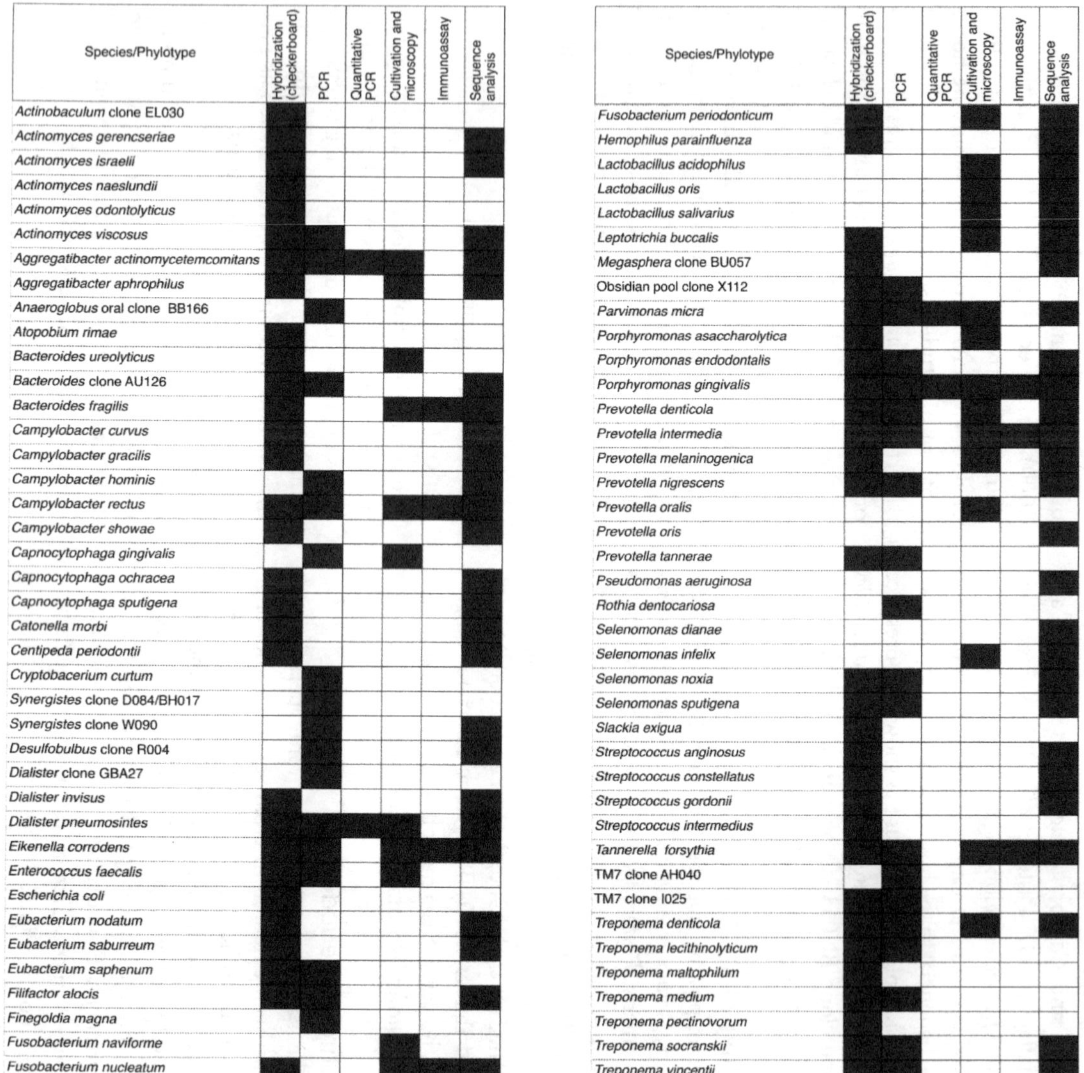


**Figure 10.2** Bacteria associated with periodontal disease. Data has been summarized from studies that are described in the text.

### Necrotizing ulcerative gingivitis

The bacterial flora of necrotizing ulcerative gingivitis has been demonstrated to be composed, for the most part, of fusobacteria and spirochetes. Recent studies have isolated previously unsuspected spirochetes, e.g. *Treponema putidum*, a proteolytic treponeme, from lesions of necrotizing ulcerative gingivitis (Wyss *et al.*, 2004). Other bacteria reported in these lesions include *Rothia dentocariosa*, *Treponema* spp., *Achromobacter* spp., *Propionibacterium acnes*, *Capnocytophaga* spp. and *P. intermedia* (Paster *et al.*, 2002).

### Periodontal abscess

A periodontal abscess is a localized purulent infection within the tissues adjacent to the periodontal pocket. *F. nucleatum*, *P. intermedia*, *P. micra*, *T. forsythia*, *C. rectus* and *P. gingivalis* have been recovered from these lesions (Newman and Sims, 1979; Herrera *et al.*, 2000).

## Summary

The bacterial aetiology of periodontal diseases has evolved over several centuries along with advances in technology and methods of bacterial identification and characterization as well as disease identification. The last decades have brought significant advances to our understanding of the pathogenesis of periodontal diseases. These include the discovery of new, uncultivated, disease-associated bacterial species, the recognition of dental plaque as a biofilm, the identification and characterization of genetic defects that predispose to periodontitis, the role of risk factors in disease susceptibility, and the discovery of new host-derived cellular and molecular mechanisms implicated in periodontal tissue destruction. Many of these discoveries hold promise for the future as foundation for the engineering of new prevention and treatment modalities.

## References


Anon. (1996). Position paper: epidemiology of periodontal diseases. American Academy of Periodontology. J. Periodontol. *67*, 935–945.

Anwar, H., Dasgupta, M.K., and Costerton, J.W. (1990). Testing the susceptibility of bacteria in biofilms to antibacterial agents. Antimicrob. Agents Chemother. *34*, 2043–2046.

Armitage, G.C. (1999). Development of a classification system for periodontal diseases and conditions. Ann. Periodontol. *4*, 1–6.

Bradshaw, D.J., and Marsh, P.D. (1999). Use of continuous flow techniques in modeling dental plaque biofilms. Methods Enzymol. *310*, 279–296.

Chen, Z., Casiano, C.A., and Fletcher, H.M. (2001). Protease-active extracellular protein preparations from *Porphyromonas gingivalis* W83 induce N-cadherin proteolysis, loss of cell adhesion, and apoptosis in human epithelial cells. J. Periodontol. 72, 641–650.

Chung, W.O., Park, Y., Lamont, R.J., McNab, R., Barbieri, B., and Demuth, D.R. (2001). Signalling system in *Porphyromonas gingivalis* based on a LuxS protein. J. Bacteriol. *183*, 3903–3909.

Costerton, J.W., Lewandowski, Z., DeBeer, D., Caldwell, D., Korber, D., and James, G. (1994). Biofilms, the customized microniche. J. Bacteriol. *176*, 2137–2142.

Curtis, M.A., Kuramitsu, H.K., Lantz, M., Macrina, F.L., Nakayama, K., Potempa, J., Reynolds, E.C., and Aduse-Opoku, J. (1999). Molecular genetics and nomenclature of proteases of *Porphyromonas gingivalis*. J. Periodontal. Res. *34*, 464–472.

Danino, J., Joachims, H.Z., and Barak, M. (1998). Predictive value of an adherence test for acute otitis media. Otolaryngol. Head Neck Surg. *118*, 400–403.

Darveau, R.P., Pham, T.T., Lemley, K., Reife, R.A., Bainbridge, B.W., Coats, S.R., Howald, W.N., Way, S.S., and Hajjar, A.M. (2004). *Porphyromonas gingivalis* lipopolysaccharide contains multiple lipid A species that functionally interact with both toll-like receptors 2 and 4. Infect. Immun. *72*, 5041–5051.

Downes, J., Munson, M.A., Spratt, D.A., Kononen, E., Tarkka, E., Jousimies-Somer, H., and Wade, W.G. (2001). Characterisation of Eubacterium-like strains isolated from oral infections. J. Med. Microbiol. *50*, 947–951.

Dowson, C.G., Hutchison, A., Woodford, N., Johnson, A.P., George, R.C., and Spratt, B.G. (1990). Penicillin-resistant viridans streptococci have obtained altered penicillin-binding protein genes from penicillin-resistant strains of *Streptococcus pneumoniae*. Proc. Natl. Acad. Sci. U.S.A. *87*, 5858–5862.

Du, L.D., and Kolenbrander, P.E. (2000). Identification of saliva-regulated genes of *Streptococcus gordonii* DL1 by differential display using random arbitrarily primed PCR. Infect. Immun. *68*, 4834–4837.

Dzink, J.L., Tanner, A.C., Haffajee, A.D., and Socransky, S.S. (1985). Gram-negative species associated with active destructive periodontal lesions. J. Clin. Periodontol. *12*, 648–659.

Dzink, J.L., Socransky, S.S., and Haffajee, A.D. (1988). The predominant cultivable microbiota of active and inactive lesions of destructive periodontal diseases. J. Clin. Periodontol. *15*, 316–323.

Dzink, J.L., Gibbons, R.J., Childs, W.C., 3rd, and Socransky, S.S. (1989). The predominant cultivable microbiota of crevicular epithelial cells. Oral Microbiol. Immunol. *4*, 1–5.

Fong, K.P., Chung, W.O., Lamont, R.J., and Demuth, D.R. (2001). Intra- and interspecies regulation of gene expression by *Actinobacillus actinomycetemcomitans* LuxS. Infect. Immun. *69*, 7625–7634.

Gafan, G.P., Lucas, V.S., Roberts, G.J., Petrie, A., Wilson, M., and Spratt, D.A. (2004). Prevalence of periodontal pathogens in dental plaque of children. J. Clin. Microbiol. *42* 4141–4146.

Gold, S.I. (1985). Paeeriodontics. The past. Microbiology. Part (III). J. Clin. Periodontol. *12*, 257–269.

Goodson, J.M., Tanner, A.C., Haffajee, A.D., Sornberger, G.C., and Socransky, S.S. (1982). Patterns of progression and regression of advanced destructive periodontal disease. J. Clin. Periodontol. *9*, 472–481.

Goulhen, F., Grenier, D., and Mayrand, D. (2003). Oral microbial heat-shock proteins and their potential contributions to infections. Crit. Rev. Oral Biol. Med. *14*, 399–412.

Haffajee, A.D., Socransky, S.S., Dzink, J.L., Taubman, M.A., Ebersole, J.L., and Smith, D.J. (1988). Clinical, microbiological and immunological features of subjects with destructive periodontal diseases. J. Clin. Periodontol. *15*, 240–246.

Haraszthy, V.I., Hariharan, G., Tinoco, E.M., Cortelli, J.R., Lally, E.T., Davis, E., and Zambon, J.J. (2000). Evidence for the role of highly leukotoxic *Actinobacillus actinomycetemcomitans* in the pathogenesis of localized juvenile and other forms of early-onset periodontitis. J. Periodontol. *71*, 912–922.

Henderson, B., Poole, S., and Wilson, M. (1996). Bacterial modulins: a novel class of virulence factors which cause host tissue pathology by inducing cytokine synthesis. Microbiol. Rev. *60*, 316–341.

Herrera, D., Roldan, S., Gonzalez, I., and Sanz, M. (2000). The periodontal abscess (I). Clinical and microbiological findings. J. Clin. Periodontol. *27*, 387–394.

Isogai, E., Isogai, H., Sawada, H., and Ito, N. (1986). Bacterial adherence to gingival epithelial cells of rats with naturally occurring gingivitis. J. Periodontol. *57*, 225–230.

Kamma, J.J., Nakou, M., Gmur, R., and Baehni, P.C. (2004). Microbiological profile of early onset/aggressive periodontitis patients. Oral Microbiol. Immunol. *19*, 314–321.

Kesavalu, L., Ebersole, J.L., Machen, R.L., and Holt, S.C. (1992). *Porphyromonas gingivalis* virulence in mice: induction of immunity to bacterial components. Infect. Immun. *60*, 1455–1464.

Kesavalu, L., Holt, S.C., and Ebersole, J.L. (1997). *Porphyromonas gingivalis* virulence in a murine lesion model: effects of immune alterations. Microb. Pathog. *23*, 317–326.

Koch, R. (1932). Die Aetiologie der Tuberkulose. J. Mol. Med. *11*, 490–492.

Kolenbrander, P.E., and Phucas, C.S. (1984). Effect of saliva on coaggregation of oral *Actinomyces* and *Streptococcus* species. Infect. Immun. *44*, 228–233.

Kolenbrander, P.E., Andersen, R.N., Blehert, D.S., Egland, P.G., Foster, J.S., and Palmer, R.J., Jr. (2002). Communication among oral bacteria. Microbiol. Mol. Biol. Rev. *66*, 486–505, table of contents.

Kolenbrander, P.E., Palmer, R.J., Jr., Rickard, A.H., Jakubovics, N.S., Chalmers, N.I., and Diaz, P.I. (2006). Bacterial interactions and successions during plaque development. Periodontol. 2000 *42*, 47–79.

Kornman, K.S., and Loesche, W.J. (1980). The subgingival microbial flora during pregnancy. J. Periodontal. Res. *15*, 111–122.

Kornman, K.S., and Robertson, P.B. (1985). Clinical and microbiological evaluation of therapy for juvenile periodontitis. J. Periodontol. *56*, 443–446.

Kremer, B.H., Loos, B.G., van der Velden, U., van Winkelhoff, A.J., Craandijk, J., Bulthuis, H.M., Hutter, J., Varoufaki, A.S., and van Steenbergen, T.J. (2000). Peptostreptococcus micros smooth and rough genotypes in periodontitis and gingivitis. J. Periodontol. *71*, 209–218.

Kumar, P.S., Griffen, A.L., Barton, J.A., Paster, B.J., Moeschberger, M.L., and Leys, E.J. (2003). New bacterial species associated with chronic periodontitis. J. Dent. Res. *82*, 338–344.

Kumar, P.S., Griffen, A.L., Moeschberger, M.L., and Leys, E.J. (2005). Identification of candidate periodontal pathogens and beneficial species by quantitative 16S clonal analysis. J. Clin. Microbiol. *43*, 3944–3955.

Kumar, P.S., Leys, E.J., Bryk, J.M., Martinez, F.J., Moeschberger, M.L., and Griffen, A.L. (2006). Changes in periodontal health status are associated with bacterial community shifts as assessed by quantitative 16S cloning and sequencing. J. Clin. Microbiol. *44* 3665–3673.

Lamont, R.J., El-Sabaeny, A., Park, Y., Cook, G.S., Costerton, J.W., and Demuth, D.R. (2002). Role of the *Streptococcus gordonii* SspB protein in the development of *Porphyromonas gingivalis* biofilms on streptococcal substrates. Microbiology *148*, 1627–1636.

Lee, J.W., Choi, B.K., Yoo, Y.J., Choi, S.H., Cho, K.S., Chai, J.K., and Kim, C.K. (2003). Distribution of periodontal pathogens in Korean aggressive periodontitis. J. Periodontol. *74*, 1329–1335.

Leys, E.J., Lyons, S.R., Moeschberger, M.L., Rumpf, R.W., and Griffen, A.L. (2002). Association of Bacteroides forsythus and a novel Bacteroides phylotype with periodontitis. J. Clin. Microbiol. *40*, 821–825.

Li, Y., and Burne, R.A. (2001). Regulation of the gtfBC and ftf genes of *Streptococcus mutans* in biofilms in response to pH and carbohydrate. Microbiology *147*, 2841–2848.

Li, Y.H., Lau, P.C., Tang, N., Svensater, G., Ellen, R.P., and Cvitkovitch, D.G. (2002a). Novel two-component regulatory system involved in biofilm formation and acid resistance in *Streptococcus mutans*. J. Bacteriol. *184*, 6333–6342.

Li, Y.H., Tang, N., Aspiras, M.B., Lau, P.C., Lee, J.H., Ellen, R.P., and Cvitkovitch, D.G. (2002b). A quorum-sensing signalling system essential for genetic competence in *Streptococcus mutans* is involved in biofilm formation. J. Bacteriol. *184*, 2699–2708.

Lindhe, J., Hamp, S., and Löe, H. (1973). Experimental periodontitis in the beagle dog. J. Periodontal. Res. *8*, 1–10.

Listgarten, M.A. (1976). Structure of the microbial flora associated with periodontal health and disease in man. A light and electron microscopic study. J. Periodontol. *47*, 1–18.

Listgarten, M.A., and Hellden, L. (1978). Relative distribution of bacteria at clinically healthy and periodontally diseased sites in humans. J. Clin. Periodontol. *5*, 115–132.

Listgarten, M.A., Schifter, C.C., and Laster, L. (1985). 3-year longitudinal study of the periodontal status of an adult population with gingivitis. J. Clin. Periodontol. *12*, 225–238.

Löe, H., Anerud, A., Boysen, H., and Morrison, E. (1986). Natural history of periodontal disease in man. Rapid, moderate and no loss of attachment in Sri Lankan laborers 14 to 46 years of age. J. Clin. Periodontol. *13*, 431–445.

Loesche, W.J. (1979). Clinical and microbiological aspects of chemotherapeutic agents used according to the specific plaque hypothesis. J. Dent. Res. *58*, 2404–2412.

McNab, R., Ford, S.K., El-Sabaeny, A., Barbieri, B., Cook, G.S., and Lamont, R.J. (2003). LuxS-based signalling in *Streptococcus gordonii*: autoinducer 2 controls carbohydrate metabolism and biofilm formation with *Porphyromonas gingivalis*. J. Bacteriol. *185*, 274–284.

Macuch, P.J., and Tanner, A.C. (2000). Campylobacter species in health, gingivitis, and periodontitis. J. Dent. Res. *79*, 785–792.

Marsh, P.D. (2004). Dental plaque as a microbial bioflim. Caries Res. *38*, 204–211.

Marsh, P.D. (2005). Dental plaque: biological significance of a biofilm and community life-style. J. Clin. Periodontol. *32 Suppl 6*, 7–15.

Marsh, P.D., McDermid, A.S., McKee, A.S., and Baskerville, A. (1994). The effect of growth rate and haemin on the virulence and proteolytic activity of *Porphyromonas gingivalis* W50. Microbiology *140*, 861–865.

Michalowicz, B.S., Ronderos, M., Camara-Silva, R., Contreras, A., and Slots, J. (2000). Human herpesviruses and *Porphyromonas gingivalis* are associated with juvenile periodontitis. J. Periodontol. *71*, 981–988.

Mombelli, A., Lang, N.P., Burgin, W.B., and Gusberti, F.A. (1990). Microbial changes associated with the development of puberty gingivitis. J. Periodontal. Res. *25*, 331–338.

Netea, M.G., van Deuren, M., Kullberg, B.J., Cavaillon, J.M., and Van der Meer, J.W. (2002). Does the shape of lipid A determine the interaction of LPS with Toll-like receptors? Trends Immunol. *23*, 135–139.

Newman, M.G., and Sims, T.N. (1979). The predominant cultivable microbiota of the periodontal abscess. J. Periodontol. *50*, 350–354.

Njoroge, T., Genco, R.J., Sojar, H.T., Hamada, N., and Genco, C.A. (1997). A role for fimbriae in *Porphyromonas gingivalis* invasion of oral epithelial cells. Infect. Immun. *65*, 1980–1984.

Nyvad, B., and Fejerskov, O. (1987). Scanning electron microscopy of early microbial colonization of human enamel and root surfaces in vivo. Scand. J. Dent. Res. *95*, 287–296.

Ohta, H., Miyagi, A., Kato, K., and Fukui, K. (1996). The relationships between leukotoxin production, growth rate and the bicarbonate concentration in a toxin-production-variable strain of *Actinobacillus actinomycetemcomitans*. Microbiology *142*, 963–970.

Page, R.C. (1999). Milestones in periodontal research and the remaining critical issues. J. Periodontal. Res. *34*, 331–339.

Paju, S., Goulhen, F., Asikainen, S., Grenier, D., Mayrand, D., and Uitto, V. (2000). Localization of heat shock proteins in clinical *Actinobacillus actinomycetemcomitans* strains and their effects on epithelial cell proliferation. FEMS Microbiol. Lett. *182*, 231–235.

Paster, B.J., Boches, S.K., Galvin, J.L., Ericson, R.E., Lau, C.N., Levanos, V.A., Sahasrabudhe, A., and Dewhirst, F.E. (2001). Bacterial diversity in human subgingival plaque. J. Bacteriol. *183*, 3770–3783.

Paster, B.J., Falkler, W.A., Jr., Enwonwu, C.O., Idigbe, E.O., Savage, K.O., Levanos, V.A., Tamer, M.A., Ericson, R.L., Lau, C.N., and Dewhirst, F.E. (2002). Prevalent bacterial species and novel phylotypes in advanced noma lesions. J. Clin. Microbiol. *40* 2187–2191.

Roberts, A.P., Cheah, G., Ready, D., Pratten, J., Wilson, M., and Mullany, P. (2001). Transfer of TN916-like elements in microcosm dental plaques. Antimicrob. Agents Chemother. *45*, 2943–2946.

Seymour, G.J., Ford, P.J., Cullinan, M.P., Leishman, S., and Yamazaki, K. (2007). Relationship between periodontal infections and systemic disease. Clin. Microbiol. Infect. *13(Suppl. 4)*, 3–10.

Socransky, S.S., and Haffajee, A.D. (1992). The bacterial etiology of destructive periodontal disease: current concepts. J. Periodontol. *63*, 322–331.

Socransky, S.S., and Haffajee, A.D. (2005). Periodontal microbial ecology. Periodontol. 2000 *38*, 135–187.

Stewart, P.S. (2003). Diffusion in biofilms. J. Bacteriol. *185*, 1485–1491.

Stingu, C.S., Eschrich, K., Rodloff, A.C., Schaumann, R., and Jentsch, H. (2008). Periodontitis is associated with a loss of colonization by *Streptococcus sanguinis*. J. Med. Microbiol. *57*, 495–499.

Takeuchi, Y., Umeda, M., Ishizuka, M., Huang, Y., and Ishikawa, I. (2003). Prevalence of periodontopathic bacteria in aggressive periodontitis patients in a Japanese population. J. Periodontol. *74*, 1460–1469.

Tanner, A., Maiden, M.F., Paster, B.J., and Dewhirst, F.E. (1994). The impact of 16S ribosomal RNA-based phylogeny on the taxonomy of oral bacteria. Periodontol. 2000 *5*, 26–51.

Theilade, E., Wright, W.H., Jensen, S.B., and Loe, H. (1966). Experimental gingivitis in man. II. A longitudinal clinical and bacteriological investigation. J. Periodontal. Res. *1*, 1–13.

Ueki, K., Tabeta, K., Yoshie, H., and Yamazaki, K. (2002). Self-heat shock protein 60 induces tumour necrosis factor-alpha in monocyte-derived macrophage: possible role in chronic inflammatory periodontal disease. Clin. Exp. Immunol. *127*, 72–77.

Van Hoogmoed, C.G., Geertsema-Doornbusch, G.I., Teughels, W., Quirynen, M., Busscher, H.J., and Van der Mei, H.C. (2008). Reduction of periodontal pathogens adhesion by antagonistic strains. Oral Microbiol. Immunol. *23*, 43–48.

Wyss, C., Moter, A., Choi, B.K., Dewhirst, F.E., Xue, Y., Schupbach, P., Gobel, U.B., Paster, B.J., and

Guggenheim, B. (2004). Treponema putidum sp. nov., a medium-sized proteolytic spirochaete isolated from lesions of human periodontitis and acute necrotizing ulcerative gingivitis. Int. J. Syst. Evol. Microbiol. *54*, 1117–1122.

Yamazaki, K., Ohsawa, Y., Tabeta, K., Ito, H., Ueki, K., Oda, T., Yoshie, H., and Seymour, G.J. (2002). Accumulation of human heat shock protein 60-reactive T cells in the gingival tissues of periodontitis patients. Infect. Immun. *70*, 2492–2501.

Zambon, J.J. (1985). *Actinobacillus actinomycetemcomitans* in human periodontal disease. J. Clin. Periodontol. *12*, 1–20.

Zambon, J.J. (1994). *Actinobacillus actinomycetemcomitans* in adult periodontitis. J. Periodontol. *65*, 892–893.

Zambon, J.J., Slots, J., and Genco, R.J. (1983). Serology of oral *Actinobacillus actinomycetemcomitans* and serotype distribution in human periodontal disease. Infect. Immun. *41*, 19–27.

# Periodontal Biofilm and Immunity: Immune Subversion by Select Pathogens as a Community Service

11

George Hajishengallis


## Abstract

The polymicrobial community that initiates periodontal disease does not represent a random compilation of bacteria. Rather, these bacteria form organized consortia that have evolved through mutually beneficial relationships. This Chapter focuses on microbial immune subversion as a means by which select pathogens may contribute to the adaptive fitness of the entire periodontal biofilm. For instance, *Porphyromonas gingivalis* expresses specialized virulence traits that undermine immunity and promote non-resolving inflammation, which, respectively, protect the bacteria and facilitate nutrient acquisition. The virulence factors involved (e.g. cysteine proteases and atypical lipopolysaccharide structures) are released as components of readily diffusible membrane vesicles, which can thus become available for the benefit of other biofilm organisms. The elucidation of immune subversion mechanisms of key periopathogens that promote the collective virulence of their communities may provide new avenues of therapeutic intervention in human periodontitis.


## Introduction

Periodontal disease is perhaps the most common chronic disorder of infectious origin in humans and is characterized by inflammatory destruction of the tooth-supporting tissues (Pihlstrom *et al.*, 2005). The disease is initiated by certain species of subgingival Gram-negative anaerobic bacteria co-existing within dynamic communities of highly organized architecture, originally named 'dental plaque' which predates the more modern term 'biofilm' (Millin and Smith, 1961; Gibbons *et al.*, 1964; Socransky *et al.*, 1998; Davey and Costerton, 2006). In fact, seldom would discussions on pathogenic biofilms omit the classical example of dental plaque. Incidentally, the first multispecies biofilm ever viewed by man (in 1683 by van Leeuwenhoek) was bacteria ('animalcules') from tooth scrapings (Dixon, 2009).

Periodontal health is characterized by the predominant presence of Gram-positive, facultative, and fermentative bacteria (e.g. streptococci and actinomyces), which form a biofilm on tooth surfaces in an orderly ecological succession (Kolenbrander *et al.*, 2010). On the other hand, the onset of periodontal disease is associated with a significant shift to Gram-negative, anaerobic, and proteolytic bacteria which begin to colonize subgingival tooth sites with greater frequencies (Socransky and Haffajee, 2005). Under periodontally healthy conditions, the microbial burden is relatively low and is estimated that $10^2$ to $10^3$ isolates may be cultured from an individual gingival crevicular site. In contrast, the microbial load becomes strikingly higher in periodontitis, where $10^5$ to $10^8$ isolates may be cultured from an individual periodontal pocket (Darveau, 2009).

Using a colour-coded system, Socransky and colleagues characterized these microbial communities as red, orange, green, purple and yellow complexes, on the basis of cluster analysis, community ordination, and associated disease severity (Socransky *et al.*, 1998). Elevated numbers of red complex members, i.e. *Porphyromonas gingivalis*, *Treponema denticola*, and *Tanerella forsythia*, correlate strongly with periodontal tissue destruction (Socransky *et al.*, 1998; Holt and Ebersole, 2005).

These species can co-adhere with each other *in vitro* and, importantly, tend to be coassociated also *in vivo* (Jenkinson and Lamont, 2005; Kolenbrander *et al.*, 2010). Other bacteria such as *Prevotella intermedia, Fusobacterium nucleatum,* and *Aggregatibacter actinomycetmecomitans* are also associated with various forms of periodontal disease (Socransky *et al.*, 1998; Ximenez-Fyvie *et al.*, 2000; Darveau, 2009). Although these putative pathogens have received particular attention, it should be noted that a great percentage (perhaps 50%) of the approximately 700 bacterial species in the human oral cavity are as yet uncultivable (Jenkinson and Lamont, 2005; Kolenbrander *et al.*, 2010).

Red or orange complex bacteria, such as *P. gingivalis* or *F. nucleatum*, are rarely seen in significant numbers in the early colonizing consortia of streptococci and actinomyces, which actually constitute a prerequisite for 'red' and 'orange' colonization (Jenkinson and Lamont, 2005). For example, *P. gingivalis* colonization appears to be dependent upon antecedent streptococci (*Streptococcus gordonii*) which provide molecular anchor sites, such as glyceraldehyde-3-phosphate dehydrogenase and antigen I/II-like adhesins that bind, respectively, *P. gingivalis* fimbrial proteins FimA and Mfa (Jenkinson and Lamont, 2005). Strains of *F. nucleatum* coaggregate with both early and late colonizers (e.g. *S. gordonii and P. gingivalis*) and are thus thought to function as 'bridge' organisms, facilitating the ordered succession of the bacterial species in the developing dental biofilm (Kolenbrander *et al.*, 2010). In addition to providing interbacterial attachment sites, early colonizers, such as facultative anaerobic streptococci, may facilitate the colonization of strictly anaerobic organisms by reducing oxygen tension to levels permissive for obligate anaerobic growth (Lamont and Jenkinson, 1998).

The mature dental biofilm displays community characteristics in terms of both architecture and function, such as morphological structures (bacterial aggregates and fluid-filled channels), adhesive matrix, quorum-sensing-based communication, and metabolic cooperation (Wood *et al.*, 2000; Jenkinson and Lamont, 2005; Darveau, 2010; Kolenbrander *et al.*, 2010). In addition, it is becoming increasingly more apparent that the undermining of host defences by key periodontal pathogens may also benefit other species that lack immune subversion strategies but share the same subgingival location (Darveau, 2009, 2010; Hajishengallis, 2009; Krauss *et al.*, 2010). Briefly stated, the microbial communities that are associated with periodontal disease are not random collections of bacteria, but rather organized consortia that have evolved through mutually beneficial interactions, securing a niche with adequate nutrients and protection against the host immunity. This chapter focuses on the importance of immune subversion as a means which contributes to the adaptive fitness of the periodontal biofilm. A case will be made that this function is contributed by select bacteria with specialized virulence traits which might act as 'keystone species'.

## Are there keystone species in the subgingival biofilm?

A keystone species is one that plays a vital role in maintaining the structure of an ecological community, analogous to the role of a keystone in an arch. Its impact on the community is greater than would be expected on the basis of its relative abundance. The concept of a keystone species was first introduced to biology by the ecologist R.T. Paine in 1969 (Paine, 1969). A classic example is the starfish *Pisaster ochraceus*; its experimental removal from a particular intertidal zone caused a dramatic reduction in the number of resident species compared to a control area (Paine, 1969; Power *et al.*, 1996). In the context of microbial ecology, a keystone species can be defined as one that 'serves an essential function for the entire community, similar to a differentiated cell serving a function for an entire tissue' (Darveau, 2010). The implication is not that the biofilm community would necessarily collapse in the absence of a single important species but that its numbers, composition, and pathogenicity could be dramatically affected. For example, *P. gingivalis* could be the species responsible for blocking or blunting Toll-like receptor (TLR)-4-dependent antimicrobial responses against all members of the community. The mechanism by which *P. gingivalis* can neutralize TLR4 is an unusual lipopolysaccharide (LPS): a TLR4 antagonist readily available in 'packages' of easily diffusible outer membrane vesicles (OMVs) that can be 'shipped' to the local environment and thereby rescue other bacteria (Darveau, 2010) (more below). The absence of this specialized function would likely result in a more effective TLR4 immunity [e.g. increased expression of β-defensins (Lu *et al.*, 2009)] leading to a better control of subgingival bacterial growth.

## Periodontal homeostasis and microbial tactics to disrupt it

A brief background on periodontal host immunity and homeostasis is necessary before one can discuss how select periodontal pathogens may provide community service through immune subversion. The clinically healthy periodontium is not free of inflammation but is under a homeostatic inflammatory state that keeps dental bacteria at bay (Darveau, 2010). Periodontal health, therefore, represents a dynamic balanced state where proinflammatory and antimicrobial activities are optimally regulated to prevent unwarranted host responses (Gaffen and Hajishengallis, 2008). This homeostasis may be disrupted, however, either by congenital immunodeficiencies or immunoregulatory defects of the host, or by pathogens that subvert the host response (Kinane *et al.*, 2006; Schenkein, 2006; Gaffen and Hajishengallis, 2008; Kumpf and Schumann, 2008).

Implicit in the 'disrupted homeostasis' concept is that the host response to periodontal bacteria is intrinsically protective. This is supported by several lines of evidence from observations in humans or experimental animals. Individuals with neutrophil disorders with respect to numbers (chronic or cyclic neutropenia), trafficking capacity (leucocyte adhesion deficiencies) or killing function (Chédiak–Higashi and Papillon–Lefèvre syndromes), all have increased susceptibility to and severity of periodontitis, including forms of the disease that appear early in life (Schenkein, 2006; Darveau, 2010). Similarly, mice with impaired mobilization of leucocytes to sites of infection, due to combined P- and E-selectin deficiency, display massive oral bacterial colonization and induction of gingival inflammation and alveolar bone loss (Niederman *et al.*, 2001). Mice whose neutrophils display impaired bacterial killing (owing to the genetic absence of the lysosomal-associated membrane protein-2) are also unable to control their indigenous oral flora and develop severe periodontitis (Beertsen *et al.*, 2008). Importantly, inflammatory bone loss in both types of immunodeficient mice could be prevented by antibiotics (Niederman *et al.*, 2001; Beertsen *et al.*, 2008). In these animals, therefore, periodontal disease is caused by the indigenous oral bacteria in the absence of sufficient host immune control.

The studies discussed above collectively suggest that normal innate immune function is important for periodontal homeostasis. In this context, healthy human gingiva display a coordinated gradient of chemokines and adhesion molecules that is conducive for the directed migration of neutrophils to the gingival crevice. Specifically, a gradient of interleukin IL-8 and cell adhesion molecules (ICAM-1 and E-selectin) was shown to be topographically associated with the pathway of neutrophil migration from the vasculature to the junctional epithelium and ultimately to the gingival sulcus (Tonetti *et al.*, 1998; Darveau, 2010). This gradient became denser towards the upper junctional epithelial cell layers and in areas exposed to bacterial challenge (Tonetti *et al.*, 1998). In periodontal health, gingival epithelial cells and leucocytes in the junctional epithelium also express an array of antibacterial factors, such as α- and β-defensins and the cathelicidin LL-37, as well as soluble or membrane-bound CD14 and soluble LPS-binding protein (Schenkein, 2006; Diamond *et al.*, 2008; Gorr, 2009; Darveau, 2010) (Fig. 11.1). It is increasingly appreciated that this controlled inflammatory state or 'armed peace' is the result of host interactions with the commensal bacteria in close contact with, but not invading, the periodontal tissues (Roberts and Darveau, 2002; Dixon *et al.*, 2004).

This armed peace may transition to open war if the homeostatic balance is compromised. Given the roles of chemokine and cell adhesion molecules in leucocyte migration, the disruption of their characteristic gradient could impair the ability of the host to effectively sense bacteria and to summon leucocytes for their control. Termed 'local chemokine paralysis', the ability of gingival epithelial cells to induce IL-8 for chemoattraction

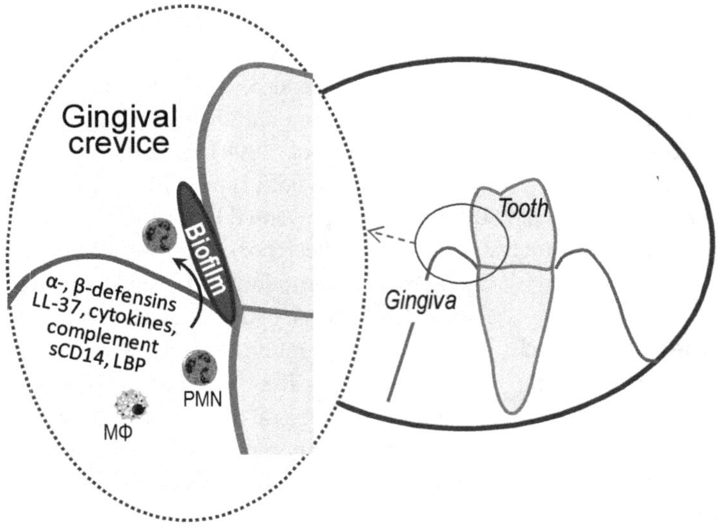


**Figure 11.1** Innate immunity in the periodontium. Gingival epithelial cells and leucocytes in the junctional epithelium or in the connective tissue produce cytokines, antimicrobial peptides (e.g. α- and β-defensins and cathelicidin LL-37), as well as soluble CD14 and LPS-binding protein, which potentially contribute to bacterial clearance. These molecules, the expression of which is inducible (e.g. following TLR activation by the dental biofilm), transit to the gingival crevice which also contains functional complement. Leucocytes, mostly neutrophils, enter the crevice to fight off bacteria. Pathogenic periodontal biofilms can resist or subvert these immune mechanisms resulting in non-resolving, chronic inflammation. Mφ, macrophages; PMN, polymorphonuclear neutrophils.

of neutrophils is proactively inhibited by *P. gingivalis,* at least *in vitro*. This inhibitory effect requires that *P. gingivalis* invade the epithelial cells (Darveau *et al.*, 1998) and secrete the serine phosphatase enzyme SerB (Hasegawa *et al.*, 2008). Accordingly, a SerB-deficient isogenic mutant induces higher levels of neutrophil recruitment to the periodontium and reduced levels of periodontal bone loss compared to wild-type *P. gingivalis* (Bainbridge *et al.*, 2010). *P. gingivalis*-infected epithelial cells are prevented from eliciting IL-8 responses even when exposed to bacteria like *F. nucleatum* that are potent inducers of this cytokine on their own (Darveau *et al.*, 1998; Hasegawa *et al.*, 2008). Moreover, *P. gingivalis* inhibits E-selectin upregulation by other periodontal bacteria (Darveau *et al.*, 1995). Inhibition of E-selectin expression results in diminished neutrophil adhesion to endothelial cells *in vitro* (Darveau *et al.*, 1995) and, along with disruption of the IL-8/ICAM-1 gradient, could presumably suppress neutrophil diapedesis and directed migration to the infected gingival crevice. In brief, disruption of focused migration of leucocytes in the early stages of biofilm formation may allow unchecked growth of the biofilm, which may subsequently cause overt inflammation and uncontrolled mobilization of inflammatory cells with destructive consequences.

## Microbial manipulation of the host response is dynamic and contextual

The capacity of *P. gingivalis* for immune evasion does not necessarily imply that periodontal bacteria induce generalized immunosuppression in the periodontium. After all, these bacteria are strongly associated with inflammatory tissue damage. It is quite likely that mechanisms for inhibition or stimulation of the host response may be employed under temporally differential conditions. In the early stages of infection, periodontal bacteria may suppress (or hide from) host defence to facilitate their colonization of the subgingival crevice. At later stages, when colonization has been established, the bacteria could perhaps become more aggressive and stimulate inflammation. In fact, inflammation can contribute to the increased demand for nutrients, such as haemin from inflammatory exudate as a source of essential iron. Moreover, the inflammatory gingival crevicular fluid (GCF) contains ample proteinaceous substrate that is both needed and can be utilized by asaccharolytic/proteolytic bacteria (Delima and Van Dyke, 2003).

Clinical and histological studies of the inflamed periodontium indicate that complement activation is a major feature, whereas more recent mechanistic studies indicate that periodontal bacteria manipulate complement to promote their adaptive fitness through both immunosuppressive and inflammatory mechanisms (Slaney and Curtis, 2008; Hajishengallis, 2010; Krauss *et al.*, 2010). The involvement of complement in inflammation does not derive solely from generation of powerful inflammatory mediators, such as the chemotactic anaphylatoxins (C3a, C5a), but also from synergistic crosstalk with TLR signalling pathways (Hajishengallis and Lambris, 2010; Ricklin *et al.*, 2010).

Activated complement components are abundant in the GCF of periodontitis patients and in inflamed gingiva, whereas they are absent or present at significantly lower levels in periodontally healthy individuals (Hajishengallis, 2010). Mechanistically, local complement activation may promote periodontal inflammation by C5a-induced vasodilation, by increased vascular permeability and flow of inflammatory exudate, and by chemotactic recruitment of inflammatory cells such as macrophages and especially neutrophils (Fig. 11.1). In periodontally diseased sites, the IL-8 gradient that characterizes healthy gingiva is disrupted, even though the expression of IL-8 per se is increased (Tonetti *et al.*, 1994). Whereas a coordinated and topographically restricted chemokine/cell adhesion molecule expression gradient could spatially limit the infiltration of inflammatory cells to sites where they are needed, the excessive production of C5a – which is chemotactic for a broad spectrum of leucocytes – may result in 'out-of-control' inflammation and collateral tissue damage.

Unwarranted and excessive recruitment of neutrophils plays a key role in host-mediated inflammatory tissue injury in periodontitis (Van Dyke and Serhan, 2003). Neutrophils can be found in great numbers in the gingival crevice and comprise ≥95% of total leucocytes (Delima and Van Dyke, 2003). As mentioned above, extravasating neutrophils enter the gingival crevice through the junctional epithelium which, under inflamed conditions, is occupied to about 60% by trafficking neutrophils (Schroeder, 1973; Delima and Van Dyke, 2003). Because the inflammatory exudate is a rich source of nutrients, this dysregulated inflammatory state seems to favour the periodontal biofilm bacteria.

## Targeted immune evasion in the context of inflammation

How could periodontal bacteria cope with complement activation and the ensuing amplification of the host response, which in large part involves crosstalk interactions between complement and TLRs? This question becomes even more puzzling considering that periodontal bacteria themselves have evolved non-immunological ways to 'voluntarily' activate specific complement components. For instance, the Arg-specific gingipains of *P. gingivalis* function as C5 convertase-like enzymes (Wingrove *et al.*, 1992) which generate high levels of C5a *in vitro* and *in vivo* (Wang *et al.*, 2010; Liang *et al.*, 2011). This is quite curious for a pathogen which goes at great lengths to inhibit all three initiation mechanisms (classic, lectin, and alternative) of complement activation (Popadiak *et al.*, 2007; Krauss *et al.*, 2010). It was shown, however, that *P. gingivalis* exploits C5a to instigate a subversive crosstalk with TLR2 (Wang *et al.*, 2010; Liang *et al.*, 2011). Mechanistically, upon binding the C5a receptor (C5aR; CD88), C5a stimulates Gαi-dependent intracellular $Ca^{2+}$ signalling which synergistically enhances an otherwise weak cAMP response induced by *P. gingivalis*-induced TLR2 activation alone. In this crosstalk pathway, sustained elevated production of cAMP leads to the activation of the cAMP-dependent protein kinase A which inactivates the glycogen synthase kinase-3β and impairs nitric oxide-dependent killing of *P. gingivalis* in macrophages (Wang *et al.*, 2010; Hajishengallis and Lambris, 2011) (Fig. 11.2). Crosstalk between C5aR and TLR2 also inhibits the ability of neutrophils to kill *P.*

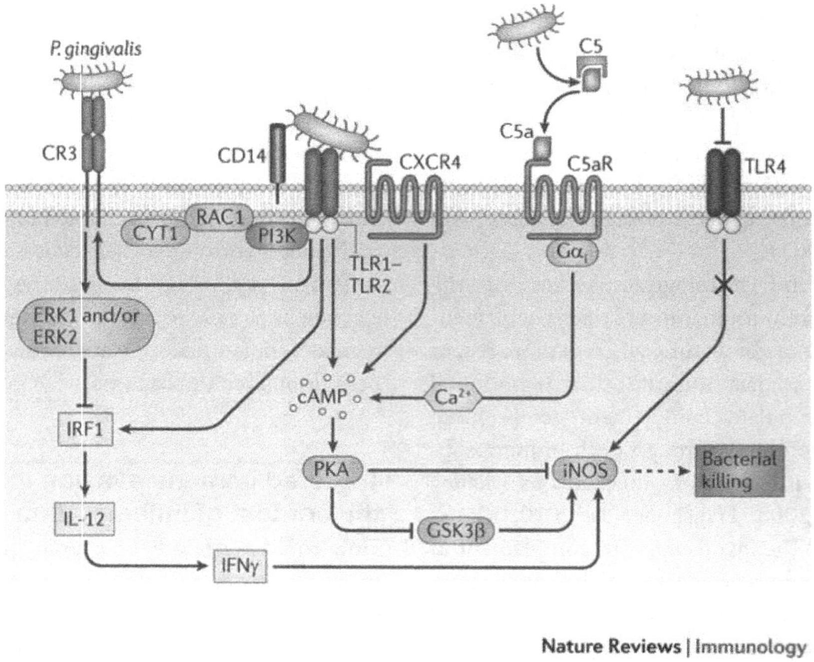


**Figure 11.2** Exploitation of innate immune receptors by *P. gingivalis* to undermine host immunity. *P. gingivalis* has surface structures that interact with Toll-like receptor (TLR)2 (specifically with the CD14–TLR2–TLR1 signalling complex) and with TLR4. The activation of TLR4, however, is blocked by the bacterium's atypical lipopolysaccharide which acts as an antagonist; therefore, TLR4 is unlikely to induce protective responses. The TLR2 response is proactively modified through crosstalk with other receptors that are under *P. gingivalis* control. *P. gingivalis* regulates C5a receptor (C5aR) by virtue of Arg-specific cysteine proteinases, which attack C5 and release biologically active C5a. C5a stimulates intracellular $Ca^{2+}$ signalling which synergistically enhances the otherwise weak cAMP responses induced by TLR2 activation alone. Maximal cAMP induction requires the participation of CXC-chemokine receptor 4 (CXCR4), which is activated directly by the bacterium's fimbriae. The resulting activation of the cAMP-dependent protein kinase A (PKA) inactivates glycogen synthase kinase-3β (GSK3β) and inhibits the inducible nitric oxide synthase (iNOS)-dependent killing of the pathogen in macrophages. An additional pathway induced downstream of TLR2 is an inside-out signalling pathway, mediated by RAC1, phosphatidylinositol-3 kinase (PI3K) and cytohesin 1 (CYT1), which transactivates complement receptor-3 (CR3). Activated CR3 binds *P. gingivalis* and induces extracellular signal-regulated kinase-1/ERK2 signalling, which in turn selectively downregulates IL-12 p35 and p40 mRNA expression through suppression of interferon regulatory factor 1 (IRF1). Inhibition of bioactive IL-12, and secondarily IFNγ, leads to impaired immune clearance of *P. gingivalis*. From Hajishengallis and Lambris (2011), with permission.

*gingivalis,* although the underlying signalling pathway is cAMP-independent (Hajishengallis *et al.,* 2010). The *P. gingivalis*-induced C5aR-TLR2 crosstalk regulates *in vivo* cytokine production in favour of the pathogen (Liang *et al.,* 2011). Specifically, this oral bacterium proactively and selectively inhibits TLR2-induced IL-12 which is required for its immune clearance, whereas the same C5aR-TLR2 crosstalk upregulates other inflammatory and bone-resorptive cytokines (IL-1β, IL-6, and TNF-α). These effects would be expected to enhance the persistence of *P. gingivalis* in the host and to create favourable conditions for inflammatory tissue damage. This notion is consistent with observations that mice deficient in either C5aR or TLR2 are protected against *P. gingivalis*-induced periodontitis (Liang *et al.,* 2011).

Are these subversive effects good solely for *P. gingivalis*? Most likely not. When neutrophils or macrophages are rendered impotent in terms of killing – even though they can still elicit inflammatory responses – the same leucocytes would also be unable to control bystander bacteria cohabiting the same region with *P. gingivalis*. For example, although *F. nucleatum* is susceptible to killing by leucocytes, it displays a dramatic resistance to leucocyte killing in the presence of *P. gingivalis* (Krauss, J.L., and Hajishengallis, G., unpublished). In the oral gavage model of mouse periodontitis, the implantation of *P. gingivalis* is accompanied by a major increase in the numbers of the indigenous oral anaerobic microbiota (Liang, S., and Hajishengallis, G., unpublished). These observations do not prove but are consistent with the notion that *P. gingivalis* might be a key pathogen capable of promoting the survival and virulence of the microbial community in which it resides. Similar conclusions can be derived from a study in the rabbit model of *P. gingivalis*-induced periodontitis (Hasturk *et al.*, 2007). In that study, the implantation of *P. gingivalis* caused the overgrowth of the dental biofilm and a shift to a more anaerobic flora. When resolvin E1 was topically applied to inhibit the inflammatory process, this treatment also resulted in the elimination of *P. gingivalis*, which correlated with the return of the indigenous oral flora to its baseline composition and numbers (Hasturk *et al.*, 2007). These findings are also consistent with the notion that *P. gingivalis* can positively impact on the welfare of additional periodontal species. Moreover, the data suggest that *P. gingivalis* cannot sustain itself under non-inflammatory conditions (inability to secure GCF-derived nutrients, such as haemin-derived iron or tissue breakdown products). Conversely, sustained inflammation and thus availability of host proteinaceous components resulting from inflammatory tissue destruction can promote the growth of proteolytic organisms (e.g. *P. gingivalis*, *T. forsythia*, and *P. intermedia*) but also of other bystander bacteria that may not be as proteolytic. This is likely a major mechanism contributing to the shift away from a healthy community to one that is pathogenic.

Histological and electron microscopy studies show that gingival crevicular neutrophils form what looks like a 'defence wall' against the dental biofilm bacteria. However, the neutrophils largely fail to control the bacteria despite being viable and capable of eliciting inflammatory responses, including release of reactive oxygen species and extracellular DNA traps, i.e. web-like scaffolds that concentrate released antimicrobial molecules resulting in the trapping and killing of extracellular bacteria (Lange and Schroeder, 1971; Newman, 1980; Schroeder and Listgarten, 1997; Delima and Van Dyke, 2003; Chapple and Matthews, 2007; Ryder, 2010; Vitkov *et al.*, 2010). Since reactive oxygen species do not discriminate between microbial and host cells, these molecules are more likely to cause collateral damage to periodontal tissues (Nathan, 2002; Chapple and Matthews, 2007; Van Dyke, 2008) than to control the infection, especially since many oral bacteria, including *P. gingivalis*, are resistant to oxidative burst killing (Mydel *et al.*, 2006; Hajishengallis *et al.*, 2008a; Krauss *et al.*, 2010). It is often assumed that biofilms are inherently resistant to phagocytosis. However, this is an unproven assumption. In fact, neutrophils can potentially recognize and respond to biofilms through phagocytosis, degranulation (e.g. release of lactoferrin and elastase) and formation of extracellular DNA traps (Meyle *et al.*, 2010). Therefore, when biofilms successfully resist elimination, this is likely to involve proactive microbial mechanisms that evade or subvert phagocytosis or other leucocyte functions. Time-lapse video microscopy and confocal laser scanning microscopy has revealed that, depending on the nature and composition of biofilms, neutrophils can move into a biofilm and eat up bacteria as they move across, or display a relatively immobile phenotype with limited phagocytosis in their immediate vicinity (Wagner *et al.*, 2007; Guenther *et al.*, 2009; Gunther *et al.*, 2009; Meyle *et al.*, 2010). Moreover, neutrophils can detect quorum-sensing molecules of bacteria which will attract them to the developing biofilm and stimulate their killing mechanisms (Wagner *et al.*, 2007). The mechanistic basis for the relative impotence of crevicular neutrophils to control periodontal infection and to become involved in non-resolving inflammation is largely unexplored, but could be due to a number of mechanisms. At least in part, this could be the result of the *P. gingivalis* capacity to generate high levels of C5a and exploit it for targeted suppression of the leucocyte killing function without affecting – in fact, enhancing – the overall inflammatory response.

## Acting at a distance

Although *P. gingivalis* is a Gram-negative bacterium that expresses LPS, its lipid A moiety presents some unusual features that render it invisible to TLR4 or, even worse for the host, antagonistic to TLR4 (Fig. 11.2). Specifically, the organism elegantly utilizes specific lipid A 1- and 4′-phosphatases and a deacylase which in concert generate a tetra-acylated and dephosphorylated lipid A structure (Coats *et al.*, 2009). This modification renders the LPS biologically inert, thereby allowing *P. gingivalis* evade TLR4-mediated immune responses (Coats *et al.*, 2009). At the same time, this LPS modification confers protection against polymyxin B and perhaps other cationic anti-microbial peptides (Coats *et al.*, 2009). Intriguingly, the presence of high concentrations of haemin suppresses lipid A 1-phosphatase activity and leads to the production of a mono-phosphorylated lipid A, which actively antagonizes TLR4 activation (Coats *et al.*, 2005, 2009). Thus, even though TLR4 could be activated by LPS from other bacteria (e.g. *A. actinomycetemcomitans*, the LPS of which is a potent TLR4 agonist), this activity could be attenuated by *P. gingivalis* LPS. In fact, experimental evidence supports this notion (Darveau *et al.*, 2002; Hajishengallis *et al.*, 2002) and the underlying mechanism involves competitive TLR4 blocking caused by the binding of the antagonistic *P. gingivalis* lipid A to MD2, a TLR4-bound protein required for responsiveness to LPS (Coats *et al.*, 2005). This mechanism may not only protect other bacteria in close vicinity to *P. gingivalis* but also those in relatively remote areas of mixed species biofilms. Indeed, *P. gingivalis* releases OMV decorated with LPS (as well as gingipains and fimbriae) that can readily diffuse in the gingival crevice or even penetrate gingival tissue (Lamont and Jenkinson, 1998; Darveau, 2010). These virulence factor-containing packages, therefore, have the potential to undermine innate immune responses for the microbial community at large. For example, the lipid A structure of *P. gingivalis* which antagonizes TLR4 (tetra-acylated monophosphorylated lipid A) reduces the expression of β-defensins in a reconstituted human gingival epithelium (Lu *et al.*, 2009). This mechanism can be lent to other biofilm bacteria as long as they encounter *P. gingivalis* OMVs.

In addition to LPS and gingipains, the fimbriae of *P. gingivalis* have also been implicated in immune evasion, although their primary role is related to adherence and colonization (Hajishengallis, 2009). At least some of the immune evasive mechanisms of *P. gingivalis* fimbriae can positively affect other periodontal bacteria in the vicinity of wandering OMVs. *P. gingivalis* uses its fimbriae to bind complement receptor-3 (CR3) which induces extracellular signal-regulated kinase 1/2 signalling that in turn selectively inhibits mRNA expression of the p35 and p40 subunits of IL-12 (Hajishengallis *et al.*, 2007) (Fig. 11.2). As alluded to above, IL-12 is a key cytokine involved in pathogen clearance through regulatory effects on the production of interferon-γ, which is a potent activator of the macrophage microbicidal capacity (Trinchieri, 2003). Consistent with the above, wild-type mice elicit lower levels of IL-12 and interferon-γ, and display impaired clearance of *P. gingivalis* infection compared to mice that lack CR3 (Hajishengallis *et al.*, 2007). Similar results are seen after pharmacological blockade of CR3 with a specific antagonist (Hajishengallis *et al.*, 2007). Importantly, the interaction of *P. gingivalis* or its purified fimbriae with CR3 inhibits the ability of LPS from other bacteria, such as *A. actinomycetemcomitans*, to induce IL-12 and IFN-γ in mouse macrophages or human monocytes.

Moreover, the fimbriae of *P. gingivalis* can co-ligate the CXC-chemokine receptor 4 (CXCR4) and TLR2 in lipid rafts. This leads to crosstalk signalling that attenuates the killing capacity of macrophages in a cAMP-dependent way reminiscent of the C5aR-mediated mechanism discussed above (Hajishengallis *et al.*, 2008b) (Fig. 11.2). Intriguingly, the same immune evasion mechanism is exploited by pancreatic adenocarcinoma which, in lieu of fimbriae, expresses a specialized antigen (PAUF) for binding the CXCR4–TLR2 complex (Park *et al.*, 2010). Macrophages can be found in relatively low numbers in the gingival crevice but they are consistently present in the junctional epithelium, regardless of periodontal health status (Tonetti *et al.*, 1994). In addition, macrophages can interact with invading periodontal bacteria in the underlying connective tissue. When *P. gingivalis* fimbriae are released as components of OMVs, these CR3- and CXCR4-dependent evasion mechanisms could impact on the survival of additional periodontal species.

## What about adaptive immunity: Should the periodontal bacteria get scared?

Inflamed periodontal tissues contain organized lymphoid/myeloid aggregates rich in $CD4^+$ T cells, B cells, and antigen-presenting cells – mostly dendritic cells (DC) – which regulate the development of adaptive immune responses (Jotwani *et al.*, 2001). The histopathology of periodontitis shows that most stages of the disease include elements of both innate and adaptive immune responses (Gemmell *et al.*, 2002; Gaffen and Hajishengallis, 2008; Garlet, 2010). Indeed, destructive inflammation and periodontal bone resorption is the outcome of a complex interplay between a persisting biofilm and innate and adaptive immune cells (Fig. 11.3). However, the evasion tactics discussed above for periodontal pathogens involve innate immunity because microbial pathogens that successfully disable host defences preferentially target innate immunity (Finlay and McFadden, 2006; Hajishengallis and Lambris, 2011). In part, this is because innate defence mechanisms are those that are first encountered by pathogens. Additionally, these microbes may also undermine the overall host defence by subverting innate immunity given the instructive role of innate immunity in the development of the adaptive response (Iwasaki and Medzhitov, 2010).

In this regard, antigen-presenting cells, such as macrophages and especially dendritic cells, regulate the development of the various T cell subsets and thus can influence the type of T cell immunity. T helper type 1 (Th1) cells are primarily responsible for cell-mediated immunity to intracellular pathogens. Th2 cells, on the other hand, mediate humoral immunity (including production of IgE) and also activate mast cells, which mediate immune responses to helminths. The newly described Th17 subset mediates responses that reinforce innate immunity and protect against extracellular pathogens (Weaver *et al.*, 2007; Khader *et al.*, 2009). In addition to their protective potential, the same T cell subsets can mediate harmful responses. In general, Th1 cells have been implicated in inflammatory diseases, whereas Th2 cells are involved in allergic reactions (Lucey *et al.*, 1996). After the discovery of the Th17 subset, however, the role of Th1 in destructive inflammation has been questioned in some diseases, including rheumatoid arthritis (Cua *et al.*, 2003; Sato *et al.*, 2006). In this regard, recent

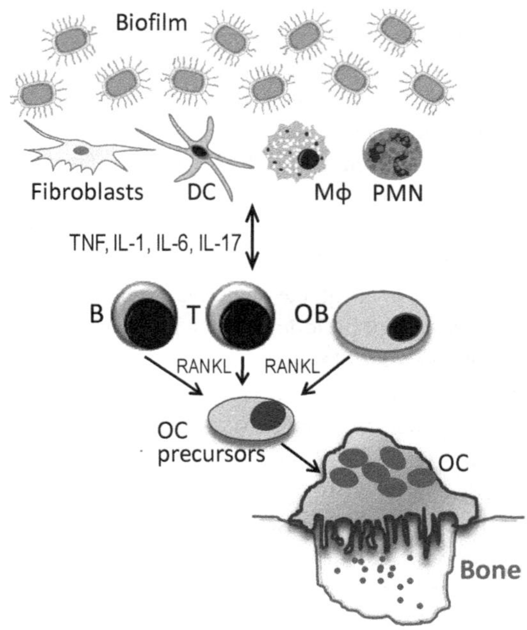


**Figure 11.3** Interplay of innate and adaptive immune players in inflammatory periodontal bone loss. The periodontal biofilm interacts with innate immune cells, including antigen-presenting cells, such as macrophages (Mϕ) and dendritic cells (DC), which in turn can activate T cells. The actual innate-adaptive cell interplay is considerably more complex than depicted and is bi-directional. For instance, IL-17 produced by T cells activates neutrophils (PMN) and fibroblasts which thereby contribute further to the inflammatory response. Inflammatory cytokines produced by innate and adaptive cells, such as TNF, IL-1, IL-6, and IL-17 stimulate the expression of receptor activator of nuclear factor-κB ligand (RANKL). Activated T and B cells, as well as osteoblasts, are major sources of RANKL by means of which they drive the maturation of osteoclast (OC) precursors into osteoclasts that cause bone erosion. These mechanisms operate in the context of unresolved periodontal infection and inflammation.

findings implicate Th17 as a specialized osteoclastogenic subset that links T cell activation to bone resorption (Sato *et al.*, 2006). Inasmuch as TLRs link innate to adaptive immunity and APCs receive cues from bacterial TLR agonists (Iwasaki and Medzhitov, 2010), periodontal bacteria can dictate the type of the T cell responses in the periodontium. Briefly stated, periodontal bacteria could also dysregulate adaptive responses and thereby disrupt periodontal host homeostasis.

The periodontal host adaptive response is highly complex and encompasses both protective and destructive elements (Gemmell *et al.*, 2007; Kinane *et al.*, 2007; Gaffen and Hajishengallis, 2008; Garlet, 2010). The destructive potential of the adaptive response in periodontitis is readily supported by animal studies. Indeed, severe combined immunodeficient (SCID) mice, which lack both T and B cells and cannot mount adaptive immune responses, develop reduced *P. gingivalis*-induced bone loss compared to immunocompetent mice (Baker *et al.*, 1994). With regard to T cells, it is the MHC class II-restricted $CD4^+$ T cells that are particularly involved, because $CD4^+$ T cell-deficient mice display greater resistance to *P. gingivalis*-induced bone loss than do normal mice or mice deficient in MHC class I-restricted $CD8^+$ T cells (Baker *et al.*, 1999). Specific implication of B cells is supported by a study showing that adoptive transfer of *A. actinomycetemcomitans*-specific B cells into rats infected with the same pathogen leads to increased periodontal bone resorption (Harada *et al.*, 2006). In principle, B cells can confer protection against periodontal biofilm bacteria through their role in humoral immunity. However, naturally induced antibodies to periodontal bacteria appear to be of low affinity and poor functionality (Gemmell *et al.*, 2002; Schenkein, 2006).

However, despite intensive research for almost four decades, the precise roles of the adaptive immune cells in periodontitis are poorly understood. Do they simply exhibit hyperactivation that is disproportionate to the bacterial challenge? Do periodontal bacteria proactively manipulate and subvert the adaptive response? Early studies in human periodontitis patients proposed that periodontal bacteria play an active role in suppressing T cell-mediated immunity (Ivanyi and Lehner, 1970; Evans *et al.*, 1989). This notion is consistent with recent experimental studies in the mouse model. Specifically, *P. gingivalis* exerts a predominantly immunosuppressive influence on gene expression in $CD4^+$ and $CD8^+$ T cells harvested from *P. gingivalis*-infected mice (Gemmell *et al.*, 2006). However, it is not known if these are direct effects on the T cells or are mediated through manipulation of APC.

The originally proposed Th1/Th2 paradigm (Mosmann and Coffman, 1989) has offered a productive conceptual framework for investigating the role of adaptive immunity in periodontitis, but is now clear that the disease cannot be adequately described in simple Th1 versus Th2 dichotomous terms (Gaffen and Hajishengallis, 2008). The recent discovery of the Th17 subset compels re-examination of the role of T cells in periodontal disease. In this context, cytokines characteristic of the Th17 subset have been found in inflamed periodontal tissue, suggesting a potential role in pathogenesis (Johnson *et al.*, 2004; Takahashi *et al.*, 2005; Vernal *et al.*, 2005; Lester *et al.*, 2007; Dutzan *et al.*, 2009; Ohyama *et al.*, 2009). Although IL-17 can stimulate short-term antimicrobial immunity (Yu *et al.*, 2007), chronic IL-17 receptor signalling can turn an acute inflammatory response into chronic immunopathology, as in rheumatoid arthritis (Gaffen and Hajishengallis, 2008; Lubberts, 2008). In this setting, IL-17 induces connective tissue destruction and bone resorption by inducing matrix metalloproteinases and the receptor activator of NF-κB ligand (RANKL), which directly stimulates maturation and activation of osteoclasts (Lubberts, 2008; Gaffen, 2009) (Fig. 11.3). These effects can be efficiently mediated by IL-17 and independent of IL-1β and TNF-α (Lubberts, 2008; Gaffen, 2009). In a periodontal context, IL-17 regulates matrix metalloproteinases and inflammatory cytokines in gingival fibroblasts (Beklen *et al.*, 2007), and *P. gingivalis* can stimulate IL-17 production from T cells *in vitro* (Oda *et al.*, 2003). The presence of IL-17 correlates with periodontal inflammation and bone loss in both humans and mouse models, although a cause-and-effect relationship has yet to be demonstrated (Lester *et al.*, 2007; Gaffen and Hajishengallis, 2008; Cardoso *et al.*, 2009; Ohyama *et al.*, 2009; Liang *et al.*, 2011).

Intriguingly, prostaglandin E2 (PGE2), which has been strongly associated with periodontal tissue destruction (Heasman *et al.*, 1993; Roberts *et al.*, 2004), enhances the expression of IL-23, a cytokine that contributes to the expansion of the Th17 subset (Sheibanie *et al.*, 2004). By contrast, PGE2 inhibits IL-12 and thus may impair Th1 development (Harris *et al.*, 2002). *P. gingivalis* is particularly potent in inducing PGE2 production (Noguchi *et al.*, 1996; Offenbacher and Salvi, 1999) and might be able to promote Th17 development at the expense of Th1. Because IL-17 upregulates COX-2 and PGE2 production (Gaffen and Hajishengallis, 2008), a positive feedback loop can be created between Th17/IL-17 and PGE2. This may cause disruption of homeostasis, if Th17 cells indeed play a destructive role through their proinflammatory and osteoclastogenic activities, whereas Th1 cells mediate protective immunity as supported by certain studies (reviewed by Gemmell *et al.*, 2002, 2007).

In summary, the adaptive response overall does not seem to control the periodontal biofilm but rather contributes to the destructive effects seen in periodontitis. This is not only supported by the mouse knockout studies discussed above (Baker *et al.*, 1994, 1999), but also by the observation that, in human periodontitis, T cells and B cells are the major contributors of RANKL in the bone resorptive lesions (Kawai *et al.*, 2006) (Fig. 11.3). Although the evidence is not as strong as with innate immunity, select periodontal pathogens may have an active role in manipulating adaptive immunity in ways that promote their adaptive fitness and collateral periodontal tissue damage.

## Summary and future perspectives

Despite its daunting complexity, periodontitis and its mechanisms can become conceptually 'simpler' if we think of the disease as disruption of homeostasis. This review has focused on the microbial factors that destabilize the homeostatic activation of periodontal immunity. Mounting evidence suggests that select periodontal pathogens, such as *P. gingivalis*, provide useful service to their residential communities by targeting and inhibiting critical antimicrobial responses, whereas they stimulate other inflammatory responses that they can tolerate. Interestingly, inhibitory as well as stimulatory actions promote the fitness of periodontal bacteria. The first is obvious and needs no additional explanation. Stimulation of inflammatory responses that do not kill (e.g. C5a-induced inflammation) may result in acquisition of essential nutrients (e.g. GCF-derived peptides and haemin). Thus, periodontal pathogens appear to have evolved in ways that allow them not only to endure inflammation, but also to exploit it for promoting their survival and, collaterally, causing periodontal tissue injury. Circumstantial evidence supports the potential of *P. gingivalis* to act as a keystone organism, i.e. one that could promote the survival and persistence of other members of the periodontal biofilm community. In this regard, the immune subversive tools of *P. gingivalis* (gingipains, atypical LPS, fimbriae) are released in readily diffusible OMVs and thereby become available to other biofilm organisms. In this way, even bacteria that lack protective mechanisms can thrive in the presence of *P. gingivalis*. A synergy of shared, communal mechanisms of immune evasion, colonization, and nutrient procurement should maximize the survival potential and virulence of the whole biofilm community. In this context, various combinations of *P. gingivalis* and other periodontal bacteria become more virulent in *in vivo* models of infection than are the individual organisms alone (Feuille *et al.*, 1996; Kesavalu *et al.*, 1998, 2007; Jenkinson and Lamont, 2005; Kinane and Hajishengallis, 2009; Polak *et al.*, 2009). In conclusion, microbial ecosystems containing *P. gingivalis* (and other potential key periopathogens) may perturb otherwise homeostatic host–bacterial interactions, thereby leading to non-protective and non-resolving chronic inflammation in the periodontium. Elucidation of the mechanisms by which select periodontal pathogens subvert immunity and promote the collective virulence of the subgingival biofilm would facilitate the rational design of therapeutic interventions against human periodontitis.

## Acknowledgements

The author is supported by grants from the U.S. National Institutes of Health DE015254, DE018292, DE017138, and DE021580.

## References

Bainbridge, B., Verma, R.K., Eastman, C., Yehia, B., Rivera, M., Moffatt, C., Bhattacharyya, I., Lamont, R.J., and Kesavalu, L. (2010). Role of *Porphyromonas gingivalis* phosphoserine phosphatase enzyme SerB in inflammation, immune response, and induction of alveolar bone resorption in rats. Infect. Immun. *78*, 4560–4569.

Baker, P.J., Evans, R.T., and Roopenian, D.C. (1994). Oral infection with *Porphyromonas gingivalis* and induced alveolar bone loss in immunocompetent and severe combined immunodeficient mice. Arch. Oral Biol. *39*, 1035–1040.

Baker, P.J., Dixon, M., Evans, R.T., Dufour, L., Johnson, E., and Roopenian, D.C. (1999). CD4(+) T cells and the proinflammatory cytokines gamma interferon and interleukin-6 contribute to alveolar bone loss in mice. Infect. Immun. *67*, 2804–2809.

Beertsen, W., Willenborg, M., Everts, V., Zirogianni, A., Podschun, R., Schroder, B., Eskelinen, E.L., and Saftig, P. (2008). Impaired phagosomal maturation in neutrophils leads to periodontitis in lysosomal-associated membrane protein-2 knockout mice. J. Immunol. *180*, 475–482.

Beklen, A., Ainola, M., Hukkanen, M., Gurgan, C., Sorsa, T., and Konttinen, Y.T. (2007). MMPs, IL-1, and TNF are regulated by IL-17 in periodontitis. J. Dent. Res. *86*, 347–351.

Cardoso, C.R., Garlet, G.P., Crippa, G.E., Rosa, A.L., Junior, W.M., Rossi, M.A., and Silva, J.S. (2009). Evidence of the presence of T helper type 17 cells in chronic lesions of human periodontal disease. Oral Microbiol. Immunol. *24*, 1–6.

Chapple, I.L., and Matthews, J.B. (2007). The role of reactive oxygen and antioxidant species in periodontal tissue destruction. Periodontol. 2000 *43*, 160–232.

Coats, S.R., Pham, T.T., Bainbridge, B.W., Reife, R.A., and Darveau, R.P. (2005). MD-2 mediates the ability of tetra-acylated and penta-acylated lipopolysaccharides to antagonize *Escherichia coli* lipopolysaccharide at the TLR4 signalling complex. J. Immunol. *175*, 4490–4498.

Coats, S.R., Jones, J.W., Do, C.T., Braham, P.H., Bainbridge, B.W., To, T.T., Goodlett, D.R., Ernst, R.K., and Darveau, R.P. (2009). Human Toll-like receptor 4 responses to *P. gingivalis* are regulated by lipid A 1- and 4'- phosphatase activities. Cell Microbiol. *11*, 1587–1599.

Cua, D.J., Sherlock, J., Chen, Y., Murphy, C.A., Joyce, B., Seymour, B., Lucian, L., To, W., Kwan, S., Churakova, T., *et al.* (2003). Interleukin-23 rather than interleukin-12 is the critical cytokine for autoimmune inflammation of the brain. Nature *421*, 744–748.

Darveau, R.P. (2009). The oral microbial consortium's interaction with the periodontal innate defense system. DNA Cell Biol. *28*, 389–395.

Darveau, R.P. (2010). Periodontitis: a polymicrobial disruption of host homeostasis. Nat. Rev. Microbiol. *8*, 481–490.

Darveau, R.P., Cunningham, M.D., Bailey, T., Seachord, C., Ratcliffe, K., Bainbridge, B., Dietsch, M., Page, R.C., and Aruffo, A. (1995). Ability of bacteria associated with chronic inflammatory disease to stimulate E-selectin expression and promote neutrophil adhesion. Infect. Immun. *63*, 1311–1317.

Darveau, R.P., Belton, C.M., Reife, R.A., and Lamont, R.J. (1998). Local chemokine paralysis, a novel pathogenic mechanism for *Porphyromonas gingivalis*. Infect. Immun. *66*, 1660–1665.

Darveau, R.P., Arbabi, S., Garcia, I., Bainbridge, B., and Maier, R.V. (2002). *Porphyromonas gingivalis* lipopolysaccharide is both agonist and antagonist for p38 mitogen-activated protein kinase activation. Infect. Immun. *70*, 1867–1873.

Davey, M.E., and Costerton, J.W. (2006). Molecular genetics analyses of biofilm formation in oral isolates. Periodontol. 2000 *42*, 13–26.

Delima, A.J., and Van Dyke, T.E. (2003). Origin and function of the cellular components in gingival crevice fluid. Periodontol. 2000 *31*, 55–76.

Diamond, G., Beckloff, N., and Ryan, L.K. (2008). Host defense peptides in the oral cavity and the lung: similarities and differences. J. Dent. Res. *87*, 915–927.

Dixon, B. (2009). Animalcules (ASM Press, Washington, DC).

Dixon, D.R., Bainbridge, B.W., and Darveau, R.P. (2004). Modulation of the innate immune response within the periodontium. Periodontol. 2000 *35*, 53–74.

Dutzan, N., Gamonal, J., Silva, A., Sanz, M., and Vernal, R. (2009). Over-expression of forkhead box P3 and its association with receptor activator of nuclear factor- B ligand, interleukin (IL)–17, IL-10 and transforming growth factor- during the progression of chronic periodontitis. J. Clin. Periodontol. *36*, 396–403.

Evans, R.I., Mikulecky, M., and Seymour, G.J. (1989). Effect of initial treatment of chronic inflammatory periodontal disease in adults on spontaneous peripheral blood lymphocyte proliferation. J. Clin. Periodontol. *16*, 271–277.

Feuille, F., Ebersole, J.L., Kesavalu, L., Stepfen, M.J., and Holt, S.C. (1996). Mixed infection with *Porphyromonas gingivalis* and *Fusobacterium nucleatum* in a murine lesion model: potential synergistic effects on virulence. Infect. Immun. *64*, 2094–2100.

Finlay, B.B., and McFadden, G. (2006). Anti-immunology: evasion of the host immune system by bacterial and viral pathogens. Cell *124*, 767–782.

Gaffen, S.L. (2009). The role of interleukin-17 in the pathogenesis of rheumatoid arthritis. Curr. Rheumatol. Rep. *11*, 365–370.

Gaffen, S.L., and Hajishengallis, G. (2008). A new inflammatory cytokine on the block: re-thinking periodontal disease and the Th1/Th2 paradigm in the context of Th17 cells and IL-17. J. Dent. Res. *87*, 817–828.

Garlet, G.P. (2010). Destructive and protective roles of cytokines in periodontitis: A re-appraisal from host defense and tissue destruction viewpoints. J. Dent. Res. *89*, 1349–1363.

Gemmell, E., Drysdale, K.E., and Seymour, G.J. (2006). Gene expression in splenic CD4 and CD8 cells from BALB/c mice immunized with *Porphyromonas gingivalis*. J. Periodontol. *77*, 622–633.

Gemmell, E., Yamazaki, K., and Seymour, G.J. (2002). Destructive periodontitis lesions are determined by the nature of the lymphocytic response. Crit. Rev. Oral Biol. Med. *13*, 17–34.

Gemmell, E., Yamazaki, K., and Seymour, G.J. (2007). The role of T cells in periodontal disease: homeostasis and autoimmunity. Periodontol. 2000 *43*, 14–40.

Gibbons, R.J., Socransky, S.S., Dearaujo, W.C., and Vanhoute, J. (1964). Studies of the predominant cultivable microbiota of dental plaque. Arch. Oral Biol. *11*, 365–370.

Gorr, S.U. (2009). Antimicrobial peptides of the oral cavity. Periodontol. 2000 *51*, 152–180.

Guenther, F., Stroh, P., Wagner, C., Obst, U., and Hansch, G.M. (2009). Phagocytosis of staphylococci biofilms by polymorphonuclear neutrophils: *S. aureus* and *S. epidermidis* differ with regard to their susceptibility towards the host defense. Int. J. Artif. Organs *32*, 565–573.

Gunther, F., Wabnitz, G.H., Stroh, P., Prior, B., Obst, U., Samstag, Y., Wagner, C., and Hansch, G.M. (2009). Host defence against *Staphylococcus aureus* biofilms infection: phagocytosis of biofilms by polymorphonuclear neutrophils (PMN). Mol. Immunol. *46*, 1805–1813.

Hajishengallis, G. (2009). *Porphyromonas gingivalis*–host interactions: open war or intelligent guerilla tactics? Microbes Infect. *11*, 637–645.

Hajishengallis, G. (2010). Complement and periodontitis. Biochem. Pharmacol. *80*, 1992–2001.

Hajishengallis, G., and Lambris, J.D. (2010). Crosstalk pathways between Toll-like receptors and the complement system. Trends Immunol. *31*, 154–163.

Hajishengallis, G., and Lambris, J.D. (2011). Microbial manipulation of receptor crosstalk in innate immunity. Nat. Rev. Immunol. *11*, 187–200.

Hajishengallis, G., Martin, M., Schifferle, R.E., and Genco, R.J. (2002). Counteracting interactions between lipopolysaccharide molecules with differential activation of Toll-like receptors. Infect. Immun. *70*, 6658–6664.

Hajishengallis, G., Shakhatreh, M.-A.K., Wang, M., and Liang, S. (2007). Complement receptor 3 blockade promotes IL-12-mediated clearance of *Porphyromonas gingivalis* and negates its virulence in vivo. J. Immunol. *179*, 2359–2367.

Hajishengallis, G., Wang, M., Bagby, G.J., and Nelson, S. (2008a). Importance of TLR2 in early innate immune response to acute pulmonary infection with *Porphyromonas gingivalis* in mice. J. Immunol. *181*, 4141–4149.

Hajishengallis, G., Wang, M., Liang, S., Triantafilou, M., and Triantafilou, K. (2008b). Pathogen induction of CXCR4/TLR2 cross-talk impairs host defense function. Proc. Natl. Acad. Sci. U.S.A. *105*, 13532–13537.

Hajishengallis, G., Wang, M., Krauss, J.L., Domon, H., Hosur, K.B., Liang, S., Magotti, P., Triantafilou, M., Triantafilou, K., and Lambris, J.D. (2010). Pathogen-instigated subversive crosstalk between complement and Toll-like receptors. Mol. Immunol. *47*, 2241.

Harada, Y., Han, X., Yamashita, K., Kawai, T., Eastcott, J.W., Smith, D.J., and Taubman, M.A. (2006). Effect of adoptive transfer of antigen-specific B cells on periodontal bone resorption. J. Periodontal. Res. *41*, 101–107.

Harris, S.G., Padilla, J., Koumas, L., Ray, D., and Phipps, R.P. (2002). Prostaglandins as modulators of immunity. Trends Immunol. 23, 144–150.

Hasegawa, Y., Tribble, G.D., Baker, H.V., Mans, J.J., Handfield, M., and Lamont, R.J. (2008). Role of *Porphyromonas gingivalis* SerB in gingival epithelial cell cytoskeletal remodeling and cytokine production. Infect. Immun. *76*, 2420–2427.

Hasturk, H., Kantarci, A., Goguet-Surmenian, E., Blackwood, A., Andry, C., Serhan, C.N., and Van Dyke, T.E. (2007). Resolvin E1 regulates inflammation at the cellular and tissue level and restores tissue homeostasis in vivo. J. Immunol. *179*, 7021–7029.

Heasman, P.A., Collins, J.G., and Offenbacher, S. (1993). Changes in crevicular fluid levels of interleukin-1 beta, leukotriene B4, prostaglandin E2, thromboxane B2 and tumour necrosis factor alpha in experimental gingivitis in humans. J. Periodontal. Res. *28*, 241–247.

Holt, S.C., and Ebersole, J.L. (2005). *Porphyromonas gingivalis, Treponema denticola,* and *Tannerella forsythia*: the 'red complex', a prototype polybacterial pathogenic consortium in periodontitis. Periodontol. 2000 *38*, 72–122.

Ivanyi, L., and Lehner, T. (1970). Stimulation of lymphocyte transformation by bacterial antigens in patients with periodontal disease. Arch. Oral Biol. *15*, 1089–1096.

Iwasaki, A., and Medzhitov, R. (2010). Regulation of adaptive immunity by the innate immune system. Science *327*, 291–295.

Jenkinson, H.F., and Lamont, R.J. (2005). Oral microbial communities in sickness and in health. Trends Microbiol. *13*, 589–595.

Johnson, R.B., Wood, N., and Serio, F.G. (2004). Interleukin-11 and IL-17 and the pathogenesis of periodontal disease. J. Periodontol. *75*, 37–43.

Jotwani, R., Palucka, A.K., Al-Quotub, M., Nouri-Shirazi, M., Kim, J., Bell, D., Banchereau, J., and Cutler, C.W. (2001). Mature dendritic cells infiltrate the T cell-rich region of oral mucosa in chronic periodontitis: In Situ, in vivo, and in vitro studies. J. Immunol. *167*, 4693–4700.

Kawai, T., Matsuyama, T., Hosokawa, Y., Makihira, S., Seki, M., Karimbux, N.Y., Goncalves, R.B., Valverde, P., Dibart, S., Li, Y.P., *et al.* (2006). B and T lymphocytes are the primary sources of RANKL in the bone resorptive lesion of periodontal disease. Am. J. Pathol. *169*, 987–998.

Kesavalu, L., Holt, S.C., and Ebersole, J.L. (1998). Virulence of a polymicrobic complex, *Treponema denticola* and *Porphyromonas gingivalis*, in a murine model. Oral Microbiol. Immunol. *13*, 373–377.

Kesavalu, L., Sathishkumar, S., Bakthavatchalu, V., Matthews, C., Dawson, D., Steffen, M., and Ebersole, J.L. (2007). Rat model of polymicrobial infection, immunity, and alveolar bone resorption in periodontal disease. Infect. Immun. *75*, 1704–1712.

Khader, S.A., Gaffen, S.L., and Kolls, J.K. (2009). Th17 cells at the crossroads of innate and adaptive immunity against infectious diseases at the mucosa. Mucosal. Immunol. *2*, 403–411.

Kinane, D.F., and Hajishengallis, G. (2009). Polymicrobial infections, biofilms, and beyond. J. Clin. Periodontol. *36*, 404–405.

Kinane, D.F., Peterson, M., and Stathopoulou, P.G. (2006). Environmental and other modifying factors of the periodontal diseases. Periodontol. 2000 *40*, 107–119.

Kinane, D.F., Demuth, D.R., Gorr, S.U., Hajishengallis, G.N., and Martin, M.H. (2007). Human variability in innate immunity. Periodontol. 2000 *45*, 14–34.

Kolenbrander, P.E., Palmer, R.J., Periasamy, S., and Jakubovics, N.S. (2010). Oral multispecies biofilm development and the key role of cell–cell distance. Nat. Rev. Microbiol. *8*, 471–480.

Krauss, J.L., Potempa, J., Lambris, J.D., and Hajishengallis, G. (2010). Complementary Tolls in the periodontium: how periodontal bacteria modify complement and Toll-like receptor responses to prevail in the host. Periodontol. 2000 *52*, 141–162.

Kumpf, O., and Schumann, R.R. (2008). Genetic influence on bloodstream infections and sepsis. Int. J. Antimicrob. Agents *32(Suppl. 1)*, S44–S50.

Lamont, R.J., and Jenkinson, H.F. (1998). Life below the gum line: Pathogenic mechanisms of *Porphyromonas gingivalis*. Microbiol. Mol. Biol. Rev. *62*, 1244–1263.

Lange, D., and Schroeder, H.E. (1971). Cytochemistry and ultrastructure of gingival sulcus cells. Helv. Odontol. Acta *15*, 65–86.

Lester, S.R., Bain, J.L., Johnson, R.B., and Serio, F.G. (2007). Gingival concentrations of interleukin-23 and -17 at healthy sites and at sites of clinical attachment loss. J. Periodontol. *78*, 1545–1550.

Liang, S., Krauss, J.L., Domon, H., McIntosh, M.L., Hosur, K.B., Qu, H., Li, F., Tzekou, A., Lambris, J.D., and Hajishengallis, G. (2011). The C5a receptor impairs IL-12-dependent clearance of *Porphyromonas gingivalis* and is required for induction of periodontal bone loss. J. Immunol. *186*, 869–877.

Lu, Q., Darveau, R.P., Samaranayake, L.P., Wang, C.Y., and Jin, L. (2009). Differential modulation of human b-defensins expression in human gingival epithelia by *Porphyromonas gingivalis* lipopolysaccharide with tetra- and penta-acylated lipid A structures. Innate Immun. *15*, 325–335.

Lubberts, E. (2008). IL-17/Th17 targeting: on the road to prevent chronic destructive arthritis? Cytokine *41*, 84–91.

Lucey, D.R., Clerici, M., and Shearer, G.M. (1996). Type 1 and type 2 cytokine dysregulation in human infectious, neoplastic, and inflammatory diseases. Clin. Microbiol. Rev. *9*, 532–562.

Meyle, E., Stroh, P., Gunther, F., Hoppy-Tichy, T., Wagner, C., and Hansch, G.M. (2010). Destruction of bacterial biofilms by polymorphonuclear neutrophils: relative contribution of phagocytosis, DNA release, and degranulation. Int. J. Artif. Organs *33*, 608–620.

Millin, D.J., and Smith, M.H. (1961). Nature and composition of dental plaque. Nature *189*, 664–665.

Mosmann, T.R., and Coffman, R.L. (1989). TH1 and TH2 cells: different patterns of lymphokine secretion lead to different functional properties. Annu. Rev. Immunol. *7*, 145–173.

Mydel, P., Takahashi, Y., Yumoto, H., Sztukowska, M., Kubica, M., Gibson, F.C., 3rd, Kurtz, D.M., Jr., Travis, J., Collins, L.V., Nguyen, K.A., *et al.* (2006). Roles of the host oxidative immune response and bacterial antioxidant rubrerythrin during *Porphyromonas gingivalis* infection. PLoS Pathog. *2*, e76 (712–725).

Nathan, C. (2002). Points of control in inflammation. Nature *420*, 846–852.

Newman, H.N. (1980). Neutrophils and IgG at the host–plaque interface on children's teeth. J. Periodontol. *51*, 642–651.

Niederman, R., Westernoff, T., Lee, C., Mark, L.L., Kawashima, N., Ullman-Culler, M., Dewhirst, F.E., Paster, B.J., Wagner, D.D., Mayadas, T., *et al.* (2001). Infection-mediated early-onset periodontal disease in P/E-selectin-deficient mice. J. Clin. Periodontol. *28*, 569–575.

Noguchi, K., Shitashige, M., Yanai, M., Morita, I., Nishihara, T., Murota, S., and Ishikawa, I. (1996). Prostaglandin production via induction of cyclooxygenase-2 by human gingival fibroblasts stimulated with lipopolysaccharides. Inflammation *20*, 555–568.

Oda, T., Yoshie, H., and Yamazaki, K. (2003). *Porphyromonas gingivalis* antigen preferentially stimulates T cells to express IL-17 but not receptor activator of NF-kappaB ligand in vitro. Oral Microbiol. Immunol. *18*, 30–36.

Offenbacher, S., and Salvi, G.E. (1999). Induction of prostaglandin release from macrophages by bacterial endotoxin. Clin. Infect. Dis. *28*, 505–513.

Ohyama, H., Kato-Kogoe, N., Kuhara, A., Nishimura, F., Nakasho, K., Yamanegi, K., Yamada, N., Hata, M., Yamane, J., and Terada, N. (2009). The involvement of IL-23 and the Th17 pathway in periodontitis. J. Dent. Res. *88*, 633–638.

Paine, R.T. (1969). The Pisaster-Tegula interaction: prey patches, predator food preference, and intertidal community structure. Ecology *50*, 950–961.

Park, H.D., Lee, Y., Oh, Y.K., Jung, J.G., Park, Y.W., Myung, K., Kim, K.H., Koh, S.S., and Lim, D.S. (2010). Pancreatic adenocarcinoma upregulated factor promotes metastasis by regulating TLR/CXCR4 activation. Oncogene *30*, 201–211.

Pihlstrom, B.L., Michalowicz, B.S., and Johnson, N.W. (2005). Periodontal diseases. Lancet *366*, 1809–1820.

Polak, D., Wilensky, A., Shapira, L., Halabi, A., Goldstein, D., Weiss, E.I., and Houri-Haddad, Y. (2009). Mouse model of experimental periodontitis induced by *P.*

*gingivalis/F. nucleatum* infection: Bone loss and host response. J. Clin. Periodontol. *36*, 406–410.

Popadiak, K., Potempa, J., Riesbeck, K., and Blom, A.M. (2007). Biphasic effect of gingipains from *Porphyromonas gingivalis* on the human complement system. J. Immunol. *178*, 7242–7250.

Power, M.E., Tilman, D., Estes, J.A., Menge, B.A., Bond, W.J., Mills, L.S., Daily, G., Castilla, J.C., Lubchenco, J., and Paine, R.T. (1996). Challenges in the quest for keystones. BioScience *46*, 609–620.

Ricklin, D., Hajishengallis, G., Yang, K., and Lambris, J.D. (2010). Complement: a key system for immune surveillance and homeostasis. Nat. Immunol. *11*, 785–797.

Roberts, F.A., and Darveau, R.P. (2002). Beneficial bacteria of the periodontium. Periodontol. 2000 *30*, 40–50.

Roberts, F.A., Houston, L.S., Lukehart, S.A., Mancl, L.A., Persson, G.R., and Page, R.C. (2004). Periodontitis vaccine decreases local prostaglandin E2 levels in a primate model. Infect. Immun. *72*, 1166–1168.

Ryder, M.I. (2010). Comparison of neutrophil functions in aggressive and chronic periodontitis. Periodontol. 2000 *53*, 124–137.

Sato, K., Suematsu, A., Okamoto, K., Yamaguchi, A., Morishita, Y., Kadono, Y., Tanaka, S., Kodama, T., Akira, S., Iwakura, Y., *et al.* (2006). Th17 functions as an osteoclastogenic helper T cell subset that links T cell activation and bone destruction. J. Exp. Med. *203*, 2673–2682.

Schenkein, H.A. (2006). Host responses in maintaining periodontal health and determining periodontal disease. Periodontol. 2000 *40*, 77–93.

Schroeder, H.E. (1973). Transmigration and infiltration of leucocytes in human junctional epithelium. Helv. Odontol. Acta *17*, 6–18.

Schroeder, H.E., and Listgarten, M.A. (1997). The gingival tissues: the architecture of periodontal protection. Periodontol. 2000 *13*, 91–120.

Sheibanie, A.F., Tadmori, I., Jing, H., Vassiliou, E., and Ganea, D. (2004). Prostaglandin E2 induces IL-23 production in bone marrow-derived dendritic cells. FASEB J. *18*, 1318–1320.

Slaney, J.M., and Curtis, M.A. (2008). Mechanisms of evasion of complement by *Porphyromonas gingivalis*. Front Biosci. *13*, 188–196.

Socransky, S.S., and Haffajee, A.D. (2005). Periodontal microbial ecology. Periodontol. 2000 *38*, 135–187.

Socransky, S.S., Haffajee, A.D., Cugini, M.A., Smith, C., and Kent, R.L., Jr. (1998). Microbial complexes in subgingival plaque. J. Clin. Periodontol. *25*, 134–144.

Takahashi, K., Azuma, T., Motohira, H., Kinane, D.F., and Kitetsu, S. (2005). The potential role of interleukin-17 in the immunopathology of periodontal disease. J. Clin. Periodontol. *32*, 369–374.

Tonetti, M.S., Imboden, M.A., Gerber, L., Lang, N.P., Laissue, J., and Mueller, C. (1994). Localized expression of mRNA for phagocyte-specific chemotactic cytokines in human periodontal infections. Infect. Immun. *62*, 4005–4014.

Tonetti, M.S., Imboden, M.A., and Lang, N.P. (1998). Neutrophil migration into the gingival sulcus is associated with transepithelial gradients of interleukin-8 and ICAM-1. J. Periodontol. *69*, 1139–1147.

Trinchieri, G. (2003). Interleukin-12 and the regulation of innate resistance and adaptive immunity. Nat. Rev. Immunol. *3*, 133–146.

Van Dyke, T.E. (2008). The management of inflammation in periodontal disease. J. Periodontol. *79*, 1601–1608.

Van Dyke, T.E., and Serhan, C.N. (2003). Resolution of inflammation: a new paradigm for the pathogenesis of periodontal diseases. J. Dent. Res. *82*, 82–90.

Vernal, R., Dutzan, N., Chaparro, A., Puente, J., Antonieta Valenzuela, M., and Gamonal, J. (2005). Levels of interleukin-17 in gingival crevicular fluid and in supernatants of cellular cultures of gingival tissue from patients with chronic periodontitis. J. Clin. Periodontol. *32*, 383–389.

Vitkov, L., Klappacher, M., Hannig, M., and Krautgartner, W.D. (2010). Neutrophil fate in gingival crevicular fluid. Ultrastruct. Pathol. *34*, 25–30.

Wagner, C., Zimmermann, S., Brenner-Weiss, G., Hug, F., Prior, B., Obst, U., and Hansch, G.M. (2007). The quorum-sensing molecule N-3-oxododecanoyl homoserine lactone (3OC12-HSL) enhances the host defence by activating human polymorphonuclear neutrophils (PMN). Anal. Bioanal. Chem. *387*, 481–487.

Wang, M., Krauss, J.L., Domon, H., Hosur, K.B., Liang, S., Magotti, P., Triantafilou, M., Triantafilou, K., Lambris, J.D., and Hajishengallis, G. (2010). Microbial hijacking of complement-Toll-like receptor crosstalk. Sci. Signal *3*, ra11.

Weaver, C.T., Hatton, R.D., Mangan, P.R., and Harrington, L.E. (2007). IL-17 family cytokines and the expanding diversity of effector T cell lineages. Annu. Rev. Immunol. *25*, 821–852.

Wingrove, J.A., DiScipio, R.G., Chen, Z., Potempa, J., Travis, J., and Hugli, T.E. (1992). Activation of complement components C3 and C5 by a cysteine proteinase (gingipain-1) from *Porphyromonas (Bacteroides) gingivalis*. J. Biol. Chem. *267*, 18902–18907.

Wood, S.R., Kirkham, J., Marsh, P.D., Shore, R.C., Nattress, B., and Robinson, C. (2000). Architecture of intact natural human plaque biofilms studied by confocal laser scanning microscopy. J. Dent. Res. *79*, 21–27.

Ximenez-Fyvie, L.A., Haffajee, A.D., and Socransky, S.S. (2000). Microbial composition of supra- and subgingival plaque in subjects with adult periodontitis. J. Clin. Periodontol. *27*, 722–732.

Yu, J.J., Ruddy, M.J., Wong, G.C., Sfintescu, C., Baker, P.J., Smith, J.B., Evans, R.T., and Gaffen, S.L. (2007). An essential role for IL-17 in preventing pathogen-initiated bone destruction: recruitment of neutrophils to inflamed bone requires IL-17 receptor-dependent signals. Blood *109*, 3794–3802.

# Oral Biofilms as a Reservoir for Extraoral Pathogens: Ventilator-associated Pneumonia

12

John G. Thomas


## Abstract

The link between oral flora and lung infections in mechanically ventilated (MV) patients (the 'oral systemic link') has always been circumstantial, based on clinical and nursing practices and preventative care. At the turn of the twentieth century, that began to change as molecular and imaging methods provided tools to better evaluate microbial exchange, refocusing on the importance of the endotracheal tube (ETT) lumen as a potential conduit, devoid of normal cellular defensive components.

We undertook the challenge in two phases, first engineering the adult-ventilator endotracheal lung (A-VEL) simulator to replicate the closed, bi-directional airway and stress of the intubated patient in the intensive care unit (ICU). Of singular importance, was the incorporation of multiple quantitative imaging techniques to define the 3D biofilm luminal development in Stages (I–IV) from single to complex microbial communities and the incorporation of dental pathogens (*Streptococcus mutans* and *Porphyromonas gingivalis*) as an endogenous event, in the preconditioning of the ETT luminal surface, an abiotic medical device.

The second phase shifted to the *in vivo* environment and in multiple clinical studies, unmasked the biphasic nature of ETT luminal colonization, oral – endogenous (early) – to systemic – exogenous (late) – at a 3–5 day 'switch'. Further, the ICU studies dramatized the shift from the infectious process in VAP to work of breathing (WOB), where the former occurred in 16%, while biofilm accretion, occlusion of the ETT lumen and increased airway resistance occurred in 100% of intubated patients.

Most recently, we have used both 16S (microarray) and 18S rRNA (pyrosequencing) to redefine the proportions of bacteria and fungi from oral reservoirs, and been astonished by the richness and diversity of the oral fungal community in the ETT accretion occlusion, often yielding > 15 species.

Endotracheal tube associated pneumonia (EAP) management continues to elude optimal strategies, but the use of selected oral probiotics coupled with better oral care in both the ICU and admitting institutions is gaining reinforcement: 'oral stewardship'. Further, the utilization of dental professionals in the ICU has importance, as has the recognition that the next fertile area of airway disease (oral to systemic) study is the neonatal intensive care unit (NICU), where 50% of newborns may be intubated and develop EAP with no teeth. How? Why?


## Introduction

Ventilator associated pneumonia (VAP) has a tortuous history, highlighting the dangers and unrecognized consequences of medical intervention. The intent of mechanical ventilation is to assist the patient in breathing, applying the physical principles of forced airway breathing for supportive care. The introduction of positive pressure ventilation, forcing air into the lungs through an endotracheal tube, marked a dramatic improvement over the iron lung. Given the ubiquitous nature of microbes and the unequalled ability to contaminate hydrated surfaces, the advantages of positive pressure mechanical ventilation were often compromised by the disadvantages of microbial intervention.

Fig. 12.1 outlines this history of VAP and its association with a rather difficult learning curve for medical and dental practitioners and health care providers, particularly in the field of respiratory therapy (RT). First reports of a ventilator associated respiratory infection were linked to aerosolization and water contamination. *Pseudomonas* was an early culprit. As improvements in ventilation and the means of carrying water vapour instead of water droplets improved, it was assumed the inherent problems with water contamination would decrease; they did not.

Hence, the focus in the middle 1980s and early 1990s was colonization from the patient's normal flora, particularly in the gastrointestinal tract. That dynamic flora became the residual target for research and the emphasis for therapy focused on reducing the gastrointestinal flora with antibiotic paste or systemic antibiotics. Attention also turned to nasogastric feeding tubes and the role of motile microbes, which paralleled the increasing use of mechanical ventilation for the at-risk population, including nursing homes (Adachi *et al.*, 2002).

Ultimately, in the early 2000s it became apparent that a major unrecognized reservoir for VAP pathogens was the oral flora and that the potential colonization of the lumen of an indwelling medical device, the endotracheal tube (ETT) (Vassilakopoulos, 2009). Indirect evidence began to support the hypothesis that the oral flora plays a major role in VAP, and preliminary investigations aimed to reduce or modify the normal oral flora through oral hygiene. Research was particularly directed

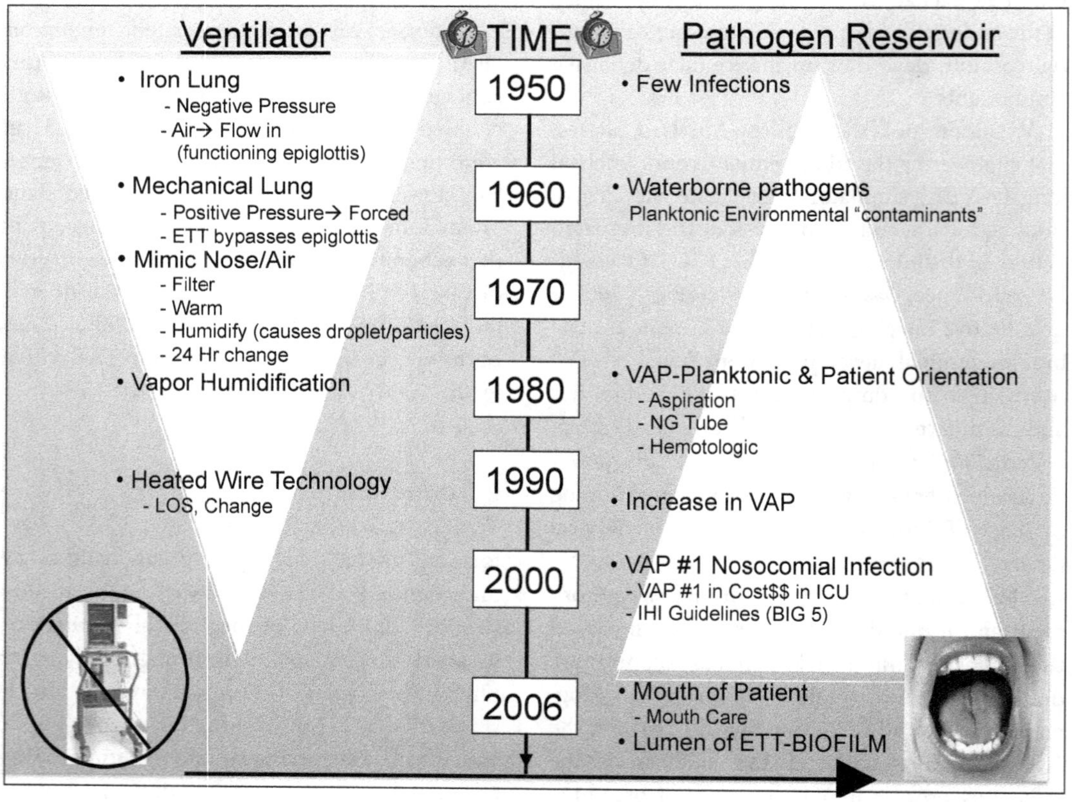


**Figure 12.1** VAP pathogenesis: transition from ventilator to oral flora over time. Early emphasis focused on ventilatory inoculum while later evolution recognized the importance of microbial colonization from the oral flora. Hence the recognition that VAP was a misnomer for EAP (endotrachea-associated pneumonia).

towards an adult ICU population, emphasizing the role of pre-colonization of oral flora in the elderly or nursing home residents prior to admission (Thomas *et al.*, 2007).

In parallel with this unmasking of the importance of oral micro-organisms, was the exponential growth of medical microbiology and the appreciation that indwelling medical devices present a significant risk of infection. Data from a variety of authors show that usage per year varied from 5 million for ventral venous catheters to 400,000 cardiac pacemakers with infections at 3–8% and 1.5% respectively; overall device associated and biomaterial associated infection range from 5–15% but was reported as high as 100%. Interestingly, the infection risk of dental implants ranges from 5% to 10% and, although not listed, VAP ranges from 3% to 19%. The numbers were astounding, the usage significant, the infection risks ranging from 1–30%, and the organisms on these materials represented an entirely new phenotype in the medical community, a biofilm. Biofilms were recalcitrant to standard therapy and were undetectable or unrecoverable by standard Pasteur driven laboratory protocols of the 1880s. Biofilms demanded an entirely new approach for medical microbiology, and uncovered the term VBNC, or viable but non-culturable.

This unparalleled rediscovery of medical microbiology also catalysed a re-description of the normal flora and its importance in colonization of devices. In its simplest form, the normal flora could be ascribed to four anatomic locations: gastrointestinal, genitourinary, oral, and skin. These reservoirs represented overlapping ecosystems, each with a unique gradient of growth dependent upon selected features of the anatomical hierarchy, particularly carbohydrate energy, stress, pH and Eh, or oxidation reduction potential. The resident flora of each reservoir was unique to an individual, but was relatively stable and represented a barrier to colonization of exogenous pathogens.

In an analogous manner, the oral cavity with its multitude of microbial 'niches' can be divided into four basic regions: buccal and lingual mucosa, teeth, dorsum of the tongue and palate. In addition, we have shown that the sinus tract is a haven for biofilm colonization, and thus may be a potential reservoir for pathogens involved in VAP.

Of immediate benefit, was the unmasking of the inhabitants of the oral flora when molecular microbiology was picked up by the dental community. It was recognized that more than 70% of the mouth was biotic with sloughing tissue, although oral hygiene was directed largely towards abiotic enamel surfaces, representing less than 30% of the total surface area. Furthermore, the communication with the sinus tract became more widely appreciated and the reinoculation from this reservoir to the oral flora was recognized by us and others (Garcia, 2005; ISI Guidelines, 2006; and see the website: http://www.hsc.wvu.edu/som/Pathology/Thomas/Videos/miniMicroOralBiofilms/miniMicroOralBiofilms.html).

The interactive flora of the sinus tract and the oral cavity is a dynamic ecosystem for as many as 900 different phylotypes. Prior to the molecular era, our understanding of the microflora was limited to organisms identified by traditional culture methods. Molecular identification techniques uncovered a plethora of new species and led to the realization that the less well-characterized organisms within the microbiota, were in fact, as important to the stability of the oral flora as any other feature.

Ventilator associated pneumonia is still the most expensive disease in the intensive care unit (ICU). Even with the best attempts at controlling infection, VAP remains a significant player in hospital costs and resource utilization (Unroe *et al.*, 2010). Approximately 50% of the ICU resources are directed towards the mechanically ventilated patient. In total, the financial burden of VAP in the US is approximately $5000 to $6000 per day. It is generally acknowledged that the average stay in the ICU with mechanical ventilation currently costs $40,000. Although exact figures depend on the method of assessment, VAP is clearly one of the four most costly diseases in the US Health Care System. Different tabulations highlighted these values: VAP ($40,212), central line associated blood strain infections ($36,000), surgical site infections ($25,546) and catheter associated UTI ($1006) (Rello *et al.*, 2002; Thomas *et al.*, 2010).

Mechanical ventilation in the ICU represents 20–30% of all hospital costs and 7–8% of total health care in the US. The Centre for Disease

Control (CDC)'s 2009 National Healthcare Safety Network (NHSN) Report stated that the incident of pneumonia in the ventilated mechanical surgical patient was as high as 53% to 64%, mortality and morbidity associated with the development of pulmonary infection was also high with the mortality rate ranging 20% to 41%.

## Biofilms, ETT and staging (I–IV)

One of the major problems with VAP is the lack of a uniform standard in classifying its characteristics and therefore tracking its frequency (Shorr *et al.*, 2011). This too has evolved. The most recent recommendations for defining VAP (available at http://www.cdc.gov/nhsn/PDFs/pscManual/6pscVAPcurrent.pdf) include key characteristics as follows:

- A diagnosis of pneumonia is made, based on a combination of radiology, clinical, and laboratory criteria (physician treatment of pneumonia is not sufficient for diagnosis)
- Patient was intubated and mechanically ventilated (MV) at time of, or within, 48 hours before onset of the event.

The cost associated with VAP is not just related to the adult population. As the supportive care for the premature and/or neonatal ICU population expanded, the cause of VAP in children demanded stabilization and tracking of numbers, and a clear definition. Today, premature infants are ventilated significantly longer than adults. Although neonates do not have an established abiotic surface or tooth microbial reservoir, there is evidence that the oral cavity serves as a reservoir for VAP pathogens here too. Clearly, therefore, biotic mixed surfaces in the oral cavity provide significant microbiota for lung inoculation. Perhaps the most important finding associated with VAP is that it is a biofilm associated disease and that the lumen of the ETT is commonly colonized by biofilm material originating in the oral cavity.

Fig. 12.2 highlights the significant features of ETT position and its significance in establishing a microbial flora. In the adult population ETTs are sized in one-half mm increments, ranging from 9.0 mm in diameter to 5.5 mm, the majority using between 7 mm and 8 mm. The ETT is positioned just 2 mm above the bifurcation of the bronchus and the cuff inflated at a pressure that maximizes resistance to fluid movement, but recognizes the

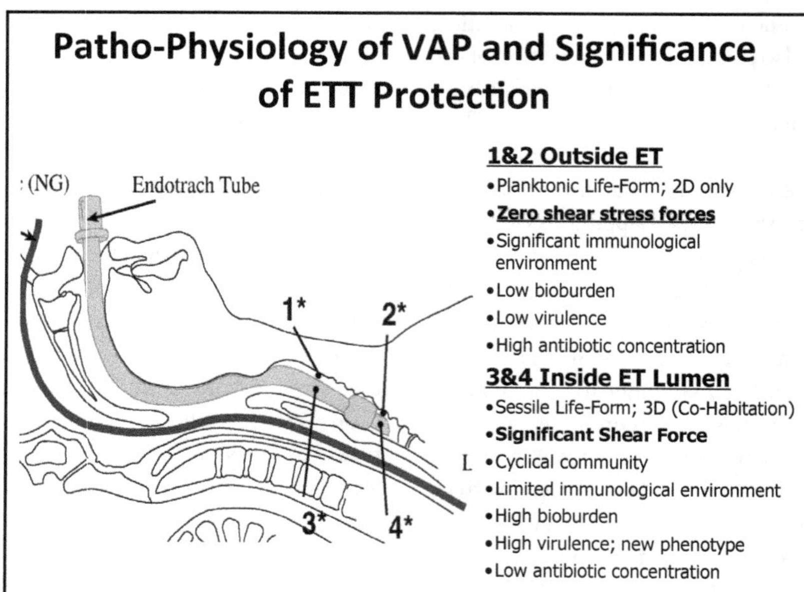


**Figure 12.2** Pathophysiology of VAP and significance of ETT. Protection: inside (3 and 4) vs. outside (1 and 2). Characteristics of microbial contamination inside and outside the ETT are described.

consequences of significant pressures on the trachea or resulting erosion. The cuff itself has been a singular activity of focused research with a variety of shapes, pressures, and materials used to achieve optimum closure.

Historically, the concept of biofilms was well recognized in the dental community as a part of dental plaque architecture and in engineering as a major influence of filtration efficiency linked to reduced energy or heat transfer in selected devices. Engineers knew of biofilm in air-conditioning units and water filtration; in fact, the word 'biofilm' was coined in the early 1980s by the engineering community and all original literature describing the properties of biofilm were in the engineering literature.

In fact, biofilms are the legacy of microbiology; the original environment of the forming earth was inhabited only by microorganisms in organized communities, where 'sociomicrobiology' was a defence mechanism that allowed for survival. Evidence of biofilms in frozen icecaps (Hall-Stoodley *et al.*, 2004) suggests that they were prevalent 4.3 billion years ago, whereas the planktonic microbes are a free floating form, or phenotype, evolved about 2.4 million years ago. Biofilms have been characterized and described for a variety of organisms and diseases in a number of reviews (Donlan and Costerton, 2002; Fux *et al.*, 2003). What makes biofilm so unique for VAP are its physical characteristics, and its ability to withstand significant pressures and stresses associated with respiratory therapy. Biofilms in VAP are usually multispecies where the sum of the whole is more dramatic than individual mono species biofilms. Hence, VAP biofilms have extracellular matrices composed of *hydrated polymers* and their physical properties include high *viscoelasticity* and *rheology*. Together these features enhance the microbes' ability to withstand and advance their population on an abiotic surface under significant stress. The property of viscoelasticity allows the biofilm to absorb shear forces and to rebound elastically, whilst over the long term, shear is dissipated through viscous flow or alternatively the biofilm is streamlined to reduce drag. Rheology is the physics underlying this flow, originally described by Greek Philosopher Herculanous Ephesus (540 to 480 BC), who stated, 'every thing flows and nothing abides; every thing gives way and nothing stays fixed.'

An important pathological feature of the biofilm is that disease progress is associated with different stages throughout the development of the biofilm. We began to recognize the importance of staging in our early research with recalcitrant colonies when, in our 50 patient assessment of luminal ETT, outcomes were associated with a Stage IV. Stages of biofilms are shown in Table 12.1 and Fig. 12.3. Further research by us highlighted the significance of staging when correlated with additional ventilatory parameters (Darouiche, 2001).

Biofilm 3D Structure analysis was enhanced by the use of a Poloxamer, a co-polymer of poloxyethylene-polyoxproylene, which undergoes a thermally reversible gelation. This material acts as a 'scaffold' supporting bacteria in three dimensions and enhancing biofilm formation due to close location of bacteria to one another. We use F-127 (BSF-Germany) at 30% varying slurry at 4°C and incubating the either single or mixed species inoculum at 37°C to induce biofilm formation. Original studies were performed at Purdue University, and focused on refractive-CLSM for systemic pathogens (Thesis, Stephen Sharpe, 2005).

We have identified four Stages in the growth pattern of the organisms within the micro-niches of the biofilm; these are defined by individuals of that community, but follow the lag-log – stationary – death batch culture growth pattern and are cyclical in nature. This growth cycle, however, clearly incorporates a planktonic form as well as the biofilm form and for simplistic sense, we have established a phase zero (T-0) for the planktonic microbe within the community (Fig. 12.3). All microbial communities have both the biofilm and planktonic organisms. Ultimately, pathogenicity is associated with prevalence of the biofilm as the major component of that microbiota. Lately, we have reported that the ratio of phenotypes, biofilm to planktonic, may be a useful biomarker for disease status (Thomas *et al.*, 2010), particularly in chronic wounds.

The ratio biofilm to planktonic can be defined by two themes. First, the ratio of the MICs of the two phenotypes can be determined by measuring

**Table 12.1** Biofilm stages of a cycling community 'form follows function'

| Stage | Form | Features of structure | Function |
|---|---|---|---|
| I | Loose association with surface | Bacteria reversibly attach to solid support | Attachment in region of survival |
| II | Robust adhesion | Surrounded by extra cellular polymeric substances (EPS)/'slime' | Enable biofilm to withstand high flow/high shear forces |
| III | Co-aggregation of cells into segregated domains with QS | Viscoelastic, cells form layers, mushrooms and quorum sensing | Nutrients diffuse in and waste products diffuse out via channels<br>Needs are communicated |
| IV | Sloughing and shedding<br>Return to motility | Maximum thickness is exceeded<br>Metastasis of the 'community' when sheer force exceeds their tensile strength | Community restarts in a new location via planktonic transmission |

Biofilm stages I–V recognize that I and IV are reversible and significant targets for intervention. This table complements Fig. 12.3.

the biofilm MIC using F-127 poloxamer-grown communities, and the planktonic MIC on Mueller–Hinton agar using a template of antimicrobials for wound infections. In preliminary studies, we have found that a ratio of greater than five is associated with poor clinical outcomes. Secondly, Congo Red can be used to measure the presence of EPS (Extra Polymeric Substances) during the analysis of *Staph. aureus* colony counts. Congo Red-positive strains compared with Congo red-negative strains of *Staph. aureus* enabled another ratio to be established. Here, a value of greater than 2 was found to correlate with poor outcomes. This approach is now being established for periodontal disease, given that the emerging opinion is that periodontal disease is an oral wound mimicking chronic wounds of systemic sites in many characteristics, or a wound is a 'toothless periodontal pocket'.

This inability to treat VAP and its biofilm compatriot has led to some unique strategies and designs for combatting the disease. One major concern is the transmission of multidrug resistance by the planktonic population; hence, one strategy involves 'crop rotation'; i.e. the sequential application of several different courses of antibiotics in order to eliminate resistant microorganisms. In our institution, we have measured the impact of 'crop rotation' and the use of limited exposure for a specific timeframe to a particular ICU antimicrobial. The uses of antimicrobial to treat VAP in the ICU is a Pharmacy and Therapeutics decision, not that of the individual physician. We reported a summary of our results in a recent publication (Sarwari *et al.*, 2008).

Complementary and/or adjunctive interventions have also been studied. The most recognized in the US is the IHI Institute of Health Care Improvement Guidelines (www.ihi.org/knowledge), which were first formulated in 2007. These emphasized not antimicrobial anti-infective approaches, but rather focused on better patient management based on the physical demands of respiratory therapy and mechanical ventilation during VAP. The four adjunctive non-medical interventions were directed towards major features of VAP and past strategies. These included:

- 30–45° head elevation
- daily 'sedation vacation'
- peptic ulcer disease prophylaxis
- deep venous thrombosis prophylaxis.

## Pathophysiology of VAP/airway disease/modelling

By 2002 it had become quite apparent that VAP was much more complicated process than that socially attributable to the ventilation; in fact, focus was addressing the lumen of the ETT as a

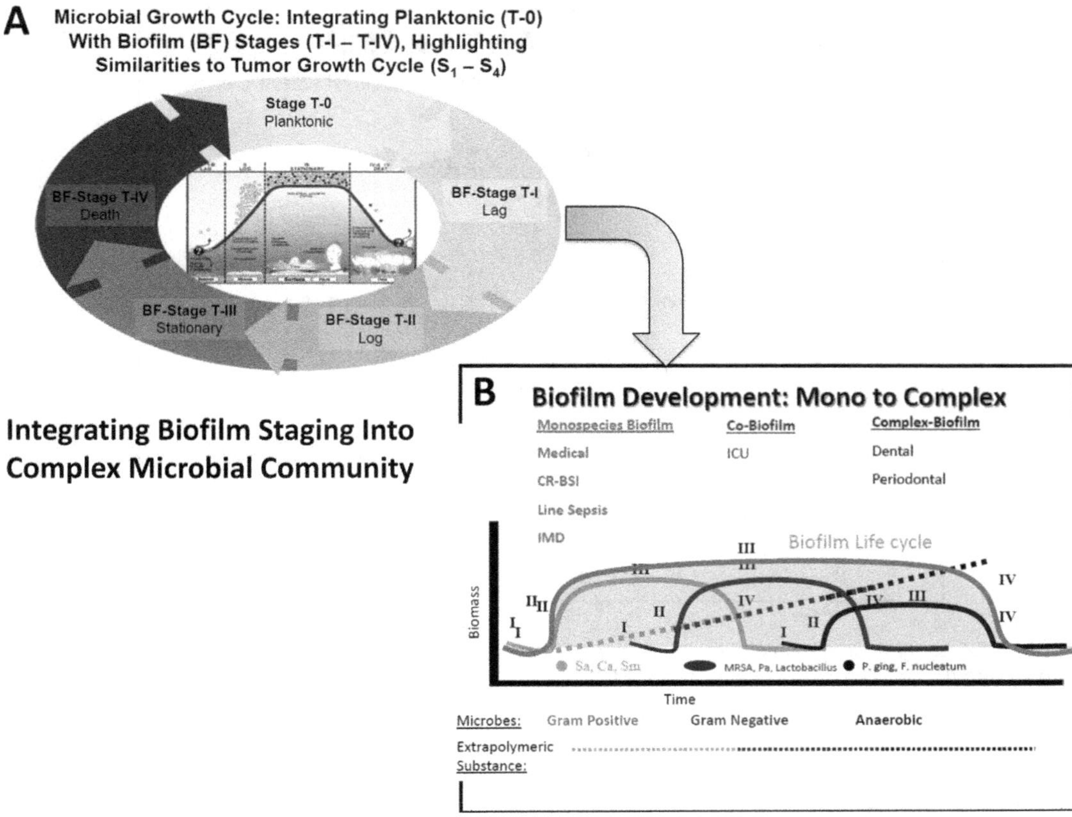


**Figure 12.3** (A) Microbial growth cycle: integrating planktonic (T-0) with biofilm stages (T-I to T-IV). (B) Biofilm development: mono to complex – integrating, multiple species. In periodontal biofilms, the microflora shifts from predominantly Gram-positive to Gram-negative. Each development integrates the life cycle T-0 to T-IV. This figure complements Table 12.1.

potential compliment or primary reservoir for microbial colonization (Kollef, 2006).

We were very interested to link the significance of the luminal airway with the development of a biofilm; and the impact of physical features in the development of that biofilm attributable to high stress of ventilation (FORM = FUNCTION or STRUCTURE = FUNCTION). Hence, in 2002, we organized a Biofilm Translational Research Laboratory for studies in medicine, dentistry, and industry with an emphasis on VAP biofilms (www.hsc.wvu.edu/som/pathology/thomas).

## Model building

It was quite apparent to us that the models evaluating stress and features associated with mechanical ventilation had not been developed. We wanted to develop a model or simulator that duplicated the closed airway environment that was bidirectional of the mechanically ventilated patient. Our initial design utilized eight characteristics (Fig. 12.4) that defined the environment in MV of a ETT biofilm, but we needed two additional items, namely (1) to provide clear evidence that the lumen is a reservoir for biofilm, and (2) to define the parameters used in ventilation, particularly associated with VAP. These could then be used to establish key features of the model, mimicking the actual forces and environment of the bidirectional mechanical ventilation.

Hence in 2003, we analysed 50 consecutive extubated ETTs from West Virginia University Hospitals. The ETTs were divided into three equal sections, A, B, and C, where A was the oral end, B the middle, and C the lung end recognizing in concert with studies at the Lane School of Engineering, WVU, that A and C forces were parallel, whereas the centre section of the ETT received

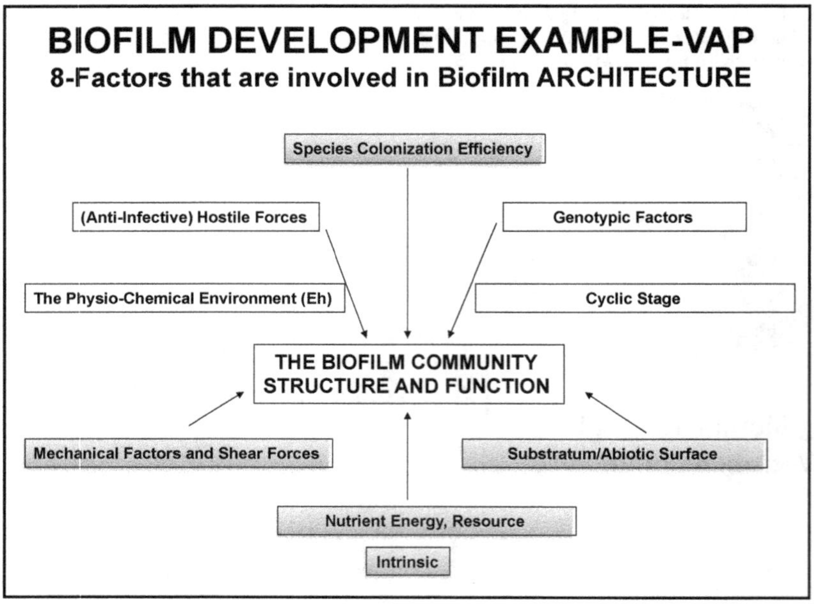


**Figure 12.4** Biofilm development in the endotracheal lumen: eight factors that are involved in biofilm architecture (the most significant are darkened).

forces attributable to vortexing. We simultaneously collected the 18 ventilatory parameters for each of the 50 patients with 22 clinical parameters.

Results were astounding and clearly defined the parameters for ventilation, the forces associated with those parameters, and the remarkable diverse three-dimensional structure of a heretofore unrecognized biofilm within the lumen. From the clinical parameters, a three-scheme event was established recognizing three patient characterizations; Category I would be acute care and short-term ventilation (primarily as a support to elective surgery); Category II ventilator parameters for a patient with chronic obstructive pulmonary disease (COPD); and, Category III, for a patient who had acute respiratory disease syndrome (ARDS).

Clinical studies and SEM highlighted another growing theme, disease outcome with VAP could be correlated with Staging of the biofilm within the lumen and it was our first attempt recognizing this important characterization with pathogenicity. The application of the Staging has further used in correlating patient outcomes of VAP. In a study of 11 woman and 21 men with a mean age of 50 and on MV in the ICU for 13 days with 8 of 10 patients with a Stage IV biofilm had pneumonia. There was a relationship between increase in biofilm staging and incidences of pneumonia ($P$ is less than 0.05). Interestingly, there was no relationship relative to duration of intubation, patient age, or hospital stay with biofilm stage (Wilson *et al.*, 2012).

The adult ventilator-ETT-lung (A-VEL) model ultimately, then, was constructed recognizing three features and zones that would allow for verification and validation (Fig. 12.5). Zone X included the parameters of ventilation utilizing the Puritan Bennett 720; Zone Y were the parameters of ETT assessment in an anatomically correct head and microbial inoculation at $10^6$ CFU/ml; and, Zone Z was the Michigan Bi-lobed Lung with adjustable resistance forces. When coupled with ventilation, it could duplicate any bidirectional airway of the ventilated patient.

Following multiple studies with the A-VEL model and assessment of inoculation, several additional features were added which included: (1) Potential nebulization at either end of the ETT allowing for dual inoculation or administration of aerosolized anti-infectives duplicating again the airway management of a VAP patient,

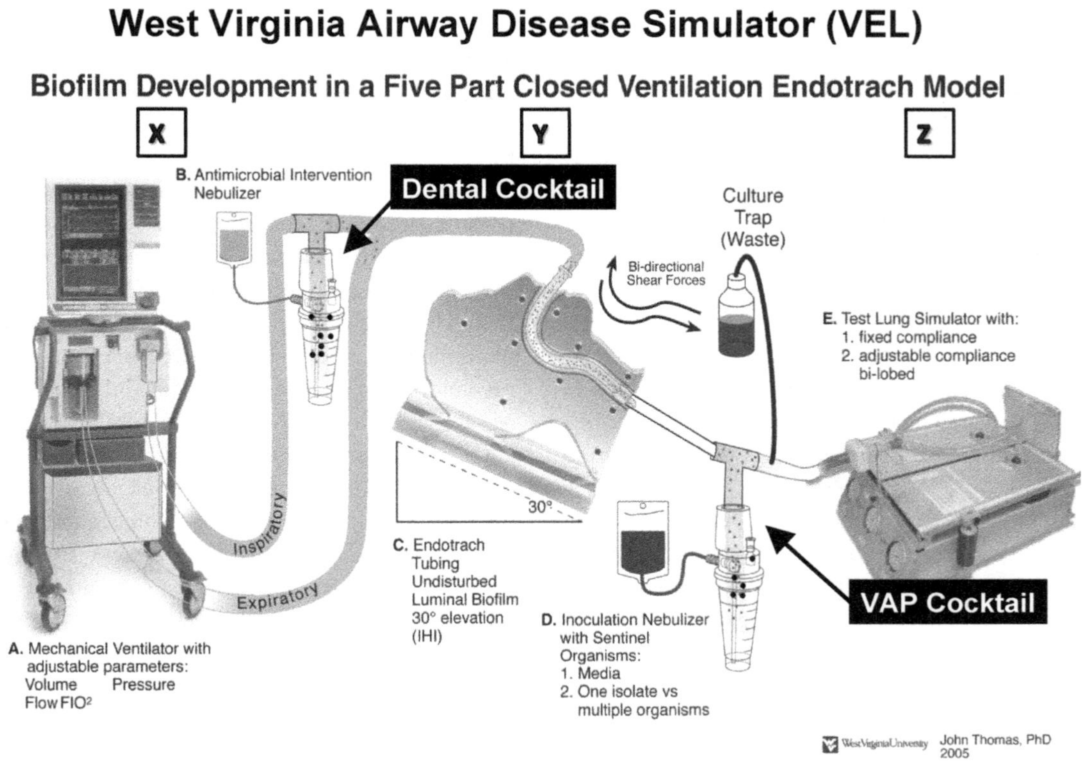


**Figure 12.5** (A) Integrated VAP model: biofilm development in a five-part engineered closed ventilator endotracheal lung model using existing MV supplies. (B) Modification and updates of VEL by zone (X/Y/Z) parallelling changes mandated by IHI and patient therapy. Inclusion of dental/oral microbes, *Strep. mutans* and *P. gingivalis* were added.

and (2) insertion of a filter distal to the end of the ETT allowed for measurement of aerosolized material addressing the potential consequence of Stage T-IV and the potential recovery of organisms that would be ventilated to the lung with consequences of alveolar colonization. We are very interested in the size of these particles given that alveolar space generally is restricted to 5 mm and it was important to define the distribution of the breakdown products, a key feature of biofilm to planktonic and back to biofilm strategy and the cyclical nature of staging of a biofilm (Krivit and Heuertz, 2011). Later modifications included, at the request of IHI Bundle, elevation of the head to 30°C, heating with a blanket of the Michigan Lung to 35°C, and five additional complete X, Y, X VEL 'stations' for simultaneous evaluations of six ETTs, which were generally run for 14 days.

Organism inoculation (CFU/ml) was also critical and we wanted to mimic the environment of the ventilated patient addressing its biphasic nature going from predominantly Gram-positive organisms to a Gram-negative-dominated microflora. Hence, initial studies utilized the concept of a simple-to-complex biofilm and the organisms used included: *Staph. aureus, Candida albicans,* MRSA, *Pseudomonas aeruginosa, Stenotrophomonas maltophila.* Other studies showed that day 5 was a significant breakpoint and the movement of the biofilm from a co to a complex was often amplified by the presence of *Candida albicans,* which we consider to be a 'universal co-aggregate' and a remarkable bridge builder in the development of a complex biofilm.

By 2005, significant publications had uncovered the importance in studies of the dental flora and its contribution via plaque accessibility to the importance of the luminal ETT microbiota. Studies by Seok-Mo *et al.* (2008) using molecular methods had highlighted the growing awareness of a link between dental flora and VAP. Using our A-VEL Model, we evaluated the significant of dental flora. This was done by colonizing the ETT for 48 hours with *Strep. mutans*, *Lactobacillus casei* and *F. nucleatum* prior to the addition of three-organism pool for 48 hours used in the initial studies with our model: *Staph. aureus*, *C. albicans*, and *P. aeruginosa*. Use of a pre-inoculum of dental flora enhanced the adherence of the traditional three-member complex biofilm.

Of added significance was the inclusion of an ETT provided by Covidien that contained various concentrations of silver as a glass anomer varying from 0.075% weight volume to 30%. A 7.5% silver inhibited the dental pre-colonization and reduced the subsequent co-biofilm formation (Fig. 12.6).

A summary of the significant results of the Covidien studies (2004–2007) using silver against the five-member complex biofilm emphasized two features. One, the growth curve of the complex biofilm occurred in essentially two phases. The silver impeded or inhibited significant growth for the first 5 days and although the growth still was reduced after 5 to 7 days compared to controls, the biofilm continued to increase in a significant manner to day 14. Further, it was first order kinetics: 0.075% to 0.75% silver had no impact, where 7.5%, 15%, and 30% had essentially the same reduction. Hence, a concentration gradient once obtained, was efficacious throughout and suggested that an optimum value could be obtained for the silver release. Based on evidence from the 50 patient study, where 90% of the patients were off ventilation by 14 days, most studies were run for the 14-day profile (Kollef *et al.*, 2008).

Our data assessment was enhanced significantly by the combination of multiple image analysis methods (Table 12.2). Image analysis included bright field microscopy, confocal laser scanning microscopy, SEM, and freeze-fracture SEM. Luminal ETT biofilms were analysed undisturbed, whereas colony counts were acquired from scraping, sonicating, and in vortexing the aggregates of the biofilm. Comstat software provided our major metric in comparing the activity of silver in the Covidien described ETTs. Comstat quantified biofilm image stacks acquired by CLSM (confocal laser scanning microscopy) (www.dtu.dk/comstat). Of the nine quantitative analyses available, we emphasized five: biovolume, biothickness, roughness coefficient, surface to volume ratio and volume of micro-colonies to the substratum.

Lastly, our most recent studies have incorporated into the A-VEL model the use of bioluminescent bacteria (Calliper Life Sciences), using the IVIS system (www.Caliperls.com). Here the goal of the studies was to pulse, for 6 hours, bioluminescent biofilm associated bacteria into pre-colonized extubated ETTs placed in the VEL Model and to determine in a naturally inoculated lumen where pulsed organisms would be detected. We have used X05 (*Pseudomonas*) and X8.1 or X29 (*Staph. aureus*). Several questions were addressed: Was the IVIS system able to capture photons emitted from bacteria within the ETT? Could Gram-positive and Gram-negative organisms be used to track early and late events? And, was there a distribution in the ETT which described a unique observation attributable to the use of bioluminescent bacterial? Preliminary results were very encouraging, and have continued to complement our earlier findings.

## Linking oral flora to airway disease, work of breathing, and systemic disease (collateral damage and antibiotic resistance)

In 2007, it became apparent that we needed to take the information from the adult, A-VEL model and substantiate with clinical findings and *in vitro* to *in vivo* switch; it was our purpose to move to a molecular method, utilize the extubated ETTs, emphasizing detection of dental oral flora as well as more traditional flora in our method of analysis. Hence, our Sabbatical in the fall of 2007 at Cardiff to University, School of Dental Medicine, Wales, UK and Wales University Hospital. Scientific staff provided the technical expertise to utilize density gradient gel electrophoresis (DGGE), and extrapolate that a Welsh population that mimicked West Virginia environment, given that Wales historically has had a coal mining industry. Our goals were to (1) define the heterogeneity of a mixed oral population within the ETT lumen, (2)

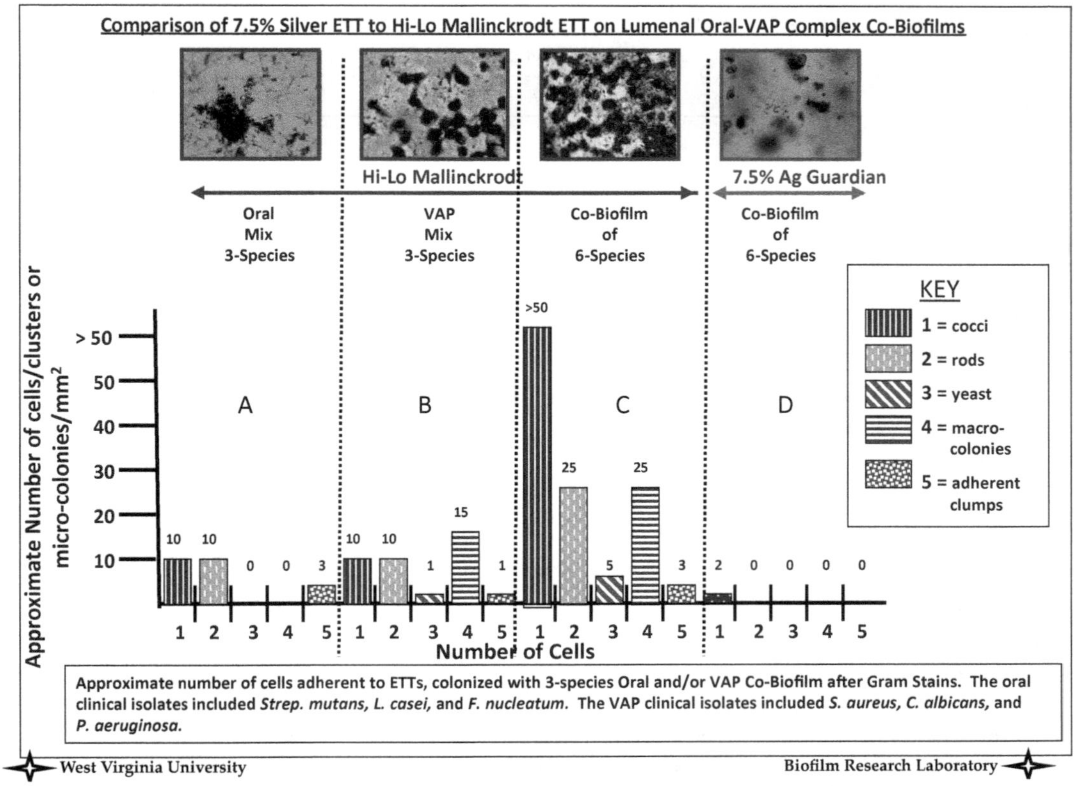


**Figure 12.6** Three-species dental/oral 'cocktail' and three-species VAP cocktail, incubated independently and together. Comparison of 7.5% silver ETT to hi-lo Mallinckrodt ETT on luminal oral-vap complex co-biofilms. Ag was effective against the combined three-species cocktail incubated for 48 hours. A colour version of this figure is available in the plate section at the back of the book.

to utilize a variety of molecular methods, and (3) use quantitative imaging with elemental analysis to further substantiate the studies utilizing our A-VEL model.

A two-pronged design was utilized (Thomas *et al.*, 2008; Williams *et al.*, 2008). First, the microbial diversity was defined using DGGE in combination with 16S-RNA amplification. Secondly, aerobic and anaerobic organisms were cultured and visualized by SEM/Elemental Analysis. Thirdly, 18 clinical parameters were assessed to define a 'predictive signature' for at-risk patients. Results of the 25-patient study emphasized several significant and key features (Cairns *et al.*, 2011). The ETT harboured a diverse population that contained significantly more species than could be identified by traditional cultures (Paster *et al.*, 2001; Park, 2005). Micro-organisms in the dental flora were early contributors to the diversity and *Strep. mutans* (orange complex) and *P. gingivalis* (*red complex*) were particularly evident using Socransky's criteria. *Strep. mutans* (*orange complex*) produces soluble and insoluble 'glucans' which make it enormously adherent to abiotic surfaces. Thirdly, based on hours of ventilation, emphasizing the 5-day break, Gram-positive isolates, particularly dental flora, had a propensity to appear earlier in the ventilation history, whereas Gram-negatives appeared later.

Fig. 12.7 schematically provides the overview of the information provided in this feasibility study. It reemphasized the recognition of 'early' and 'late' sequence in ventilator associated colonization, Gram-positive first, Gram-negative second, but it clearly identified an endogenous early phase and an exogenous late phase. SEMs (Fig. 12.8) emphasize the diversity of the ETT luminal biofilm.

**Table 12.2** Imaging techniques for biofilms

| Biofilm architecture | Analytical imaging | | | | | | |
|---|---|---|---|---|---|---|---|
| | X-ray | BF | FC | CLSM | FISH | SEM | FF |
| Microcolony | | | | | | | |
| Size | | X | | | | | X |
| % confluency | | X | | | | | X |
| Pattern of organization | | | | | | | |
| Open/closed | X | X | | X | | | X |
| Loose and channels/tightly bound (compact) | | X | | X | | | X |
| Spatial arrangement | | | | | | | |
| No bridging/bridging | | | | X | | X | |
| Average distance | | | | X | | X | |
| No domain/domains (grouping) | | | | X | X | X | |
| Complexity or heterogeneity (visual peaks/bonds) | X | | | | | | X |
| Single species/multispecies consortia | | X | | | | | |
| G+/G– morphotype | | | X | X | X | | |
| Viability/susceptibility–resistance | | | | | | | |
| Non-vital/vital role | | | X | X | | | |
| Biomass/bioburden/biodiversity of consortia | X | | X | | X | | |
| Organism size/shape (individually) | | X | X | | X | | |

X-ray, zinc dust radiography; BF, bright field; FC, flow cytometry; CSLM, confocal laser scanning microscopy; SEM, scanning electron microscopy; FISH, fluorescent *in situ* hybridization; FF, freeze fracture SEM; G+, Gram-positive; G–, Gram-negative.

We found the recovery of *Candida albicans* particularly meaningful. Sissons *et al.* had earlier reported the frequent occurrence of *Candida* in caries utilizing a chequerboard analysis (Filoche, 2008). More recently Ghannoum *et al.* (2010) added credibility to the initial findings and enhanced our statement that *Candida albicans* is a universal co-aggregate. They analysed 20 healthy patients and utilized pyro-sequencing with a pan-fungal internal transcribed spacer (ITS). He focused on 18-S RNA and found that a total of 74 culturable and 11 non-culturable genera were recovered. Each individual had between 9 and 23 species and *Candida* was the most frequent at 75%, followed by *Cladosporium* at 65% and *Aurebasidium* and *Saccharomyces* at 50%.

We felt our data strongly suggested that the importance of dental flora could be symbolized as indicated in Fig. 12.9. The ETT is an extension of the abiotic surface, it adds to every patient essentially 15% more of the surface to be colonized by, first, dental exogenous flora and second by exogenous or hospital infection control flora (Coppadoro *et al.*, 2011).

This also underscored another perspective: dental plaque provided a reservoir for endogenous colonization of the ETT, but that the structure and composition of the ETT biofilm mimicked the structure of dental plaque. The SEM characterization of the luminal biofilm was reminiscent of dental plaque architecture (Fig. 12.10). The same 3D features that enhanced dental plaque and its recalcitrant nature were evident in the luminal biofilm, hence our name 'bio-plaque' within the lumen.

We addressed this further by using SEM elemental analysis in our 25 patient study and subsequently, at the WVU Lane School of Engineering. The results were most interesting, similar profiles of elements were clustered between 0 and 5 KEV. X-ray analysis detected a mean of 14 elements, of which elements 12 through 32 were commonly found in plaque and in ETT biofilm.

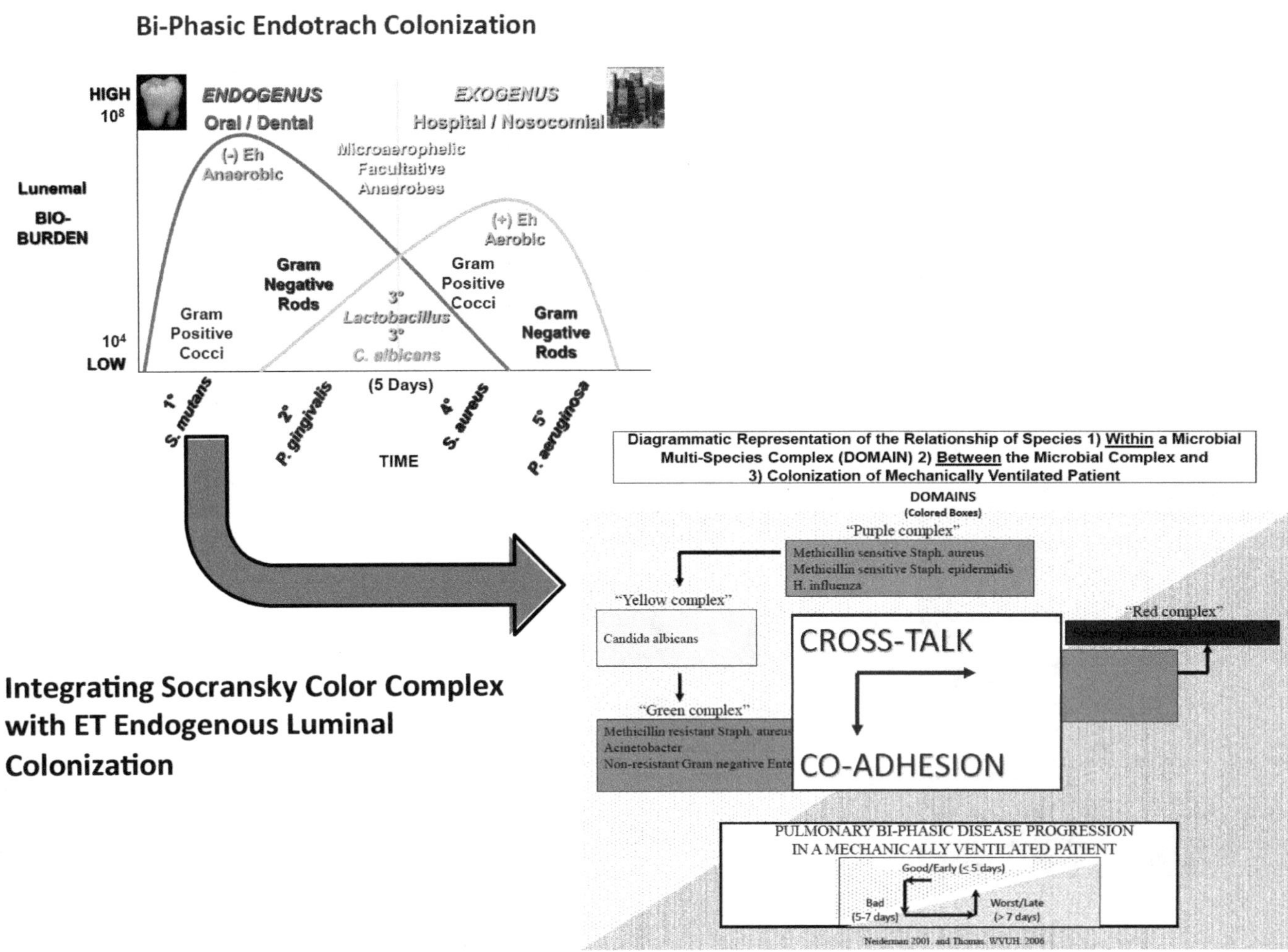


**Figure 12.7** (A) Diagrammatic representation of luminal endotracheal colonization – endogenous (oral) to exogenous (environmental), based on Cardiff study: emphasizing Gram-positive to Gram-negative in each. (B) Extrapolating the significance of VAP isolates, both oral and exogenous using the Socransky colour code template.

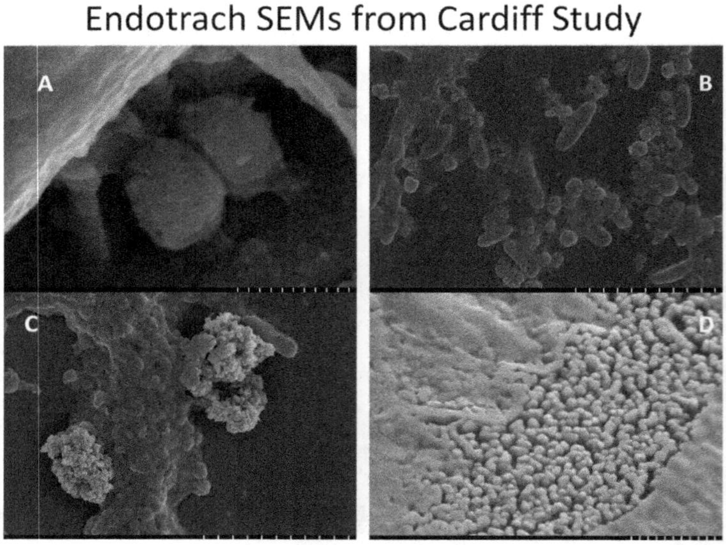


**Figure 12.8** SEM of undisturbed ETT lumen on section B highlighting diversity of colonized microbiota/biofilm. Sections A, B, C, D reflect different patients and establish an OMS (oral microbial signature).

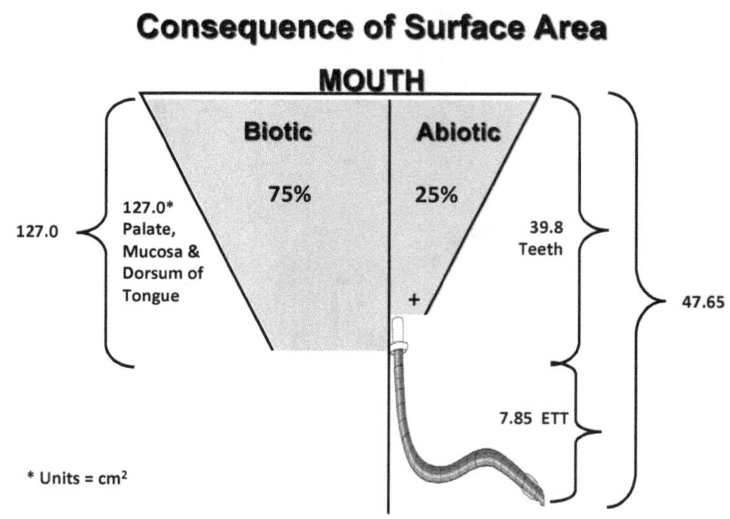


**Figure 12.9** Recognition that the ETT effectively adds three to four teeth of abiotic surface to the mouth in the ventilated patient. This new oral-like abiotic surface is undetected by the immune response and systemic antimicrobials.

These included Ca, Se, Na, Ps, Ka and Ni. Semi-quantitative ratios of elements help highlight similarities between dental plaque and ETT matrix. We were astonished and surprised by the variability of the 3D architecture of plaque like material in the ETT (Fig. 12.11).

Ultimately, though, the most enlightened feature of the Cardiff studies and subsequent evaluation of data at West Virginia University Hospital was the revelation that luminal colonization with mucus adherence resulted in a significantly altered airway resistance, or work of breathing. Our earlier studies highlighted this potential 'biphasic' nature by suggesting that ventilatory biofilms could be defined 'microscopically' and 'macroscopically'. Microscopic was the more

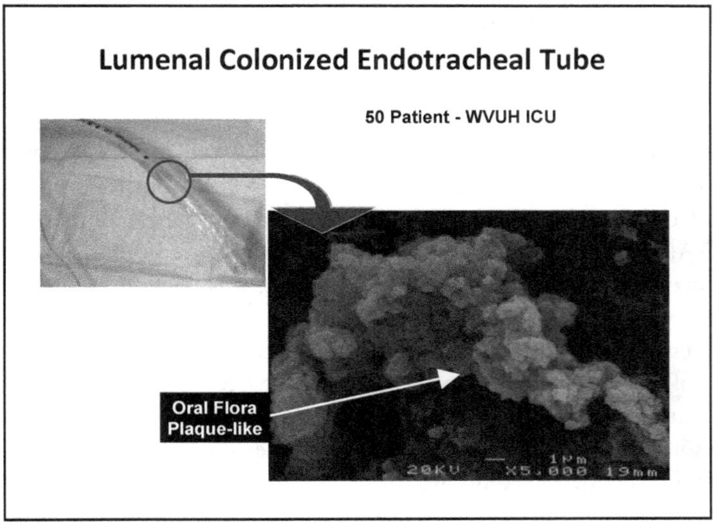


**Figure 12.10** Luminal colonized endotracheal tube in 3D architecture defined by SEM. Presence of oral flora by traditional culture and PCR methods.

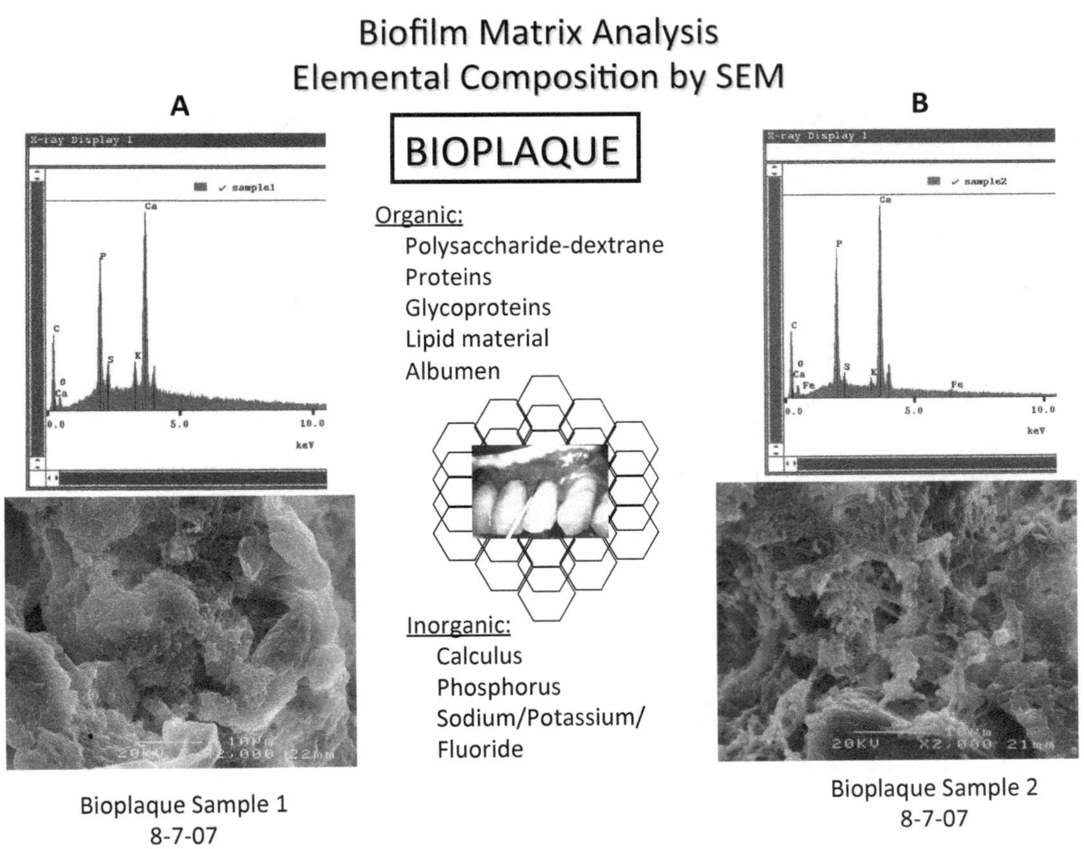


**Figure 12.11** BioPlaque and endotracheal biofilm matrix comparison using SEM and SEM elemental analysis. Elemental composition is similar between the two patients (A and B) but the 3D biofilm dimension defines a different OMS (Oral Microbial Signature).

traditional assessment of the consequences of ETT colonization: VAP. However, it was also apparent that the biofilm acting as an adherent bridge to mucus altered dramatically the work of breathing, as was noted in 'Weaning Techniques' used in the ICU. In fact, it became apparent that an altered work of breathing occurred 100% of the time whereas VAP has a resultant colonization phenomenon occurring between 3% and 16% of the time.

Our thoughts were organized in Fig. 12.12, by the biphasic nature of luminal biofilm, microscopic and macroscopic, hindering the recognition that maintenance of 'nominal ETT function' should be the goal of any therapeutic interventions. We highlighted the significance of airway resistance in an 87 patient study (Wilson *et al.*, 2009); here, utilizing the PTS 2000 we evaluated the airway resistance as a function of luminal size and ventilation days. Our results were most startling, as they revealed there was no 'predictability' with the length of time of ventilation, some airway resistance occurring within 8 hours. Startling was the magnitude of the impact, where clearly 50% of the extubated ETTs had an increased resistance associated with a functional airway of an ETT by at least one-half size smaller (Mahul *et al.*, 1992). These findings were considered benchmark and changed significantly the manner in which the patients were evaluated prior to being weaned and extubated (Maselli *et al.*, 2011).

We evaluated further the ETT occlusion by studying the location at which the occlusion was greatest. Earlier studies with our A-VEL model indicated that stress in vortexing caused the thickest biofilm in section B. We were very interested in defining a signature or profile of the occlusions within the ETTs under ventilation. We approached this in two ways. First, patient ETTs were extubated, coated with zinc dust, and placed into a CT scan; 500 sections at a 0.5 mm were evaluated utilizing VITAL software. Fig. 12.13 highlights the signature of one ETT. Generally occlusion was highest in the B section, which complimented the findings of our SEMs. There was less evidence of biofilm in the distal end (C) suggesting higher turnover (Stage IV) or inoculation of the lung and alveolar sacs from this region.

Our present strategy of VAP management focuses on maintaining the 'nominal function' of the ETT, reducing the two major features, microbial colonization and work of breathing. Here, we are using an optically directed clearing device that has the potential of, on a daily basis, maintaining nominal function of the airway while

**PARADIGM SHIFT: Management and Intervention**

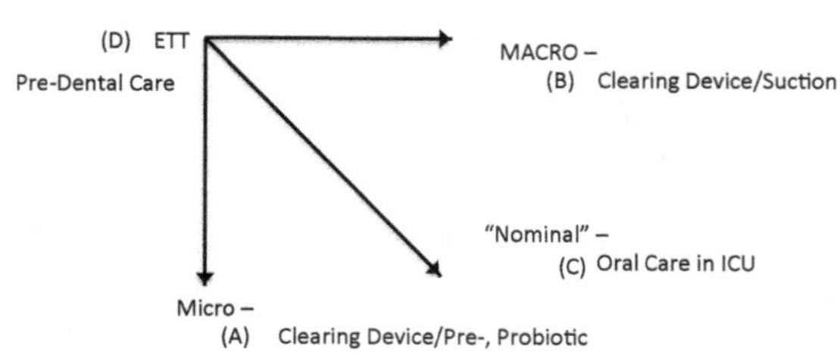


Figure 1B

1. VAP and Endotrach.
   Recognizing that balanced oral and systemic normal flora or Microbiota is a defensive "Organ System", 1) reduce the use of systemic antibiotics (A) use a "green" clearing device daily adjunctive to routine suctioning (B) to maintain nominal airway function of the endotrach.
2. Complement the ICU Oral Care (C) with preventative dental care (D) prior to intubation or hospital admission.
3. Maximize the benefit of reduced ventilation hours with earlier Extubation (weaning), and reduced *C. difficile* with concomitant oral use of Pre-Probiotics (B).

**Figure 12.12** Two linked pathways to patient disease – integration of 'microscopic' and 'macroscopic' pathways to effectively reduce ETT function.

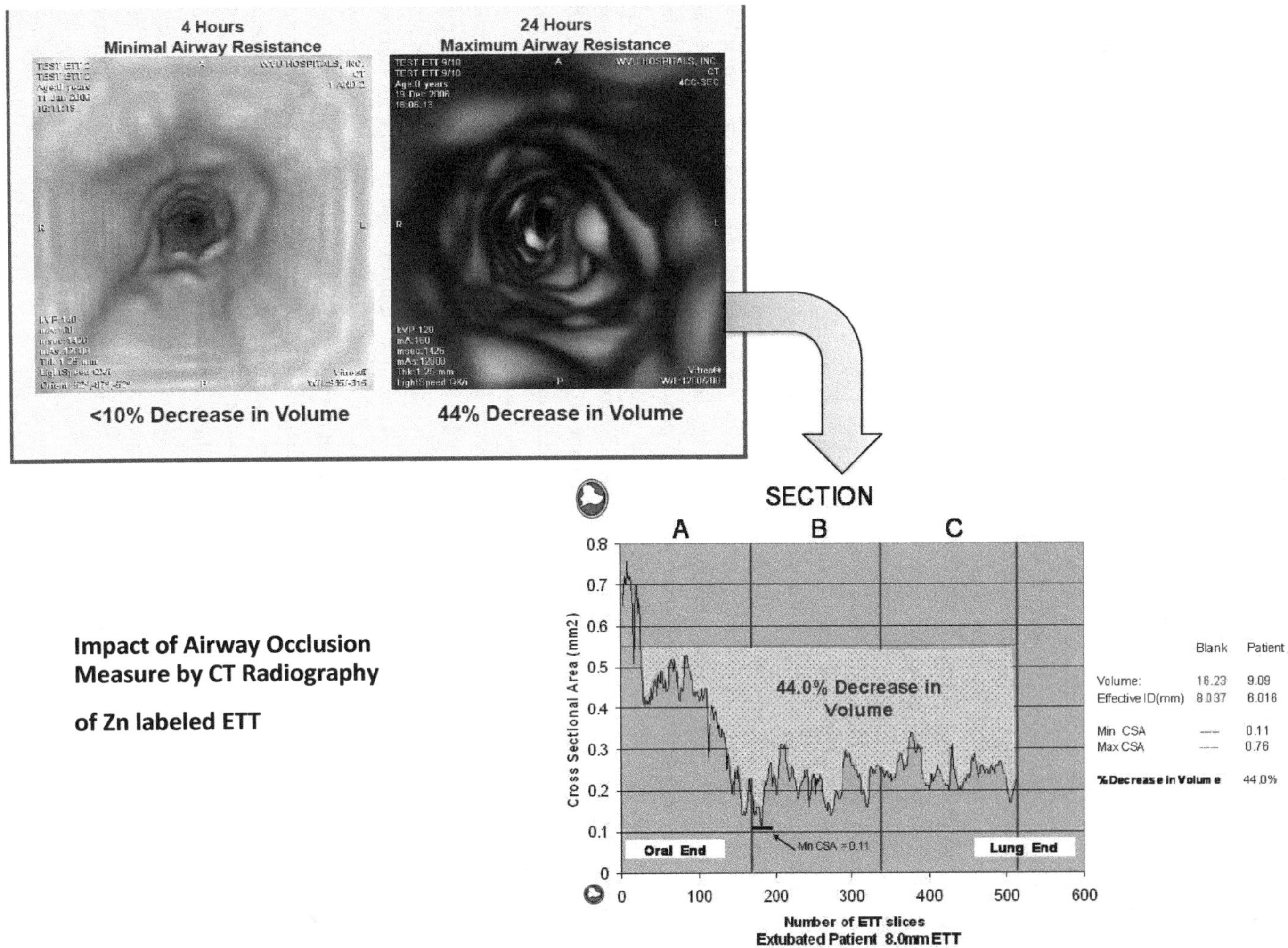


**Figure 12.13** Biofilm mucus endotracheal signature. (Lower-right image) CT radiography of zinc dust-coated endotrachea showing distribution and occlusion based on 500 0.5-mm sections. (Upper-left image) Fly-thru 3D reconstruction of luminal biofilm mucus matrix comparing 4-hour and 24-hour mechanical ventilation.

simultaneously reducing the bioburden. It also does so in a 'green' fashion, that is, without the use of anti-infectives (www.virtual-ports.com). Preliminary results using the clearing device (Endoclear) and measurements of work of breathing using the PTS-2000 have indicated a greater than 90–95% return to nominal ETT function and a greater than $10^3$ log reduction of microbial challenge (Waters, 2011).

Application of our growing information on the importance of biofilms, dental disease, and the oral systemic link lead us to investigate, organize, and implement in 2002 the Mountain State Oral Microbiological Laboratory (www.hsc.wvu.edu/som/Pathology/Thomas/PDFS/Educational-Resources/Oral-infections-antiinfective-mngt-&-Cases.pdf). This was a natural extension of our original studies with low dose doxycycline (PerioSTAT) (Thomas *et al.*, 2000) and our continual evolution of studies on biofilms. The Laboratory was uniquely affiliated with both the WVU School of Dentistry and WVU School of Medicine. Given its microbial intent, it also had the unique organization of being a unit within the clinical microbiology laboratory at WVUH, thereby linking traditional clinical microbiology with not so traditional dental microbiology. It amplified the strengths of both. It was tailored for detection, identification, and resistance testing of viable oral/dental microbiota emphasizing rapid microscopy and culture techniques differentiating biofilm and planktonic phenotypes. Its two-pronged approach not only identified dental pathogens, but the importance of sentinel organisms involved in healthy oral biofilm. Six FDA 'approved for laboratory testing' antibiotics were evaluated in pure culture, both planktonic and biofilm (F-127) including amoxicillin, tetracycline, clindamycin, erythromycin, ciprofloxacin, and metronidazole. To enhance turn-around time and to parallel traditional methods in clinical microbiology, the dental laboratory employed real-time web based algorithms for comparison of organism characteristics to known controls (www.anaerobicsystems.com) (California). Antimicrobial susceptibilities were compared to a national data base using The Surveillance Network (TSN), a national electronic surveillance system having access to greater than 500 laboratories downloading information daily (www.MRLworld.com). This data mining allowed us to establish an oral dental antibiogram and allowed for continual tracking of periopathogens particularly associated with antibiotic resistance.

## Conclusion

Clinical translational research is the opportunity to transcribe science from the 'bench to the bedside'. Our focus was on the disease ventilator associated pneumonia (VAP), the most expensive disease for hospitalized patients on mechanical ventilation and the unrecognized contribution of endogenous oral dental flora.

We reported in this text on the evolution of four integrated steps over 12 years to better define first, the science and then its application to clinical dentistry. The architecture of our studies involved the '*in vitro* to the *in vivo* switch' emphasizing in our initial *in vitro* studies the adult ventilator ETT lung model (A-VEL) emphasizing the cascade of microbes from a simple to a co, to a complex biofilm. The organism selection was based on a 50 patient study that ultimately recognized in the *in vivo* extension the importance of three oral endogenous flora. These included *Strep. mutans*, (orange complex), *P. gingivalis*, (red complex), and the 'universal co-aggregate' (as we have named it), *Candida albicans*.

A key feature was the revelation that the luminal ETT biofilm was the intermediate reservoir that allowed for co-colonization and ultimate metastases to the lung of the oral-systemic Microbiota. Further, it was the Staging (I-IV) which was key in defining the risk for this lung colonization.

Staging also highlighted the fact that 'structure equals function'. Multispecies or complex biofilm represent a more recalcitrant nature than did biofilms with one or two microorganisms. This concept was independent of whether one dealt with traditional medical diseases such as VAP or oral disease such as periodontal disease (form equals function).

In our opinion, future studies should continue to unravel the pathophysiology of a recognized endogenous oral flora with an exogenous flora, often associated with the environment of that patient. Further, it is critical to recognize that this endogenous, exogenous combination should dictate the management of the oral-systemic disease; this should highlight the importance of 'Stewardship' of the normal oral flora and perhaps its re-establishment using probiotics colonizing the oral mucosa and/or the gastrointestinal flora. Clearly in the management of VAP, the impact of a poor oral flora maximized by poor oral hygiene, either in the hospital, outpatient, or institutional setting, is critical. 'Pay me now or pay me later.'

## Addendum

Our on-going translational research has resulted in two major studies that emphasize the continuing importance and contribution of oral flora to VAP.

In the first study, we analysed endotracheal luminal biofilm, Stages (I-IV), to clinical pneumonia. Thirty-two extubated endotracheae were analysed by this author for stage (Fig. 12.3A and B), blinded to all patient data. Clinical data included pneumonia, duration of intubation, comorbidities and microbiology. Staging was defined using characteristics listed in Table 12.1. The scanning electron microscopy analysis was performed by NIOSH Laboratories and visualized by both transmission and scanning gold-palladium images from a JOEL microscope of 2000 to 5000 magnification. Importantly, there was a statistically valid relationship between increasing biofilm stage and incidence of pneumonia ($P < 0.05$), and Stage IV biofilm with pneumonia ($P < 0.02$). Simultaneously, there was no correlation between duration of intubation and prediction of biofilm stage (Wilson *et al.*, 2012).

Our next study in these observations is to correlate plaque and/or periodontal biofilm stage with endotracheal luminal stage and pneumonia. Staging is a key pathologic characterization of a biofilm and re-emphasizes 'Structure = Function' or 'Form = Function'. We are presently using staging as a criterion of periodontal analysis (www.hsc.wvu.edu/som/Pathology/Thomas/PDFS/Educational-Resources/Oral-infections-antiinfective-mngt-&-Cases.pdf), when paperpoint samples are received from the periodontal clinic.

Secondly, we emphatically continued the work-of-breathing (WOB) studies initially reported in 2009 and complemented recently by our published data emphasizing non-culture techniques and detection of dental/oral flora (Williams *et al.*, 2008; Cairns *et al.*, 2011).

This has involved a uniquely engineered, optically directed, endotracheal clearing device that focuses on both pathways (micro and macro) defined by Fig. 12.12, while emphasizing returning the endotrachea to its 'nominal function' (www.hsc.wvu.edu/som/pathology/thomas). This 43-patient study has allowed us to critically reassess the oral/dental contribution using non-culture techniques employing both 16S (bacterial) and 18S (fungi), pre- and post cleaning with the device. The 16S DNA focus utilized the HOMIM Protocol (http://mim.forsyth.org/) and resources of Bruce Paster and Forsyth Dental Institute, Boston, USA.

Preliminary results were remarkable. A total of 101 phylotypes were detected emphasizing rich dental/oral microbiota; they were extremely adherent, reflecting a modest reduction in numbers and intensity in pre versus post cleaning. The top 10, usually Gram-positive streptococci, were also consistent pre- and post-cleaning, complementing our previous work of breathing studies where the occlusion was not influenced by length of intubation and size of the endotrachea. The top three organisms were *Streptococcus* Cluster III, *Lactobacillus* Cluster I and *Streptococcus anginosus* and *S. intermedius* (ASM), in both pre and post cleaning.

## Acknowledgements

We acknowledge the support of Hamed M. Motlagh, MPH, Sara B. Posey, MPH, Linda Corum, MS and Christopher L. Waters, MS.

## Web-based resources

www.hsc.wvu.edu/som/pathology/thomas/educational-resources. (Click on 'Educational resources'.)

### Mini-Micro Educational Series

- Seven 30 minute PowerPoint, with voice, presentations.

1 Biofilms: Architect of Survival: http://www.

hsc.wvu.edu/som/Pathology/Thomas/Videos/thomas_060707_biofilms1/thomas_060707_biofilms1.html

2 Biofilms: Architect of Chronic Diseases Head to Toe: http://www.hsc.wvu.edu/som/Pathology/Thomas/Videos/thomas_060707_biofilms2/thomas_060707_biofilms2.html

3 A Comprehensive Engineered Ventilator-Endotrach-Lung (VEL) Model: http://www.hsc.wvu.edu/som/Pathology/Thomas/Videos/thomas_060707_biofilms3/thomas_060707_biofilms3.html

4 Diagnostic Microbiology, Pathology and Biofilms: http://www.hsc.wvu.edu/som/Pathology/Thomas/Videos/thomas_061807_biofilms4/thomas_061807_biofilms4.html

5 Rediscovering the Benefit of 'Microbial Engineering': Probiotics: http://www.hsc.wvu.edu/som/Pathology/Thomas/Videos/microMiniProbiotics/microMiniProbiotics.html

6 Putting the Pieces Together: The Link between Oral Care, Biofilms, and Pneumonia: http://www.hsc.wvu.edu/som/Pathology/Thomas/Videos/miniMicroOralBiofilms/miniMicroOralBiofilms.html

7 Defining the Impact of Microbes in Chronic Wounds: A Biofilm Reactor: http://www.hsc.wvu.edu/som/Pathology/Thomas/Videos/miniMicroWounds/miniMicroWounds.html

8 Prebiotics & Probiotics: Old Fad with New Interventions: http://www.hsc.wvu.edu/som/Pathology/Thomas/Videos/probiotics_ii_qtp.mp4

9 A Biofilms Procaryotic Tumours? Or... Are Tumours Eucaryotic Biofilms?: http://www.hsc.wvu.edu/som/Pathology/Thomas/Videos/tumours_qtp.mp4

10 Ten Universal Microbial Principles: Anti-Koch (Under Construction).

## Biofilms Made Easy: 'A Picture Tutorial'

- An Introduction to Biofilms: Structure, Medical/Dental/Microbiological impacts, via 100 PowerPoint images with corresponding description.
- http://www.hsc.wvu.edu/som/Pathology/Thomas/PDFS/Educational-Resources/
- Biofilms-For-Everyone-Booklet.pdf

## Microbiology Pocket Electronic Handbook (PDF)

- A Microbiology Internet Guide, includes: antibiotics and pharmacology, susceptibility profiles, oral microbiology, and normal flora.
- http://www.hsc.wvu.edu/som/Pathology/Thomas/Educational-Resources/Microguide.pdf

## Dental Microbiology Pocket Electronic Handbook with Instructional Dental Case Presentations

- A computation of published reviews for management, utilizing Dental Antibiograms from the WV Mountain State Oral-Facial Microbiology Laboratory (Oral Infections Anti-Infective Management).
- Six Instructive Cases for Periodontal Disease (P. 5).
- Five Instructive Cases for Endodontic Disease (P.17).

http://www.hsc.wvu.edu/som/Pathology/Thomas/PDFS/Educational-Resources/Oral-Infections-AntiInfective-Mngt-&-Cases.pdf

## Online courses (with certificate)

- John G. Thomas's West Virginia Microbiology Annual Fall Symposium (Movie).
- The annual complete 2-day Course, including all five speakers through PowerPoint and audio presentations for each faculty, with sort options per presentation.

- http://www.hsc.wvu.edu/som/Pathology/Thomas/Online-Courses/
- Featured Workshops:
  - 19th Annual Clinical Microbiology Seminar (September 2008)
  - 20th Annual Clinical Microbiology Seminar (September 2009)
  - 21st Annual Clinical Microbiology Seminar (September 2010).
- Cost is $70 for a 2 week availability http://wvuecommerce.wvu.edu.
- Select Clinical Microbiology Seminars: Fill Basket.

## References


Adachi, M., Ishihara, K., Abe, S., Okida, K., and Ishihara, T. (2002). Effect of professional oral health care on elderly living in nursing homes. Oral Surg. Oral Med. Oral Pathol. Oral Radiol. Endo. *94*, 191–195.

Cairns, S., Thomas, J.G., Hooper, S.J., Wise, M.P., and Frost, P.J. (2011). Molecular analysis of microbial communities in endotracheal tube biofilms. PLoS One *6*, e14759.

Coppadoro, A., Berra, L., and Bigatello, L.M. (2011). Modifying endotracheal tubes to prevent ventilator associated pneumonia. Curr. Opin. Infect. Dis. *24*, 157–162.

Darouiche, R.O. (2001), Device-associated infections: a macro problem that starts with microadherence. Clin. Infec. Dis. *33*, 1567.

Dewar, D., Lambrinos, J., Mallick, R., and Zhong, Y. (2000). Cost benefit analysis of mechanical ventilation. Int. J. Technol. Assess. Health Care *16*, 148–164.

Donlan, R.M., and Costerton, J.W. (2002). Biofilms: survival mechanisms of clinically relevant microorganisms. Clin. Microbiol. Rev. *15*, 167–193.

Filoche, S.K. (2008). Plaques from different individuals yield different microbiota responses to oral-antiseptic treatment. FEMS Immunol. Med. Microbiol. *54*, 27–36.

Forestier, C., Guelon, D., Cluytens, V., Gillart, T., Sirot, J., and De Champs, C. (2008). Oral probiotic and prevention of *Pseudomonas aeruginosa* infections: a randomized, double-blind, placebo-controlled pilot study in intensive care unit patients. Crit. Care *12*, R69.

Fux, C.A., Stoodley, P., Hall-Stoodley, L., and Costerton, J.W. (2003). Bacterial biofilms: a diagnostic and therapeutic challenge. Expert Rev. Anti-Infect. Ther. *1*, 667–683.

Garcia, R. (2005). A review of the possible role of oral and dental colonization on the occurrence of health care-associated pneumonia: underappreciated risk and a call for interventions. Am. J. Infect. Control 33, 527–541.

Ghannoum, M.A., Jurevic, R.J., Mukherjee, P.K., Cui, F., Sikaroodi, M., Naqvi, A., and Gillevet, P.M. (2010). Characterization of the oral fungal microbiome (mycobiome) in healthy individuals. PLoS Pathog. *6*, e1000713.

Hall-Stoodley, L., Costerton, J.W., and Stoodley, P. (2004). Bacterial biofilms: from the natural environment to infectious diseases. Nat. Rev. Microbiol. *2*, 95–108.

Höffken, G., and Niederman, M.S. (2002) Nosocomial pneumonia: the importance of a de-escalating strategy for antibiotic treatment of pneumonia in the ICU. Chest *122*, 2183–2196.

IHI Guidelines (2006). Getting Started Kit: Prevent ventilator-associated pneumonia: How-to Guide. Crit Care Nursing Quarterly *29*, 157–173.

Klevens, R.M., Edwards, J.R., Richards, C.L. Jr, Horan, T.C., Gaynes, R.P., Pollack, D.A., and Cardo, D.M. (2007). Estimating health care-associated infections and deaths in U.S. hospitals, 2002. Public Health Rep. *122*, 160–166.

Kollef, M.H. (2006). The prevention of ventilator-associated pneumonia. N. Eng. J. Med. *340*, 627.

Kollef, M.H., Afessa, B., Anzueto, A., Veremakis, C., Kerr, K.M., Margolis, B.D., Craven, D.E., Roberts, P.R., Arroliga, A.C., Hubmayr, R.D., *et al.* (2008). Silver-coated endotracheal tubes and incidence of ventilator associated pneumonia: The NASCENT randomized trial. J. Am. Med. Assoc. *300*, 805–813.

Krivit, B., and Heuertz, R. (2011). Bacterial biofilms and healthcare-associated infections. MLO Med. Lab. Obs. *43*, 38–39.

Levy, S.B. (2002). Factors impacting on the problem of antibiotic resistance. J. Antimicrob. Chemother. *49*, 25–30.

Mahul, P., Auboyer, C., Jospe, R., Ros, A., Guerin, C., el Khouri, Z., Galliez, M., Dumont, A., and Gaudin, O. (1992). Prevention of nosocomial pneumonia in intubated patients: respective role of mechanical subglottic secretions drainage and stress ulcer prophylaxis. Intensive Care Med. *18*, 20–25.

Maselli, D.J., and Restrepo, M.I. (2011). Strategies in the prevention of ventilator-associated pneumonia. Ther. Adv. Respir. Dis. *5*, 131–141.

Morrow, L.E., Kollef, M.H., and Casale, T.B. (2010). Probiotic prophylaxis of ventilator-associated pneumonia: a blinded, randomized, controlled trial. Am. J. Respir. Crit. Care Med. *182*, 1058–1064.

Park, D.R. (2005). The microbiology of ventilator-associated pneumonia. Respir. Care. *50*, 742–765.

Paster, B.J., Boches, S.K., Galvin, J.L., Ericson, R.E., Lau, C.N., Levanos, V.A., Sahasrabudhe, A., and Dewhirst, F.E. (2001). Bacterial diversity in human subgingival plaque. J. Bacteriol. *183*, 3770–3783.

Rello, J., Ollendorf, D.A., Oster, G., Vera-Llonch, M., Bellm, L., Redman, R., and Kollef, M.H. (2002). VAP Outcomes Scientific Advisory Group. Epidemiology and outcomes of ventilator-associated pneumonia in a large US database. Chest. *122*, 2115–2121.

Safdar, N., Dezfulian, C., Collard, H.R., and Saint, S. (2005). Clinical and economic consequences of ventilator associated pneumonia: A systematic review. Crit. Care Med. 33, 2184–2193.

Sarwari, A.R., Petros, K.O., Sager, R.B., Mullett, C.J., Thomas, J.G., *et al.* (2008). Efforts to Influence Antimicrobial Use and Resistance at a Tertiary

Care Institution Using Collaborative Multifaceted Interventions. S10 – Oral Presentation, Annual Conference on Antimicrobial Resistance. June 23–25, Bethesda, Maryland.

Seok-Mo, H., Haase, E.M., Lesse, A.J., Gill, S.R., and Scannapieco, F.A. (2008). Genetic relationships between respiratory pathogens isolated from dental plaque and bronchoalveolar lavage fluid from patients in the intensive care unit undergoing mechanical ventilation. Clin. Infect. Dis. *47*, 1562–1570.

Shorr, A.F., Chan, C.M., and Zilberberg, M.D. (2011). Diagnostics and epidemiology in Ventilator-associated pneumonia. Ther. Adv. Respir. Dis. *5*, 121–130.

Thomas, J., Nakaishi, L., and Corum, L. (2007). Consequences of biofilms on indwelling medical devices: costs and prevention. In Disinfection and Decontamination: Principles, Applications, and Related Issues. Mannivannan, G., ed. (CRC Press, Boca Raton, FL), pp. 289–338.

Thomas, J.G., Walker, C., and Bradshaw, M. (2000). Long-term use of subantimicrobial dose doxycycline does not lead to changes in antimicrobial susceptibility. J. Periodontol. *71*, 1472–1483.

Thomas, J.G., Litton, S., and Corum, L. (2006). Economic impact of biofilms on treatment costs. In Biofilms, Infections, and Antimicrobial Therapy, Pace, J.L., Rupp, M.E., and Finch, R.G., eds. (CRC Press, Boca Raton, FL), pp. 21–38.

Thomas, J.G., Corum, L., and Miller, K. (2009). Biofilms and ventilation. In The Role of Biofilms in Device-Related Infections Biofilm Ventilation, Shirtliff, M., and Leid, J.. eds. (Springer, Berlin, Germany), pp. 75–107.

Thomas, J.G., Motlagh, H.M., and Percival, S. (2010). Establishing the 'Critical Ratio' as a Microbial BioMarker for Chronic Wound Classification/Status. Poster session presented at SAWC Atlanta, GA.

Thomas, J.G., Williams, D.W., Hooper, S.J., Malic, S., and Cairns, S. (2008). PCR DGGE to unmask *Streptococcus mutans* in endotracheal biofilms. J. Dent. Res. *87*. Available at: www.dentalresearch.org

Unroe, M., Kahn, J., and Carson, S., Govert, J.A., Martinu, T., Sathy, S.J., Clay, A.S., Chia, J., Gray, A., Tulsky, J.A., *et al.* (2010). One year trajectories of care and resource utilization for recipients of prolonged MV: a cohort study. Ann. Intern. Med. *153*, 167–175.

Vassilakopoulos, T. (2009). The patient–ventilator interaction has a third player: the endotracheal tube. Chest *136*, 957–959.

Waters, C. (2011). Use of an Endotrachael Tube optically directed clearing device to remove airway biofilm occlusion. Poster session presented at ASM/ICAAC Annual Meeting, Chicago, Ill.

Williams, D.W., Thomas, J.G., Hooper, S.J., *et al.* (2008). Profiling microbial diversity in endotracheal tube biofilms. J. Dent. Res. *87*. Available from: www.dentalresearch.org

Wilson, A., Gray, D., and Thomas, J.G. (2009). Increases in endotracheal tube resistance are unpredictable relative to duration of intubation. Chest *136*, 1006–1013.

Wilson, A., Gray, D., Karakiozis, J., and Thomas, J.G. (2012). advanced endotracheal tube biofilm stage, not duration of intubation, is related to pneumonia. J. Trauma Acute Care Surg. 72, 916–923.

# Oral Biofilm as a Vehicle for Chemotherapeutic Agents

# 13

Marieke P.T. Otten, Henk J. Busscher, Chris G. van Hoogmoed, Frank Abbas and Henny C. van der Mei


## Abstract

An overview is presented on oral biofilm formation and recent developments in oral biofilm control using mechanical devices (manual or powered toothbrushes and interdental cleaning devices) and biofilm control based on oral chemotherapeutics (antibacterial toothpastes and mouthrinses). For clinical efficacy of oral chemotherapeutics, it is important that the antibacterial remains active in the oral cavity for periods longer than the actual brushing or rinsing time, a characteristic called 'substantivity'. Substantivity can be achieved by adsorption of antibacterials to oral hard and soft tissues followed by release. Mechanical cleaning never results in complete removal of oral biofilm: most notably in fissures, interproximal spaces, gingival pockets and around orthodontic appliances. Recently, it has been demonstrated that this residual biofilm can act as a reservoir for oral chemotherapeutics that are slowly released over time in bio-active concentrations. This function of oral biofilm was already known for fluoride and has been demonstrated to aid in preventing caries. Results from our laboratory showed that residual biofilm after mechanical cleaning can release absorbed antibacterial agents from toothpastes and mouthrinses in bio-active concentrations.


## The oral biofilm

The oral biofilm, or dental plaque, is a complex microbial community embedded in a polymeric matrix adherent to a surface. The development of an oral biofilm can be described in a few basic steps. The first step in biofilm formation is the adsorption of a salivary conditioning film on surfaces: teeth, restorations, prosthetic devices and soft tissue surfaces. In the second step, individual bacteria adhere to these surfaces followed by co-adhesion of other strains and species. However, in this initial phase, adhesion is still reversible and bacteria may actually detach quite easily from the surface. When adherent bacteria produce extracellular polymeric substances (EPS), adhesion becomes irreversible. The next step is maturation of the biofilm, including the development of microcolonies to form a large, matrix enclosed structure (Stoodley *et al.*, 2002): the oral biofilm (Brecx, 1997; Baehni and Takeuchi, 2003; Hojo *et al.*, 2009). Fig. 13.1 schematically presents the initial steps in the formation of this complex, multispecies biofilm, including the clinical appearance of a nine days old oral biofilm on a tooth surface.

The resident oral microflora lives in harmony with the host, helps to maintain oral health and harbour low numbers of pathogenic microorganisms, as e.g. veillonella consuming short-chain acids such as lactates that are secreted by Gram-positive bacteria like mutans streptococci (Hojo *et al.*, 2009). Pathogenic bacteria can develop in high numbers under appropriate circumstances and cause oral diseases like caries, gingivitis and periodontitis (Brecx, 1997; Rosan and Lamont, 2000). Key environmental factors, like diet and the performance of oral hygiene can trigger shifts in the balance of the resident oral microflora, a process which has been termed the 'ecological plaque hypothesis' (Marsh, 1994). The goal of oral health care is to maintain the equilibrium between the resident flora and the host. This can be done either by reducing the total biofilm

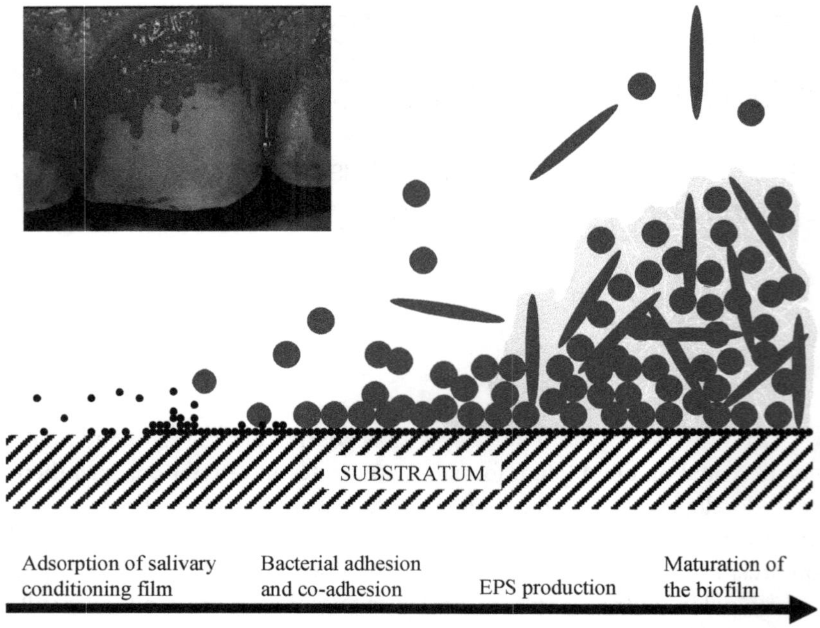


**Figure 13.1** Schematic, sequential presentation of the initial steps in biofilm formation. In the left top corner a photograph of a 9 days old, undisturbed oral biofilm on a human front incisor *in vivo*, stained by 0.5% aqueous neutral red solution is shown (Quirynen *et al.*, 1989). A colour version of this figure is available in the plate section at the back of the book.

mass, by targeted reduction of the prevalence of specific pathogens, or by interfering with the environmental factors responsible for the shift of the resident flora to a more pathogenic composition. Currently, the most wide-spread and accepted way for removal of oral biofilm is toothbrushing with a toothpaste, advisably supported by the use of mouthrinses and interdental cleaning aids. A large cohort study in adults (Axelsson *et al.*, 2004) studied the effects of a 30-year plaque control program on tooth mortality, caries and periodontal disease; self-performed biofilm control led to adequate maintenance of oral health. Nevertheless, a small number of subjects included in this study suffered from caries and periodontal diseases and had an average plaque coverage of 20% despite the extensive preventative program and supplementary use of interdental cleaning devices. In children however, adequate daily oral hygiene is much more difficult to achieve than in adults. Moreover, orthodontic appliances have become popular in juvenile and, recently, in adult patient populations, thereby creating numerous retention sites in the oral cavity that are difficult to clean. Thus, the need for improved plaque control programs has increased greatly over the past decade.

Out of the thousands of different bacterial strains and species that inhabit the human oral cavity (Kolenbrander *et al.*, 2006; Ten Cate, 2006; Keijser *et al.*, 2008), a subset of causative microorganisms exists for caries and periodontal diseases. Caries develops when specific types of acid producing bacteria, such as mutans streptococci and lactobacilli, convert fermentable carbohydrates into acids that dissolve the enamel surface (demineralization) and finally result in cavities. For about half a century, caries was regarded as an 'infectious and transmittable' disease caused by a particular microbial strain (Fejerskov, 2004). This would suggest that caries can be prevented by vaccination. Research on anti-caries vaccination with mutans streptococcal cells or antigens (Balakrishnan *et al.*, 2000; Chen and Wang, 2010) showed that these vaccines can prevent colonization by mutans streptococci, but clinical studies have failed to prove that vaccination is effective against caries. Currently, caries is considered to be a 'multifactorial' disease, caused by a complex interplay between saliva, diet, biofilm composition, and without a simple causative pathway (Fejerskov, 2004) the idea that caries is an infectious, transmittable disease has been abandoned. Plaque-induced gingivitis is due to the accumulation of bacteria in the biofilm around the gingiva, which then promotes an inflammatory response by the host and results in red, swollen gingiva. In some patients, especially those with high levels of the pathogens *Aggregatibacter actinomycetemcomitans* and *Porphyromonas gingivalis*, gingivitis progresses to periodontitis: destruction of the supporting fibres and surrounding alveolar bone which eventually results in tooth loss. In addition to the dental diseases caries and gingivitis, an association between oral focal infections and the risk for cardiovascular diseases (Seymour *et al.*, 2007; de Oliviera *et al.*, 2010), and for preterm low-birthweight babies (Scannapieco *et al.*, 2003; Lopez, 2008) has been reported.

Oral diseases remain a public health problem despite major achievements in oral health care (WHO Oral Healthcare Report, 2003). Because a close relationship exists between oral biofilm and the occurrence of caries and periodontitis (Hardie, 1992), mechanical removal (Haffajee *et al.*, 2006; Corby *et al.*, 2008) and chemical plaque control (Sekino *et al.*, 2004) are still widely applied.

## Controlling oral biofilms: mechanical removal

The toothbrush is the most widely employed tool to remove oral biofilm, although its proper use is not trivial and requires quite some skill. When performed with an adequate technique and for an appropriate duration, manual brushing is highly effective. However, for most patients, neither of these criteria are fulfilled. Biofilm removal from pits, fissures, interproximal spaces, and around orthodontic appliances and imperfect restorations is never achieved by manual toothbrushing only. Therefore a number of tools have been advocated to assist in removal of biofilm from difficult to reach places: dental floss, toothpicks, mini-brushes and most importantly, interdental brushes. Risk analysis for the development of caries and periodontitis has shown that the interproximal area is most susceptible. Therefore, it is of utmost importance that interdental cleaning devices are used in combination with mechanical brushing. The choice for an interdental cleaning device can differ from patient to patient, and should be based on the specific dental status of each patient (Claydon, 2008; Van der Weijden *et al.*, 2008). Powered toothbrushes have been developed to compensate for a poor brushing technique and to facilitate biofilm removal from hard to reach places. Powered toothbrushes with a rotating, oscillating or sonic action remove biofilm and reduce gingivitis significantly better than do manual brushes (Tritten and Armitage, 1996; Ho and Niederman, 1997; Moritis *et al.*, 2002; Biesbrock *et al.*, 2008; Rosema *et al.*, 2008), and it is suggested that biofilm removal may even extend beyond the reach of the bristles (Hope and Wilson, 2003; Busscher *et al.*, 2010). Nevertheless, complete removal of oral biofilm by means of habitual brushing and use of interdental cleaning aides remains difficult (Nyvad, 2003; Claydon, 2008; Van der Weijden *et al.*, 2008). Reduction of biofilm by only 50–60% can be achieved by one self-performed manual brushing (Paraskevas *et al.*, 2006; Van der Weijden *et al.*, 2008), i.e. 40–50% of the biofilm is left behind.

The resident microflora contributes to oral health and is well tolerated by the oral cavity (Hojo *et al.*, 2009) and by the human body in general. Therefore, the challenge is to prevent a shift in the commensal microbial composition to a pathogenic flora. Untargeted removal of oral biofilm every 48 hours is sufficient to prevent the development of gingivitis and caries (Lang *et al.*, 1973; Claydon, 2008). However, twice-daily brushing is usually recommended by dental professionals. This high brushing frequency is justified by the idea that it increases the efficacy of biofilm removal and enhances the beneficial effects of therapeutic toothpaste components such as fluoride (Claydon, 2008) and stain-removing abrasives (Addy, 1997). It also meets the patients' desire for fresh breath. Note that toothpastes with moderate or high abrasivity do not necessarily yield more efficient removal of oral biofilm (Paraskevas *et al.*, 2006).

## Controlling oral biofilms: chemotherapeutic approaches

When delivered by toothpastes or mouthrinses, chemotherapeutic adjuncts such as fluoride and antibacterial agents can help to prevent oral diseases by altering the pathogenicity of oral biofilms (Scheie, 2003). A wide range of chemotherapeutic agents is deliverable in products like toothpastes, gels and mouthrinses (see Table 13.1). In general, chemotherapeutic agents can control biofilm formation by reducing accumulation of new biofilm, by reducing or removing existing biofilm, by suppressing growth and development of pathogenic bacteria and by inhibiting production of virulence factors (Marsh, 1992). It is important that antibacterials not disrupt the healthy oral microbiome even though their action is typically non-specific (Marsh, 1992). Antibacterial agents can perform their action in different specific ways, i.e. causing leakage of cellular contents or affecting microbial metabolism (Baehni and Takeuchi, 2003; Scheie, 2003). Compared with planktonic or free-floating bacteria, organisms in a biofilm mode of growth are less sensitive to antibacterial agents (Sedlacek and Walker, 2007) and take advantage of the protective functions of the biofilm (Hojo *et al.*, 2009). Biofilm bacteria differentiate themselves from planktonic ones by producing EPS. Bacterial EPS is comprised of biosynthetic polymers that can be highly diverse in chemical composition and may include polysaccharides, proteins, nucleic acids and phospholipids (Stoodley *et al.*, 2002). Apart from acting as a glue and providing structural support to the biofilm, EPS also acts as an extremely protective slime encasing. Antibacterials often bind to, or are inactivated by, the EPS matrix of the biofilm. As a result, the agents do not reach the deeper layers of a biofilm, as was first shown for vinegar by Antonie van Leeuwenhoek in the seventeenth century and as later confirmed for chlorhexidine using confocal laser scanning microscopy (see Fig. 13.2) (Zaura-Arite *et al.*, 2001).

Numerous studies have shown that in general up to 1000-fold higher concentrations of chemotherapeutics are needed to kill bacteria in biofilms compared to free-floating or planktonic ones (Wilson, 1996; Ten Cate, 2006). Chlorhexidine is the most employed antibacterial in the oral cavity. It has a broad spectrum of activity against Gram-negative and Gram-positive bacteria and yeast, and proven anti-plaque activity. At high concentrations it is bactericidal by damaging the bacterial cell membrane (Baehni and Takeuchi, 2003). At lower concentrations, it is bacteriostatic by inhibiting sugar transport and membrane functions (Baehni and Takeuchi, 2003; Marsh, 2010). Fluoride inhibits the metabolism of bacteria (Wiegand *et al.*, 2007; Stoodley *et al.*, 2008) by affecting cell glycolysis (Marquis *et al.*, 2003), thereby inhibiting bacterial growth (Hamilton and Bowden, 1988).

Inhibition of bacterial growth by fluoride ions is most effective under acidic conditions (Hamilton and Bowden, 1988). Fluoride can also reduce demineralization and enhance remineralization (Ten Cate *et al.*, 2003). Essential oils, like thymol and eucalyptol, can reduce the biomass of Gram-negative anaerobic bacteria, resulting in a reduction in oral biofilm mass and gingivitis (Marsh, 2010). Triclosan has a broad spectrum of antibacterial activity (Baehni and Takeuchi, 2003). And, at sublethal levels, it can inhibit acid production by oral streptococci (Marsh, 2010). Detergents like sodium lauryl sulphate can damage cell membranes, inhibit enzymes and kill bacteria. Metal salts, like stannous fluoride are very well retained in the oral cavity and possess bactericidal activity against both Gram-positive and Gram-negative bacteria (Baehni and Takeuchi, 2003; Marsh, 2010).

Antibacterial photodynamic therapy involves activation of photosensitizers added to the biofilm to create a reactive state by formation of oxygen-derived free radicals that lead to bacterial cell death (Braham *et al.*, 2009; Fontana *et al.*, 2009). However, the antibacterial effect of photodynamic therapy using methylene blue on biofilm bacteria was less than on bacteria in a planktonic state (Fontana *et al.*, 2009). Yet, the beneficial effects of photodynamic therapy can include not only antibacterial effects but also inactivation of proteases and inflammatory cytokines (Braham *et al.*, 2009).

It is not always clear in what concentration oral chemotherapeutics are actually applied in

**Table 13.1** Chemotherapeutic agents used in oral health care products

| Class of inhibitor | Active ingredient | Modes of action |
|---|---|---|
| Amine alcohol | Octapinol<br>Delmopinol | Plaque inhibition (Eley, 1999; Addy and Moran, 2008; Scheie and Petersen, 2008), by interfering with plaque matrix formation and reducing bacterial adherence |
| Bisbiguanide | Chlorhexidine | Antibacterial (Eley, 1999; Addy and Moran, 2008)<br>Cell wall damage (Marsh, 1992; Eley, 1999; Scheie and Petersen, 2008)<br>Plaque inhibition by binding to cell membranes (Eley, 1999; Scheie and Petersen, 2008) |
| Enzymes | Lactoperoxidase<br>Lysozyme<br>Glucose oxidase<br>Amyloglucosidase | Antibacterial (Addy and Moran, 2008)<br>Enhances host defence mechanisms (Marsh, 1992; Addy and Moran, 2008; Scheie and Petersen, 2008) |
| Essential Oils | Thymol<br>Eucalyptol | Antibacterial (Eley, 1999; Addy and Moran, 2008)<br>Antioxidative activity (Eley, 1999)<br>Inhibition of enzyme activity (Marsh, 1992; Mandel, 1994), reducing glycolysis (Marsh, 1992), reducing bacterial adherence (Marsh, 1992) |
| Fluorides | Sodium fluoride<br>Stannous fluoride<br>Amine fluoride<br>Monofluorophosphate | Prevents demineralization (Ten Cate *et al.*, 2003)<br>Enhances remineralization (Ten Cate *et al.*, 2003)<br>Antibacterial effects derived from non-fluoride portion (Mandel, 1994; Addy and Moran, 2008; Scheie and Petersen, 2008) |
| Metal ions | Stannous<br>Zinc<br>Copper | Antibacterial (Marsh, 1992; Addy and Moran, 2008; Scheie and Petersen, 2008)<br>Plaque inhibition (Marsh, 1992; Eley, 1999; Addy and Moran, 2008)<br>Inhibiting enzyme systems and glycolysis (Marsh, 1992; Eley, 1999; Scheie and Petersen, 2008) |
| Oxygenating agents | Hydrogen peroxide<br>Sodium peroxycarborate | Antibacterial (Mandel, 1994; Addy and Moran, 2008) |
| Plant extracts/ Natural products | Sanguinarine extracts | Antibacterial (Addy and Moran, 2008)<br>Plaque inhibition (Mandel, 1994; Eley, 1999) by suppression growth of bacterial strains and enzyme activity (Scheie and Petersen, 2008) |
| Phenols | Triclosan | Antibacterial (Marsh, 1992; Eley, 1999; Addy and Moran, 2008; Scheie and Petersen, 2008)<br>Plaque inhibition (Marsh, 1992; Mandel, 1994; Eley, 1999; Addy and Moran, 2008)<br>Interference with plaque metabolism (Marsh, 1992)<br>Disruption of bacterial cell (Scheie and Petersen, 2008) |
| Quaternary ammonium compounds | Cetylpyridium chloride | Antibacterial (Eley, 1999; Addy and Moran, 2008; Scheie and Petersen, 2008)<br>Plaque inhibition (Mandel, 1994; Eley, 1999; Addy and Moran, 2008) by interaction with microorganisms (Mandel, 1994; Scheie and Petersen, 2008) |
| Surfactants | Sodium lauryl sulfate | Antibacterial (Addy and Moran, 2008; Scheie and Petersen, 2008)<br>Inactivate bacterial enzymes (Marsh, 1992; Scheie and Petersen, 2008) |

the oral cavity, although this can greatly affect their efficacy. Large individual variations exist, for instance, in the amount of toothpaste used during brushing and in the amount of water rinsing afterwards. Most people add water to the toothbrush with toothpaste on it, thereby diluting the active ingredients in the toothpaste. Further dilution takes place in the oral cavity due to salivation and swallowing. Also post-brushing rinsing with water will reduce the concentration of actives in saliva. A water rinse after use of a toothpaste containing 5000 ppm F decreased

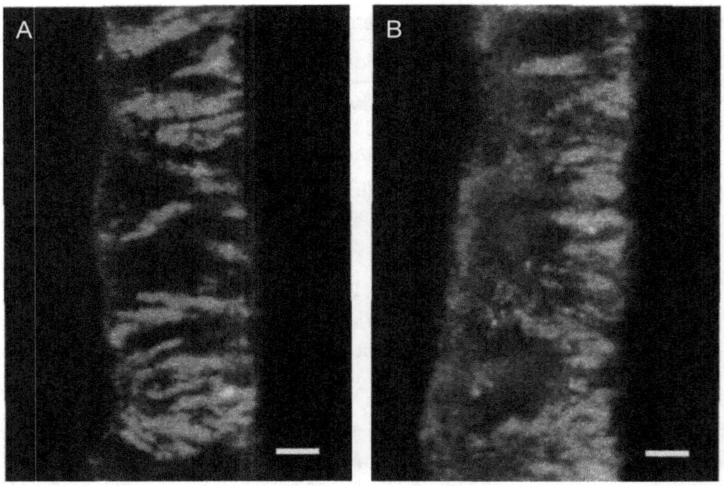


**Figure 13.2** Confocal laser scanning micrographs of a 24 hours old biofilm (Zaura-Arite *et al.*, 2001) grown on an *in situ* surface before (A) and after (B) rinsing with a chlorhexidine mouthrinse. Green and red areas represent live and dead bacteria. Note, that oral biofilm is only killed on the outside, and not on the substratum side (right). A colour version of this figure is available in the plate section at the back of the book.

the fluoride retention in oral biofilm to levels comparable to those after using a 1450 ppm toothpaste without water rinsing (Nordstrom and Birkhed, 2009). Rinsing with toothpaste-foam for 1 minute after brushing resulted in a high fluoride concentration in both saliva (Sjogren and Birkhed, 1994) and in oral biofilm (Sjogren *et al.*, 1996), which is more effective than rinsing with a NaF-containing mouthrinse (Sjogren *et al.*, 1996) after toothbrushing. A similar mechanism can be found for post-rinsing behaviour: extrinsic factors (e.g. eating, drinking and chewing) reduced the activity of a chlorhexidine mouthrinse (Tomas *et al.*, 2010). Also, the concentrations used (Cousido *et al.*, 2010), the frequency of application (Cousido *et al.*, 2010), or the age of the product (Weiland *et al.*, 2008) all influence the efficacy of antibacterial mouthrinses and toothpastes.

## Retention of chemotherapeutic agents in the oral cavity

An important property of effective oral antibacterial agents is their ability to be retained in the oral cavity by interaction with oral surfaces, and their subsequent slow release in an active concentration at relevant places. This process is called substantivity and is defined by the Oxford Dictionary of Dentistry (Ireland, 2010) as a characteristic of an antibacterial product whereby it remains active in the oral cavity for a longer period than the average brushing or rinsing time. Substantivity leads to the long-lasting effectiveness of oral hygiene products, and is especially important because the concentration of toothpaste and mouthrinse components in the oral cavity decreases rapidly through rinsing and swallowing.

Examples of substantive action are numerous. Antibacterial effects on bacterial viability in saliva e.g. of a 0.2% chlorhexidine rinse (Cousido *et al.*, 2010) could be perceived up to 7 hours post-use, and included significant reductions in plaque re-growth and biofilm viability until 24 hours post-use (Arweiler *et al.*, 2002). Increased percentages of dead bacteria have been found in oral biofilm until at least 6 hours after last use of a chlorhexidine mouthrinse (see Fig. 13.3). Effects of mouthrinse and toothpaste both containing amine fluoride/stannous fluoride (Weiland *et al.*, 2008) on bacterial viability in a biofilm can be also found up to 7 hours after its application. A Triclosan containing toothpaste had a substantive effect on bacterial viability in oral biofilm until 24 hours after the last use (Arweiler *et al.*, 2002).

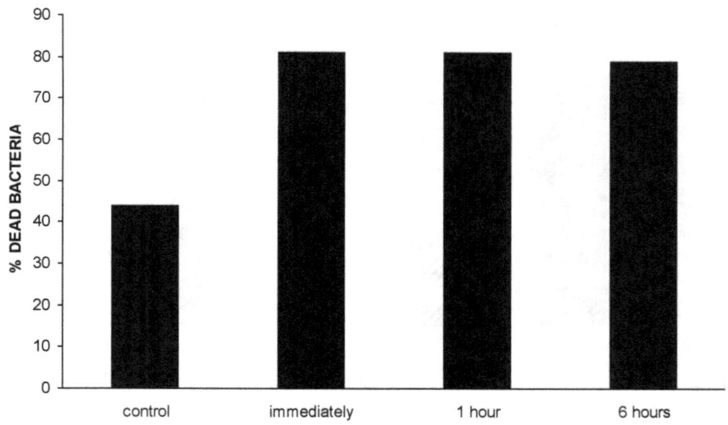


**Figure 13.3** Percentage dead bacteria in 9 hours old oral biofilm formed *in vivo*, at different times after the *in vivo* use of a chlorhexidine mouthrinse. Adapted from Van der Mei *et al.* (2006).

## The reservoir function of residual biofilm

Fluoride is much smaller than most antibacterial agents, and it has been shown in several studies that, after exposure to fluoride, biofilms can absorb fluoride ions to become a fluoride reservoir (Sjogren *et al.*, 1996; Nordstrom and Birkhed, 2009). Increased fluoride retention in biofilms can be found until 12 hours after the last brushing with a 1030 ppm fluoride toothpaste (Pessan *et al.*, 2008). Depending on the concentration of fluoride included in a toothpaste, which ranges from 500 ppm to 5000 ppm, only 0.4 ppm fluoride has been demonstrated to accumulate per mg biofilm (Nordstrom *et al.*, 2009) which, based on *in vitro* research, may be expected to promote remineralization of enamel (Lynch *et al.*, 2006). In general, a fluoride gel is more effective than are fluoride-containing toothpastes in maintaining high levels of fluoride in both oral biofilm and saliva; and increased fluoride retention in oral biofilm can be found until 24 hours after application of a fluoride gel (Zero *et al.*, 1992). Also, fluoride-containing toothpastes as well as fluoride-releasing restorations maintain increased fluoride levels in oral biofilm (Cenci *et al.*, 2008). It has been demonstrated recently that fluoride uptake in residual biofilm is very important for the anticaries effect of fluoride containing toothpaste (Tenuta *et al.*, 2009). This means that the biofilm is not only a reservoir for fluoride, but also contributes to the clinically relevant effect of fluoride by promotion of enamel remineralization and anticaries efficacy.

Recently, it was shown that biofilm viability and re-attachment *in vitro* on top of a biofilm that was exposed to antibacterial chemotherapeutics tended to decrease as compared to an untreated biofilm (Verkaik *et al.*, 2011), which suggests that the oral biofilm may also act as a reservoir for oral antibacterials. This was confirmed *in vitro* (see Fig. 13.4a) and *in vivo* (see Fig. 13.4b), by showing that after mechanical cleaning, residual biofilm can release bio-active concentrations of antibacterial agents absorbed from toothpastes and mouthrinses. Dead biofilm exposed *in vitro* to antibacterial agents (Otten *et al.*, 2011) reduced the growth of freshly added salivary bacteria to the exposed dead biofilm by 50% to 70%. Biofilm collected from the oral cavity up to 6 hours after use of an antibacterial was also able to kill unexposed oral biofilm (Fig. 13.4b) (Otten *et al.*, 2010).

## Summary and future perspectives

This chapter provides a brief overview on oral biofilm formation and control, focusing on the reservoir function of oral biofilm. Evidence is provided that oral biofilm can absorb antibacterials and subsequently release these antibacterials in bio-active concentrations. These data support the use of antibacterial toothpastes and mouthrinses

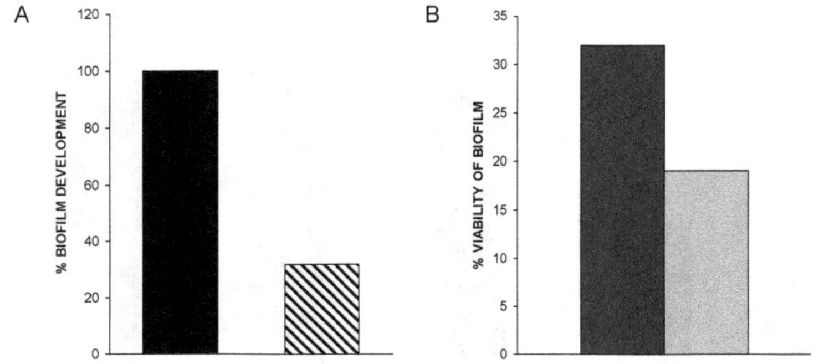


**Figure 13.4** Illustrations of the reservoir function of oral biofilm for antibacterial agents. (A) The percentage biofilm developed on top of an *in vitro* grown 4 hours old biofilm that has been exposed to sterile water (black bar) or 25% Crest Pro Health mouthrinse (hatched area). Adapted from Otten *et al.* (2011). (B) The theoretically expected initial viability (black) and clinically measured viability after 2 hours (grey) of *in vivo* grown biofilm samples, consisting of control biofilm and biofilm collected after using Crest Pro Health mouthrinse. Adapted from Otten *et al.* (2010).

in oral health care. Most people are unable to remove oral biofilm completely, especially in vulnerable places such as interproximally or around orthodontic appliances and restorations. The substantive release of absorbed antibacterial from residual biofilm can thus be of great value because the antibacterials are released just there where plaque control is most needed. However, this should not be taken as a license for incomplete oral hygiene, but rather as a suggestion that negative effects of residual biofilm can be reduced by the use of antibacterial health care products. To what extent this reservoir function contributes towards the control of caries and gingivitis remains to be determined clinically. Clinical benefits of the absorptive role of residual biofilm, may possibly be enhanced by a toothbrush that introduces energy into the biofilm (Hope and Wilson, 2003; Busscher *et al.*, 2010) and thus leads to fluffing up of the biofilm (Busscher *et al.*, 2010). Fluffed up biofilms may have an increased capacity to absorb oral antibacterials. Further clinical studies are needed to demonstrate any additional effect of an antibacterial reservoir in residual biofilm after powered brushing.

## References

Addy, M., and Moran, J. (2008). Chemical supragingival plaque control. In Clinical Periodontology and Implant Dentistry, Lang, N.P., and Lindhe, J., eds. (Blackwell Munskgaard, Oxford, UK), pp. 734–765.

Addy, M. (1997). Antiseptics in periodontal therapy. In Clinical Periodontology and Implant Dentistry, Lindhe, J., Karring, T., and Lang, N.P., eds. (Munksgaard, Copenhagen, Denmark), pp. 461–482.

Arweiler, N.B., Auschill, T.M., Reich, E., and Netuschil, L. (2002). Substantivity of toothpaste slurries and their effect on reestablishment of the dental biofilm. J. Clin. Periodontol. *29*, 615–621.

Axelsson, P., Nystrom, B., and Lindhe, J. (2004). The long-term effect of a plaque control program on tooth mortality, caries and periodontal disease in adults. Results after 30 years of maintenance. J. Clin. Periodontol. *31*, 749–757.

Baehni, P.C., and Takeuchi, Y. (2003). Anti-plaque agents in the prevention of biofilm-associated oral diseases. Oral Dis. *9(Suppl. 1)*, 23–29.

Balakrishnan, M., Simmonds, R.S., and Tagg, J.R. (2000). Dental caries is a preventable infectious disease. Aust. Dent. J. *45*, 235–245.

Biesbrock, A.R., Walters, P.A., Bartizek, R.D., Goyal, C.R., and Qaqish, J.G. (2008). Plaque removal efficacy of an advanced rotation-oscillation power toothbrush versus a new sonic toothbrush. Am. J. Dent. *21*, 185–188.

Braham, P., Herron, C., Street, C., and Darveau, R. (2009). Antimicrobial photodynamic therapy may promote periodontal healing through multiple mechanisms. J. Periodontol. *80*, 1790–1798.

Brecx, M. (1997). Strategies and agents in supragingival chemical plaque control. Periodontol. 2000 *15*, 100–108.

Busscher, H.J., Jager, D., Finger, G., Schaefer, N., and Van der Mei, H.C. (2010). Energy transfer, volumetric expansion, and removal of oral biofilms by non-contact brushing. Eur. J. Oral Sci. *118*, 177–182.

Cenci, M.S., Tenuta, L.M., Pereira-Cenci, T., Del Bel Cury, A.A., Ten Cate, J.M., and Cury, J.A. (2008). Effect of microleakage and fluoride on enamel-dentine demineralization around restorations. Caries Res. *42*, 369–379.

Chen, F., and Wang, D. (2010). Novel technologies for the prevention and treatment of dental caries: a patent survey. Expert. Opin. Ther. Pat. *20*, 681–694.

Claydon, N.C. (2008). Current concepts in toothbrushing and interdental cleaning. Periodontol. 2000 *48*, 10–22.

Corby, P.M., Biesbrock, A., Bartizek, R., Corby, A.L., Monteverde, R., Ceschin, R., and Bretz, W.A. (2008). Treatment outcomes of dental flossing in twins: molecular analysis of the interproximal microflora. J. Periodontol. *79*, 1426–1433.

Cousido, M.C., Tomas Carmona, I., Garcia-Caballero, L., Limeres, J., Alvarez, M., and Diz, P. (2010). *In vivo* substantivity of 0.12% and 0.2% chlorhexidine mouthrinses on salivary bacteria. Clin. Oral Investig. *14*, 397–402.

De Oliviera, C., Watt, R., and Hamer, M. (2010). Toothbrushing, inflammation, and risk of cardiovascular disease: results from Scottish Health Survey. BMJ *340*, c2451.

Duckworth, R.M., and Gao, X.J. (2006). Plaque as a reservoir for active ingredients. In The Teeth and Their Environment, Duckworth, R.M., ed. (Karger, Basel, Switzerland), pp. 132–149.

Eley, B.M. (1999). Antibacterial agents in the control of supragingival plaque – a review. Br. Dent. J. *186*, 286–296.

Fejerskov, O. (2004). Changing paradigms in concepts on dental caries: consequences for oral health care. Caries Res. *38*, 182–191.

Fontana, C.R., Abernethy, A.D., Som, S., Ruggiero, K., Doucette, S., Marcantonio, R.C., Boussios, C.I., Kent, R., Goodson, J.M., Tanner, A.C., *et al.* (2009). The antibacterial effect of photodynamic therapy in dental plaque-derived biofilms. J. Periodontal Res. *44*, 751–759.

Haffajee, A.D., Teles, R.P., and Socransky, S.S. (2006). The effect of periodontal therapy on the composition of the subgingival microbiota. Periodontol. 2000 *42*, 219–258.

Hamilton, I., and Bowden, G. (1988). Effect of fluoride on oral microorganisms. In Fluoride in Dentistry, Ekstrand, J., Fejerskov, O., and Silverstone, L.M., eds. (Munskgaard, Copenhagen, Denmark), pp. 77–103.

Hardie, J.M. (1992). Oral microbiology: current concepts in the microbiology of dental caries and periodontal disease. Br. Dent. J. *172*, 271–278.

Ho, H.P., and Niederman, R. (1997). Effectiveness of the Sonicare sonic toothbrush on reduction of plaque, gingivitis, probing pocket depth and subgingival bacteria in adolescent orthodontic patients. J. Clin. Dent. *8*, 15–19.

Hojo, K., Nagaoka, S., Ohshima, T., and Maeda, N. (2009). Bacterial interactions in dental biofilm development. J. Dent. Res. *88*, 982–990.

Hope, C.K., and Wilson, M. (2003). Effects of dynamic fluid activity from an electric toothbrush on *in vitro* oral biofilms. J. Clin. Periodontol. *30*, 624–629.

Ireland, R. (2010). A Dictionary of Dentistry (Oxford University Press, Oxford, UK).

Keijser, B.J., Zaura, E., Huse, S.M., Van der Vossen, J., Schuren, F.H., Montijn, R.C., Ten Cate, J.M., and Crielaard, W. (2008). Pyrosequencing analysis of the oral microflora of healthy adults. J. Dent. Res. *87*, 1016–1020.

Kolenbrander, P.E., Palmer, R.J., Jr., Rickard, A.H., Jakubovics, N.S., Chalmers, N.I., and Diaz, P.I. (2006). Bacterial interactions and successions during plaque development. Periodontol. 2000 *42*, 47–79.

Lang, N.P., Cumming, B.R., and Loe, H. (1973). Toothbrushing frequency as it relates to plaque development and gingival health. J. Periodontol. *44*, 396–405.

Lopez, R. (2008). Periodontal disease and adverse pregnancy outcomes. Evid. Based. Dent. *9*, 48.

Lynch, R.J., Mony, U., and Ten Cate, J.M. (2006). The effect of fluoride at plaque fluid concentrations on enamel de- and remineralisation at low pH. Caries Res. *40*, 522–529.

Mandel, I.D. (1994). Antimicrobial mouthrinses: overview and update. J. Am. Dent. Assoc. *125(Suppl. 2)*, 2S–10S.

Marquis, R.E., Clock, S.A., and Mota-Meira, M. (2003). Fluoride and organic weak acids as modulators of microbial physiology. FEMS Microbiol. Rev. *26*, 493–510.

Marsh, P.D. (1992). Microbiological aspects of the chemical control of plaque and gingivitis. J. Dent. Res. *71*, 1431–1438.

Marsh, P.D. (1994). Microbial ecology of dental plaque and its significance in health and disease. Adv. Dent. Res. *8*, 263–271.

Marsh, P.D. (2010). Controlling the oral biofilm with antimicrobials. J. Dent. *38(Suppl. 1)*, S11–S15.

Moritis, K., Delaurenti, M., Johnson, M.R., Berg, J., and Boghosian, A.A. (2002). Comparison of the Sonicare Elite and a manual toothbrush in the evaluation of plaque reduction. Am. J. Dent. *15(Spec)*, 23B–25B.

Nordstrom, A., and Birkhed, D. (2009). Fluoride retention in proximal plaque and saliva using two NaF dentifrices containing 5,000 and 1,450 ppm F with and without water rinsing. Caries Res. *43*, 64–69.

Nordstrom, A., Mystikos, C., Ramberg, P., and Birkhed, D. (2009). Effect on *de novo* plaque formation of rinsing with toothpaste slurries and water solutions with a high fluoride concentration (5,000 ppm). Eur. J. Oral Sci. *117*, 563–567.

Nyvad, B. (2003). The role of oral hygiene. In Dental Caries, The Disease and its Clinical Management, Fejerskov, O., and Kidd, E., eds. (Blackwell Munksgaard, Oxford, UK), pp. 171–177.

Otten, M.P., Busscher, H.J., Van der Mei, H.C., Abbas, F., and Van Hoogmoed, C.G. (2010). Retention of antimicrobial activity in plaque and saliva following mouthrinse use *in vivo*. Caries Res. *44*, 459–464.

Otten, M.P., Busscher, H.J., Van der Mei, H.C., Van Hoogmoed, C.G., and Abbas, F. (2011). Acute and substantive action of antimicrobial toothpastes and mouthrinses on oral biofilm *in vitro*. Eur. J. Oral Sci. *119*, 151–155.

Paraskevas, S., Timmerman, M.F., Van der Veen, M., and Van der Weijden, G.A. (2006). Additional effect of dentifrices on the instant efficacy of toothbrushing. J. Periodontol. *77*, 1522–1527.

Paraskevas, S., Rosema, N.A., Versteeg, P., Timmerman, M.F., Van der Veen, M., and Van der Weijden, G.A. (2007). The additional effect of a dentifrice on the instant efficacy of toothbrushing: a crossover study. J. Periodontol. *78*, 1011–1016.

Pessan, J.P., Silva, S.M., Lauris, J.R., Sampaio, F.C., Whitford, G.M., and Buzalaf, M.A. (2008). Fluoride uptake by plaque from water and from dentifrice. J. Dent. Res. *87*, 461–465.

Quirynen, M., Marechal, M., Busscher, H.J., Weerkamp, A.H., Arends, J., Darius, P.L., and Van Steenbergen, D. (1989). The influence of surface free-energy on planimetric plaque growth in man. J. Dent. Res. *68*, 796–799.

Rosan, B., and Lamont,R.J. (2000). Dental plaque formation. Microbes Infect. *2*, 1599–1607.

Rosema, N.A., Timmerman, M.F., Versteeg, P.A., Van Palenstein Helderman, W.H., Van der Velden, U., and Van der Weijden, G.A. (2008). Comparison of the use of different modes of mechanical oral hygiene in prevention of plaque and gingivitis. J. Periodontol. *79*, 1386–1394.

Scannapieco, F.A., Bush, R.B., and Paju, S. (2003). Periodontal disease as a risk factor for adverse pregnancy outcomes. A systematic review. Ann. Periodontol. *8*, 70–78.

Scheie, A.A. (2003). The role of antimicrobials. In Dental Caries: The Disease and its Clinical Management, Fejerskov, O., and Kidd, E., eds. (Blackwell Munksgaard, Oxford, UK), pp. 179–188.

Scheie, A.A., and Petersen, F.C. (2008). Antimicrobials in caries control. In Dental Caries: The Disease and its Clinical Management, Fejerskov, O., and Kidd, E., eds. (Blackwell Munskgaard, Oxford, UK), pp. 265–277.

Sedlacek, M.J., and Walker, C. (2007). Antibiotic resistance in an *in vitro* subgingival biofilm model. Oral Microbiol. Immunol. *22*, 333–339.

Sekino, S., Ramberg, P., Uzel, N.G., Socransky, S., and Lindhe,J. (2004). The effect of a chlorhexidine regimen on *de novo* plaque formation. J. Clin. Periodontol. *31*, 609–614.

Seymour, G.J., Ford, P.J., Cullinan, M.P., Leishman, S., and Yamazaki, K. (2007). Relationship between periodontal infections and systemic disease. Clin. Microbiol. Infect. *13(Suppl. 4)*, 3–10.

Sjogren, K., and Birkhed, D. (1994). Effect of various post-brushing activities on salivary fluoride concentration after toothbrushing with a sodium fluoride dentifrice. Caries Res. *28*, 127–131.

Sjogren, K., Birkhed, D., Rangmar, S., and Reinhold, A.C. (1996). Fluoride in the interdental area after two different post-brushing water rinsing procedures. Caries Res. *30*, 194–199.

Stoodley, P., Sauer, K., Davies, D.G., and Costerton, J.W. (2002). Biofilms as complex differentiated communities. Annu. Rev. Microbiol. *56*, 187–209.

Stoodley, P., Nguyen, D., Longwell, M., Nistico, L., von Ohle, C., Milanovich, N., and de Jager, M. (2007). Effect of the sonicare flexcare power toothbrush on fluoride delivery through *Streptococcus mutans* biofilms. Compend. Contin. Educ. Dent. *28(9 suppl. 1)*, 15–22.

Stoodley, P., Wefel, J., Gieseke, A., Debeer, D., and Von Ohle, C. (2008). Biofilm plaque and hydrodynamic effects on mass transfer, fluoride delivery and caries. J. Am. Dent. Assoc. *139*, 1182–1190.

Ten Cate, J.M. (2006). Biofilms, a new approach to the microbiology of dental plaque. Odontology *94*, 1–9.

Ten Cate, J.M., Larsen, M.J., Pearce, E.I.F., and Fejerskov, O. (2003). Chemical interactions between the tooth and oral fluids. In Dental Caries: The Disease and its Clinical Management, Fejerskov, O., and Kidd, E., eds. (Blackwell Munksgaard, Oxford, UK), pp. 49–69.

Tenuta, L.M., Zamataro, C.B., Del Bel Cury, A.A., Tabchoury, C.P., and Cury, J.A. (2009). Mechanism of fluoride dentifrice effect on enamel demineralization. Caries Res. *43*, 278–285.

Tomas, I., Cousido, M.C., Garcia-Caballero, L., Rubido, S., Limeres, J., and Diz, P. (2010). Substantivity of a single chlorhexidine mouthwash on salivary flora: Influence of intrinsic and extrinsic factors. J. Dent. *38*, 541–546.

Tritten, C.B., and Armitage, G.C. (1996). Comparison of a sonic and a manual toothbrush for efficacy in supragingival plaque removal and reduction of gingivitis. J. Clin. Periodontol. *23*, 641–648.

Van der Mei, H.C., White, D.J., Atema-Smit, J., Van de Belt-Gritter, E., and Busscher, H.J. (2006). A method to study sustained antimicrobial activity of rinse and dentifrice components on biofilm viability *in vivo*. J. Clin. Periodontol. 33, 14–20.

Van der Weijden, G.A., Echeverria, J.J., Sanz, M., and Lindhe, J. (2008). Mechanical supragingival plaque control. In Clinical Periodontology and Implant Dentistry, Lindhe, J., Lang, N.P., and Karring, T., eds. (Blackwell Munskgaard, Copenhagen), pp. 705–733.

Verkaik, M.J., Busscher, H.J., Jager, D., Slomp, A.M., Abbas, F., and Van der Mei, H.C. (2011). Efficacy of natural antimicrobials in toothpaste formulations against oral biofilms *in vitro*. J. Dent. *39*, 218–224.

Weiland, B., Netuschil, L., Hoffmann, T., and Lorenz, K. (2008). Substantivity of amine fluoride/stannous fluoride following different modes of application: a randomized, investigator-blind, placebo-controlled trial. Acta Odontol. Scand. *66*, 307–313.

Wiegand, A., Buchalla, W., and Attin, T. (2007). Review on fluoride-releasing restorative materials – fluoride release and uptake characteristics, antibacterial activity and influence on caries formation. Dent. Mater. *23*, 343–362.

Wilson, M. (1996). Susceptibility of oral bacterial biofilms to antimicrobial agents. J. Med. Microbiol. *44*, 79–87.

The World Oral Health Report – Continuous improvement of oral health in the 21st century – the approach of the WHO Global Oral Health Programme. (2003).

Zaura-Arite, E., Van Marle, J., and Ten Cate, J.M. (2001). Confocal microscopy study of undisturbed and chlorhexidine-treated dental biofilm. J. Dent. Res. *80*, 1436–1440.

Zero, D.T., Raubertas, R.F., Fu, J., Pedersen, A.M., Hayes, A.L., and Featherstone, J.D. (1992). Fluoride concentrations in plaque, whole saliva, and ductal saliva after application of home-use topical fluorides [published erratum appears in J. Dent. Res. 1993, *72*, 87]. J. Dent. Res. *71*, 1768–1775.

# Probiotics: a Possible Tool in Oral Health Care?

14

Christof Godts, Gitte Loozen, Marc Quirynen and Wim Teughels


## Abstract

The human oral cavity is colonized by a wide variety of bacteria, which form very complex and dynamic biofilms on hard and soft tissues. Certain members of these microbiological communities are associated with oral infections, such as caries and periodontal diseases. New treatment approaches, which do not rely on conventional antibiotic therapies, are emerging since complete eradication of pathogenic bacteria from oral biofilms is impossible and antibiotic resistance is becoming problematic. For example, attempts have been made to reduce the overall pathogenicity of tissue-associated biofilms by introducing live beneficial bacteria. Early successes, primarily in the field of gastrointestinal microbiology, have paved the way for the introduction of probiotics in oral health care. These new anti-/pro-microbial therapies are considered very promising for prevention and treatment of plaque-related oral diseases.

In this chapter, the concept of probiotics for oral healthcare is introduced, followed by an overview of the diverse mechanisms of probiotic action in the oral cavity. Since the benefits of probiotics will ultimately be revealed by clinical studies, the clinical outcomes of probiotic applications for combating dental caries and periodontal diseases are addressed. Finally the interactions of probiotics with the oral microbial ecosystem are discussed and future perspectives regarding the oral probiotic concept are presented.


## Introduction

### Time for change

As described in the previous chapters, the oral cavity is densely populated with many microorganisms living in complex microbiological communities. It has been known for many decades that certain members of these microbial ecosystems are associated with oral diseases such as caries and periodontal disease. Prevention and treatment of these diseases is primarily targeted at the reduction of this bacterial threat. For periodontal diseases the world-wide 'gold standard' treatment strategy is based on mechanical subgingival debridement (rootplanning), if necessary including periodontal surgery to reduce the depth of the periodontal pocket, in combination with improved oral hygiene. Additionally, chemotherapeutics are widely used to prevent and treat infections by indigenous and exogenous microbes. The availability of effective and cheap antibiotics in the latter half of the twentieth century revolutionized the treatment of infectious diseases and, for developed countries at least, dramatically reduced the impact of infectious diseases on the death rate. The immunologist and Nobel laureate, Macfarlane Burnett, stated in 1962 that 'by the late twentieth century, we can anticipate the virtual elimination of infectious diseases as a significant factor in social life' (Burnett, 1962). However, the development of resistance to a range of antibiotics by some important pathogens has raised the possibility of a return to the pre-antibiotic 'dark-ages'. In the oral cavity, the widespread use of antibiotics is for example, reflected in the level of resistance in the subgingival microbiota of adult periodontitis patients (Van Winkelhoff *et al.*, 2000). These developments have encouraged researchers in various fields of healthcare to develop alternative approaches that do not involve the use of conventional antibiotics (Teughels *et al.*, 2000).

Recent insights in oral microbiology and its relation to the development of oral disease have resulted in a paradigm shift away from treating oral diseases by targeting specific oral pathogens towards ecological and microbiological community-based approaches (Marsh, 2003; Socransky and Haffajee, 2005). These approaches recognize the importance of maintaining a natural balance of the resident commensal oral microbiota and the need to carefully modulate host immune responses to the microbiota at a site.

The application of 'health promoting' bacteria for therapeutic purposes is one of the most promising emerging techniques focused on manipulation of microbial communities. Because the oral microbiota is at least as complex as other human-associated microbial communities, such as the gastro-intestinal or vaginal microbiota, dental biofilms are considered to be difficult therapeutic targets (Socransky and Haffajee, 2002). However, the encouraging effects of health-promoting bacteria in different fields of healthcare have resulted recently in the introduction of probiotics for oral healthcare (Meurman, 2005; Teughels *et al.*, 2008). Several clinical studies on the effects of probiotics in different fields of oral healthcare, such as halitosis (Henker *et al.*, 2001; Burton *et al.*, 2006; Kang *et al.*, 2006b), oral candidiasis (Ahola *et al.*, 2002; Hatakka *et al.*, 2007a), tooth decay (Stecksen-Blicks *et al.*, 2009) and periodontal disease (Shimauchi *et al.*, 2008), have been published. Although the use of such bacteria to specifically improve oral health is still in its infancy, the widespread oral intake of probiotics as preventative and therapeutic products for gastro-intestinal health may have major implications for the composition of the oral microflora. Therefore, probiotics are already of considerable interest for oral healthcare workers.

## The probiotic concept

The term 'probiotic' is a relatively new word and is currently used for bacteria with beneficial effects on humans and animals. As an antonym of the term 'antibiotics' it was introduced by Lilly and Stillwell (1965) as 'Substances produced by micro-organisms which promote the growth of other micro-organisms'. However, the use of micro-organisms to promote health originates from ancient times, and can be traced back to the classical Roman literature where food fermented by micro-organisms was used as a therapeutic agent. Since its introduction in 1965, the term 'probiotics' has been given several different definitions (Parker, 1974; Fuller, 1989; Havenaar and Huis In't Veld, 1992; Schaafsma, 1996; Naidu *et al.*, 1999; Schrezenmeir and de Vrese, 2001) leading to the final one, which is the currently used consensus definition, put forward by the World Health Organization and the Food and Agriculture Organization of the United States (Table 14.1). Probiotics are defined as 'Live micro-organisms, that when administered in adequate amounts, confer health benefits upon the host'. The dynamics in the definition mirrors the rapid developments in the field. Still some slight modifications in this definition are likely to occur as scientists rediscover events taking place at the interface between mucosal surfaces and the microbiota, and interactions of probiotic microorganisms with the host (Böhm and Kruis, 2006). Since Saito (2004) observed beneficial effects on human health using inactivated micro-organisms or their cell compounds, even the restriction to 'live' organisms is questioned (Stamatova and Meurman, 2009). The term 'replacement therapy' (also called 'bacteriotherapy' or 'bacterial interference') is sometimes used interchangeably with 'probiotics' (for review see Wilson, 2005). Although both approaches use live bacteria for the prevention or treatment of infectious disease, there are some slight differences (Table 14.2). Since there is much confusion about the terminology, we did not specifically differentiate between probiotic therapies and replacement therapies.

In contrast, 'prebiotics' are generally defined as 'not digestible food ingredients that beneficially affect the host by selectively stimulating the growth and/or activity of one or a limited number of bacterial species already established in the colon, and thus in effect improve host health' (Gibson and Roberfroid, 1995). These are inulin,

**Table 14.1** Changes in the definition of 'probiotics'

| Year | Definition | Reference |
|---|---|---|
| 1965 | Substances produced by micro-organisms which promote the growth of other microorganisms | Lilly and Stillwell (1965) |
| 1974 | Organisms and substances which contribute to intestinal microbial balance | Parker (1974) |
| 1989 | A live microbial feed supplement which beneficially affects the host animal by improving its intestinal microbial balance | Fuller (1989) |
| 1992 | A viable mono- or mixed-culture of micro-organisms which applied to animal or man, beneficially affects the host by improving the properties of the indigenous microflora | Havenaar and Huis In't Veld (1992) |
| 1996 | Living microorganisms, which upon ingestion in certain numbers, exert health benefits beyond inherent basic nutrition | Schaafsma (1996) |
| 1999 | A microbial dietary adjuvant that beneficially affects the host physiology by modulating mucosal and systemic immunity, as well as improving nutritional and microbial balance in the intestinal tract | Naidu *et al.* (1999) |
| 2001 | A preparation of or a product containing viable, defined microorganisms in sufficient numbers, which alter the microflora (by implantation or colonization) in a compartment of the host and by that exert beneficial health effects in this host | Schrezenmeir and de Vrese (2001) |
| 2001 | Live micro-organisms which, when administered in adequate amounts, confer a health benefit to the host | FAO/WHO (2002) |

**Table 14.2** Differences between 'replacement' and 'probiotic' therapy

| Replacement therapy | Probiotic therapy |
|---|---|
| Effector strain is not ingested and is applied directly on the site of infection | Probiotics are generally used as dietary supplements |
| Colonization of the site by the effector strain is essential | Probiotics are able to exert a beneficial effect without permanently colonizing the site |
| Involves dramatic and long-term change in the indigenous microbiota | Rarely a dramatic and long-term microbiological change |
| Directed at displacing or preventing colonization of a pathogen | |
| Has a minimal immunological impact | Exerts beneficial effects by influencing the immune system |

fructo-oligosaccharides, galacto-oligosaccharides and lactulose. The concept of prebiotics essentially has the same aim as probiotics, which is to improve host health via modulation of the intestinal flora, although by a different mechanism.

However, there are some cases in which prebiotics may be beneficial for the probiotic, especially with regard to bifidobacteria, a case that is described as the synbiotic concept. Synbiotics are defined as 'mixtures of probiotics and prebiotics that beneficially affect the host by improving the survival and implantation of live micro microbial dietary supplements in the gastrointestinal tract of the host' (Andersson *et al.*, 2001).

Probiotics are often brought in connection with 'functional foods'. This term comprises the knowledge of the relationship between foods and health and the effect of food ingredients on physiological functions.

The history of the term 'probiotics' mirrors the rapid developments in our understanding and use of micro-organisms in human conditions and diseases. The definition will surely have to be further adapted as we learn even more about actions of probiotic micro-organisms and their interaction with the host (Bohm and Kruis, 2006).

## Mechanisms of probiotic action

The diversity of conditions that may benefit from ingestion of probiotics (Table 14.3) illustrates the

**Table 14.3** Well-established probiotic effects

| Effect | References |
|---|---|
| Prevention and/or reduction of duration and complaints of rotavirus-induced or antibiotic-associated diarrhoea as well as alleviation of complaints due to lactose intolerance | Majamaa *et al.* (1995), Buydens and Debeuckelaere (1996), Cremonini *et al.* (2002), Hawrelak *et al.* (2005) |
| Decrease of the concentration of cancer-promoting enzymes and/or putrefactive (bacterial) metabolites in the gut | Goldin and Gorbach, (1984), Haskard *et al.* (2001), Ouwehand *et al.* (2002) |
| Prevention and alleviation of unspecific and irregular complaints of the gastrointestinal tracts in healthy people | de Vrese and Schrezenmeir (2008) |
| Beneficial effects on microbial aberrancies, inflammation and other complaints in connection with: inflammatory diseases of the gastrointestinal tract, *Helicobacter pylori* infection or bacterial overgrowth | Vanderhoof *et al.* (1998), Gionchetti *et al.* (2000), Felley *et al.* (2001), Gaon *et al.* (2002), Ishikawa *et al.* (2003) |
| Normalization of passing stool and stool consistency in subjects suffering from constipation or an irritable colon | Niedzielin *et al.* (2001), Ouwehand *et al.* (2002), Koebnick *et al.* (2003) |
| Prevention or alleviation of allergies and atopic diseases in infants | Isolauri *et al.* (2000), Kalliomaki *et al.* (2001), Isolauri (2003), Ogawa *et al.* (2006), Forsythe *et al.* (2007) |

variety of mechanisms that may be involved in their actions and that some effects are systemic rather than just local. There are common themes emerging in studies of the modes of action of probiotics and numerous mechanisms related to oral microbiological ecology have been proposed to explain their beneficial effects, including (i) competition with potential pathogens for nutrients or adhesion sites, (ii) degradation of toxins and production of antimicrobial substances and (iii) local or systemic immuno-modulation (Silva *et al.*, 1987; Lewis and Freeman, 1998; Isolauri, 2001; Meurman and Stamatova, 2007). In all likelihood, there does not exist a single probiotic bacterium exhibiting all three functions; at least not to the extent that it could be a remedy for prevention or therapy of all types of diseases (Bonifait *et al.*, 2009; Oelschlaeger, 2010). Therefore, probiotic strains are often used in combination with each other in order to increase the number of beneficial effects. However, it is important to realize that probiotic bacterial strains can behave differently or induce completely opposite effects under different conditions or when combined. This makes generalizations of strain effects or extrapolating effects from single probiotic strains speculative. Also, extrapolating effects observed on the gastro-intestinal microbiota to effects on the oral microbiota is inappropriate.

Ultimately, evidence supporting the benefits of probiotics must emerge from clinical studies. While certain modes of actions shown *in vitro* look promising, these action mechanisms might be altered when the probiotic is in the oral cavity and thereby may have little chance of conferring oral health benefits (Reid *et al.*, 2006)

### Competitive exclusion

According to the 'competitive exclusion principle', also referred to as the Gause's law, two species in competition for the same resources (nutrients, space,...) cannot stably co-exist. One species will always have a slight advantage over the other, leading towards a suppression or a shift of the latter species to another niche. The competitive exclusion principle used by beneficial bacteria can occur on two levels: (1) hindering the adhesion of pathogenic bacteria or (2) competing for the same nutrients.

The literature points out that antagonistic strains are better adapted to their niche than potential pathogens, and can therefore interfere in diseases by passively occupying the niche or actively restricting the adhesion capability of pathogens to surfaces. However, definitive proof that any of these mechanisms occur for probiotic strains introduced *in vivo*, has seldom been shown. It has been revealed that several bacterial strains, mainly streptococci can hinder colonization of periodontopathogens to hard and soft tissue surfaces *in vitro* (Teughels *et al.*, 2007a; Sliepen *et al.*, 2008; Van Hoogmoed *et al.*, 2008; Sliepen *et al.*, 2009b). Through the production of biosurfactants the probiotic agent is able to reduce the adhesion capacity of specific pathogens. Van Hoogmoed *et al.* (2004, 2008) observed that biosurfactants produced by *Streptococcus mitis* strains decreased adhesion of *Streptococcus mutans* and several periodontal pathogens. Alternatively probiotics have been shown to inhibit adhesion by modifying protein composition at the binding site. Certain probiotic strains are able to modify the protein composition in the salivary pellicle by removing salivary agglutinin (gp340), which is an important host receptor for *S. mutans* adhesion (Haukioja, 2008).

In addition micro-organisms can compete for certain essential nutrients or chemicals, and in doing so can inhibit growth of the pathogen (Elli *et al.*, 2000). For example, *Prevotella intermedia* utilizes vitamin K as a growth resource, which can be substituted by progesterone or oestrogen. Levels of progesterone and oestrogen in gingival crevicular fluid are greatly increased during pregnancy. This may explain the transition from a healthy microbiota to a pathogenic during pregnancy gingivitis. Probiotics, able to outcompete periodontal pathogens on nutrient uptake, could potentially improve oral health (Wang *et al.*, 1990; Smith and Pippin, 1998).

### Degradation of toxins and production of antimicrobial substances

Probiotic bacteria are able to produce a diverse range of compounds acting as antimicrobial agents, such as lactic acid, hydrogen peroxide, bacteriocins and bacteriocin like substances (Gillor *et al.*, 2008; Gordon, 2009; Oelschlaeger, 2010).

Short-chain fatty acids, such as lactic acids, are able to pass across bacterial cell membranes. Subsequent acidification of the cytoplasm will lead to inhibition of bacterial proliferation. In this respect, Sookkhee *et al.* (2001) showed antimicrobial activity and inhibition of *Porphyromonas gingivalis* and *S. mutans* by oral isolates of lactic acid bacteria. Activity was higher at an acidic pH range, indicating that the antimicrobial effect was partially due to involvement of organic acids. Kõll-Klais *et al.* (2005) largely confirm this observation by showing higher prevalence of homofermentative lactobacilli among healthy persons compared to patients with periodontitis. Since production of lactic acid in homofermentative lactobacilli exceeds production in heterofermentative lactobacilli, it may be expected that homofermentative lactobacilli would display a more pronounced inhibition of *P. gingivalis* and *P. intermedia*.

Various *in vitro* and *in vivo* studies have shown inhibition of pathogenic bacterial species through production of hydrogen peroxide (Mashimo *et al.*, 1985; Tompkins and Tagg, 1986; Makras and De Vuyst, 2006; Falagas *et al.*, 2007). In this regard, Hillman and Shivers (1988) observed a 45-fold lower appearance of *Aggregatibacter actinomycetemcomitans* in gnotobiotic rats infected with a hydrogen peroxide producing *S. sanguinis* strain than in rats monoinfected with *A. actinomycetemcomitans*. Vanderhoeven and Camp (1993) showed inhibition of *S. mutans* in co-cultures with *S. sanguinis* when hydrogen peroxide was added to the mixture.

Bacteriocins are ribosomally synthesized cationic peptides with a narrow spectrum of antimicrobial activity compared with bacteriocin-like inhibitory substances, which have a broader spectrum of action (Cintas *et al.*, 2001). Several bacteriocins derived from indigenous oral bacteria have been described (Oliveira *et al.*, 1998; Teanpaisan *et al.*, 1998: Hillman *et al.*, 2000, 2007; Hillman, 2002; Lima *et al.*, 2002). In *S. salivarius* production of two potent bacteriocins, salivaricin types A and B, was demonstrated. This particular species can be used in the prevention of dental caries caused by *S. sobrinus* and *S. mutans*. Salivaricin B was successfully used in the treatment of *Prevotella* spp.- and *Parvimonas micra*-induced halitosis (Balakrishan *et al.*, 2000; Burton *et al.*, 2005, 2006). Additionally, a *Lactobacillus paracasei* HL32 produced bacteriocin was shown to be able to kill *P. gingivalis* by changing the cell envelope of the pathogen (Pangsomboom *et al.*, 2006).

### Immune modulation

In addition to the clear anti-microbial actions of probiotics, these strains can also act on a wide variety of cells to modulate the immune system towards anti-inflammatory action. Probiotic bacteria or their products (e.g. metabolites, cell wall components and DNA) can be recognized by host cells such as epithelial cells and immune cells (Delcenserie *et al.*, 2008; Oelschlaeger, 2010). Increased phagocytic capacity of macrophages when challenged with *Lactobacillus acidophilus* and *Lactobacillus casei* has been reported (Perdigon *et al.*, 2002). It is known that probiotics can regulate the expression of phagocytosis receptors in the neutrophils of healthy individuals (Pelto *et al.*, 1998) and can enhance natural killer cell activity (Takeda *et al.*, 2006). They have also been shown to modulate the adaptive immune response (Link-Amster *et al.*, 1994; Braat *et al.*, 2004).

Few studies have been conducted to determine whether immunomodulation by so-called beneficial bacteria also applies to the oral environment. In this respect, several publications have shown that certain streptococci, such as *S. cristatus, S. salivarius, S. mitis* and *S. sanguinis* can attenuate the interleukin IL-8 response induced by periodontopathogens such as *Fusobacterium nucleatum* and *A. actinomycetemcomitans* on epithelial cells (Cosseau *et al.*, 2008; Zhang *et al.* 2008; Sliepen *et al.* 2009a). The exact regulatory systems are still unclear, although there are indications that these streptococci can inhibit the nuclear factor κB-pathway (Cosseau *et al.*, 2008; Zhang *et al.*, 2008). Recently, Della Riccia *et al.* (2007) tested *in vivo* the immunomodulatory effects of *Lactobacillus brevis* on periodontal disease. The *in vivo* use of this probiotic led to a significant decrease in inflammatory markers in the saliva, such as metalloproteinase and nitric oxide synthase activity, prostaglandin $E_2$ ($PGE_2$) and interferon *g* (IFN-*g*) levels. No effect was observed on IgA levels.

### Other mechanisms of probiotic action

The above outlined mechanisms of probiotic action are numerous; however, other modes of action exist. These are either not applicable for the oral situation or studies have not yet been conducted in this area. One example of a probiotic mechanism that is relevant for improving gastro-intestinal health is the enhancement of the mucosal barrier function. Probiotics can influence mucosal cell–cell interactions by the enhancement of the intestinal barrier function. Disruption of this epithelial barrier is encountered in several conditions including inflammatory bowel disease and autoimmune diseases such as Type 1 diabetes. Enhancement of the barrier by probiotics can benefit the host in such diseases (Ng *et al.*, 2009). Additionally, invasion of epithelial cells is an important mode by which bacteria exert their pathogenicity. For gut epithelial cells, anti-invasive properties of probiotic bacteria have already been established. Secreted factors of *Bifidobacterium bifidum* strain Bb12 interfere with the invasion of epithelial cells by *Salmonell,* Typhimurium (Botes *et al.*, 2008). These effects have not yet been explored in relation to the oral cavity.

## Effects of oral probiotics

### Probiotics and dental caries

The fact that caries is a bacterially mediated process was postulated more than 115 years ago (Miller, 1890). Since then, research has refined our understanding of caries development to show that it is a multifaceted disease process. Today we know that host factors, bacteria and nutrients all contribute to the production of organic acids and the subsequent demineralization activity (Keyes, 1968). Since all three elements are involved in disease initiation and progression, removal of any one element leads to the interception of the disease process (Anderson and Shi, 2006). To overcome the limitations of the traditional disease management strategies, a number of researchers are developing 'probiotic' methods to prevent caries by interfering with the oral colonization of cariogenic pathogens.

Many studies have concentrated on probiotic bacteria that express bacteriocins or bacteriocin like inhibitory substances that specifically prevent the growth of cariogenic bacteria (Devine and Marsh, 2009). Several clinical studies have reported the outcomes of probiotic interventions for caries prevention (Näse *et al.*, 2001; Montalto *et al.*, 2004; Nikawa *et al.*, 2004; Caglar *et al.*, 2005, 2006, 2008, 2009; Cildir *et al.*, 2009; Stecksen-Blicks *et al.*, 2009). However, the extreme heterogeneity between these studies makes it difficult to draw strong conclusions. This heterogeneity is reflected in the set-up of the studies, the probiotic strains employed (lactobacilli such as *L. sporogens, L. bifidum, L. bulgaricus, L. termophilus,*

*L. acidophilus, L. casei, L. rhamnosus, L. reuteri* or *Streptococcus thermophilus* or *Bifidobacteria*), the used vehicula (milk, cheese, yogurt, ice cream, tablets, solutions) and outcome measures (caries incidence, *S. mutans* numbers, *Lactobacillus* spp numbers). Caries incidence, which is the ultimate outcome measure, was reported in 2 studies, one of which did not show statistically significant differences (Näse *et al.*, 2001). The most recent double blind placebo controlled study showed a significantly lower DMFT index after 21 months of probiotic usage compared with a control population (Stecksen-Blicks *et al.*, 2009). However, next to the probiotic, the experimental group also received fluoride, in contrast to the control group. Therefore it cannot clearly be derived whether these differences were attributable to either the probiotic or the fluoride. In terms of *S. mutans* numbers, the majority of studies showed moderate reductions when a probiotic was used (Ahola *et al.*, 2002; Nikawa *et al.*, 2004; Caglar *et al.*, 2005, 2006, 2007a,b, 2008, 2009; Cildir *et al.*, 2009; Chuang *et al.*, 2010). However, some studies showed no effect of a probiotic intervention (Näse *et al.*, 2001; Montalto *et al.*, 2004; Stecksen-Blicks *et al.*, 2009; Lexner, 2010; Petersson *et al.*, 2011).

All the studies which monitored *Lactobacillus* numbers during a probiotic intervention report no statistically significant effects in this parameter. (Ahola, 2002; Caglar, 2005, 2006, 2007a,b, 2008, 2009; Cildir *et al.*, 2009, Stecksen-Blicks, 2009; Chuang *et al.*, 2010; Lexner, 2010; Singh *et al.*, 2011; Petersson *et al.*, 2011). These data might indicate that the used 'probiotic lactobacilli' are not colonizing the oral cavity or at least total number of lactobacilli in the oral cavity did not increase dramatically. However none of these studies specifically determined numbers of 'probiotic lactobacilli'. This might be reassuring since, in general, increased numbers of lactobacilli have been historically related to increased caries susceptibility.

Although this more classical approach where probiotic bacteria are used that express bacteriocins or bacteriocin like inhibitory substances that specifically prevent the growth of cariogenic bacteria, some researchers have started to develop 'designer probiotics'. Clancy *et al.* (2000) developed a recombinant strain of *S. mutans* expressing urease, which was shown to reduce the cariogenicity of plaque in an animal model.

Hillman and coworkers (1994) isolated a strain of *S. mutans*, from a human subject, that produces a bacteriocin called mutacin 1140 that is capable of killing virtually all other strains of mutans streptococci against which it was tested. Apparently this strain persistently colonizes the oral cavities of human subjects and aggressively displaces indigenous mutans streptococci (Hillman *et al.*, 1987). Three years following a single, 3 minutes infection regimen involving brushing and flossing of a concentrated cell suspension onto and between the teeth, all of the subjects remained colonized. No other strains of mutans streptococci were observed in saliva and plaque samples of these colonized volunteers. The same results were found recently, 14 years after colonization, for at least two of three subjects who were still available for testing (Hillman, 2002). Consequently, *S. mutans* strain JH1140, which has a spontaneous mutation resulting in 3-fold elevated production of mutacin 1140, served as the starting strain for the construction of a clinically applicable effector strain for replacement therapy against caries. In accord with the acidogenic theory of dental caries, lactic acid production by *S. mutans* has long been considered to be the main pathogenic mechanism for production of caries lesions. Consequently, recombinant DNA methods were used to replace essentially the entire open reading frame (ORF) for lactic acid dehydrogenase (LDH) with the ORF for alcohol dehydrogenase B from *Zymomonas mobilis*. The resulting BCS3-L1 effector strain lacks any measurable LDH activity and shows ca. 10-fold elevated levels of ADH activity relative to its parent. The ability of BCS3-L1 to serve as an effector strain in the replacement therapy of dental caries was extensively tested in the laboratory and in animal models (Hillman *et al.*, 2000). The strain proved to have significantly reduced pathogenic potential; it persistently and pre-emptively colonized the niche on the tooth surface normally occupied by wild-type strains of *S. mutans*; it was genetically stable; and it showed no ill effects in acute or chronic toxicity studies. If BCS3-L1 could be shown to have similar properties in humans, it would serve as an ideal effector strain. Similarly, genetically modified probiotics with enhanced properties are being developed. For example, a recombinant *Lactobacillus* strain that expressed antibodies targeting one of the major adhesins of *S. mutans* was able to reduce both the viable counts of *S. mutans* and the caries score in a rat model (Kruger *et al.*, 2002).

Unfortunately, although the *in vitro* and animal data with designer probiotics appear promising, no human trials have been conducted yet. This is due to the fact that the use of a persistently implanted, genetically modified bacterium has not been allowed for any purpose by governmental safety institutions.

## Probiotics and plaque related periodontal diseases

The limited knowledge regarding the effect of probiotics on oral health is even more striking for plaque related periodontal diseases. The current view on the aetiology of plaque related periodontal inflammation considers three factors that determine whether disease will develop in a subject (Slots and Rams, 1991; Socransky and Haffajee, 1992; Wolff *et al.*, 1994): a susceptible host, the presence of pathogenic species and the reduction or absence of so-called 'beneficial bacteria'. Since it is difficult to influence the host response without the risk for serious side effects (e.g. COX-2 inhibitors), periodontal therapy especially aims for a reduction of the bacterial threat (Salvi and Lang, 2005). Mechanical subgingival debridement in combination with an improved oral hygiene shifts the subgingival microbiota to a less pathogenic composition, characterized by high proportions of Gram-positive aerobic species (Ximenez-Fyvie *et al.*, 2000; Roberts and Darveau, 2002). Although reductions in the total subgingival microbiota up to 2-log values can easily be achieved, a recolonization primarily by less pathogenic bacteria towards baseline numbers occurs within 1 to 2 weeks (Harper and Robinson, 1987; Goodson *et al.*, 1991; Maiden *et al.*, 1991). The shift towards a less pathogenic microbiota is only temporary, with a re-establishment of a more aggressive microbiota within weeks to months (Mousques *et al.*, 1980; Magnusson *et al.*, 1984; van Winkelhoff *et al.*, 1988; Wade *et al.*, 1992; Quirynen *et al.*, 2005). The dynamics of this recolonization depends on the level of oral hygiene, the efficacy of the subgingival debridement and the residual probing depth (Magnusson *et al.*, 1984; Petersilka *et al.*, 2002). The use of antibiotics or antiseptics, either locally or systemically, does not really improve the long-term effect of periodontal therapy (Quirynen *et al.*, 2002). Therefore, some investigators have started to focus on the third aetiological factor for plaque related periodontal inflammation: 'the reduction or absence of so-called beneficial bacteria'. From a theoretical point of view, restoring these reduced numbers of beneficial bacteria via probiotics might be of considerable interest in the treatment of plaque related periodontal diseases. Probiotics might not only suppress the emergence of endogenous pathogens or prevent the superinfection with exogenous pathogens, but they might also protect us through the promotion of a beneficial host response (Roberts and Darveau, 2002).

Although the use of probiotics in prevention and treatment of plaque-related periodontal disease is widely regarded to be a newly developing treatment approach, the first scientific report on the beneficial effect of lactic acid bacteria on infections of the oral mucosa dates back to 1954 (Kragen, 1954). In the 1990s, several Russian studies on the use of *Lactobacillus* species in the treatment and prevention of periodontal infections were reported (Pozharitskaia *et al.*, 1994; Grudianov *et al.*, 2002). In the past few years, several larger scale randomized control clinical trials have been published that describe the use of probiotics for the treatment or prevention of plaque-related periodontal disease. These studies have generally expressed great optimism based on the obtained results. However, there has been a marked lack of consistency between the studies regarding the used set-up, the probiotic agent (*Lactobacillus salivarius, Weisella cibaria, Lactobacillus reuteri*), the mode of application (tablets, chewing gum, oral rinses), and the outcome measure (probing pocket depth, gingivitis index, plaque index, bleeding on probing, microbiology).

Six studies have reported the microbiological effects of oral probiotics (Hillman and Shivers, 1988; Ishikawa *et al.*, 2003; Teughels *et al.*, 2007a; Sugano *et al.*, 2007; Mayanagi, 2009; Zahradnik, 2009). In general, all studies showed at least that probiotic application resulted in microbiological changes with up to 0.55 log reductions in total anaerobic bacteria, up to 0.25 log increases in total aerobic bacteria, up to 1.3 log reductions in black pigmented bacteria, up to 1.8 log reductions in *A. actinomycetemcomitans* numbers, up to 2.6 log reductions in *Campylobacter rectus* numbers, up to 2.1 log reductions in *P. gingivalis* numbers, up to 1 log reductions in *P. intermedia* numbers and up to 0.17 log reductions in *Tannerella forsythia*. Additionally, significant clinical changes in gingivitis index (Krasse *et al.*, 2006; Della Riccia *et al.*, 2007; Shimauchi *et al.*, 2008), bleeding upon probing (Della Riccia *et al.*, 2007; Twetman *et al.*, 2009), and probing pocket depth (Shimauchi *et al.*, 2008) have been reported. In relation to the potentially anti-inflammatory properties of oral probiotics, clinical studies have shown that the use of oral probiotics can decrease the crevicular fluid, TNF-α, IL-8, polymorphonuclear elastase, matrix metalloproteinase-3, prostaglandin $E_2$ and IFN-gamma production (Della Riccia *et al.*, 2007; Twetman *et al.*, 2009; Staab *et al.*, 2009).

## Considerations regarding the interaction of probiotics with the oral ecology

It can be concluded that most of the results from probiotic intervention studies for the management of caries and plaque related periodontal diseases show no extreme effects. It is rather surprising that the above mentioned studies tested primarily lactobacilli as possible probiotics for oral bacterial infections since lactic acid bacteria are of considerable interest for oral healthcare due to their cariogenic potential. They are highly acidogenic and produce short chain carboxylic acids from the fermentation of sucrose. These organisms consequently lower the pH, which potentially can dissolve hard tissues such as enamel and dentine (Toi *et al.*, 2000). However, Fitzgerald *et al.* (1980) showed that only 3 of 50 lactobacillus strains isolated from dental plaque of school children induced significant caries activity in conventional hamsters. Additionally, in an artificial caries model, lactobacilli produced significantly shallower caries lesions than *S. mutans* and *Actinomyces israelii*, although a synergistic effect on growth of *S. mutans* and *A. israelii* was observed in mixed cultures where *L. acidophilus* was present (Shen *et al.*, 2004). Therefore it has been speculated that, in contrast to mutans streptococci, lactobacilli are more related to caries progression than to the initiation of a caries lesion (Edwardsson, 1974; Maltz *et al.*, 2002). By no means does this imply that lactobacilli are not capable of inducing caries under favourable environmental conditions. Out of 32 *Lactobacillus* strains, comprising eight different species obtained from human dental plaque or other sources, 17 were moderately to highly cariogenic in animals receiving a cariogenic diet (Fitzgerald *et al.*, 1981). Only one, a *L. lactis* strain, was scored as non-cariogenic. A similar observation was recently made for *Lactobacillus salivarius* (Matsumoto *et al.*, 2005). *Lactobacillus* strains are however also found in orally healthy persons. Therefore one might speculate that *Lactobacillus* strains from the oral cavity of caries free persons possess inhibitory properties towards cariogenic bacteria. Such inhibitory properties are however rather uncommon. Sookkhee *et al.* (2001) isolated 3790 lactic acid bacteria from 130 orally healthy volunteers. These strains were all screened on their ability to inhibit the growth on agar of several oral pathogens such as *S. mutans*, *A. viscosus*, but also *Porphyromonas gingivalis* and *Candida albicans*. Only five oral *Lactobacillus* isolates were good antimicrobial producers which could inhibit a number of oral pathogens. The strains with the most omnipotent antimicrobial activity were *L. paracasei* and *L. rhamnosus* species. The latter species is a well known gastro-intestinal probiotic. Additionally, the contribution of these lactobacilli to periodontal health is far from elucidated. However, Koll-Klais *et al.* (2005) noted that periodontal destruction and inflammation is closely associated with decreased levels of certain lactic acid bacteria. Looking more closely, they showed that the *Lactobacillus* species composition in the oral cavity differs in respect to periodontal health and habitat. In their study, using bacterial culturing, it seems that lactobacilli rarely colonize subgingival sites whereas saliva was significantly more colonized. One might question if the subgingival region is a common habitat for lactobacilli, and if not, this might implicate that lactobacilli might not be the bacterial species of choice as a probiotic? There were no differences in overall counts of salivary lactobacilli between healthy and chronic periodontitis patients. However, significantly higher proportions of homofermentative lactobacilli, especially *Lactobacillus gasseri,* were present in saliva of healthy persons. The presence of this species was also associated with less dental plaque and inflammation. As a whole, the homofermentative lactobacilli were associated with the absence of subgingival periodontal pathogens. Finally, as in the gastro-intestinal microbiota, lactobacilli make up for only a relatively small percentage of the cultivable oral microbiota. This all brings us to the question whether lactobacilli are the first choice probiotics in the treatment and prevention of plaque-related oral diseases. In order to answer this question we might look at other studies in the oto-oro-pharyngeal region. We can compare a study of Roos and coworkers (2001) and Hattaka and coworkers (2007), focusing on the effects of probiotics on the occurrence of otitis media in otitis media-prone children. Both studies were double-blinded randomized placebo-controlled studies with a follow-up time of 3 to 6 months. In the study of Roos and coworkers two *Streptococcus sanguinis,* two *Streptococcus mitis* and one *Streptococcus oralis* strain were used. The authors noticed a significantly reduced recurrence of otitis media in children treated with probiotics. However, in the Hatakka study, *Lactobacillus rhamnosus GG, Lactobacillus rhamnosus* LC705, *Bifidobacterium breve* 99 and *Propionibacterium freudenreichii* JS were used, and no statistically significant differences could be detected between the placebo treated group and the probiotic treated group. This makes us wonder whether streptococci would be more effective probiotics than lactobacilli in the oral cavity.

The second important question in relation to oral ecology is: 'What is the effect of pretreatment?'. It is very well known that it is difficult to introduce new species in an already established microbiological ecology. Although it is conceivable that suppression or removal of the existing microbial ecology in the gastro-intestinal tract is difficult, in the oral cavity this might be easier to achieve. The existing microbial ecology can be suppressed or removed by scaling and root planing, oral hygiene measures and antiseptics. However, none of the existing studies took this factor into consideration. The importance of the suppression or the removal of an established ecology might be derived from several lines of evidence. First of all, the aforementioned study of Roos *et al.* (2001) can be compared to a study of Tano *et al.* (2002). These studies had similar set-ups, and looked at the recurrence of recurrence of otitis media in otitis media-prone children. In both studies a mixture of two *Streptococcus sanguinis*, two *Streptococcus mitis* and one *Streptococcus oralis* strain were used as probiotics. In contrast to the study of Tano and coworkers (2002), Roos and coworkers used a 10 day antibiotic pretreatment prior to the use of a probiotic sodium chlorite nasal spray. This approach resulted in a significant lower recurrence of otitis media in contrast to the study of Tano and coworkers were no significant difference between the placebo and the probiotic group was detected.

Additionally, it has been documented that the indigenous microbiota in premature infants is more easily influenced by a probiotic than in humans with an established microbial ecology (Sprunt *et al.*, 1980). In the field of oral microbiology, it has been documented that there are windows of infectivity for species such as *Streptococcus mutans* and *Streptococcus sanguinis* (Caufield *et al.*, 1993, 2000). Perhaps these windows of infectivity offer opportunities for long-term influences on resident bacterial populations.

There has been little proof that the use of probiotics orally will lead to the permanent colonization of the oral cavity by the administered probiotic strain. However, one case has been described where a female person has been permanently colonized by *Lactobacillus rahmnosus* GG (Busscher *et al.*, 1999). This could be attributed to a 1-year intensive use of *Lactobacillus rahmnosus* GG during the childhood of this patient. This opens up the discussion whether also for oral probiotics, there might not be a window for infectivity at young age.

## Conclusions

Probiotics have made their way into oral health care and are more likely our friend than our enemy. Despite our rapidly increasing knowledge of host–pathogens interactions, the role of beneficial bacteria and probiotics in preventing the emergence of oral pathogens and oral health remains obscure. There is a great need for elucidating the role of the oral beneficial microbiota and probiotics, for identifying beneficial bacteria and for properly conducted large scale studies on the usefulness of probiotics to maintain or improve oral health.

## References

Ahola, A.J., Yli-Knuuttila, H., Suomalainen, T., Poussa, T., Ahlstrom, A., Meurman, J.H., and Korpela, R. (2002). Short-term consumption of probiotic-containing cheese and its effect on dental caries risk factors. Arch. Oral Biol. *47*, 799–804.

Andersson, H., Asp, N.G., Bruce, A., Roos, S., Wadstrom, T., and Wold, A.E. (2001). Health effects of probiotics and prebiotics: a literature review on human studies. Scand. J. Nutr. *45*, 58–75.

Anderson, M.H., and Shi, W.A. (2006). Probiotic approach to caries management. Pediatr. Dent. *28*, 151–153.

Balakrishnan, M., Simmonds, R.S., and Tagg, J.R. (2000). Dental caries is a preventable infectious disease. Aust. Dent. J. *45*, 235–245.

Braat, H., van den Brande, J., van Tol, E., Hommes, D., Peppelenbosch, M., and van Deventer, D.S. (2004). *Lactobacillus rhamnosus* induces peripheral hyporesponsiveness in stimulated CD4+ T cells via modulation of dendritic cell function. Am. J. Clin. Nutr. *80*, 1618–1625.

Bohm, S.K., and Kruis, W. (2006). Probiotics: do they help to control intestinal inflammation? Ann. N. Y. Acad. Sci. *1072*, 339–350.

Bonifait, L., Chandad, F., and Grenier, D. (2009). Probiotics for oral health: myth or reality? J. Can. Dent. Assoc. *75*, 585–590.

Botes, M., Loos, B., van Reenen, C.A., and Dicks, L.M.T. (2008). Adhesion of the probiotic strains *Enterococcus mundtii* ST4SA and *Lactobacillus plantarum* 423 to Caco-2 cells under conditions simulating the intestinal tract, and in the presence of antibiotics and anti-inflammatory medicaments. Arch. Microbiol. *190*, 573–584.

Burnett, F.M. (1962). Natural History of Infection Diseases, Third Edition (Cambridge University Press, Cambridge, UK).

Burton, J.P., Chilcott, C.N., and Tagg, J.R. (2005). The rationale and potential for the reduction of oral malodour using *Streptococcus salivarius* probiotics. Oral Dis. *11(Suppl. 1)*, 29–31.

Burton, J.P., Chilcott, C.N., Moore, C.J., Speiser, G., and Tagg, J.R. (2006). A preliminary study of the effect of probiotic *Streptococcus salivarius* K12 on oral malodour parameters. J. Appl. Microbiol. *100*, 754–764.

Busscher, H.J., Mulder, A.F., and van der Mei, H.C. (1999). In vitro adhesion to enamel and in vivo colonization of tooth surfaces by lactobacilli from a bio-yoghurt. Caries Res. 33, 403–404.

Buydens, P., and Debeuckelaere, S. (1996). Efficacy of SF 68 in the treatment of acute diarrhea. A placebo-controlled trial. Scand. J. Gastroenterol *31*, 887–891.

Caglar, E., Sandalli, N., Twetman, S., Kavaloglu, S., Ergeneli, S., and Selvi, S. (2005). Effect of yogurt with *Bifidobacterium* DN-173 010 on salivary mutans streptococci and lactobacilli in young adults. Acta Odont. Scand. *63*, 317–320.

Caglar, E., Cildir, S.K., Ergeneli, S., Sandalli, N., and Twetman, S. (2006). Salivary mutans streptococci and lactobacilli levels after ingestion of the probiotic bacterium *Lactobacillus reuteri* ATCC 55730 by straws or tablets. Acta Odont. Scand. *64*, 314–318.

Caglar, E., Kuscu, O.O., Cildir, S.K., Kuvvetli, S.S., and Sandalli, N. (2008). A probiotic lozenge administered medical device and its effect on salivary mutans Streptococci and Lactobacilli. Int. J. Paediatr. Dent. *18*, 35–39.

Caglar, E., Topcuoglu, N., Cildir, S.K., Sandalli, N., and Kulekci, G. (2009). Oral colonization by *Lactobacillus reuteri* ATCC 55730 after exposure to probiotics. Int. J. Paediatr. Dent. *19*, 377–381.

Caufield, P.W., Cutter, G.R., and Dasanayake, A.P. (1993). Initial acquisition of mutans streptococci by infants: evidence for a discrete window of infectivity. J. Dent. Res. *72*, 37–45.

Caufield, P.W., Dasanayake, A.P., Li, Y., Pan, Y., Hsu, J., and Hardin, J.M. (2000). Natural history of *Streptococcus sanguinis* in the oral cavity of infants: evidence for a discrete window of infectivity. Infect. Immun. *68*, 4018–4023.

Chuang, L.C., Huang, C.S., Ou-Yang, L.W., and Lin. S.Y. (2010). Probiotic *Lactobacillus paracasei* effect on cariogenic bacterial flora. Clin. Oral. Investig. *15*, 471–476.

Cildir, S.K., Germec, D., Sandalli, N., Ozdemir, F.I., Arun, T., Twetman, S., and Caglar, E. (2009). Reduction of salivary mutans Streptococci in orthodontic patients during daily consumption of yoghurt containing probiotic bacteria. Eur. J. Orthod. *31*, 407–411.

Cintas, L.M., Casaus, M.P., Herranz, C., Nes, I.F., and Hernandez, P.E. (2001). Review: Bacteriocins of lactic acid bacteria. Food Sci. Tech. Int. *7*, 281–305.

Clancy, K.A., Pearson, S., Bowen, W.H., and Bume, R.A. (2000). Characterization of recombinant, ureolytic *Streptococcus mutans* demonstrates an inverse relationship between dental plaque ureolytic capacity and cariogenicity. Infect. Immun. *68*, 2621–2629.

Cosseau, C., Devine, D.A., Dullaghan, E., Gardy.J.L., Chikatamarla, A., Gellatly, S., Yu, L.L., Pistolic, J., Falsafi, R., Tagg, J., and Hancock, R.E.W. (2008). The commensal *Streptococcus salivarius* K12 downregulates the innate immune responses of human epithelial cells and promotes host-microbe homeostasis. Infect. Immun. *76*, 4163–4175.

Cremonini, F., Di Caro, S., Nista, E.C., Bartolozzi, F., Capelli, G., Gasbarrini, G., and Gasbarrini, A. (2002). Meta-analysis: the effect of probiotic administration on antibiotic-associated diarrhoea. Aliment. Pharmacol. Ther. *16*, 1461–1467.

Delcenserie, V., Martel, D., Lamoureux, M., Amiot, J., Boutin, Y., and Roy, D. (2008). Immunomodulatory effects of probiotics in the intestinal tract. Curr. Issues Mol. Biol. *10,* 37–54.

Della Riccia, D.N., Bizzini, F., Perilli, M.G., Polimeni, A., Trinchieri, V., Amicosante, G., and Cifone, M.G. (2007). Anti-inflammatory effects of *Lactobacillus brevis* (CD2) on periodontal disease. Oral Dis. *13,* 376–385.

Devine, D.A., and Marsh, P.D. (2009). Prospects for the development of probiotics and prebiotics for oral applications. J. Oral Microbiol. *1,* DOI: 10.3402/jom.v1i0.1949.

Edwardsson, S. (1974). Bacteriological studies on deep areas of carious dentine. Odontol. Revy. Suppl. *32,* 1–143.

Elli, M., Zink, R., Rytz, A., Reniero, R., and Morelli, L. (2000). Iron requirement of Lactobacillus spp. in completely chemically defined growth media. J. Appl. Microbiol. *88,* 695–703.

FAO/WHO (2002). Guidelines for the evaluation of probiotics in food. FAO/WHO report, London, Ontario, Canada, april 30 and May 1 2002, 11p.

Falagas, M.E., Betsi, G.I., and Athanasiou, S. (2007). Probiotics for the treatment of women with bacterial vaginosis. Clin. Microbiol. Infec. *13,* 657–664.

Felley, C.P., Corthesy-Theulaz, I., Rivero, J.L.B., Sipponen, P., Kaufmann, M., Bauerfeind, P., Wiesel, P.H., Brassart, D., Pfeifer, A., Blum, A.L., *et al.* (2001). Favourable effect of an acidified milk (LC-1) on *Helicobacter pylori* gastritis in man. Eur. J. Gastroen. Hepat. *13,* 25–29.

Fitzgerald, R.J., Fitzgerald, D.B., Adams, B.O., and Duany, L.F. (1980). Cariogenicity of human oral lactobacilli in hamsters. J. Dent. Res. *59,* 832–837.

Fitzgerald, R.J., Adams, B.O., Fitzgerald, D.B., and Knox, K.W. (1981). Cariogenicity of human plaque lactobacilli in gnotobiotic rats. J. Dent. Res. *60,* 919–926.

Forsythe, P., Inman, M.D., and Bienenstock, J. (2007). Oral treatment with live *Lactobacillus reuteri* inhibits the allergic airway response in mice. Am. J. Resp. Crit. Care. *175,* 561–569.

Fuller, R. (1989). Probiotics in man and animals. J. Appl. Bacteriol. *66,* 365–378.

Gaon, D., Garmendia, C., Murrielo, N.O., Games, A.D., Cerchio, A., Quintas, R., Gonzalez, S.N., and Oliver, G. (2002). Effect of Lactobacillus strains (*L. casei* and *L. acidophillus* strains CERELA) on bacterial overgrowth-related chronic diarrhea. Medicina-Buenos Aires *62,* 159–163.

Genco, R.J., Ho, A.W., Grossi, S.G., Dunford, R.G., and Tedesco, L.A. (1999). Relationship of stress, distress and inadequate coping behaviours to periodontal disease. J. Periodontol. *70,* 711–723.

Gibson, G.R., and Roberfroid, M.B. (1995). Dietary modulation of the human colonic microbiota: introducing the concept of prebiotics. J. Nutr. *125,* 1401–1412.

Gillor, O., Etzion, A., and Riley, M.A. (2008). The dual role of bacteriocins as anti- and probiotics. Appl. Microbiol. Biotechnol. *81,* 591–606.

Gionchetti, P., Rizzello, F., Venturi, A., and Campieri, M. (2000). Probiotics in infective diarrhoea and inflammatory bowel diseases. J. Gastroen. Hepat. *15,* 489–493.

Goodson, J.M., Tanner, A., McArdle, S., Dix, K., and Watanabe, S.M. (1991). Multicenter evaluation of tetracycline fiber therapy. III. Microbiological response. J. Periodontal. Res. *26,* 440–451.

Gordon, D.M. (2009). The potential of bacteriocin-producing probiotics and associated caveats. Future Microbiol. *4,* 941–943.

Grudianov, A.I., Dmitrieva, N.A., and Fomenko, E.V. (2002). Use of probiotics Bifidumbacterin and Acilact in tablets in therapy of periodontal inflammations. Stomatologiia (Mosk) *81,* 39–43.

Haffajee, A.D., Arguello, E.I., Ximenez-Fyvie, L.A., and Socransky, S.S. (2003). Controlling the plaque biofilm. Int. Dent. J. *53,* 191–199.

Haskard, C.A., El-Nezami, H.S., Kankaanpaa, P.E., Salminen, S., and Ahokas, J.T. (2001). Surface binding of aflatoxin B(1) by lactic acid bacteria. Appl. Environ. Microb. *67,* 3086–3091.

Hatakka, K., Ahola, A.J., Yli-Knuuttila, H., Richardson, M., Poussa, T., Meurman, J.H., and Korpela, R. (2007a). Probiotics reduce the prevalence of oral candida in the elderly: a randomized controlled trial. J. Dent. Res. *86,* 125–130.

Hatakka, K., Blomgren, K., Pohjavuori, S., Kaijalainen, T., Poussa, T., Leinonen, M., Korpela, R., and Pitkaranta, A. (2007b). Treatment of acute otitis media with probiotics in otitis-prone children: A double-blind, placebo-controlled randomised study. Clin. Nutr. *26,* 314–321.

Haukioja, A., Loimaranta, V., and Tenovuo, J. (2008). Probiotic bacteria affect the composition of salivary pellicle and Streptococcal adhesion in vitro. Oral Microbiol. Immun. *23,* 336–343.

Havenaar, R., and Huis In't Veld, J.M.J. (1992). Probiotics: a general view. In Lactic Acid Bacteria in Health and Disease, Vol. 1, Wood, B.J., ed. (Elsevier Applied Science Publishers, New York), pp. 151–170.

Hawrelak, J.A., Whitten, D.L., and Myers, S.P. (2005). Is *Lactobacillus rhamnosus* GG effective in preventing the onset of antibiotic-associated diarrhoea: A systematic review. Digestion *72,* 51–56.

Henker, J., Schuster, F., and Nissler, K. (2001). Successful treatment of gut-caused halitosis with a suspension of living non-pathogenic *Escherichia coli* bacteria: a case report. Eur. J. Pediatr. *160,* 592–594.

Hillman, J.D. (2002). Genetically modified *Streptococcus mutans* for the prevention of dental caries. Antonie Van Leeuwenhoek Int. J. Gen. Mol. Microbiol. *82,* 361–366.

Hillman, J.D., and Shivers, M. (1988). Interaction between wild-type, mutant and revertant forms of the bacterium *Streptococcus sanguis* and the bacterium *Actinobacillus actinomycetemcomitans* in vitro and in the gnotobiotic rat. Arch. Oral Biol. *33,* 395–401.

Hillman, J.D., Yaphe, B.I., and Johnson, K.P. (1985). Colonization of the human oral cavity by a strain of *Streptococcus mutans.* J. Dent. Res. *64,* 1272–1274.

Hillman, J.D., Dzuback, A.L., and Andrews, S.W. (1987). Colonization of the human oral cavity by a *Streptococcus mutans* mutant producing increased bacteriocin. J. Dent. Res. *66,* 1092–1094.

Hillman, J.D., Chen, A., Duncan, M., and Lee, S.W. (1994). Evidence that L-(+)-lactate dehydrogenase deficiency is lethal in *Streptococcus mutans*. Infect. Immun. *62:* 60–64.

Hillman, J.D., Brooks, T.A., Michalek, S.M., Harmon, C.C., Snoep, J.L., and Van Der Weijden, C.C. (2000). Construction and characterization of an effector strain of *Streptococcus mutans* for replacement therapy of dental caries. Infect. Immun. *68,* 543–549.

Hillman, J.D., Mo, J., McDonell, E., Cvitkovitch, D., and Hillman, C.H. (2007). Modification of an effector strain for replacement therapy of dental caries to enable clinical safety trials. J. Appl. Microbiol. *102,* 1209–1219.

Ishikawa, H., Aiba, Y., Nakanishi, M., Oh-hashi, Y., and Koga, Y. (2003). Suppression of periodontal pathogenic bacteria in the saliva of humans by the administration of *Lactobacillus salivarius* TI 2711. J. Jpn. Soc. Periodontol. *45,* 105–112.

Isolauri E. (2001). Probiotics in the prevention and treatment of allergic disease. Pediatr. Allergy Immunol. *12,* 56–59.

Isolauri, E., Arvola, T., Sutas, Y., Moilanen, E., and Salminen, S. (2000). Probiotics in the management of atopic eczema. Clin. Exp. Allergy *30,* 1604–1610.

Kalliomaki, M., Salminen, S., Arvilommi, H., Kero, P., Koskinen, P., and Isolauri, E. (2001). Probiotics in primary prevention of atopic disease: a randomized placebo-controlled trial. Lancet *357,* 1076–1079.

Kang, M.S., Chung, J., Kim, S.M., Yang, K.H., and Oh, J.S. (2006a). Effect of *Weissella cibaria* isolates on the formation of *Streptococcus mutans* biofilm. Caries Res. *40,* 418–425.

Kang, M.S., Kim, B.G., Chung, J., Lee, H.C., and Oh, J.S. (2006b). Inhibitory effect of *Weissella cibaria* isolates on the production of volatile sulphur compounds. J. Clini. Periodontol. *33,* 226–232.

Keyes P.H. (1968). Research in dental caries. J. Am. Dent. Assoc. *76,* 1357–1373.

Koebnick, C., Wagner, I., Leitzmann, P., Stern, U., and Zunft, H.J.F. (2003). Probiotic beverage containing *Lactobacillus casei* Shirota improves gastrointestinal symptoms in patients with chronic constipation. Can. J. Gastroenterol. *17,* 655–659.

Kõll-Klais, P., Mandar, R., Leibur, E., and Mikelsaar, M. (2005). Oral microbial ecology in chronic periodontitis and periodontal health. Microb. Ecol. Health D. *17,* 146–155.

Kragen, H. (1954). The treatment of inflammatory infections of the oral mucosa with a lactic acid bacterial culture preparation. Zahnarztl Welt *9,* 306–308.

Krasse, P., Carlsson, B., Dahl, C., Paulsson, A., Nilsson, A., and Sinkiewicz, G. (2006). Decreased gum bleeding and reduced gingivitis by the probiotic *Lactobacillus reuteri*. Swed. Dent. J. *30,* 55–60.

Kruger, C., Hu, Y.Z., Pan, Q., Marcotte, H., Hultberg, A., and Delwar, D. (2002). In situ delivery of passive immunity by lactobacilli producing single-chain antibodies. Nat. Biotech. *20,* 702–706.

Lexner, M.O., Blomqvist, S., Dahlen, G., and Twetman, S. (2010). Microbiological profiles in saliva and supragingival plaque from caries-active adolescents before and after a short-term daily intake of milk supplemented with probiotic bacteria-a pilot study. Oral Health Prev. Dent. *8,* 383–388.

Lilly, D.M., and Stillwell, R.H. (1965). Probiotics: growth-promoting factors produced by microorganisms. Science *147,* 747–748.

Lima, F.L., Farias, F.F., Carvalho, M.A.R., Alviano, C.S., and Farias, L.M. (2002). Influence of abiotic factors on the bacteriocinogenic activity of *Actinobacillus actinomycetemcomitans*. Res. Microbiol. *153,* 249–252.

Link-Amster, H., Rochat, F., Saudan, K.Y., Mignot, O., and Aeschlimann, J.M. (1994). Modulation of a specific humoral immune response and changes in intestinal flora mediated through fermented milk intake. Fems Immunol. Med. Mic. *10,* 55–63.

Magnusson, I., Lindhe, J., Yoneyama, T., and Liljenberg, B. (1984). Recolonization of a subgingival microbiota following scaling in deep pockets. J. Clin. Periodontol. *11,* 193–207.

Maiden, M.F., Tanner, A., McArdle, S., Najpauer, K., and Goodson, J.M. (1991). Tetracycline fiber therapy monitored by DNA probe and cultural methods. J. Periodontal Res. *26,* 452–459.

Majamaa, H., Isolauri, E., Saxelin, M., and Vesikari, T. (1995). Lactic acid bacteria in the treatment of acute *Rotavirus gastroenteritis*. J. Pediatr. Gastr. Nutr. *20,* 333–338.

Makras, L., and De Vuyst, L. (2006). The in vitro inhibition of Gram-negative pathogenic bacteria by Bifidobacteria is caused by the production of organic acids. Int. Dairy J. *16,* 1049–1057.

Maltz, M., de Oliveira, E.F., Fontanella, V., and Bianchi, R.A. (2002). Clinical, microbiologic, and radiographic study of deep caries lesions after incomplete caries removal. Quintessence Int. *33,* 151–159.

Marsh, P.D. (2003). Are dental diseases examples of ecological catastrophes? Microbiology *149,* 279–294.

Mashimo, P.A., Yamamoto, Y., Nakamura, M., Reynolds, H.S., and Genco, R.J. (1985). Lactic acid production by oral *Streptococcus mitis* inhibits the growth of oral Capnocytophaga. J. Periodontol. *56,* 548–552.

Matsumoto, M., Tsuji, M., Sasaki, H., Fujita, K., Nomura, R., Nakano, K., Shintani, S., and Ooshima, T. (2005). Cariogenicity of the probiotic bacterium *Lactobacillus salivarius* in rats. Caries Res. *39,* 479–483.

Mayanagi, G., Kimura, M., Nakaya, S., Hirata, H., Sakamoto, M., Benno, Y., and Shimauchi, H. (2009). Probiotic effects of orally administered *Lactobacillus salivarius* WB21-containing tablets on periodontopathic bacteria: a double-blinded, placebo-controlled, randomized clinical trial. J. Clin. Periodontol. *36,* 506–513.

Meurman, J.H. (2005). Probiotics: contributions to oral health. Oral Diseases *13,* 443–451.

Meurman, J.H., Antila, H., Korhonen, A., and Salminen, S. (1995). Effect of *Lactobacillus rhamnosus* strain GG (ATCC 53103) on the growth of *Streptococcus sobrinus* in vitro. Eur. J. Oral Sci. *103,* 253–258.

Miller, W.D. (1890). Micro-organisms of the Human Mouth (SS White, Philadelphia, PA).

Montalto, M., Vastola, M., Marigo, L., Covino, M., Graziosetto, R., Curigliano, V., Santoro, L., Cuoco, L., Manna, R., and Gasbarrini, G. (2004). Probiotic treatment increases salivary counts of lactobacilli: a double-blind, randomized, controlled study. Digestion *69*, 53–56.

Moore, W.E. (1987). Microbiology of periodontal disease. J. Periodontal Res. *22*, 335–341.

Mousques, T., Listgarten, M.A., and Phillips, R.W. (1980). Effect of scaling and root planing on the composition of the human subgingival microbial flora. J. Periodontal Res. *15*,144–151.

Naidu, A.S., Bidlack, W.R., and Clemens, R.A. (1999). Probiotic spectra of lactic acid bacteria (LAB). Crit. Rev. Food Sci. Nutr. *39*, 13–126.

Näse, L., Hatakka, K., Savilahti, E., Saxelin, M., Ponka, A., Poussa, T., Korpela, R., and Meurman, J.H. (2001). Effect of long-term consumption of a probiotic bacterium, *Lactobacillus rhamnosus* GG, in milk on dental caries and caries risk in children. Caries Res. *35*, 412–420.

Ng, S.C., Hark, A.L., Kamm, M.A., Stagg, A.J., and Knight, S.C. (2009). Mechanisms of action of probiotics: recent advances. Inflamm. Bowel Dis. *15*, 300–310.

Niedzielin, K., Kordecki, H., and Birkenfeld, B. (2001). A controlled, double-blind, randomized study on the efficacy of *Lactobacillus plantarum* 299V in patients with irritable bowel syndrome. Eur. J. Gastroent. Hepatol. *13*, 1143–1147.

Nikawa, H., Makihira, S., Fukushima, H., Nishimura, H., Ozaki, Y., Ishida, K., Darmawan, S., Hamada, T., Hara, K., Matsumoto, A., *et al.* (2004). *Lactobacillus reuteri* in bovine milk fermented decreases the oral carriage of mutans streptococci. Int. J. Food Microbiol. *95*, 219–223.

Oelschlaeger, T.A. (2010). Mechanisms of probiotic actions – a review. Int. J. Med. Microbiol. *300*, 57–62.

Ogawa, T., Hashikawa, S., Asai, Y., Sakamoto, H., Yasuda, K., and Makimura, Y. (2006). A new synbiotic, *Lactobacillus casei* subsp *casei* together with dextran, reduces murine and human allergic reaction. Fems Immunol. Med. Mic. *46*, 400–409.

Oliveira, A.A.P., Farias, L.M., Nicoli, J.R., Costa, J.E., and Carvalho, M.A.R. (1998). Bacteriocin production by Fusobacterium isolates recovered from the oral cavity of human subjects with and without periodontal disease and of marmosets. Res. Microbiol. *149*, 585–594.

Ouwehand, A.C., Lagstrom, H., Suomalainen, T., and Salminen, S. (2002). Effect of probiotics on constipation, fecal azoreductase activity and fecal mucin content in the elderly. Ann. Nutr. Metab. *46*, 159–162.

Parker, R.B. (1974). Probiotics, the other half of the antibiotic story. Anim. Nutr. Health *29*, 4–8.

Pelto, L., Isolauri, E., Lilius, E.M., Nuutila, J., and Salminen, S. (1998). Probiotic bacteria down-regulate the milk-induced inflammatory response in milk-hypersensitive subjects but have an immunostimulatory effect in healthy subjects. J. Brit. Soc. Allergy Clin. Immunol. *28*, 1474–1479.

Perdigon, G., Maldonado, G.C., Valdez, J.C., and Medici, M. (2002). Interaction of lactic acid bacteria with the gut immune system. Eur. J. Clin. Nutri. *56*, S21–S26.

Petersilka, G.J., Ehmke, B., and Flemmig, T.F. (2002). Antimicrobial effects of mechanical debridement. Periodontol. 2000 *28*, 56–71.

Petersson, L.G., Magnusson, K., Hakestam, U., Baigi, A., and Twetman, S. (2011). Reversal of primary root caries lesions after daily intake of milk supplemented with fluoride and probiotic lactobacilli in older adults. Acta Odontol. Scand. PMID:21563871.

Pozharitskaia, M.M., Morozova, L.V., Mel'nichuk, G.M., and Mel'nichuk, S.S. (1994). The use of the new bacterial biopreparation Acilact in the combined treatment of periodontitis. Stomatologiia (Mosk) 73, 17–20.

Quirynen, M., Teughels, W., De Soete, M., and van Steenberghe, D. (2002). Topical antiseptics and antibiotics in the initial therapy of chronic adult periodontitis: microbiological aspects. Periodontol. 2000 *28*, 72–90.

Quirynen, M., Vogels, R., Pauwels, M., Haffajee, A.D., Socransky, S.S., and Uzel, N.G. (2005). Initial subgingival colonization of 'pristine' pockets. J. Dent. Res. *84*, 340–344.

Quirynen, M., Teughels, W., Kinder Haake, S.A., and Newman, M.G. (2006). Microbiology of periodontal diseases. In Carranza's Clinical Periodontology, Newman, M.G., Takei, H., Klokkevold, P., and Carranza, F.A., eds. (Elsevier, St. Louis, MO), pp. 134–169.

Roberts, F.A., and Darveau, R.P. (2002). Beneficial bacteria of the periodontium. Periodontol. 2000 *30*, 40–50.

Roos, K., Grahn, E., Holm, S.E., Johansson, H., and Lind, L. (1993a). Interfering alpha-streptococci as a protection against recurrent streptococcal tonsillitis in children. Int. J. Pediatr. Otorhinolaryngology 25, 141–148.

Roos, K., Holm, S.E., Grahn, E., and Lind, L. (1993b). Alpha-streptococci as supplementary treatment of recurrent streptococcal tonsillitis: a randomized placebo-controlled study. Scand. J. Infect. Dis. 25, 31–35.

Roos, K., Holm, S.E., Grahn-Hakansson, E., and Lagergren, L. (1996). Recolonization with selected alpha-streptococci for prophylaxis of recurrent streptococcal pharyngotonsillitis: a randomized placebo-controlled multicentre study. Scand. J. Infect. Dis. 28, 459–462.

Roos, K., Hakansson, E.G., and Holm, S.E. (2001). Effect of recolonisation with 'interfering' alpha streptococci on recurrences of acute and secretory otitis media in children: randomised placebo controlled trial. Brit. Med. J. *322*, 210–212.

Saito, T. (2004). Selection of useful probiotic lactic acid bacteria from the *Lactobacillus acidophilus* group and their applications to functional foods. Anim. Sci. J. 75: 1–13.

Salvi, G.E., and Lang, N.P. (2005). The effects of non-steroidal anti-inflammatory drugs (selective and non-selective) on the treatment of periodontal diseases. Curr. Pharm. Des. *11*, 1757–1769.

Schaafsma, G. (1996). State of art concerning probiotic strains in milk products. IDF Nutr. News Lett. *5*, 23–24.

Schrezenmeir, J., and de Vrese, M. (2001). Probiotics, prebiotics, and synbiotics: approaching a definition. Am. J. Clin. Nutr. *73*, 361S–364S.

Shen, S., Samaranayake, L.P., and Yip, H.K. (2004). In vitro growth, acidogenicity and cariogenicity of predominant human root caries flora. J. Dent. *32*, 667–678.

Shimauchi, H., Mayanagi, G., Nakaya, S., Minamibuchi, M., Ito, Y., Yamaki, K., and Hirata, H. (2008). Improvement of periodontal condition by probiotics with *Lactobacillus salivarius* WB21: a randomized, double-blind, placebo-controlled study. J. Clin. Periodontol. *35*, 897–905.

Singh, R.P., Damle, S.G., and Chawla, A. (2011). Salivary mutans streptococci and lactobacilli modulations in young children on consumption of probiotic ice-cream containing Bifidobacterium lactis Bb12 and *Lactobacillus acidophilus* La5. Acta Odontol. Scand. *69*, 389–394.

Sliepen, I., Hofkens, J., Van, E.M., Quirynen, M., and Teughels, W. (2008). *Aggregatibacter actinomycetemcomitans* adhesion inhibited in a flow cell. Oral Microbiol. Immun. *23*, 520–524.

Sliepen, I., Van Damme, J., Van Essche, M., Loozen, G., Quirynen, M., and Teughels, W. (2009a). Microbial interactions influence inflammatory host cell responses. J. Dent. Res. *88*, 1026–1030.

Sliepen, I., Van, E.M., Loozen, G., Van, E.J., Quirynen, M., and Teughels, W. (2009b). Interference with *Aggregatibacter actinomycetemcomitans*: colonization of epithelial cells under hydrodynamic conditions. Oral Microbiol. Immun. *24*, 390–395.

Slots, J. (2005). Herpesviruses in periodontal diseases. Periodontol. 2000 *38*, 33–62.

Slots, J., and Rams, T.E. (1991). New views on periodontal microbiota in special patient categories. J. Clin. Periodontol. *18*, 411–420.

Smith, V.H., and Pippin, D.J. (1998). Implications of resource-ratio theory for oral microbial ecology. Eur. J. Oral Sci. *106*, 605–615.

Socransky, S.S., and Haffajee, A.D. (1992). The bacterial etiology of destructive periodontal disease: current concepts. J. Periodontol. *63*, 322–331.

Socransky, S.S., and Haffajee, A.D. (2002). Dental biofilms: difficult therapeutic targets. Periodontol. 2000 *28*, 12–55.

Socransky, S.S., and Haffajee, A.D. (2005). Periodontal microbial ecology. Periodontol. 2000 *38*, 135–187.

Sookkhee, S., Chulasiri, M., and Prachyabrued, W. (2001). Lactic acid bacteria from healthy oral cavity of Thai volunteers: inhibition of oral pathogens. J. Appl. Microbiol. *90*, 172–179.

Sprunt, K., Leidy, G., and Redman, W. (1980). Abnormal colonization of neonates in an ICU: conversion to normal colonization by pharyngeal implantation of alpha hemolytic *Streptococcus* strain 215. Pediatr. Res. *14*, 308–313.

Staab, B., Eick, S., Knofler, G., and Jentsch, H. (2009). The influence of a probiotic milk drink on the development of gingivitis: a pilot study. J. Clin. Periodontol. *36*, 850–856.

Stamatova, I.S., and Meurman, J.H. (2009). Probiotics and periodontal disease. Periodontol. 2000 *51*, 141–151.

Stecksen-Blicks, C., Sjostrom, I., and Twetman, S. (2009). Effect of long-term consumption of milk supplemented with probiotic lactobacilli and fluoride on dental caries and general health in preschool children: a cluster-randomized study. Caries Res. *43*, 374–381.

Sugano, N., Matsuoka, T., Koga, Y., and Ito, K. (2007). Effects of probiotics on periodontal disease. Dentistry in Japan *43*, 123–126.

Takeda, K., Suzuki, T., Shimada, S.I., Shida, K., Nanno, M.,and Okumura, K. (2006). Interleukin-12 is involved in the enhancement of human natural killer cell activity by Lactobacillus casei Shirota. Clin. Exp. Immunol. *146*, 109–115.

Tano, K., Grahn, H.E., Holm, S.E., and Hellstrom, S. (2002). A nasal spray with alpha-haemolytic streptococci as long term prophylaxis against recurrent otitis media. Int. J. Pediatric Otorhinolaryngology *62*, 17–23.

Teanpaisan, R., Baxter, A.M., and Douglas, C.W.I. (1998). Production and sensitivity of bacteriocin-like activity among *Porphyromonas gingivalis*, *Prevotella intermedia* and *Pr-nigrescens* strains isolated from periodontal sites. J. Med. Microbiol. *47*, 585–589.

Teughels, W., Kinder Haake, S.A., Sliepen, I., Pauwels, M., Van Eldere, J., Cassiman, J.J., and Quirynen, M. (2007a). Bacteria interfere with *A. actinomycetemcomitans* colonization. J. Dent. Res. *87*, 611–617.

Teughels, W., Newman, M.G., Coucke, W., Haffajee, A.D., van der Mei, H.C., Haake, S.K., Schepers, E., Cassiman, J.J., Van Eldere, J., van Steenberghe, D., *et al.* (2007b). Guiding periodontal pocket recolonization: a proof of concept. J. Dent. Res. *86*, 1078–1082.

Teughels, W., Van, E.M., Sliepen, I., and Quirynen, M. (2008). Probiotics and oral healthcare. Periodontol. 2000 *48*, 111–147.

Teughels, W., Loozen, G., and Quirynen, M. (2011). Do probiotics offer opportunities to manipulate the periodontal microbiota? J. Clin. Periodontol. *38*, 159–177.

Toi, C.S., Mogodiri, R., and Cleaton-Jones, P.E. (2000). Mutans streptococci and lactobacilli on healthy and carious teeth in the same mouth of children with and without dental cariës. Microb. Ecol. Health Dis. 2000 *12*, 227–233.

Tompkins, G.R., and Tagg, J.R. (1986). Incidence and characterization of anti-microbial effects produced by *Actinomyces viscosus* and *Actinomyces naeslundii*. J. Dent. Res. *65*, 109–112.

Twetman, S., Derawi, B., Keller, M., Ekstrand, K., Yucel-Lindberg, T., and Stecksen-Blicks, C. (2009). Short-term effect of chewing gums containing probiotic *Lactobacillus reuteri* on the levels of inflammatory mediators in gingival crevicular fluid. Acta. Odontol. Scand. *67*, 19–24.

Vanderhoeven, J.S., and Camp, P.J.M. (1993). Mixed continuous cultures of *Streptococcus mutans* with

*Streptococcus sanguis* or with *Streptococcus oralis* as a model to study the ecological effects of the lactoperoxidase system. Caries Res. *27*, 26–30.

Vanderhoof, J.A., Young, R.J., Murray, N., and Kaufmann, S.S. (1998). Treatment strategies for small bowel bacterial overgrowth in short bowel syndrome. J. Pediatr. Gastr. Nutr. *27*, 155–160.

Van Hoogmoed, C.G., Van der Mei, H.C., and Busscher, H.J. (2004).The influence of biosurfactants released by *S. mitis* BMS on the adhesion of pioneer strains and cariogenic bacteria. Biofouling *20*, 261–267.

Van Hoogmoed, C.G., Geertsema-Doornbusch, G.I., Teughels, W., Quirynen, M., Busscher, H.J., and van der Mei, H.C. (2008). Reduction of periodontal pathogens adhesion by antagonistic strains. Oral Microbiol. Immun. *23*, 43–48.

de Vrese, M., and Schrezenmeir, J. (2008). Probiotics, prebiotics, and synbiotics. Food Biotech. *111*, 1–66.

van Winkelhoff, A.J., van, d., V., and de Graaff, J. (1988). Microbial succession in recolonizing deep periodontal pockets after a single course of supra- and subgingival debridement. J. Clin. Periodontol. *15*, 116–122.

van Winkelhoff, A.J., Herrera, G.D., Winkel, E.G., Dellemijn-Kippuw, N., Vandenbroucke-Grauls, C.M., and Sanz, M. (2000). Antimicrobial resistance in the subgingival microflora in patients with adult periodontitis. A comparison between The Netherlands and Spain. J. Clini. Periodontol. *27*, 79–86.

Wade, W.G., Moran, J., Morgan, J.R., Newcombe, R., and Addy, M. (1992). The effects of antimicrobial acrylic strips on the subgingival microflora in chronic periodontitis. J. Clin. Periodontol. *19*, 127–134.

Wang, H.L., Greenwell, H., and Bissada, N.F. (1990). Crevicular fluid iron changes in treated and untreated periodontally diseased sites. Oral Surg. Oral Med. O. *69*, 450–456.

Wilson M. (2005). Manipulation of the indigenous microbiota. In Microbial Inhabitants of Humans, Wilson, M., ed. (Cambridge University Press, New York), pp. 395–416.

Wolff, L., Dahlen, G., and Aeppli, D. (1994). Bacteria as risk markers for periodontitis. J. Periodontol. *65*, 498–510.

Ximenez-Fyvie, L.A., Haffajee, A.D., Som, S., Thompson, M., Torresyap, G., and Socransky, S.S. (2000). The effect of repeated professional supragingival plaque removal on the composition of the supra- and subgingival microbiota. J. Clin. Periodontol. 27, 637–647.

Zahradnik, R.T., Magnusson, I., Walker, C., McDonell, E., Hillman, C.H., and Hillman, J.D. (2009). Preliminary assessment of safety and effectiveness in humans of ProBiora(3), a probiotic mouthwash. J. Appl. Microbiol. *107*, 682–690.

Zhang, G., Chen, R., and Rudney, J.D. (2008). *Streptococcus cristatus* attenuates *Fusobacterium nucleatum*-induced interleukin-8 expression in oral epithelial cells. J. Periodontal Res. *43*, 408–416.

# Index

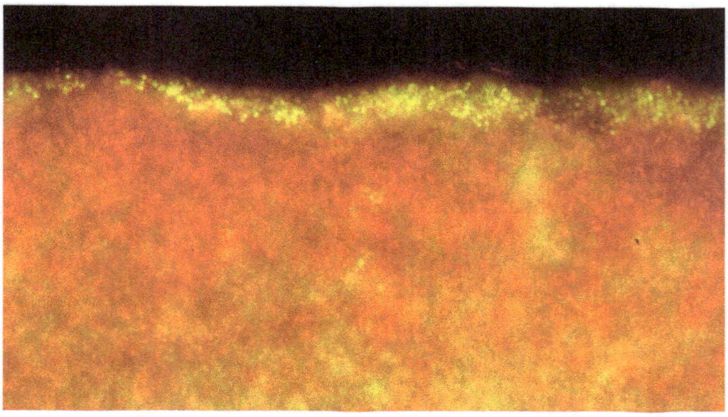

**Figure 4.3** An example of supragingival plaque with a metabolic active layer of streptococci (yellow-green) on top of a less active older biofilm composed of rods and cocci. The red colour is the signal from the red bacterial probe and the yellow-green and orange from a double labelling of a red bacterial probe and a green streptococcal probe (Zijnge *et al.*, 2010).

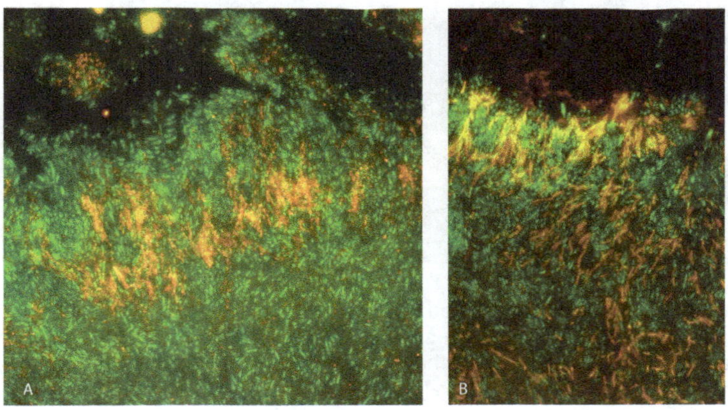


**Figure 4.4** The intermediate layer of the biofilm is dominated by fusobacteria (A) and *Tannerella forsythia* (B). In both panels the yellow-orange bacteria are of interest, which is a double labelling of a green bacterial probe and a red fusobacterial probe (A), or a red *Tannerella* probe (B). Modified from Zijnge *et al.* (2010).

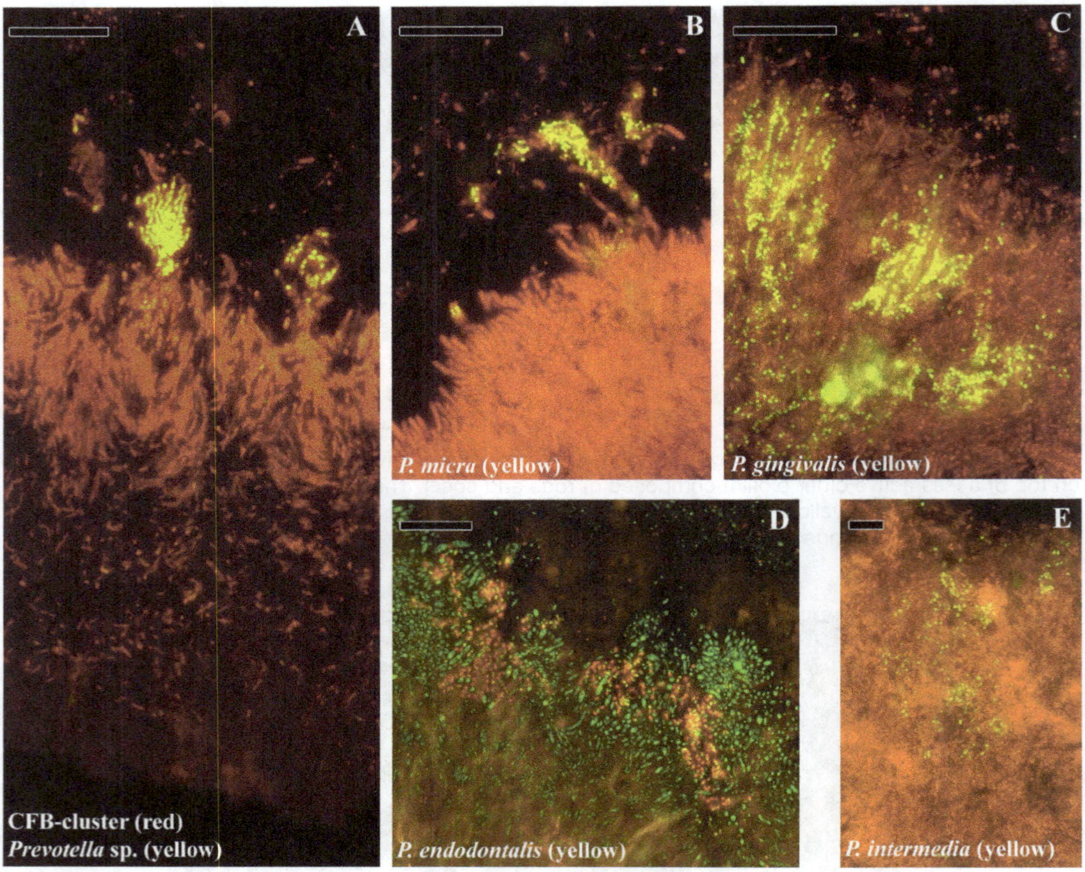


**Figure 4.5** Localization of species associated with periodontitis. (A) Overview of the subgingival biofilm with CFB-cluster species (red) and *Prevotella* sp. (yellow). Since *Prevotella* sp. are part of the CFB-cluster of bacteria, cells appear in yellow. (B) Top of the biofilm with a micro-colony of *P. micra* (yellow). (C) Micro-colonies of *P. gingivalis* (yellow) in the top layer. (D) Micro-colonies of *P. endodontalis* (yellow) in the top layer. (E) Microcolonies of *P. intermedia* in the top layer. Bars are 10 µm.

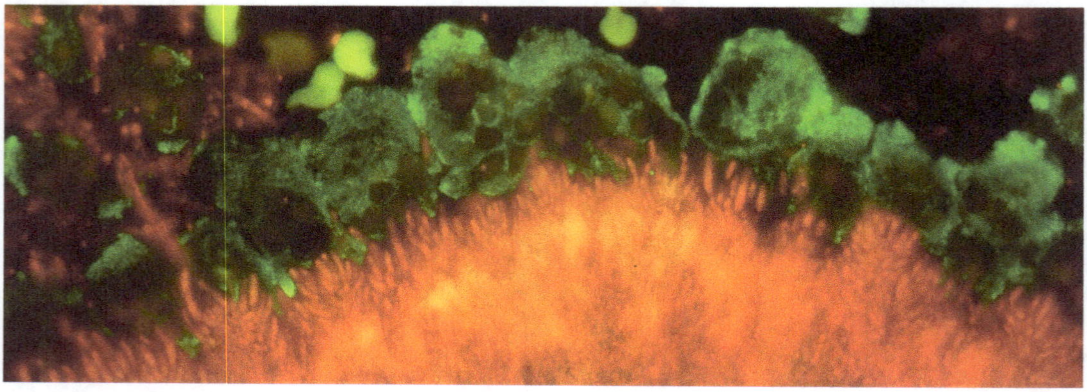

**Figure 4.6** Detail showing a subgingival biofilm with bacteria (red) and eukaryotic cells (green) (probably PMN). Note the suggestive palisade lining of bacteria and the numerical difference between bacteria and immune cells.

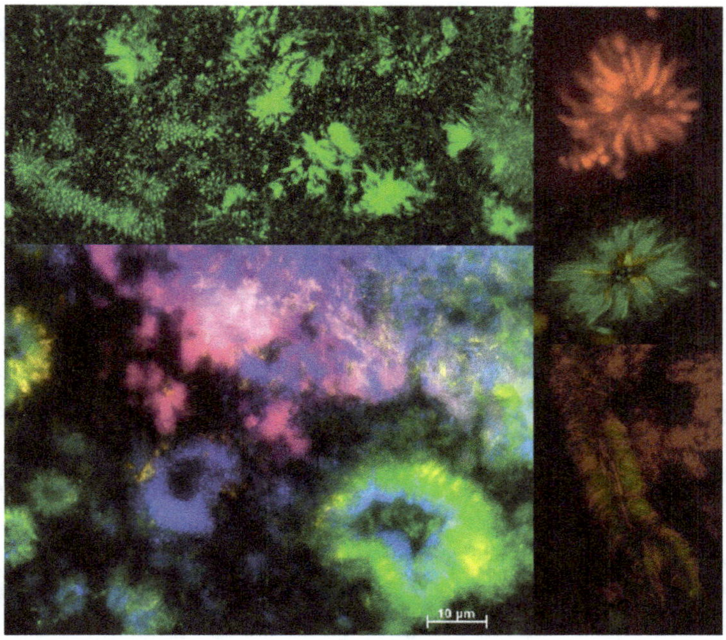


**Figure 4.7** Examples of test tube brushes in the outer layer of subgingival biofilms.

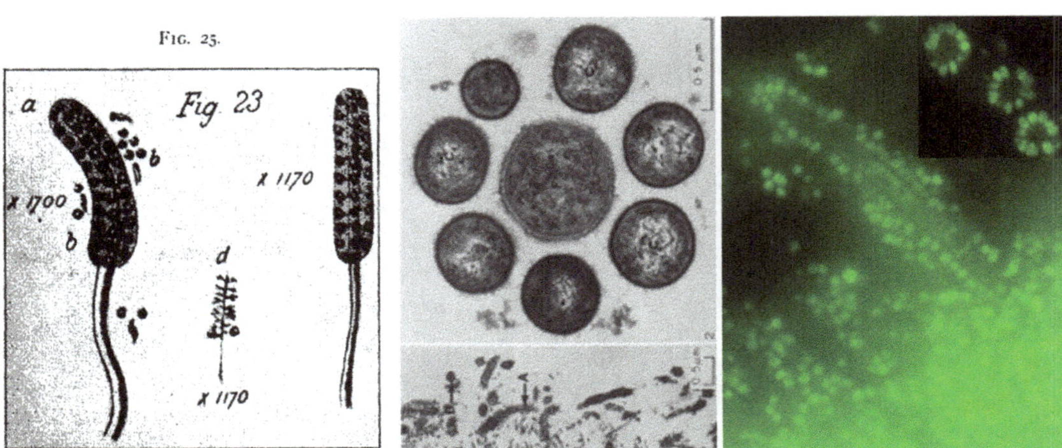


**Figure 4.8** Descriptions of corncob structures through the centuries. The left panel is a graphical representation by Vincentini in 1890 (Vincentini, 1987), the middle panel has been obtained by Listgarten in 1973 (Listgarten *et al.*, 1973) and the right panel shows corncobs as observed by FISH in 2010 (Zijnge *et al.*, 2010).

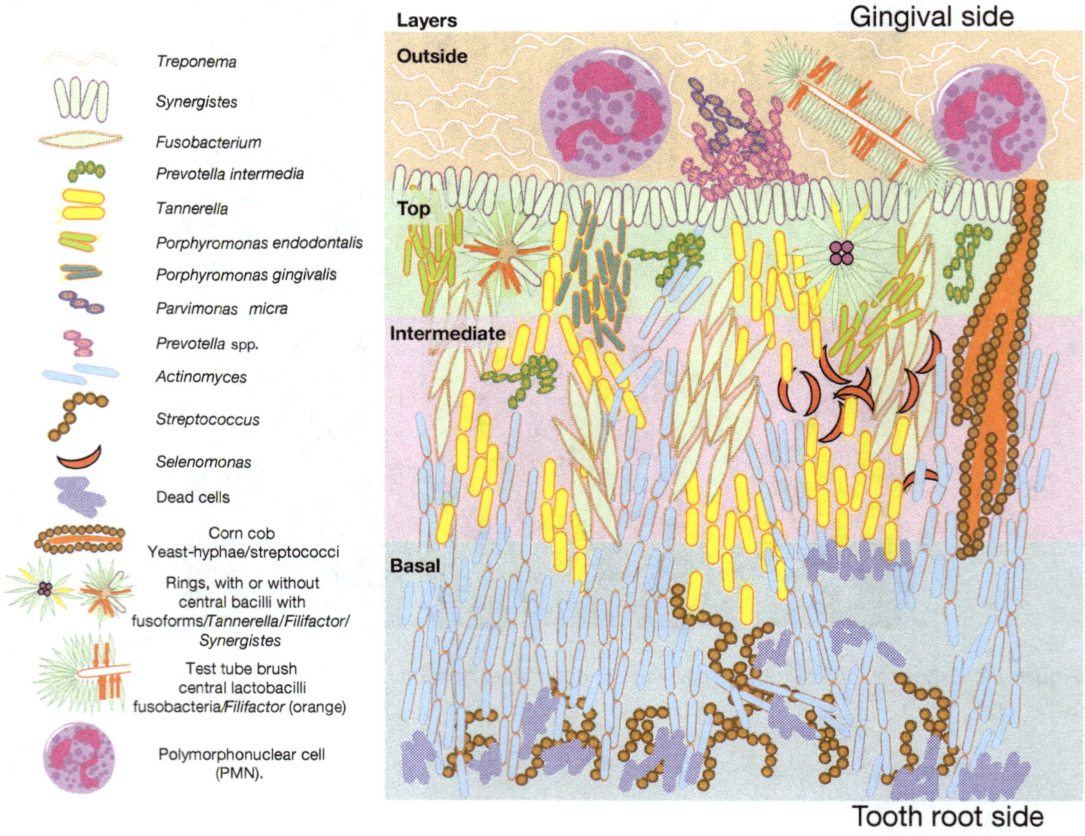


**Figure 4.9** A schematic model of a mature biofilm architecture. This drawing schematically represents the observations that are described in recent literature. It shows the three-layer architecture, with the additional loose outer layer and shows the key players in the architecture of the biofilm.

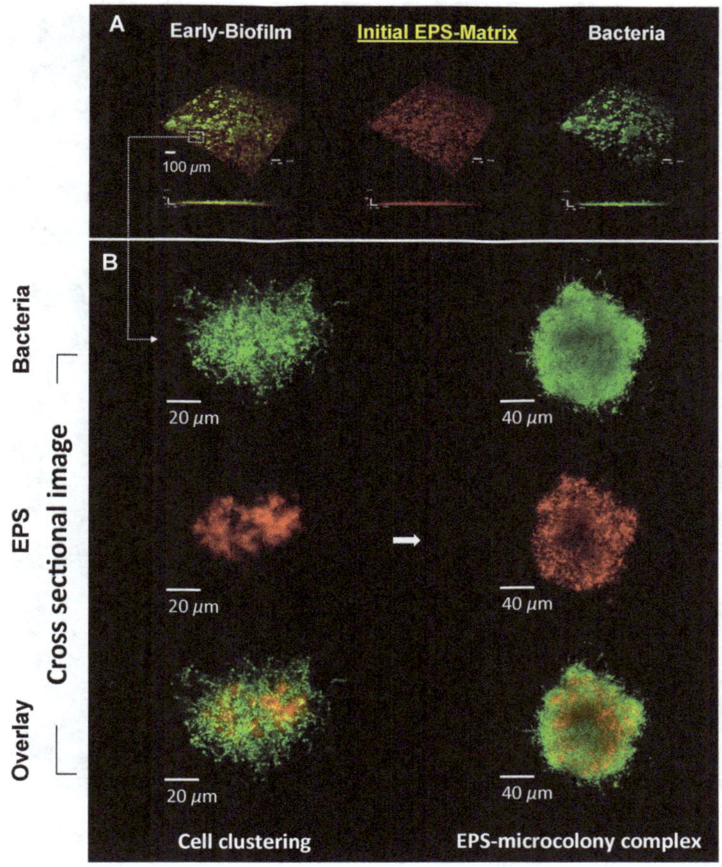


**Figure 5.3** The initial EPS-matrix and EPS-microcolony complex. (A) 3D reconstruction of EPS-matrix (in red) and bacterial cells (in green) within early-formed biofilm. (B) EPS-mediated cell clustering and further development into localized EPS-microcolony complexes. (Adapted from Xiao *et al.*, 2012.)

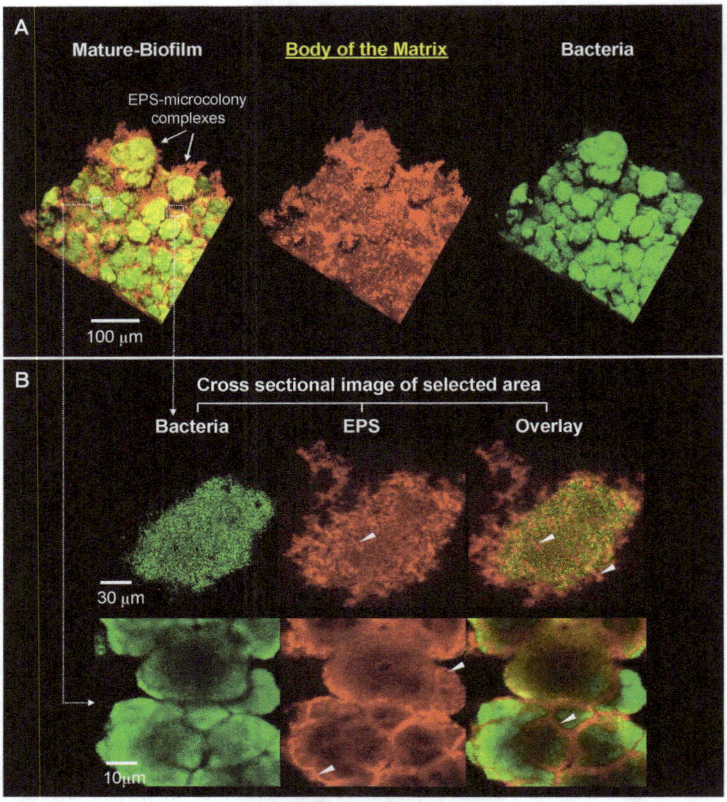


**Figure 5.4** The body of the EPS-matrix. (A) 3D reconstruction of EPS-matrix (in red) and bacterial cells (in green) within mature biofilm. (B) Close-up view of cell-matrix structural organization. Arrowheads show the intricate architecture of the matrix providing a 3D web-like meshwork surrounding and enmeshing the bacterial cells, forming compartmentalized microcolony complexes. (Bowen and Koo, 2011. Image reproduced with the permission of S. Karger AG, Basle.)

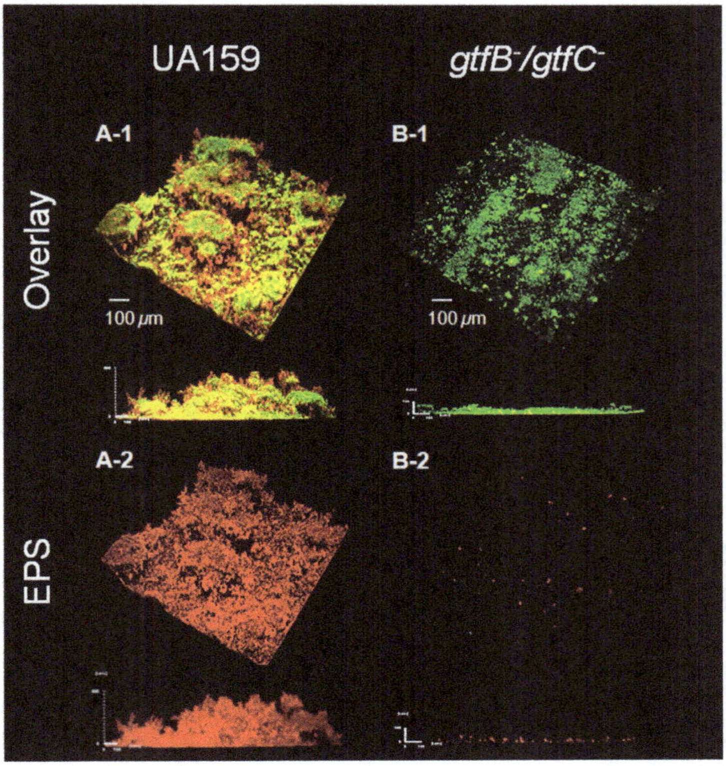


**Figure 5.5** Mixed-species biofilm formed by parental strain of *S. mutans* (UA159) and *gtfB⁻/gtfC⁻* mutant strain (Δ*gtfBC*::kan). (A-1, B-1) Representative 3D reconstruction of the biofilms. (A-2, B-2) 3D reconstruction of EPS-matrix in the same biofilms. (Adapted from Xiao *et al.*, 2012.)

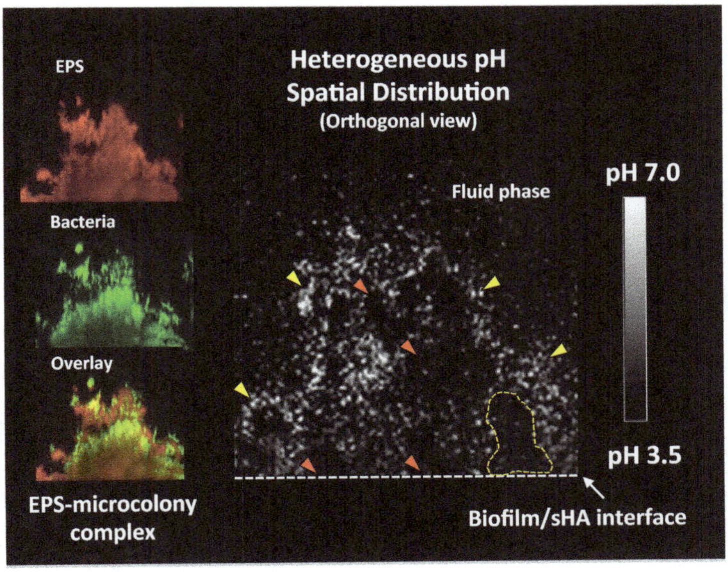


**Figure 5.6** Mapping of *in situ* pH of intact mixed-species biofilm. Orthogonal view of a representative image of the spatial distribution of pH within a EPS-microcolony complex. Red arrows indicate acidic pH regions (i) across the structure and (ii) at the microcolony/sHA interface (also highlighted with orange lines). Yellow arrows indicate pH close to neutral at the microcolony/fluid phase. (Adapted from Xiao *et al.*, 2012.)

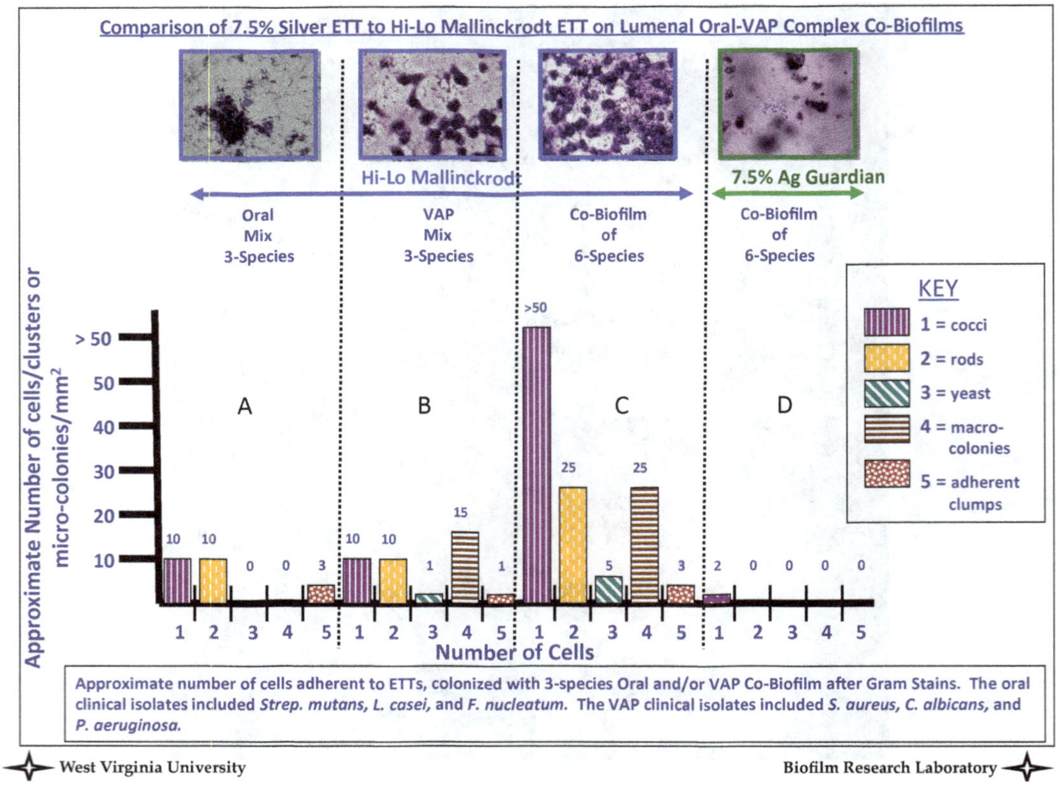


**Figure 12.6** Three-species dental/oral 'cocktail' and three-species VAP cocktail, incubated independently and together. Comparison of 7.5% silver ETT to hi-lo Mallinckrodt ETT on luminal oral-vap complex co-biofilms. Ag was effective against the combined three-species cocktail incubated for 48 hours.

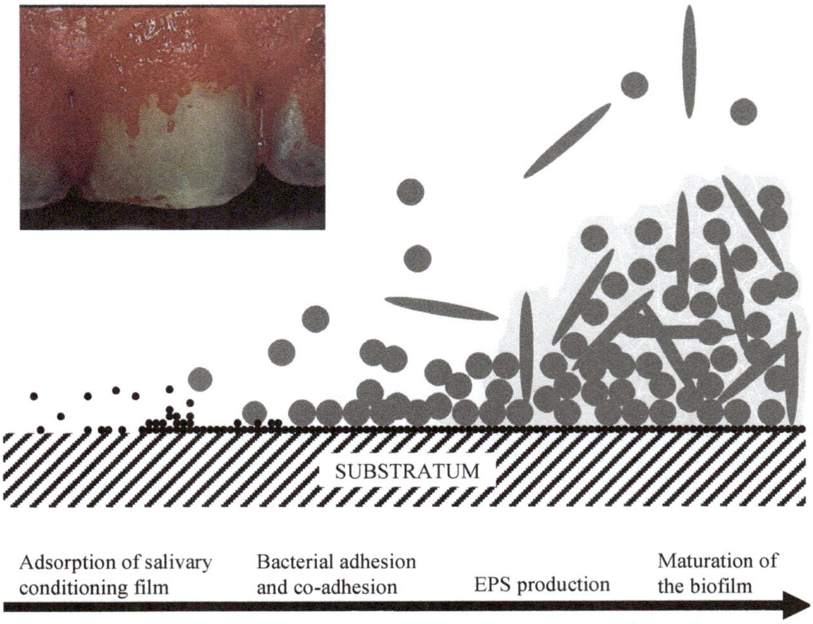


**Figure 13.1** Schematic, sequential presentation of the initial steps in biofilm formation. In the left top corner a photograph of a 9 days old, undisturbed oral biofilm on a human front incisor *in vivo*, stained by 0.5% aqueous neutral red solution is shown (Quirynen *et al.*, 1989).

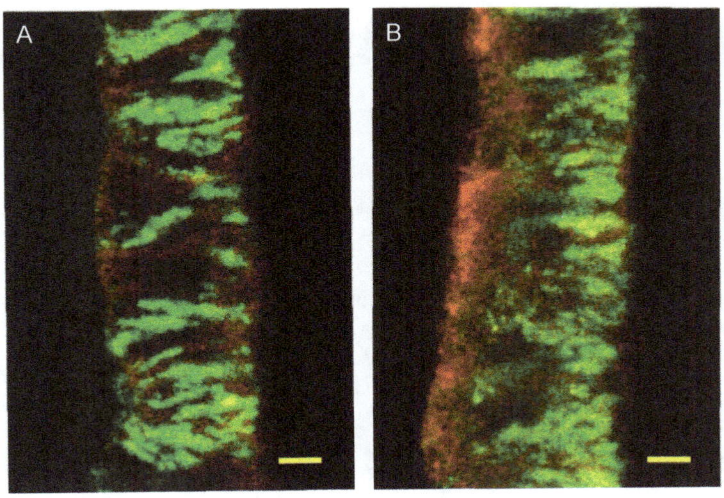


**Figure 13.2** Confocal laser scanning micrographs of a 24 hours old biofilm (Zaura-Arite *et al.*, 2001) grown on an *in situ* surface before (A) and after (B) rinsing with a chlorhexidine mouthrinse. Green and red areas represent live and dead bacteria. Note, that oral biofilm is only killed on the outside, and not on the substratum side (right).

www.ingramcontent.com/pod-product-compliance
Lightning Source LLC
LaVergne TN
LVHW062232200926
848484LV00006B/1519